Grundlehren der mathematischen Wissenschaften 245

A Series of Comprehensive Studies in Mathematics

I. P. Cornfeld
S. V. Fomin
Ya. G. Sinai

Ergodic Theory

Springer-Verlag
New York Heidelberg Berlin

I. P. Cornfeld
S. V. Fomin
Ya. G. Sinai
Landau Institute of Theoretical Physics
Academy of Sciences
Vorobiew Chasse-2
Moscow V-334
USSR

Translator
A. B. Sossinskii

AMS Subject Classification (1980): 47A35, 54H20, 58F11

Library of Congress Cataloging in Publication Data
Fomin, S. V. (Sergeĭ Vasil′evich).
Ergodic theory.
(Grundlehren der mathematischen Wissenschaften; 245)
Bibliography: p.
Includes index.
1. Ergodic theory. 2. Differentiable dynamical systems. I. Cornfeld, I. P. II. Sinai, I͡Akov Grigor′evich, 1935– III. Title. IV. Series.
QA611.5.F65 515.4′2 81-5355
AACR2

Printed in the United States of America.

9 8 7 6 5 4 3 2 1

ISBN 0-387-90580-4 Springer-Verlag New York Heidelberg Berlin
ISBN 3-540-90580-4 Springer-Verlag Berlin Heidelberg New York

Preface

Ergodic theory is one of the few branches of mathematics which has changed radically during the last two decades. Before this period, with a small number of exceptions, ergodic theory dealt primarily with averaging problems and general qualitative questions, while now it is a powerful amalgam of methods used for the analysis of statistical properties of dynamical systems. For this reason, the problems of ergodic theory now interest not only the mathematician, but also the research worker in physics, biology, chemistry, etc.

The outline of this book became clear to us nearly ten years ago but, for various reasons, its writing demanded a long period of time. The main principle, which we adhered to from the beginning, was to develop the approaches and methods or ergodic theory in the study of numerous concrete examples. Because of this, Part I of the book contains the description of various classes of dynamical systems, and their elementary analysis on the basis of the fundamental notions of ergodicity, mixing, and spectra of dynamical systems. Here, as in many other cases, the adjective "elementary" is not synonymous with "simple."

Part II is devoted to "abstract ergodic theory." It includes the construction of direct and skew products of dynamical systems, the Rohlin–Halmos lemma, and the theory of special representations of dynamical systems with continuous time. A considerable part deals with entropy. We have included here the proof of Ornstein's theorem on the isomorphism of Bernoulli automorphisms with the same entropy due to Keane and Smorodinski; this proof is nearer to information theory than Ornstein's original proof.

Before the appearance of the entropy theory of dynamical systems, the principal invariant of a dynamical system was thought to be its spectrum. Problems of the spectral theory are developed in Part III. First, we present the theory of dynamical systems with discrete spectrum due to von Neumann, then we construct various examples of spectra occurring for dynamical systems, some of which resemble discrete spectra while others have little in common with them. A special chapter is devoted to the spectral analysis of dynamical systems corresponding to the Gauss stationary random processes of probability theory.

Part IV is concerned with the approximation of dynamical systems by periodic systems, and the application of approximation theory to the analysis of dynamical systems corresponding to smooth vector fields on the two-dimensional torus.

It is clear from the above that entropy methods in the study of classical dynamical systems have not been sufficiently developed in this book. We intend to make these questions the topic of a separate monograph.

The present book appears to be somewhat uneven: together with relatively simple parts, it contains sections including the proofs of deep and difficult theorems. We have attempted to make the various sections of this book independent of one another.

Our point of view and attitude to the problems of the theory of dynamical systems developed to a great extent under the influence of the ideas and works of A. N. Kolmogorov, who initiated the rapid development of ergodic theory. Numerous conversations with V. A. Rohlin were of great importance to us. During the entire period of our work on this book, we benefited from the help of many participants of the Ergodic Theory Seminar at Moscow University. A. N. Zemlyakov helped write Chaps. 5 and 6, and A. M. Stepin helped in writing a number of sections. We had numerous useful discussions on certain questions of spectral theory with V. I. Oseledets. B. M. Gurevich carefully read the manuscript and made many useful remarks. It gives us great pleasure to express our gratitude to our teachers and colleagues for their support and assistance.

Ergodic theory was one of Sergei Vasilyevich Fomin's favorite branches of mathematics. His unfailing optimism and kindness always stimulated enthusiasm in our joint work. It is difficult to express the sorrow and grief we experienced upon Sergei Vasilyevich's untimely death, which occurred in August 1975, while the work on this book was in full progress.

Moscow, 1980

I. P. Cornfeld
Ya. G. Sinai

Contents

Part I

Ergodicity and Mixing. Examples of Dynamical Systems

Chapter 1

Basic Definitions of Ergodic Theory

§1. Definition of Dynamical Systems

Ergodic theory studies motion in a measure space. Therefore we begin by considering the notion of measure space.

We assume that we are given an abstract space M, whose points are denoted by $x, y, z, \ldots$. Further this space will turn out to be the phase space of a dynamical system.

We assume that a σ-algebra $\mathfrak{S}$ of subsets of the space M has been chosen and that a measure μ is defined on it. As a rule, we further assume that the measure is normalized ($\mu(M) = 1$) and complete, i.e., all the subsets of vanishing measure belong to $\mathfrak{S}$.

Sometimes we shall consider spaces M for which the measure μ is infinite ($\mu(M) = \infty$), but which can be represented as the union of a countable number of subsets of finite measure. Such spaces are called spaces with σ-finite measure or $\mathfrak{S}$-finite measure spaces.

Measure spaces will usually be denoted by $(M, \mathfrak{S}, \mu)$. If only the σ-algebra $\mathfrak{S}$ has been chosen on M, while the measure μ is not necessarily given, then $(M, \mathfrak{S})$ is said to be a measurable space.

At this point it is already useful to give some examples of measure spaces which often appear in ergodic theory.

Examples of measure spaces

1. M is a compact topological group with a normalized Haar measure μ. For the σ-algebra $\mathfrak{S}$ we take the completion of the σ-algebra of Borel sets of the space M with respect to the measure μ. This σ-algebra is said to be the Borel σ-algebra.

In particular, M may be an m-dimensional torus, $M = \text{Tor}^m = S^1 \times \cdots \times S^1$, on which the natural cyclic coordinates $x_1, \ldots, x_m$ have been introduced. In these coordinates $d\mu = dx_1 \cdots dx_m$.

2. M is an m-dimensional compact closed oriented manifold of class C^∞. The differential of the measure μ is given in the form $d\mu = p(x)\, dx_1 \cdots dx_m$, where $x_1, \ldots, x_m$ is a system of local coordinates and $p(x)$ is an infinitely differentiable function.

Here again $\mathfrak{S}$ is the Borel σ-algebra, i.e., the completion of the σ-algebra of Borel sets with respect to the measure μ.

To make this example more specific, assume that M is an m-dimensional compact Riemann manifold of class C^∞. In local coordinates, the metric can be written in the form $ds^2 = \sum_{i,j=1}^m g_{ij}\,dx_i\,dx_j$. The differential of the measure induced by the Riemann metric is of the form $d\mu_0 = \text{const.}\ \sqrt{\det\|g_{ij}\|} \times dx_1 \ldots dx_m$. We shall often consider measures μ for which $d\mu = p(x)\,d\mu_0$, i.e., measures given by the density with respect to the measure μ_0.

3. M is the space of sequences $x = \{x_i\}$, infinite in one or in both directions, where each coordinate x_i assumes values from a fixed finite or countable set I.

The space M may be interpreted as countable Cartesian product of spaces isomorphic to I. The discrete topology on I induces a topology on M which, in its turn, induces the Borel σ-algebra of subsets of M.

From the point of view of probability theory, M may be naturally interpreted as the space of realizations of a random process and the measure μ on M as the probability distribution for this process. According to Kolmogorov's theorem, in order to define the measure it suffices to indicate its values on all possible finite dimensional cylinders $\{x: x_{i_1} \in A_1, \ldots, x_{i_r} \in A_r\}$, where $A_1, \ldots, A_r$ are arbitrary subsets of I, and $i_1, \ldots, i_r$ are arbitrary numbers of coordinates.

4. M is the space of real-valued functions $x(s)$, $-\infty < s < \infty$. As in the previous example, the measure μ on M is defined by its values on finite dimensional cylinders

$$\{x: x_{s_1} \in A_1, \ldots, x_{s_r} \in A_r\},$$

where $A_1, \ldots, A_r$ are Borel subsets of the line, $-\infty < s_1, \ldots, s_r < \infty$. The measure itself is defined on the minimal σ-algebra generated by the cylinders.

One of the fundamental notions of ergodic theory is that of an automorphism of a measure space.

Definition 1. An *automorphism* of the measure space $(M, \mathfrak{S}, \mu)$ is a one-to-one map T of the space M onto itself such that for all $A \in \mathfrak{S}$ we have $TA, T^{-1}A \in \mathfrak{S}$ and

$$\mu(A) = \mu(TA) = \mu(T^{-1}A).$$

The measure μ is said to be an *invariant measure* for the automorphism T.

EXAMPLES

1. M is the m-dimensional torus Tor^m with a normalized Lebesgue measure ρ; the transformation T, in the cyclic coordinates $x_1, \ldots, x_m$, is given by the formula

$$T(x_1, \ldots, x_m) = (x_1 + \alpha_1 \pmod 1), \ldots, x_m + \alpha_m \pmod 1)),$$

where $\alpha_1, \ldots, \alpha_m$ are real numbers. Clearly T is an automorphism (i.e., the measure ρ is invariant). This automorphism is called a *translation* on the torus. In the particular case when $m = 1$, i.e., $M = S^1$, the automorphism T is called a *rotation* of the circle.

2. M is an m-dimensional compact closed oriented manifold of class C^∞, T is a diffeomorphism of class C^∞. If $x_1, \ldots, x_m$ are local coordinates in the neighborhood of the point $x^{(0)} \in M$ and $y_1, \ldots, y_m$ are local coordinates in the neighborhood of the point $y^{(0)} = Tx^{(0)}$, then T is locally determined by the C^∞-functions f_k, $k = 1, 2, \ldots, m$:

$$\begin{aligned} y_1 &= f_1(x_1, \ldots, x_m), \\ &\vdots \\ y_m &= f_m(x_1, \ldots, x_m). \end{aligned}$$

If the differential of the measure $d\mu$ is given by the density $p > 0$ (see Example 2), then the invariance condition of the measure μ may be written in the form:

$$\det \left\| \frac{\partial f_k}{\partial x_l} \right\|_{x = x^{(0)}} = \frac{p(y^{(0)})}{p(x^{(0)})}.$$

3. The space M of sequences, infinite in both directions (Example 3), possesses a natural *shift* T: $Tx = x'$, where $x_i' \stackrel{\text{def}}{=} x_{i+1}$, $-\infty < i < \infty$. The measure μ invariant with respect to this shift is said to be *stationary* (sometimes *stationary in the narrow sense*). This terminology arose in the theory of probability; to be more precise, in the theory of stationary random processes, with which ergodic theory is intimately connected. The invariance condition of the measure is written in the form

$$\mu(\{x: x_{i_1+k} \in A_1, \ldots, x_{i_r+k} \in A_r\}) = \mu(\{x: x_{i_1} \in A_1, \ldots, x_{i_r} \in A_r\}), \quad -\infty < k < \infty.$$

Definition 2. An *endomorphism* of the space M is a surjective (not necessarily one-to-one) map T of the space M onto itself such that for any $A \in \mathfrak{S}$ we have $T^{-1}A \in \mathfrak{S}$ and

$$\mu(A) = \mu(T^{-1}A),$$

where $T^{-1}A$ is the inverse image of the set A.

EXAMPLES

1. M is the unit circle S^1 with Lebesgue measure; again we identify M with the semi-interval $0 \leq x < 1$. The transformation T is given by the formula $Tx = 2x \pmod 1$. Under this transformation each closed interval $\Delta \subset S^1$ of length $< 1/2$ is mapped onto a interval of double length, and for

each closed interval $\Delta' \subset S^1$ there exist two intervals whose length is a half of the length of the given one, which are mapped onto Δ' by T. It follows from this last property that the Lebesgue measure is invariant with respect to T.

2. M is the space of one-sided infinite sequences $x = \{x_i\}$, $i = 0, 1, \ldots$. The transformation T is the shift $Tx = x'$ where $x_i' = x_{i+1}$. The invariance condition of the measure μ is written in the form:

$$\mu(\{x: x_{i_1+k} \in A_1, \ldots, x_{i_r+k} \in A_r\}) = \mu(\{x: x_{i_1} \in A_1, \ldots, x_{i_r} \in A_r\}),$$

for all $k \geq 0$.

Definition 3. Suppose $\{T^t\}$ is a one-parameter group of automorphisms of the measure space $(M, \mathfrak{S}, \mu)$, $t \in \mathbb{R}^1$, i.e., $T^{t+s}(x) = T^t(T^s(x))$ for all $t, s \in \mathbb{R}^1$ and $x \in M$. Then $\{T^t\}$ is said to be a *flow* if for any measurable function $f(x)$ on M the function $f(T^t x)$ is measurable on the Cartesian product $M \times \mathbb{R}^1$.

The measurability condition appearing in this definition may also be stated in the following (equivalent) form: the map $\psi: M \times \mathbb{R}^1 \to M$ given by the formula $\psi(x, t) = T^t x$ is measurable.

Definition 4. Suppose $\{T^t\}$ is a one-parameter semigroup of endomorphisms of the measure space $(M, \mathfrak{S}, \mu)$, $t \in \mathbb{R}^1_+ \stackrel{\text{def}}{=} \{s: s \geq 0\}$, i.e., $T^{t+s}x = T^t(T^s x)$ for all $t, s \in \mathbb{R}^1_+$, $x \in M$. Then $\{T^t\}$ is said to be a *semiflow* if for any measurable function $f(x)$ on M the function $f(T^t x)$ is measurable on the Cartesian product $M \times \mathbb{R}^1_+$.

In Definitions 1–4 we have introduced four fundamental objects studied in ergodic theory: automorphisms, endomorphisms, flows, and semiflows in measure spaces. Further the expression "dynamical system" stands for any of these objects. The measure space itself is said to be the *phase space* of the dynamical system.

The cyclic group (semigroup) generated by one automorphism (endomorphism) is often called a dynamical system with discrete time, while a flow (semiflow) is said to be a dynamical system with continuous time.

Sometimes we consider transformations, as well as one-parameter groups and semigroups of transformations, on a measurable space $(M, \mathfrak{S})$ for which the measure μ is not given *a priori*. If the measurability conditions contained in Definitions 1–4 are satisfied for these transformations (groups, semigroups) then they are called *measurable dynamical systems*.

In certain cases, the notion of "continuous flow," which is more general than the notion of flow introduced above, turns out to be useful. Let us call two automorphisms T_1, T_2 on the space $(M, \mathfrak{S}, \mu)$, coinciding mod 0, if there is a set $M' \in \mathfrak{S}$, $\mu(M') = 1$, such that $T_1 x = T_2 x$ for all $x \in M'$.

Definition 3′. A family of automorphisms $\{T^t\}$, $t \in \mathbb{R}^1$ of the measure space $(M, \mathfrak{S}, \mu)$ is said to be a *continuous flow*, if for all $t_1, t_2 \in \mathbb{R}^1$ the automorphisms $T^{t_1+t_2}$ and $T^{t_1}T^{t_2}$ coincide mod 0.

EXAMPLES

1. M is an m-dimensional compact closed orientable manifold of class C^∞ and X is a vector field of class C^∞ on M. Consider the system of differential equations $dx/dt = X(x)$, or in local coordinates

$$\frac{dx_1}{dt} = X_1(x_1, \ldots, x_m), \ldots, \frac{dx_m}{dt} = X_m(x_1, \ldots, x_m). \tag{1}$$

Suppose the action of the automorphism T^t, $-\infty < t < \infty$, consists in sending each point $x \in M$ into the point $x(t)$ which is the value at the moment t of the solution of the system (1) with initial condition $x(0) = 0$.

The fact that the group property holds is obvious since the right-hand side of equations (1) is independent of t.

It is clear that $\{T^t\}$ is a flow. The invariance condition of the measure μ for this flow is given by the following theorem.

Liouville's Theorem. *The measure μ with density p of class C^∞, i.e., the measure with differential $d\mu = p(x)\, dx_1 \ldots dx_m$, is invariant with respect to $\{T^t\}$ if and only if we have*

$$\sum_{k=1}^{m} \frac{\partial}{\partial x_k}(pX_k) = 0.$$

The proof of this theorem will be given in §2, Chap. 2.

Consider the important particular case of systems of the form (1). Suppose M is the m-dimensional torus Tor^m and the system of differential equation (1) in the cyclic coordinates $x_1, \ldots, x_m$ is written as follows

$$\frac{dx_1}{dt} = \alpha_1, \ldots, \frac{dx_m}{dt} = \alpha_m, \tag{2}$$

where $\alpha_1, \ldots, \alpha_m$ are real numbers. The invariant measure of the corresponding flow $\{T^t\}$ will obviously be the Lebesgue measure. Motion on the torus corresponding to the system of equation (2) is called *conditionally periodic motion* and the numbers $\alpha_1, \ldots, \alpha_m$ are the *frequencies* of this motion.

2. Suppose M is the space of all real-valued functions $x(s)$, $s \in \mathbb{R}^1$ and $\{T^t\}$ is the one-parameter group of translations, i.e., $T^t x(s) = x(s + t)$. We can introduce a topology into M by taking open sets to be sets of the form

$$\{x(s) \colon a_1 < x(s_1) < b_1, \ldots, a_r < x(s_r) < b_r\},$$

where $s_1, \ldots, s_r, a_1, \ldots, a_r, b_1, \ldots, b_r \in \mathbb{R}^1$.

This topology defines a Borel σ-algebra $\mathfrak{S}$ of subsets of M. The measure μ given on $\mathfrak{S}$ and invariant with respect to the group $\{T^t\}$ is said to be *stationary* (or *stationary in the narrow sense*).

It is easy to verify that $\{T^t\}$ with the invariant measure μ is a flow.

3. If in the previous example, instead of functions $x(s)$ given on $\mathbb{R}^1$, we consider functions defined only on $\mathbb{R}^1_+ \overset{\text{def}}{=} \{s: s \geq 0\}$ and, instead of the group of translations $\{T^t\}$, we take the semigroup $t \geq 0$, then we obtain a semiflow.

The notion of dynamical system is so general that in order to obtain meaningful results it is necessary, as a rule, to impose certain additional conditions on the system. However, there is a simple but important statement which is valid in a general case—the so-called Poincaré recurrence theorem.

Definition 5. Suppose T is an endomorphism of the space $(M, \mathfrak{S}, \mu)$ and $A \in \mathfrak{S}$. Then the point $x \in A$ is said to be a *recurrence point* (in the set A) if $T^n x \in A$ for at least one $n > 0$.

Theorem 1 (The Poincaré Recurrence Theorem). *For any endomorphism T and any $A \in \mathfrak{S}$ almost all (with respect to the measure μ) points $x \in A$ are recurrence points.*

Proof. Denote by N the subset of the set A consisting of all points which do not return to A. Then $N \in \mathfrak{S}$, since

$$N = A \cap \left(\bigcup_{n=1}^{\infty} T^{-n}(M \setminus A)\right).$$

If $x \in N$, then all points of the form $T^n x$, $n = 1, 2, \ldots$ do not belong to A, and hence $T^n x \notin N$, i.e., $x \notin T^{-n}N$. Therefore $N \cap T^{-n}N = \varnothing$, $n = 1, 2, \ldots$. This implies that the sets $N, T^{-1}N, T^{-2}N, \ldots$ are disjoint. Indeed, for $0 \leq n_1 < n_2$ we have

$$T^{-n_1}N \cap T^{-n_2}N = T^{-n_1}(N \cap T^{-(n_2-n_1)}N) = \varnothing.$$

Therefore

$$1 \geq \mu\left(\bigcup_{n=0}^{\infty} T^{-n}N\right) = \sum_{n=0}^{\infty} \mu(T^{-n}N) = \sum_{n=0}^{\infty} \mu(N).$$

The last inequality can hold only if $\mu(N) = 0$. This proves the theorem. □

It follows from this theorem that in fact almost all points $x \in A$ return to A an infinite number of times. Indeed if $T^k x \notin A$ for all $k \geq p$, then x is a nonrecurrence point for the endomorphism T^p. By Theorem 1 the measure of all such points vanishes.

Using this remark let us prove the following useful statement.

Lemma 1. *If $f(x)$ is a positive measurable function on M, then for almost all $x \in M$ we have the equality*

$$\sum_{k=1}^{\infty} f(T^k x) = \infty. \tag{3}$$

Proof. Consider the sets $A_m = \{x \in M: f(x) \geq 1/m\}$, $m = 1, 2, \ldots$ for any point $x \in A_m$ returning to A_m an infinite number of times we obviously have (3). But almost all points $x \in A_m$ possess this property. Since $\bigcup_{m=1}^{\infty} A_m = M$, the lemma is proved. □

In abstract ergodic theory the main problem is the classification of dynamical systems. Sometimes it is useful to have the possibility of identifying dynamical systems defined on various phase spaces. In this connection let us introduce the following definition.

Definition 6. Suppose $\{T_1^t\}$, $\{T_2^t\}$ are two dynamical systems with similar time (discrete or continuous) acting in the spaces $(M_1, \mathfrak{S}_1, \mu_1)$ and $(M_2, \mathfrak{S}_2, \mu_2)$ respectively. Such systems are said to be *metrically isomorphic* if there exist sets $M_1' \in \mathfrak{S}_1$, $\mu_1(M_1') = 1$, $M_2' \in \mathfrak{S}_2$, $\mu_2(M_2') = 1$ as well as an isomorphism $\varphi: (M_1', \mathfrak{S}_1, \mu_1) \to (M_2', \mathfrak{S}_2, \mu_2)$ of measure spaces such that

$$T_2^t \varphi x^{(1)} = \varphi T_1^t x^{(1)}, \qquad T_1^t \varphi^{-1} x^{(2)} = \varphi^{-1} T_2^t x^{(2)},$$

for all t, $x^{(1)} \in M_1'$, $x_2 \in M_2'$.

Examples

1. Suppose M_1 is the space of sequences infinite in both directions $x = (\ldots, y_1, y_0, y_1, \ldots)$ whose coordinates y_i equal 0 or 1, the measure μ on M_1 being the direct product of the countable number of identical measures σ on $\{0, 1\}$: $\sigma(\{0\}) = \sigma(\{1\}) = \frac{1}{2}$. The automorphism T_1 is the shift

$$T_1(\ldots, y_{-1}, y_0, y_1, \ldots) = (\ldots, y'_{-1}, y'_0, y'_1, \ldots),$$

where $y_i' = y_{i+1}$, $-\infty < i < \infty$.

For M_2 take the unit square $0 \leq x_1, x_2 < 1$ (with the normalized Lebesgue measure) on which the automorphism T_2 acts according to the formula

$$T_2(x_1, x_2) = \begin{cases} (2x_1 \ (\mathrm{mod}\ 1), \frac{1}{2}x_2) & \text{if } x_1 \in [0, \frac{1}{2}) \\ (2x_1 \ (\mathrm{mod}\ 1), \frac{1}{2}(x_2 + 1)) & \text{if } x_1 \in [\frac{1}{2}, 1). \end{cases}$$

(This automorphism is sometimes referred to as the "baker's transformation.")

It is easy to check directly that the transformation $\varphi: M_1 \to M_2$, where

$$\varphi(\ldots, y_{-1}, y_0, y_1, \ldots) = \left(\sum_{k=0}^{\infty} \frac{y_k}{2^{k+1}}, \sum_{k=1}^{\infty} \frac{y_{-k}}{2^k}\right),$$

which is defined everywhere on M_1, except a countable number of points, is one-to-one and establishes a metric isomorphism of the automorphisms T_1 and T_2.

2. Suppose M_1 is the space of one-sided sequences $x = (y_0, y_1, \ldots)$ with coordinates y_i equalling 0 or 1 and the measure μ is the direct product of a countable number of identical measures σ on $\{0, 1\}$: $\sigma(\{0\}) = \sigma(\{1\}) = \frac{1}{2}$. The endomorphism T_1 is the one-sided shift:

$$T_1(y_0, y_1, \ldots) = (y_1, y_2, \ldots).$$

Suppose further the endomorphism T_2 acts on $M_2 = [0, 1)$ with normalized Lebesgue measure according to the formula $T_2 x = 2x \pmod 1$. It is easy to see that T_1 and T_2 are metrically isomorphic, the isomorphism $\varphi: M_1 \to M_2$ being determined by the formula

$$\varphi(y_0, y_1, \ldots) = \sum_{k=0}^{\infty} \frac{y_k}{2^{k+1}}.$$

An important class of nontrivial examples of metric isomorphisms will be considered in Part II of this book.

If some property, some numerical characteristic, etc., defined for any dynamical system is the same for metrically isomorphic systems, then this property, number, etc., is said to be a metric invariant of the dynamical system. In this chapter we shall introduce the notion of ergodicity and mixing, which are the most important metric invariants of a dynamical system.

Generalizations. The definitions of this section may be generalized in two directions. Note however, that in the present book these generalizations shall not in fact be needed.

1. Suppose G is a topological group (or semigroup) with a fixed Borel σ-algebra of subsets. Then for any measure space M there is a natural σ-algebra subset structure in the Cartesian product $M \times G$.

Definition 7. The measure space M is said to be a *G-space* if we have a measurable map $\varphi: G \times M \to M$ with the following properties:

(1) for any $g \in G$ the map $T_g: M \to M$ defined by the relation $T_g x = \varphi(g, x)$ is an endomorphism of the space M;
(2) the transformations T_g determine a representation of the group G, i.e., $T_{g_1} T_{g_2} = T_{g_1 g_2}$ for all $g_1, g_2 \in G$.

If G is a group, then the group property implies that the transformations T_g are automorphisms of the space M.

The group (semigroup) of transformations $\{T_g\}$, $g \in G$ is called a *G-flow* (a *G-semiflow*) on M.

2. Assume that M, besides the σ-algebra structure $\mathfrak{S}$, possesses some additional structure; for example, M is a compact metric space. Then every measure on M generates a linear positive normed functional on the commutative normed ring $C(M)$ of continuous functions on M while each automorphism T generates an automorphism of the ring $C(M)$ which preserves this functional.

If as our starting point we take a noncommutative normed ring, for example a noncommutative C^*-algebra and a linear positive normed functional on it, and then consider automorphisms of this C^*-algebra, we then come to a theory called noncommutative ergodic theory. This theory is important for certain branches of theoretical physics. At the present time it is rapidly developing.

§2. The Birkhoff–Khinchin Ergodic Theorem. Ergodicity

The theorem stated below is one of the most important in ergodic theory.

Theorem 1 (The Birkhoff–Khinchin Ergodic Theorem). *Suppose $(M, \mathfrak{S}, \mu)$ is a space with normalized measure and $f \in L^1(M, \mathfrak{S}, \mu)$. Then for almost every (in the sense of the measure μ) $x \in M$ the following limits exist and are equal to each other*

$$\lim_{n\to\infty} \frac{1}{n} \sum_{k=0}^{n-1} f(T^k x) = \lim_{n\to\infty} \frac{1}{n} \sum_{k=0}^{n-1} f(T^{-k} x) = \frac{1}{2n+1} \sum_{k=-n}^{n} f(T^k x) \overset{\text{def}}{=} \bar{f}(x),$$

in the case of an automorphism T; for almost every $x \in M$ the following limit exists

$$\lim_{n\to\infty} \frac{1}{n} \sum_{k=0}^{n-1} f(T^k x) \overset{\text{def}}{=} \bar{f}(x),$$

in the case of an endomorphism T; for almost every $x \in M$ the following limits exist and are equal to each other

$$\lim_{t\to\infty} \frac{1}{t} \int_0^t f(T^\tau x)\, d\tau = \lim_{t\to\infty} \frac{1}{t} \int_0^t f(T^{-\tau} x)\, d\tau$$

$$= \lim_{t\to\infty} \frac{1}{2t} \int_{-t}^{t} f(T^\tau x)\, d\tau \overset{\text{def}}{=} \bar{f}(x),$$

in the case of a flow $\{T^t\}$; *for almost every* $x \in M$ *the following limit exists*

$$\lim_{t\to\infty} \frac{1}{t} \int_0^t f(T^\tau x)\, d\tau \stackrel{\text{def}}{=} \bar{f}(x)$$

in the case of a semiflow $\{T^t\}$.

Further $\bar{f}(Tx) = \bar{f}(x)$ *or* $\bar{f}(T^t x) = \bar{f}(x)$ *whenever the right-hand sides of these equations exist. Moreover*

$$\bar{f} \in L^1(M, \mathfrak{S}, \mu) \quad \text{and} \quad \int_M \bar{f}(x)\, d\mu = \int_M f(x)\, d\mu.$$

The limits which appear in the Birkhoff–Khinchin ergodic theorem are called *time means* or *means along the trajectory* (the *trajectory of the point* x is the set of all points of the form $T^k x$, in the case of discrete time, and of the form $T^t x$, in the case of continuous time).

If, for example, $\chi_A(x)$ is the indicator of the set $A \subset M$ then $(1/n)\sum_{k=0}^{n-1} \chi_A(T^k x)$ is the mean number of times that the trajectory of the point x passes through the set A from time 0 to time $n - 1$. Thus the Birkhoff–Khinchin ergodic theorem guarantees the existence, for almost every point x, of the mean number of occurrences in any measurable set A.

The proof of the Birkhoff–Khinchin ergodic theorem is far from trivial. It is hard to simplify even in particular special cases. At the same time this proof by itself is not used anywhere except in problems directly connected with the existence of time means. For this reason we do not give the proof of the Birkhoff–Khinchin theorem now; it is provided in Appendix 1.

Definition 1. The measurable function g is called *invariant* with respect to the automorphism T (endomorphism T, flow $\{T^t\}$, semiflow $\{T^k\}$) if for all $x \in M$ we have

$$g(Tx) = g(x) = g(T^{-1}x)$$

$$(g(Tx) = g(x);\ g(T^t x) = g(x) \text{ for all } t \in \mathbb{R}^1;$$

$$g(T^t x) = g(x) \text{ for all } t \in \mathbb{R}^1_+).$$[1]

In other words, an invariant function assumes constant values on every trajectory of the dynamical system, i.e., an invariant function is a function on trajectories.

[1] If some endomorphism T is actually an automorphism (i.e., is one-to-one mod 0), then any set, invariant with respect to T viewed as an endomorphism, will also be invariant with respect to T regarded as an automorphism. A similar remark is valid for all the notions which shall be introduced further for automorphisms as well as for endomorphisms (and also for flows and semiflows).

Definition 1′. The set $A \in \mathfrak{S}$ is said to be *invariant* with respect to the automorphism T (the endomorphism T, the flow or semiflow $\{T^t\}$), if its indicator χ_A is an invariant function.

In the case of an automorphism T the invariance of the set A means that $TA = A = T^{-1}A$, in the case of an endomorphism T or a flow (semiflow) $\{T^t\}$ the invariance means that $T^{-1}A = A$ or $T^tA = A$ for all $t \in \mathbb{R}^1 (t \in \mathbb{R}^1_+)$.

Definition 2. The measurable function g is said to be *invariant* mod 0 with respect to the automorphism T (the endomorphism T, the flow or semiflow $\{T^t\}$), if $g(Tx) = g(x) = g(T^{-1}x)$ almost everywhere in the case of automorphisms; $g(Tx) = g(x)$ almost everywhere in the case of endomorphisms; $g(T^tx) = g(x)$ for any $t \in \mathbb{R}^1(\mathbb{R}^1_+)$ for almost all x, which of course depend on t, in the case of a flow (semiflow).

Lemma 1. *If g is an invariant* mod 0 *function, then there exists an invariant function g_1 such that $g = g_1$ almost everywhere.*

Proof. First consider the case of discrete time. Denote

$$N = \{x \in M : g(x) \neq g(Tx)\},$$

$$A = \begin{cases} \bigcup_{n=0}^{\infty} T^{-n}N & \text{for an endomorphism} \\ \bigcup_{n=-\infty}^{\infty} T^{n}N & \text{for an automorphism.} \end{cases}$$

Then $\mu(A) = 0$ and the function

$$g_1(x) = \begin{cases} g(x) & \text{for } x \in M \setminus A \\ 0 & \text{for } x \in A, \end{cases}$$

is the one we need.

In the case of a flow, consider the set

$$E = \{(x, t) \in M \times \mathbb{R}^1 : g(x) \neq g(T^tx)\}.$$

Clearly $\nu(E) = 0$ where $\nu = \mu \times \rho$ and ρ is the Lebesgue measure on $\mathbb{R}^1$. By Fubini's theorem, there is a set $N \subset M$, $\mu(N) = 0$, and sets $N_x \subset \mathbb{R}^1$, $\rho(N_x) = 0$ $(x \in M \setminus N)$ such that for $x \in M \setminus N$, $t \in \mathbb{R}^1 \setminus N_x$ we have $g(x) = g(T^tx)$.

Let us show that for any pair of points $x_1, x_2 \in M \setminus N$ contained in the same trajectory of the flow $\{T^t\}$ we have $g(x_1) = g(x_2)$. Suppose $x_2 = (T^{t_0})x_1$,

$t_0 \in \mathbb{R}^1$. Consider the set $N_{x_1} - t_0 = \{t - t_0 : t \in N_{x_1}\}$. Clearly $\rho(N_{x_1} - t_0) = 0$ so that $\rho((N_{x_1} - t_0) \cup N_{x_2}) = 0$. Hence we can find a

$$\tau \in \mathbb{R}^1 \backslash (N_{x_1} - t_0) \cup N_{x_2}),$$

obtaining $g(x_1) = g(T^{\tau + t_0} x_1) = g(T^{\tau} x_2) = g(x_2)$. Now let us define the required function g_1. For $x \in M \backslash N$ put $g_1(x) = g(x)$. If $x \in N$ satisfies $T^t x \in N$ for all t, put $g_1(x) = 0$. For all other $x \in N$ take any point $x_1 \in M \backslash N$ on the trajectory of the point x and put $g_1(x) = g(x_1)$. It follows from what we have proved above that $g_1(x)$ does not depend on the choice of x_1. The function g_1 is invariant; the fact that it is measurable follows from the completeness of the measure μ. In the case of a semiflow the argument is similar. The lemma is proved. □

Definition 2′. The set $A \in \mathfrak{S}$ is called *invariant* mod 0 with respect to the automorphism T (the endomorphism T; the flow or semiflow $\{T^t\}$) if its indicator χ_A is an invariant function mod 0 with respect to the automorphism T (the endomorphism T, the flow or the semiflow $\{T^t\}$).

Lemma 1 implies that for any invariant mod 0 set A there exists an invariant set A_1 such that $\mu(A \triangle A_1) = 0$.

Clearly, if a dynamical system has an invariant set A, $0 < \mu(A) < 1$, then we can assume that it acts independently on the set A and its complement $M \backslash A = \bar{A}$. Then we can find invariant sets inside A and $\bar{A}$ and consider our system on these smaller subsets, etc. As a result we would like to obtain a partition of the phase space into separate elements inside of which there are no longer any nontrivial invariant sets. The difficulty here is that the final partition may turn out to be continuous and its elements may be of zero measure. We can overcome this difficulty by means of the theory of measurable partitions; to apply it, however, we need certain additional assumptions concerning the measure space. Note that these assumptions always hold in all more or less natural cases. Now we shall give the definition of ergodic or (which is the same thing) undecomposable dynamical systems.

Definition 3. A dynamical system is said to be *ergodic* if the measure $\mu(A)$ of any invariant set A equals 0 or 1.

Lemma 2. *If a dynamical system is ergodic, then any invariant function is constant on any set of full measure.*

Proof. If $g(x)$ is an invariant function, then for any a the set $C_a = \{x : g(x) < a\}$ is invariant. Therefore $\mu(C_a)$ equals 0 or 1. The lemma is proved. □

Suppose T is an ergodic system. Then for any function $f \in L^1(M, \mathfrak{S}, \mu)$ the function $\bar{f}(x)$ appearing in the Birkhoff–Khinchin ergodic theorem is invariant and therefore by Lemma 2 it is constant almost everywhere. Since $\int_M f(x)\, d\mu = \int_M \bar{f}(x)\, d\mu$, we obtain $\bar{f} = \int_M f(x)\, d\mu$ almost everywhere. The integral $\int_M f(x)\,d\mu$ is the mean value of the function f over the space M, or the space means. Hence the ergodic theorem, in the case of an ergodic dynamical system, may be stated as follows:

For almost every point $x \in M$ the time means equals the space means.

It is in this form that the ergodic theorem usually appears in the physics literature.

Suppose $f = \chi_A$ for some $A \in \mathfrak{S}$. As we have already pointed out, $(1/n)\sum_{k=0}^{n-1}\chi_A(T^k A)$ equals the mean number of times that the trajectory of the point x passes through the set A in time 0 to $n-1$. In the case of an ergodic system, the last expression tends to $\int_M \chi_A(x)\,d\mu = \mu(A)$ if $n \to \infty$. Thus in the case of an ergodic system the trajectory of almost every point x passes through every set of positive measure and, moreover, remains in the set for a length of time asymptotically proportional to the measure of the set. This property is called the property of asymptotic equidistribution of trajectories.

Lemma 3. *If a dynamical system with discrete time is ergodic and $\mu(A) > 0$, then $\mu(\bigcup_{n>0} T^{-n}A) = 1$.*

Proof. Set $B = \bigcup_{n>0} T^{-n}A$. Then $T^{-1}B \subset B$ and $\mu(T^{-1}B) = \mu(B) \geq \mu(A)$. Therefore the indicator χ_B of the set B is mod 0 invariant. According to Lemma 1, we can find an invariant set B_1 satisfying $\mu(B \triangle B_1) = 0$. It follows from ergodicity that $\mu(B_1) = 1$ and therefore $\mu(B) = 1$. The lemma is proved. □

This lemma may be restated as follows:

In the case of an ergodic dynamical system with discrete time the trajectory of any set of positive measure is the entire phase space with the exception of a set of zero measure.

If we are given a dynamical system T on the space $(M, \mathfrak{S}, \mu)$, then other normalized measures (possibly different from the given measure μ) defined on $\mathfrak{S}$ and invariant with respect to T may exist. The Birkhoff–Khinchin ergodic theorem gives us some information on the structure of the set of invariant measures.

Theorem 2. *Suppose we are given a dynamical system T on the measurable space $(M, \mathfrak{S})$ along with two normalized measures μ_1, μ_2 defined on $\mathfrak{S}$ and invariant with respect to T. Then*

(1) *if the measure μ_1 is ergodic with respect to T while μ_2 is absolutely continuous with respect to μ_1 (μ_2 is not assumed ergodic a priori), then $\mu_1 = \mu_2$;*

(2) *if both measures μ_1, μ_2 are ergodic with respect to T, then either $\mu_1 = \mu_2$ or μ_1, μ_2 are singular with respect to each other.*

Proof. We shall only consider the case of dynamical systems with discrete time. The necessary modification for the case of continuous time are straightforward.

(1) For any set $A \in \mathfrak{S}$, according to the Birkhoff–Khinchin ergodic theorem, we have

$$\lim_{n\to\infty} \frac{1}{n} \sum_{k=0}^{n-1} \chi_A(T^k x) = \mu_1 A,$$

almost everywhere with respect to the measure μ_1. Since μ_2 is absolutely continuous with respect to μ_1, this relation also holds almost everywhere with respect to the measure μ_2. According to the Lebesgue dominated convergence theorem, we have

$$\lim_{n\to\infty} \int_M \frac{1}{n} \sum_{k=0}^{n-1} \chi_A(T^k x)\, d\mu_2 = \int_M \mu_1(A)\, d\mu_2 = \mu_1(A).$$

But for any n

$$\int_M \sum_{k=0}^{n-1} \chi_A(T^k x)\, d\mu_2 = \mu_2(A).$$

Therefore $\mu_1(A) = \mu_2(A)$. Since A was arbitrary, statement (1) is proved.

(2) Assume that $\mu_1 \neq \mu_2$, i.e., $\mu_1(A) \neq \mu_2(A)$ for some $A \in \mathfrak{S}$. Denote

$$A_i = \left\{x \in M\colon \lim_{n\to\infty} \frac{1}{n} \sum_{k=0}^{n-1} \chi_A(T^k x) = \mu_i(A)\right\}, \qquad i = 1, 2.$$

By the Birkhoff–Khinchin ergodic theorem $\mu_i(A_i) = 1$, $i = 1, 2$. But by construction $A_1 \cap A_2 = \varnothing$. Therefore the measures μ_1 and μ_2 are singular with respect to each other. The theorem is proved. □

§3. Nonergodic Systems. Decomposition into Ergodic Components

In order to give an example of a nonergodic system, let us return to Example 1 of a dynamical system on page 7. In this example, the phase space is a smooth compact manifold of class C^∞, while the dynamical system is generated by translations along the trajectories of the system of differential equations

$$\frac{dx_k}{dt} = X_k(x_1, \ldots, x_m), \qquad 1 \le k \le m. \tag{1}$$

In the theory of ordinary differential equations a differentiable function $h(x_1, \ldots, x_m)$ is called a *first integral* if it preserves its value along the trajectories of the system (1). In differential form this condition may be written as follows

$$\frac{dh}{dt} = \sum_{k=1}^{m} \frac{\partial h}{\partial x_k} \cdot X_k = 0.$$

It is clear that every first integral h is an invariant function in the sense of Definition 1 in section 2. Thus we can say that invariant functions are "measurable first integrals." The ergodicity property thus means that our dynamical system has no "first integrals" given by measurable functions other than constants.

Now suppose $(M, \mathfrak{S}, \mu, \{T^t\})$ is an arbitrary dynamical system, not necessarily ergodic.

Consider all possible sets invariant mod 0 with respect to $\{T^t\}$. It is clear that the intersection as well as the union of a countable number of such sets is also invariant mod 0. Thus the family of invariant mod 0 sets constitutes a σ-algebra of invariant mod 0 sets $\mathfrak{S}^{\text{inv}}$.

First consider the case when the σ-algebra $\mathfrak{S}^{\text{inv}}$ is generated by a countable number of atoms, i.e., there exists a sequence of nonintersecting sets C_1, $C_2, \ldots$ such that $C_i \in \mathfrak{S}^{\text{inv}}$, $\mu(C_i) > 0$, $\sum_i \mu(C_i) = 1$ and every set from the σ-algebra $\mathfrak{S}^{\text{inv}}$ coincides, up to a set of vanishing measure, with the union of a finite or countable number of elements of this sequence. According to Lemma 1, for every C_i we can find an invariant set C_i' satisfying $\mu(C_i \triangle C_i') = 0$. It is clear from the description of $\mathfrak{S}^{\text{inv}}$ that there is no invariant set A of positive measure contained within some C_i' and satisfying $\mu(A) < \mu(C_i')$. Therefore, if we consider C_i' by itself as a measure space, the σ-algebra of measurable sets of which is the system of sets of the form $B \cap C_i$, $B \in \mathfrak{S}$, while the measure is the function $\mu_i(B) = \mu(B \cap C_i')/\mu(C_i')$, then the dynamical system $\{T^t\}$ restricted to C_i' is ergodic. Thus the entire space M, up to a set of zero measure, has been represented as the union of a countable number of subsets on each of which the dynamical system is ergodic. This representation is said to be the *decomposition of the dynamical system $\{T^t\}$ into ergodic components.*

Consider another particular case of a nonergodic dynamical system. Suppose that the measure space $(M, \mathfrak{S}, \mu)$ is the Cartesian product of two measure spaces $(M_1, \mathfrak{S}_1, \mu_1)$ and $(M_2, \mathfrak{S}_2, \mu_2)$. To be definite, let us assume that we are dealing with a flow. Suppose that for every $x_1 \in M_1$ we are given an ergodic flow $\{T_{x_1}^t\}$ on the space M_2. Assume that $\{T_{x_1}^t\}$ depends on x_1 measurably in the following sense: for any measurable function $f(x_1, x_2)$ the function $f(x_1, T_{x_1}^t x_2)$ is measurable on the Cartesian product $M_1 \times M_2 \times \mathbb{R}^1$. Consider the flow $\{T^t\}$ on the space M defined by the formula $T^t(x_1, x_2) = (x_1, T_{x_1}^t x_2)$. Clearly $\{T^t\}$ preserves the measure $\mu = \mu_1 \times \mu_2$. Let us study the σ-algebra $\mathfrak{S}^{\text{inv}}$ for this flow.

For any set A invariant with respect to the entire flow $\{T^t\}$ consider the set $A_{x_1} = \{x_2 : (x_1, x_2) \in A\}$. It follows from the definition of $\{T^t\}$ that A_{x_1} for any x_1 is an invariant set with respect to $\{T^t_{x_1}\}$. For all x_1 such that A_{x_1} is measurable, we have $\mu_2(A_{x_1}) = 0$ or 1 in view of the ergodicity of $\{T^t_{x_1}\}$. Suppose $B \in \mathfrak{S}_1$ is the set of all x_1 satisfying $\mu_2(A_{x_1}) = 1$. Then $\mu_1(B) = \mu(A)$, and for the set $A_1 = B \times M_2$ we obviously have $\mu(A \triangle A_1) = 0$. It also follows from the definition of the flow $\{T^t\}$ that A_1 is invariant. Thus every invariant set $A \in \mathfrak{S}^{\text{inv}}$, up to a set of zero measure, can be presented in the form $A = B \times M_2$, where $B \in \mathfrak{S}_1$. In other words, the entire measure space M can be decomposed into elements $M_{x_1} = \{x_1\} \times M_2$ on which the flow $\{T^t\}$ is ergodic. It is natural to call this decomposition, just as above, *the decomposition into ergodic components*. Under sufficiently general conditions the whole space may be represented in the form of the sum of two subsets, on the first of which the decomposition into ergodic components is of the form described in the first example; and on the second—of the form described in the second example. However, we shall not provide the proof of this statement.

§4. Averaging in the Ergodic Case

It follows from the Birkhoff–Khinchin theorem that for an ergodic system T and any function $f \in L^2(M, \mathfrak{S}, \mu)$, in the case of a dynamical system with discrete time, we have

$$\lim_{n\to\infty} \frac{1}{n} \sum_{k=0}^{n-1} f(T^k x) = \int_M f(x)\, d\mu.$$

Multiplying the right- and left-hand sides by an arbitrary function $g \in L^2(M, \mathfrak{S}, \mu)$, we get $f(T^k x) \cdot g(x) \in L^1(M, \mathfrak{S}, \mu)$ for all k. Integrating with respect to x, we obtain

$$\lim_{n\to\infty} \frac{1}{n} \sum_{k=0}^{n-1} \int_M f(T^k x) g(x)\, d\mu = \int_M f(x)\, d\mu \int_M g(x)\, d\mu. \tag{1}$$

This last equality is equivalent to ergodicity. Indeed it implies that for any two invariant sets A_1, A_2

$$\lim_{n\to\infty} \frac{1}{n} \sum_{k=0}^{n-1} \int_M \chi_{A_1}(T^k x) \chi_{A_2}(x)\, d\mu = \int_M \chi_{A_1}(x)\chi_{A_2}(x)\, d\mu$$

$$= \mu(A_1 \cap A_2) = \mu(A_1)\mu(A_2). \tag{2}$$

Therefore for $A_1 = A_2$ we get $\mu(A_1) = \mu^2(A_1)$, i.e., $\mu(A_1) = 0$ or 1.

In particular, if we put $f = \chi_{A_1}, g = \chi_{A_2}$, then (2) implies

$$\lim_{n\to\infty} \frac{1}{n} \sum_{k=0}^{n-1} \mu(T^k A_1 \cap A_2) = \mu(A_1) \cdot \mu(A_2).$$

We can say that any two sets A_1 and A_2 are statistically independent (in the mean). Of course this is a very weak form of independence, but under simple ergodicity conditions more cannot be expected. This will be visible on very simple examples in the sequel.

For a flow $\{T^t\}$, we have similarly

$$\lim_{t\to\infty} \frac{1}{t} \int_0^t d\tau \int_M f(T^\tau x)g(x)\, d\mu = \lim_{t\to\infty} \frac{1}{t} \int_0^t d\tau \int_M f(T^{-\tau} x)g(x)\, d\mu$$
$$= \int_M f\, d\mu \cdot \int_M g\, d\mu.$$

Suppose that in a measure space $(M, \mathfrak{S}, \mu)$, supplied with an ergodic flow $\{T^k\}$, we are given another measure μ_0 which is absolutely continuous with respect to the measure μ and the density $d\mu_0/d\mu = p(x) \neq \text{const}$. In the physics literature μ_0 is sometimes called a nonequilibrium distribution, and μ is an equilibrium distribution. For example, if the dynamical system is generated by the motion of material points (molecules) in a fixed volume V, then the phase space is the set of coordinates and impulses of all the points. If we assume the system closed, i.e., fix the total energy of the particles, then the phase space often turns out to be compact. In this case, under natural assumptions, as we shall see later, there exists a measure invariant with respect to the dynamical system. If we choose a part of the volume V_1 and assume that at the initial moment all the particles were concentrated in V_1, then the set of all admissible coordinates and impulses of molecules constitutes a subset A of the phase space. The formula $\mu_0(B) = \mu(A \cap B)/\mu(A)$ determines a nonequilibrium distribution μ_0.

If we are given the measure μ_0, we can construct the measure μ_t, where $\mu_t(B) = \mu_0(T^{-t}B)$. For any set A, in view of the ergodicity of the flow $\{T^t\}$, we have

$$\lim_{t\to\infty} \frac{1}{t} \int_0^t \mu_\tau(A)\, d\tau = \lim_{t\to\infty} \frac{1}{t} \int_0^t d\tau \int_M \chi_A(T^\tau x)\, d\mu = \mu(A).$$

Thus the measures μ_t converge to μ in the mean. This convergence does not exclude the possibility of having, at far-off moments t, a measure μ_t which does not resemble μ, i.e., ergodicity alone does not imply that a nonequilibrium distribution will tend to an equilibrium one with time. In order to have such a convergence, it is necessary that the dynamical system possess the mixing properties which shall be discussed in the following sections (see §6).

§5. Integral and Induced Automorphisms

In this section we describe two constructions which enable us to determine new automorphisms from a given one. Suppose T is an automorphism of the measure space $(M, \mathfrak{S}, \mu)$ and $E \in \mathfrak{S}$ satisfies $\mu(E) > 0$. Let us transform E into a space with normalized measure by choosing the σ-algebra $\mathfrak{S}_E$ consisting of all subsets $A \subset E$, $A \in \mathfrak{S}$ and setting $\mu_E(A) = \mu(A)/\mu(E)$.

Introduce the subset $E' \subset E$ consisting of all x such that $T^n x \in E$ for an infinite set of positive and an infinite set of negative values of n. Clearly if $x \in E'$, then $T^n x \in E'$ for all such n. It follows from the Poincaré recurrence theorem that $\mu(E') = \mu(E)$. Therefore we further assume that $E' = E$. Now for E there is a function $k_E(x) = \min\{n \geq 1 : T^n x \in E\}$. It is natural to call this function the *return time* in E.

Lemma 1.

$$\int_E k_E(x)\, d\mu = \mu\left(\bigcup_{n \geq 0} T^n E\right).$$

Proof. Suppose $E_n = \{x \in E : k_E(x) = n\}$. Then the sets $T^i E_n$, $n = 1, 2, \ldots,$ $0 \leq i < n$, are pairwise disjoint and $\bigcup_{n \geq 0} T^n E = \bigcup_{n \geq 1} \bigcup_{i=0}^{n-1} T^i E_n$. Therefore

$$\mu\left(\bigcup_{n \geq 0} T^n E\right) = \mu\left(\bigcup_{n \geq 1} \bigcup_{i=0}^{n-1} T^i E_n\right) = \sum_{n=1}^{\infty} \sum_{i=0}^{n-1} \mu(T^i E_n) = \sum_{n=1}^{\infty} n\mu(E_n)$$

$$= \int_E k_E(x)\, d\mu.$$

The lemma is proved. □

It follows from the lemma that $k_E(x) \in L^1(E, \mathfrak{S}_E, \mu_E)$.

Put $T_E x = T^{k_E(x)} x$, $x \in E$. Then clearly the transformation T_E is one-to-one. Let us show that T_E preserves the measure μ_E. Any $A \subset E$ may be presented in the form $A = \bigcup_{n=1}^{\infty} A_n$ where $A_n = A \cap \{x : k_E(x) = n\}$. Then $T_E A = \bigcup_n T^n A_n$ and the sets $T^n A_n$ are obviously disjoint. Therefore

$$\mu_E(T_E A) = \sum_{n=1}^{\infty} \mu_E(T^n A_n) = \sum_{n=1}^{\infty} \mu_E(A_n) = \mu_E(A).$$

Corollary. *Suppose T is an ergodic automorphism and $E \subset M$, $\mu(E) > 0$. Then* $\int_E k_E(x)\, d\mu_E = 1/\mu(E)$.

Indeed, the ergodicity of T implies $\mu(\bigcup_{n>0} T^n E) = 1$; by Lemma 1 we have $\sum_{n=1}^{\infty} n\mu(E_n) = 1$. Therefore

$$\frac{1}{\mu(E)} = \sum_{n=1}^{\infty} n \frac{\mu(E_n)}{\mu(E)} = \sum_{n=1}^{\infty} n\mu_E(E_n) = \int_E k_E(x)\, d\mu_E$$

This corollary is known as the *Kac lemma.*

Definition 1. The automorphism T_E of a measure space $(E, \mathfrak{S}_E, \mu_E)$ is said to be the *induced automorphism* constructed from the automorphism T and the set E.

Suppose $E_2 \supset E_1$. The definition immediately implies $T_{E_2} = (T_{E_1})_{E_2}$.

Now let us describe the "dual" construction, and then study the properties of both constructions simultaneously. Assuming that we are given an automorphism T of the measure space $(M, \mathfrak{S}, \mu)$, consider the measurable integer valued positive function $f \in L^1(M, \mathfrak{S}, \mu)$. By using this function construct a new measure space M^f, whose points are of the form (x, i), where $x \in M$, $1 \le i \le f(x)$ and i is an integer. The σ-algebra of measurable sets in M^f is constructed in an obvious way. The measure μ^f is defined as follows: for any subset of the form (A, i), $A \in \mathfrak{S}$ we put

$$\mu^f((A, i)) = \frac{\mu(A)}{\int_M f(x)\, d\mu}$$

Let

$$T^f(x, i) = \begin{cases} (x, i+1) & \text{if } i+1 \le f(x), \\ (Tx, 1) & \text{if } i+1 > f(x). \end{cases}$$

It is easy to check that T^f preserves the measure μ^f.

Definition 2. The automorphism T^f of the space M^f is said to be the *integral automorphism* corresponding to the automorphism T and the function f.

The space M^f can naturally be visualized as a "tower" whose foundation is the space M and which has $f(x)$ floors over the point $x \in M$. Under the action of T^f the point $(x, i) \in M^f$ is lifted vertically up one floor, if this is possible, and, if this is impossible, it is lowered down to the ground floor where it takes the position of the point $(Tx, 1)$. The space M is identified with the set of points $(x, 1)$.

It easily follows from Definitions 1 and 2 that the given automorphism T is the induced automorphism of the automorphism T^f; more precisely,

$T = (T^f)_M$. Let us prove the converse statement: if T_E is the induced automorphism of T and $\bigcup_{n\geq 0} T^n E = M$, then T may be represented as the integral automorphism corresponding to the automorphism T_E and to the function $k_E(x)$.

First of all, by the remark which follows the proof of Lemma 1, we have $k_E \in L^1(E, \mathfrak{S}_E, \mu_E)$. Further, if $E_n = \{x \in E, k_E(x) = n\}$ then each point x belonging to $T^i E_n$ for some i, $0 \leq i < n$ may be written in the form $(T^{-i}x, i)$, $T^{-i}x \in E_n$. This shows that M may be represented as the space E^{k_E} and for this representation T acts as the integral automorphism.

Theorem 1. (1) *If the automorphism T is ergodic, then any of its induced automorphisms T_E is also ergodic.*

(2) *If T^f is the integral automorphism corresponding to the automorphism T and the function f, and T is ergodic, then so is T^f.*

Proof. (1) Let A be an invariant set for T_E, with $\mu(A) > 0$. Then $(\bigcup_{n\geq 0} T^n A) \cap E = A$. But the ergodicity of T implies $\bigcup_{n\geq 0} T^n A = M$ (mod 0), therefore $A = M \cap E$ (mod 0), i.e., $A = E$ (mod 0).

(2) If A is an invariant set of positive measure for T^f, then $A \cap M$ is an invariant set of positive measure for T. In view of the ergodicity of T, we have $A \cap M = M$ (mod 0). But then, obviously, $A = M^f$ (mod 0). The theorem is proved. □

§6. Weak Mixing, Mixing, Multiple Mixing

Definition 1. The dynamical system T possesses the *weak mixing property* or, simply, is *weak mixing*, if for all $f, g \in L^2(M, \mathfrak{S}, \mu)$ we have

$$\lim_{n\to\infty} \frac{1}{n} \sum_{k=0}^{n-1} \left[\int_M f(T^k x) g(x)\, d\mu - \int_M f\, d\mu \cdot \int_M g\, d\mu\right]^2$$

$$= \lim_{n\to\infty} \frac{1}{n} \sum_{k=0}^{n-1} \left[\int_M f(T^{-k} x) g(x)\, d\mu - \int_M f\, d\mu \cdot \int_M g\, d\mu\right]^2$$

$$= \lim_{n\to\infty} \frac{1}{2n+1} \sum_{k=-n}^{n} \left[\int_M f(T^k x) g(x)\, d\mu - \int_M f\, d\mu \cdot \int_M g\, d\mu\right]^2 = 0,$$

in the case of an automorphism T;

$$\lim_{n\to\infty} \frac{1}{n} \sum_{k=0}^{n-1} \left[\int_M f(T^k x) g(x)\, d\mu - \int_M f\, d\mu \cdot \int_M g\, d\mu\right]^2 = 0,$$

in the case of an endomorphism T;

$$\lim_{t\to\infty}\frac{1}{t}\int_0^t\left[\int_M f(T^\tau x)g(x)\,d\mu-\int_M f\,d\mu\cdot\int_M g\,d\mu\right]^2 d\tau$$

$$=\lim_{t\to\infty}\frac{1}{t}\int_0^t\left[\int_M f(T^{-\tau}x)g(x)\,d\mu-\int_M f\,d\mu\cdot\int_M g\,d\mu\right]^2 d\tau$$

$$=\lim_{t\to\infty}\frac{1}{2t}\int_{-t}^t\left[\int_M f(T^\tau x)g(x)\,d\mu-\int_M f\,d\mu\cdot\int_M g\,d\mu\right]^2 d\tau=0,$$

in the case of a flow $\{T^t\}$;

$$\lim_{t\to\infty}\frac{1}{t}\int_0^t\left[\int_M f(T^\tau x)g(x)\,d\mu-\int_M f\,d\mu\cdot\int_M g\,d\mu\right]^2=0,$$

in the case of a semiflow $\{T^t\}$.

Let us show that any dynamical system possessing the weak mixing property is ergodic. We shall carry out the argument only in the case of an endomorphism. The other cases are similar.

Suppose the endomorphism T is weak mixing. Choose two functions $f, g \in L^2(M, \mathfrak{S}, \mu)$. Then

$$\lim_{n\to\infty}\frac{1}{n}\sum_{k=0}^{n-1}\left|\int_M f(T^k x)g(x)\,d\mu-\int_M f\,d\mu\cdot\int_M g\,d\mu\right|$$

$$=\lim_{n\to\infty}\frac{1}{n}\sum_{k=0}^{n-1}\left[\int_M f(T^k x)g(x)\,d\mu-\int_M f\,d\mu\cdot\int_M g\,d\mu\right]^2=0.$$

(This follows from the fact that any bounded sequence $\{c_n\}$ of complex numbers the conditions

$$\lim_{n\to\infty}\frac{1}{n}\sum_{k=0}^{n-1}|c_k|\to 0 \quad\text{and}\quad \lim_{n\to\infty}\sum_{k=0}^{n-1}|c_k|^2\to 0$$

are equivalent.) Thus

$$\lim_{n\to\infty}\left|\frac{1}{n}\sum_{k=0}^{n-1}\int_M f(T^k x)g(x)\,d\mu-\int_M f\,d\mu\cdot\int_M g\,d\mu\right|$$

$$\leq\lim_{n\to\infty}\frac{1}{n}\sum_{k=0}^{n-1}\left|\int_M f(T^k x)g(x)\,d\mu-\int_M f\,d\mu\cdot\int_M g\,d\mu\right|=0.$$

As was pointed out in §4 (see formula (1) in §4) this last relation is equivalent to the ergodicity of T.

Definition 2. A dynamical system possesses *the mixing property* or, simply, is *mixing*, if for any two functions $f, g \in L^2(M, \mathfrak{S}, \mu)$ we have

$$\lim_{n \to \pm\infty} \int_M f(T^n x) g(x)\, d\mu = \int_M f(x)\, d\mu \cdot \int_M g(x)\, d\mu,$$

in the case of an automorphism;

$$\lim_{n \to \infty} \int_M f(T^n x) g(x)\, d\mu = \int_M f(x)\, d\mu \cdot \int_M g(x)\, d\mu,$$

in the case of an endomorphism;

$$\lim_{t \to \pm\infty} \int_M f(T^t x) g(x)\, d\mu = \int_M f(x)\, d\mu \cdot \int_M g(x)\, d\mu,$$

in the case of a flow;

$$\lim_{t \to \infty} \int_M f(T^t x) g(x)\, d\mu = \int_M f(x)\, d\mu \cdot \int_M g(x)\, d\mu,$$

in the case of a semiflow.

Putting $f = \chi_{A_1}$, $g = \chi_{A_2}$ in these relations, we see that in the case of a mixing, for any two sets $A_1, A_2 \in \mathfrak{S}$ and any endomorphism or automorphism T, we have

$$\lim_{n \to \infty} \mu(T^{-n} A_1 \cap A_2) = \mu(A_1) \cdot \mu(A_2).$$

The last relation means the following: if we take any set A_2 of positive measure, then any set of positive measure A_1 in the process of its motion beginning with some moment of time will always intersect the set A_2, and the measure of that part of A_1 which is contained in A_2 at the moment n is asymptotically proportional (for $n \to \infty$) to the measure of A_2. It is precisely this property which explains the origin of the following expression: "A set A_1 of positive measure in its motion mixes uniformly in the phase space." In the physics literature the term "relaxation" is used to describe processes under which the system passes to a certain stationary state independently of its original state. Using this term, we can say that systems with the mixing property are relaxation systems in the following sense: if μ_0 is an arbitrary normalized measure absolutely continuous with respect to the measure μ and μ_n is the translate of this measure under the action of T^n (i.e., $\mu_n(A) = \mu_0(T^{-n}A)$, then $\mu_n(A) \to \mu(A)$ for any measurable A in the case of a mixing system. Indeed if we denote $p(x) = (d\mu_0/d\mu)(x)$, we obtain

$$\mu_n(A) = \int_M \chi_A(T^n x)\, d\mu_0 = \int_M \chi_A(T^n x)p(x)\, d\mu$$

$$\to \int_M \chi_A(x)\, d\mu \cdot \int_M p(x)\, d\mu = \mu(A).$$

Thus we can say that it is precisely in dynamical systems with the mixing property that any nonequilibrium distribution tends to an equilibrium one with time.

Since mixing obviously implies weak mixing, any dynamical system with the mixing property is ergodic. We shall see further that there exist ergodic systems without the weak mixing property, systems with weak mixing without the mixing property, as well as systems with the mixing property.

We can introduce notions which characterize "a stronger mixing" than the one considered above.

Definition 3. A dynamical system has the mixing property of multiplicity $r \geq 1$ or, simply, is mixing of multiplicity r, if for any functions $f_0, f_1, \ldots, f_r \in L^{r+1}(M, \mathfrak{S}, \mu)$ we have

$$\lim_{k_1, \ldots, k_r \to \infty} \int_M f_0(x) f_1(T^{k_1}x) f_2(T^{k_2+k_2}x) \cdot \ldots \cdot f_r(T^{k_1+k_2+\ldots+k_r}x)\, d\mu$$

$$= \prod_{i=0}^{r} \int_M f_i(x)\, d\mu,$$

in the case of an automorphism or endomorphism;

$$\lim_{t_1, \ldots, t_r \to \infty} \int_M f_0(x) f_1(T^{t_1}x) \cdot \ldots \cdot f_r(T^{t_1+\cdots+t_r}x)\, d\mu = \prod_{i=0}^{r} \int_M f_i(x)\, d\mu,$$

in the case of a flow or semiflow.

For measurable sets $A_0, A_1, \ldots, A_r$ in the case of a mixing of multiplicity r we have

$$\lim_{k_1, \ldots, k_r \to \infty} \mu(A_0 \cap T^{-k_1}A_1 \cap \cdots \cap T^{-(k_1+\cdots+k_r)}A_r) = \mu(A_0) \cdot \ldots \cdot \mu(A_r)$$

It is clear that ordinary mixing is a mixing of multiplicity 1, while for $r > 1$ mixing of multiplicity r implies ordinary mixing.

The K-mixing introduced further is convenient in that it can be effectively checked in a number of important classes of dynamical systems. We give the definition in the case of automorphisms.

Definition 4. The automorphism T is *K-mixing*, if for any sets $A_0, A_1, \ldots, A_r \in \mathfrak{S}$, $r \geq 0$, we have

$$\lim_{n \to \infty} \sup_{B \in \mathfrak{S}_n^\infty(A_1, \ldots, A_r)} |\mu(A_0 \cap B) - \mu(A_0)\mu(B)| = 0,$$

where $\mathfrak{S}_n^\infty(A_1, \ldots, A_r)$ is the minimal σ-algebra generated by the set $T^k A_i$ for $k \geq n$, $i = 1, \ldots, r$.

K-mixing means that the motion of the set A_0 does not depend on any events taking place on the sufficiently far off part of the trajectory of the sets A_i, $i \geq 1$. It is easy to see that K-mixing implies mixing of any multiplicity. At the same time there exist dynamical systems which are mixings of any multiplicity but do not possess the K-mixing property. The notion of K-mixing will be considered again in Chap. 10 when we study the entropy of dynamical system.

§7. Unitary and Isometric Operators Adjoint to Dynamical Systems

Any measurable map T of the measurable space $(M, \mathfrak{S})$ into itself generates the operator U_T which acts in the space of measurable complex functions on M according to the formula $(U_T f)(x) = f(Tx)$. U_T is said to be the operator adjoint to the transformation T.

In a more general situation, for a one-parameter group or semigroup $\{T^t\}$ of measurable transformations of the measurable space $(M, \mathfrak{S})$ we can introduce the adjoint group or semigroup of operators $\{U^t\} = \{U_{T^t}\}$.

Any operator U_T of the form considered has eigen-functions $f \equiv \text{const}$ corresponding to the eigen-value 1.

In this section we consider operators adjoint to dynamical systems.

Lemma 1. (1) *If $\{T^t\}$ is a one-parameter group of automorphisms (or a semigroup of endomorphisms) of the measure space $(M, \mathfrak{S}, \mu)$, then the complex Hilbert space $L^2(M, \mathfrak{S}, \mu)$ is invariant with respect to the adjoint group (or semigroup) of operators $\{U^t\}$. In this case the group (or semigroup) $\{U^t\}$, considered on $L^2(M, \mathfrak{S}, \mu)$, is the group of unitary operators (or the semigroup of isometric operators) of the space $L^2(M, \mathfrak{S}, \mu)$.*

(2) *If $\{T^t\}$ is a flow and the Hilbert space $L^2(M, \mathfrak{S}, \mu)$ is separable, then the adjoint group $\{U^t\}$ is continuous.*

Proof. (1) The invariance of the measure μ implies that $\int_M |f(T^t x)|^2 \, d\mu = \int_M |f(x)|^2 \, d\mu$ for any function $f \in L^2(M, \mathfrak{S}, \mu)$ and any $t \in \mathbb{R}^1$, hence the space $L^2(M, \mathfrak{S}, \mu)$ is invariant with respect to $\{U^t\}$. Further for all $f, g \in L^2(M, \mathfrak{S}, \mu)$ we have

$$(U^t f, U^t g)_{L^2} = \int_M f(T^t x)\overline{g(T^t x)} \, d\mu = \int_M f(x)\overline{g(x)} \, d\mu = (f, g)_{L^2},$$

i.e., all the operators U^t on $L^2(M, \mathfrak{S}, \mu)$ are isometric.

If $\{T^t\}$ is the automorphism group, then $(U^t)^{-1} = U^{-t}$, i.e., the operators U^t are invertible and therefore are unitary operators in $L^2(M, \mathfrak{S}, \mu)$.

(2) If $\{T^t\}$ is a flow, then for any $f, g \in L^2(M, \mathfrak{S}, \mu)$ the function $b(t) = (U^t f, g) = \int_M f(T^t x)\overline{g(x)} \, d\mu$ is a measurable function in t, i.e., $\{U^t\}$ is a measurable group of unitary operators. But in separable Hilbert spaces any measurable group of unitary operators is continuous (see Appendix 2). The lemma is proved. □

In the sequel, operators, as well as groups of operators, adjoint to dynamical systems will as a rule be considered in the space $L^2(M, \mathfrak{S}, \mu)$.

If the dynamical systems $\{T_1^t\}$, $\{T_2^t\}$ acting on the measure spaces $(M_1, \mathfrak{S}_1, \mu_1)$, $(M_2, \mathfrak{S}_2, \mu_2)$, respectively, are metrically isomorphic, then their isomorphism φ induces an isomorphism V_φ of the Hilbert spaces $L^2(M_1, \mathfrak{S}_1, \mu_1)$ and $L^2(M_2, \mathfrak{S}_2, \mu_2)$ which acts according to the formula $(V_\varphi f)(x) = f(\varphi(x))$. In this case if $\{U_1^t\}$, $\{U_2^t\}$ are the adjoint groups of unitary operators, then $U_2^t = VU_1^t V^{-1}$. In the spectral theory of operators this last relation is called unitary equivalence. Thus if dynamical systems are metrically isomorphic then the corresponding adjoint groups (or semigroups) of unitary operators are unitarily equivalent. (The converse statement, as we shall see later, is false).

Spectral invariants of a dynamical system are by definition spectral invariants of the group (or semigroup) adjoint to it $\{U^t\}$. The properties of a dynamical system which can be expressed in terms of its spectral invariants are said to be *spectral properties*. In Part III of the book we will study the corresponding spectral properties for certain dynamical systems.

Suppose T is an automorphism of the measure space $(M, \mathfrak{S}, \mu)$. For any element $f \in L^2(M, \mathfrak{S}, \mu)$ we can consider the scalar products

$$b_n = b_n(f) = (U_T^t f, f), \qquad -\infty < n < \infty.$$

The sequence $\{b_n\}$ is positive definite. Indeed for any family of complex numbers $\xi_0, \ldots, \xi_m$, since U_T is isometric, we have the inequality

$$0 \le \left\| \sum_{k=0}^{m} \xi_k U_T^k f \right\|^2 = \left(\sum_{i=0}^{m} \xi_i U_T^i f, \sum_{j=0}^{m} \xi_j U_T^j f \right) = \sum_{i,j=0}^{m} b_{i-j} \xi_i \bar{\xi}_j. \tag{1}$$

By Bohner's theorem, $\{b_n\}$ can be presented in the form

$$b_n(f) = \int_0^1 \exp(2\pi i n\lambda)\, d\sigma_f(\lambda), \tag{2}$$

where σ_f is a finite Borel measure on the unit circle S^1 and

$$\sigma_f(S^1) = (f, f) = \|f\|^2.$$

In the case of the endomorphism T the sequence $b_n(f) = (U_T^n f, f)$ is defined only for $n \geq 0$. For negative n put $b_n = \bar{b}_{-n}$. Then the equality (1) remains valid, so that we also have the representation (2).

Now suppose $\{T^t\}$ is a flow on the space $(M, \mathfrak{S}, \mu)$ and assume that the Hilbert space $L^2(M, \mathfrak{S}, \mu)$ is separable. For any element $f \in L^2(M, \mathfrak{S}, \mu)$, consider the scalar products

$$b(t) = b_f(t) = (U^t f, f), \qquad t \in \mathbb{R}^1.$$

By Lemma 1, $b(t)$ is continuous. Also $b(t)$ is a positive definite function, i.e., for any $(m + 1)$-tuple of complex numbers $\xi_0, \ldots, \xi_m$ and any $(m + 1)$-tuple of real numbers $t_0, \ldots, t_m$ we have

$$0 \leq \left\| \sum_{k=0}^{m} \xi_k U^{t_k} f \right\|^2 = \left(\sum_{i=0}^{m} \xi_i U^{t_i} f, \sum_{j=0}^{m} \xi_j U^{t_j} f \right) = \sum_{i,j=0}^{m} b(t_i - t_j)\xi_i \bar{\xi}_j. \tag{3}$$

By Bohner's theorem, $b(t)$ can be presented in the form

$$b_f(t) = \int_{-\infty}^{\infty} \exp(it\lambda)\, d\sigma_f(\lambda), \tag{4}$$

where σ_f is a finite Borel measure on $\mathbb{R}^1$, and

$$\sigma_f(\mathbb{R}^1) = (f, f) = \|f\|^2.$$

The measures σ_f which appear in the integrals (2) and (4) are said to be *spectral measures* of the element f with respect to the corresponding dynamical system (see Appendix 2). The spectral theory of dynamical systems studies the properties of dynamical systems which can be expressed in terms of the properties of the measures σ_f. These properties are also spectral properties.

Let us illustrate (by two examples) the type of information concerning dynamical systems which can be obtained from its spectral properties.

1. Suppose $\{T^t\}$ is a flow, $\{U^t\}$ is a corresponding one-parameter group of unitary operators. Assume that the group has an eigen-function $f(x)$ corresponding to a nonzero eigen-value, i.e., $U^t f = \exp(i\alpha t) f$ for some $\alpha \neq 0$.

Consider the level sets of this function: $C_a = \{x : f(x) = a\}$. Then $T^t C_a = C_{e^{i\alpha t}a}$. For $t_0 = 2\pi/\alpha$, we obtain $T^{t_0} C_a = C_a$. This means that the decomposition of the space M into the sets C_a is a "measurable fibering" over the circle, for which the C_a are the fibers. Under the action of the dynamical system these fibers are mapped one into the other, and each fiber returns in its place after the period t_0. Of course the map T^{t_0} of the fiber C_a onto itself will not necessarily be the identity.

2. Suppose M is the space of sequences infinite in both directions with stationary measure μ (see Example 3, p. 5), T is a translation. From the point of view of probability theory the element $f(x)$ of the Hilbert space $L^2(M, \mathfrak{S}, \mu)$ is in general a nonlinear square integrable functional defined on the realizations x of the stationary process considered. Stationarity implies that the family of random variables $y_k = f(T^k x) = U_{T^k} f$, $-\infty < k < \infty$, again constitutes a stationary random process. We can construct the correlation function of this process $b_{k-l} = E y_k y_l = \int y_k(x) \bar{y}_l(x)\, d\mu(x)$ and its spectrum decomposition $b_r = \int_0^{2\pi} \exp(-2\pi i \lambda r)\, d\sigma_f(\lambda)$. The measure σ_f depends on the functional f. The spectral theory of dynamical system studies the set of measures σ_f which can arise in this way.

The Theorems 1, 2, and 3 given below in this section, show that ergodicity, weak mixing, and mixing are spectral properties. We shall prove this for automorphisms, since in the other cases the argument is similar.

Theorem 1. *Suppose T is an automorphism of the measure space $(M, \mathfrak{S}, \mu)$. The following conditions are equivalent*:

(i) *T is ergodic*;
(ii) *any function $f \in L^2(M, \mathfrak{S}, \mu)$, invariant with respect to U_T, is a constant* (mod 0);
(iii) *any function $f \in L^2(M, \mathfrak{S}, \mu)$ such that the measure σ_f is concentrated at the point $\lambda = 0$ is a constant* (mod 0).

Theorem 2. *Suppose T is an automorphism of the measure space $(M, \mathfrak{S}, \mu)$. The following conditions are equivalent*:

(i) *T possesses a weak mixing*;
(ii) *U_T has no eigenfunctions which are not constants* (mod 0);
(iii) *for any function*

$$f \in L_0^2(M, \mathfrak{S}, \mu) \stackrel{\text{def}}{=} \left\{ f \in L^2(M, \mathfrak{S}, \mu) : \int f\, d\mu = 0 \right\},$$

the measure σ_f is continuous.

Theorem 3. *The automorphism T of the measure space $(M, \mathfrak{S}, \mu)$ possesses a mixing if and only if for any function $f \in L_0^2(M, \mathfrak{S}, \mu)$ the Fourier coefficients $\{b_n\}$ of the spectral measure σ_f tend to zero when $|n| \to \infty$.*

For the proof we shall need the following two lemmas:

Lemma 2. *The function* $f \in L^2(M, \mathfrak{S}, \mu)$ *is an eigenfunction of the operator* U_T *corresponding to the eigen-value* $\exp(2\pi i\lambda_0)$ *if and only if the measure* σ_f *is concentrated at the point* λ_0, *i.e.*, $\sigma_f = \|f\|^2\,\delta(\lambda - \lambda_0)$.

Lemma 3 (Wiener's Lemma). *Suppose* σ *is a finite Borel measure on the unit circle* $S^1 = \{\lambda: 0 \le \lambda < 1\}$, $b_n = \int_0^1 \exp(2\pi i n\lambda)\,d\sigma$, $-\infty < n < \infty$, *are its Fourier coefficients,* $\lambda_1, \lambda_2, \dots$ *is the sequence of all atoms of the measure* σ, *i.e.*, $\sigma(\{\lambda_n\}) > 0$, $n = 1, 2, \dots$, $\sigma(\{\lambda\}) = 0$ if $\lambda \ne \lambda_n$, $n = 1, 2, \dots$. *Then*

$$\lim_{n\to\infty} \frac{1}{n}\sum_{k=0}^{n-1} |b_k|^2 = \sum_{n=1}^{\infty} \sigma^2(\lambda_n).$$

The proof of the lemmas will be given below.

Proof of Theorem 1. If T is ergodic, then, as was shown in §2, any invariant measurable function with respect to T is a constant (mod 0). This of course is also true for $f \in L^2(M, \mathfrak{S}, \mu)$. If T is not ergodic and the set $E \in \mathfrak{S}$ is invariant, $0 < \mu(E) < 1$, then its indicator $\chi_E \in L^2(M, \mathfrak{S}, \mu)$ is an invariant function and $\chi_E \ne \text{const}$ (mod 0). Therefore, (i) $\Leftrightarrow$ (ii). The equivalence (ii) $\Leftrightarrow$ (iii) follows from Lemma 2. The theorem is proved. □

Proof of Theorem 2. Suppose T possesses a weak mixing, but in contradiction to our statement U_T has an eigen-function $f \ne \text{const}$ (mod 0) corresponding to the eigen-value λ, $|\lambda| = 1$. It is clear that $\lambda \ne 1$ since T is ergodic. Therefore the element $f \in L^2(M, \mathfrak{S}, \mu)$ is orthogonal to the subspace of constants so that we have $f \in L_0^2(M, \mathfrak{S}, \mu)$. But then

$$\frac{1}{n}\sum_{k=0}^{n-1}\left|\int_M f(T^k x)\overline{f(x)}\,d\mu - \left|\int_M f\,d\mu\right|^2\right|^2 = \frac{1}{n}\sum_{k=0}^{n-1}\left|\int_M f(T^k x)\overline{f(x)}\,d\mu\right|^2$$

$$= \frac{1}{n}\sum_{k=0}^{n-1}|(U_T^k f, f)|^2 = \frac{1}{n}\sum_{k=0}^{n-1}|\lambda^{2k}|(f, f)^2 = \|f\|^2 \ne 0,$$

which contradicts the definition of a weak mixing. Thus we have proved the implication (i) $\Rightarrow$ (ii). Let us show that (ii) $\Rightarrow$ (iii). Assume the converse, i.e., that for any element $f \in L_0^2(M, \mathfrak{S}, \mu)$ the spectral measure is not continuous, i.e., $\sigma_f(\{\lambda_0\}) > 0$, $\lambda_0 \in S^1$, then the normalized measure $\delta(\lambda - \lambda_0)$ concentrated at the point λ_0 is absolutely continuous with respect to σ_f. Hence in a subspace of $L_0^2(M, \mathfrak{S}, \mu)$, invariant with respect to U_T, there is an element g such that $\sigma_g = \delta(\lambda - \lambda_0)$ (see Appendix 2). By Lemma 2, g is an eigen-function, which contradicts (ii). Hence (ii) $\Rightarrow$ (iii). It remains to prove that (iii) $\Rightarrow$

(i). Take an arbitrary function $f \in L_0^2(M, \mathfrak{S}, \mu)$. Since its spectral measure is continuous, we have by Lemma 3

$$\lim_{n\to\infty} \frac{1}{n} \sum_{k=0}^{n-1} |(U^k f, f)|^2 = \lim_{n\to\infty} \frac{1}{n} \sum_{k=0}^{n-1} \left| \int_0^1 \exp(2\pi i k\lambda)\, d\sigma_f(\lambda) \right|^2 = 0. \qquad (5)$$

Now consider two functions $f, g \in L_0^2(M, \mathfrak{S}, \mu)$. It is easy to verify the identity

$$(U_T^n f, g) = \tfrac{1}{4}[(U_T^n(f + g), f + g) + i(U_T^n(f + ig), f + ig)$$
$$- (U_T^n(f - g), f - g) - i(U_T^n(f - ig), f - ig)]. \qquad (6)$$

The functions $\psi_1 = f + g$, $\psi_2 = f - g$, $\psi_3 = f + ig$, $\psi_4 = f - ig$ belong to $L_0^2(M, \mathfrak{S}, \mu)$, hence each of them satisfies a relation similar to (5).

We will use the following simple statement which concerns numerical sequences: *for any bounded sequence $\{c_n\}$, $n = 0, 1, 2, \ldots$ of complex numbers the relation $\lim_{n\to\infty}(1/n)\sum_{k=0}^{n-1}|c_k|^2 = 0$ is equivalent to the existence of a set N_0 of natural numbers of zero density* (i.e., $\lim_{n\to\infty}(1/n)\,\mathrm{card}(N_0 \cap [1, n]) = 0$) *such that* $\lim_{n\to\infty, n\notin N_0}|c_n| = 0$.

It follows from this statement that for the functions ψ_i, $0 \le i \le 4$, we can find exceptional sets N_i, $1 \le i \le 4$ of zero density such that the scalar product $(U_T^n\psi_i, \psi_i)$ tends to zero when $n \to \infty$ outside of the exceptional set N_i. The set $N = \bigcup_{i=1}^4 N_i$ is also of zero density and if $n \to \infty$ outside of N, we have $\lim(U_T^n f, g) = 0$. Using the statement formulated above once again we obtain

$$\lim_{n\to\infty} \frac{1}{n} \sum_{k=0}^{n-1} |(U_T^k f, g)|^2 = 0.$$

In the general case, when $f, g \in L^2(M, \mathfrak{S}, \mu)$ we put

$$a = \int_M f\, d\mu, \qquad b = \int_M g\, d\mu, \qquad f_0 = f - a, \qquad g_0 = g - b.$$

Then $f_0, g_0 \in L_0^2(M, \mathfrak{S}, \mu)$ and

$$(U_T^k f, g) = (U_T^k(f_0 + a), g_0 + b) = (U_T^k f_0, g_0) + a \cdot \bar{b}$$
$$= (U_T^k f_0, g_0) + \int_M f\, d\mu \cdot \int_M \bar{g}\, d\mu.$$

Hence

$$\frac{1}{n} \sum_{k=0}^{n-1} \left| \int_M f(T^k x) g(x)\, d\mu - \int_M f\, d\mu \int_M g\, d\mu \right|^2 = \frac{1}{n} \sum_{k=0}^{n-1} |(U_T^n f_0, \bar{g}_0|^2 \to 0, \quad (7)$$

for $n \to \infty$. In a similar way

$$\frac{1}{n}\sum_{k=0}^{n-1}\left|\int_M f(T^{-k}x)g(x)\,d\mu - \int_M f\,d\mu \int_M g\,d\mu\right|^2 \to 0 \quad \text{when } n \to \infty. \tag{8}$$

It follows from (7) and (8) that T possesses a weak mixing. The theorem is proved. □

Proof of Theorem 3. If T possesses a mixing, then, for any function $f_0 \in L_0^2(M, \mathfrak{S}, \mu)$,

$$\lim_{|n|\to\infty} \int_0^1 \exp(2\pi i n\lambda)\,d\sigma_{f_0}(\lambda) = \lim_{|n|\to\infty} (U_T^n f_0, f_0)$$

$$= \lim_{|n|\to\infty} \int_M f_0(T^n x)\overline{f_0(x)}\,d\mu = 0. \tag{9}$$

Conversely, if (9) holds, we obtain, by using the identity (6),

$$\lim_{|n|\to\infty} (U_T^n f_0, g_0) = 0 \quad \text{for all } f_0, g_0 \in L_0^2(M, \mathfrak{S}, \mu).$$

In the general case, for $f, g \in L^2(M, \mathfrak{S}, \mu)$,

$$(U_T^n f, g) = (U_T^n f_0, g_0) + \int_M f\,d\mu \cdot \int_M g\,d\mu,$$

where

$$f_0 = f - \int f\,d\mu, \qquad g_0 = g - \int g\,d\mu; \qquad f_0, g_0 \in L_0^2(M, \mathfrak{S}, \mu).$$

Therefore $\lim_{|n|\to\infty}(U_T^n f, g) = \int_M f\,d\mu \cdot \overline{\int_M g\,d\mu}$, which is equivalent to the mixing property. The theorem is proved. □

It remains to prove Lemmas 2 and 3.

Proof of Lemma 2. If f is an eigen-function, then

$$(U_T^n f, f) = \exp(2\pi i n\lambda_0)\cdot(f, f) = \|f\|^2 \int_0^1 \exp(2\pi i n\lambda)\,d\delta(\lambda - \lambda_0),$$

can be checked directly. Conversely, if $\sigma_f = \|f\|^2 \delta(\lambda - \lambda_0)$ then

$$(U_T^n f, f) = \int_0^1 \exp(2\pi i n \lambda_0)\, d\sigma_f = \|f\|^2 \exp(2\pi i n \lambda_0). \tag{10}$$

For $n = 1$, we get

$$|(U_T f, f)| = \|f\|^2 = \|U_T f\| \cdot \|f\|,$$

i.e., for the functions f, $U_T f$ the Cauchy–Buniakowski inequality becomes an equality. This is possible only if $U_T f = c \cdot f$ where c is a complex number. By (10), $c = \exp(2\pi i \lambda_0)$. The lemma is proved. □

Proof of Lemma 3. We have

$$|b_n|^2 = b_n \bar{b}_n = \int_0^1 \exp(2\pi i n \lambda^{(1)})\, d\sigma(\lambda^{(1)}) \cdot \overline{\int_0^1 \exp(2\pi i n \lambda^{(2)})\, d\sigma(\lambda^{(2)})}$$

$$= \int_0^1 \int_0^1 \exp[2\pi i n(\lambda^{(1)} - \lambda^{(2)})]\, d(\sigma(\lambda^{(1)}) \times \sigma(\lambda^{(2)})).$$

Therefore

$$\frac{1}{n} \sum_{k=0}^{n} |b_k|^2 = \int_0^1 \int_0^1 \varphi_n(\lambda^{(1)}, \lambda^{(2)})\, d(\sigma(\lambda^{(1)}) \times \sigma(\lambda^{(2)})),$$

where

$$\varphi_n(\lambda^{(1)}, \lambda^{(2)}) = \frac{1}{n} \sum_{k=0}^{n-1} \exp[2\pi i k(\lambda^{(1)} - \lambda^{(2)})].$$

The functions φ_n are uniformly bounded: $|\varphi_n(\lambda^{(1)}, \lambda^{(2)})| \leq 1$, $n = 1, 2, \ldots$. Moreover

$$\lim_{n\to\infty} \varphi_n(\lambda^{(1)}, \lambda^{(2)}) = \begin{cases} \lim\limits_{n\to\infty} \dfrac{1}{n} \dfrac{\exp[2\pi i n(\lambda^{(1)} - \lambda^{(2)})] - 1}{\exp[2\pi i(\lambda^{(1)} - \lambda^{(2)})] - 1} = 0 & \text{if } \lambda^{(1)} \neq \lambda^{(2)}, \\ 1 & \text{if } \lambda^{(1)} = \lambda^{(2)}. \end{cases} \tag{11}$$

By the Lebesgue dominated convergence theorem, we have

$$\lim_{n\to\infty} \frac{1}{n} \sum_{k=0}^{n=1} |b_k|^2 = \int_0^1 \int_0^1 \lim_{n\to\infty} \varphi_n(\lambda^{(1)}, \lambda^{(2)})\, d(\sigma(\lambda^{(1)}) \times \sigma(\lambda^{(2)})).$$

By (11) it suffices to take this last integral not on the entire torus $S^1 \times S^1$, but only on its diagonal $D = \{(\lambda^{(1)}, \lambda^{(2)}): \lambda^{(1)} = \lambda^{(2)}\}$. The atoms of the measure $\sigma \times \sigma$ are only the points of the diagonal of the form (λ_n, λ_n), $n = 1, 2, \ldots$, where $(\sigma \times \sigma)(\{\lambda_n, \lambda_n\}) = \sigma^2(\lambda_n)$; moreover, it is easy to see that $(\sigma \times \sigma)(D \setminus \bigcup_{n=1}^{\infty}(\{\lambda_n, \lambda_n\}) = 0$. This implies

$$\lim_{n\to\infty} \frac{1}{n} \sum_{k=0}^{n-1} |b_k^2| = \iint_D d(\sigma(\lambda^{(1)}) \times \sigma(\lambda^{(2)})) = \sum_{n=1}^{\infty} \sigma^2(\lambda_n).$$

The lemma is proved. □

In terms of isometric or unitary operators adjoint to the dynamical system, it is possible to prove a statement similar to the Birkhoff–Khinchin theorem. This statement is known as the von Neumann ergodic theorem. Unlike the Birkhoff–Khinchin theorem, it considers convergence with respect to the norm of the Hilbert space, rather than convergence almost everywhere. We shall prove it for arbitrary isometric operators (not necessarily adjoint to some dynamical system).

Theorem 4 (The von Neumann Ergodic Theorem). *Suppose U is an isometric operator in complex Hilbert space H; H_U^{inv} is the subspace of vectors $f \in H$ invariant with respect to U i.e., $H_U^{\text{inv}} = \{f \in H: Uf = f\}$; P_U is the operator of orthogonal projection on H_U^{inv}. Then*

$$\lim_{n\to\infty} \left\| \frac{1}{n} \sum_{k=0}^{n-1} U^k f - P_U f \right\|_H = 0,$$

for any $f \in H$.

For the proof we shall need the following lemma.

Lemma 4. *If U^* is the operator adjoint to U, then $H_{U^*}^{\text{inv}} = H_U^{\text{inv}}$.*

The proof of the lemma will be given later.

Proof of Theorem 4. If $f \in H_U^{\text{inv}}$, then

$$\lim_{n\to\infty} \frac{1}{n} \sum_{k=0}^{n-1} U^k f = f = Pf,$$

and therefore the theorem is valid for such f.

Suppose further that f is of the form

$$f = Ug - g \quad \text{for some } g \in H \tag{12}$$

(such f are sometimes said to be *cohomologic to zero*). Then

$$\left\| \frac{1}{n} \sum_{k=0}^{n-1} U^k f \right\| = \left\| \frac{1}{n} \sum_{k=0}^{n-1} (U^{k+1}g - U^k g) \right\| = \left\| \frac{1}{n} (U^n g - g) \right\| \leq \frac{2}{n} \|g\| \to 0,$$

for $n \to \infty$.

On the other hand, if f is of the form (12), then $P_U f = 0$. Indeed, for any $h \in H_U^{\text{inv}}$ we have

$$(f, h) = (Ug - g, h) = (Ug, h) - (g, h) = (Ug, Uh) - (g, h) = 0 \quad (13)$$

Therefore $f \perp H_U^{\text{inv}}$, i.e.,

$$P_U f = 0 \tag{14}$$

and therefore the statement of the theorem is valid for all f of the form (12).

Suppose G_U is the subspace of H spanning the vectors of the form (12). Since those vectors form a linear manifold, G_U is the closure of the set of all such vectors. For arbitrary $f \in G_U$ and $\varepsilon > 0$, we can find a vector f_ε of the form $f_\varepsilon = Ug_\varepsilon - g_\varepsilon$ such that $\| f - f_\varepsilon \| < \varepsilon$. Then

$$\frac{1}{n} \sum_{k=0}^{n-1} U^k f = \frac{1}{n} \sum_{k=0}^{n-1} U^k f_\varepsilon + \frac{1}{n} \sum_{k=0}^{n-1} U^k (f - f_\varepsilon) = \textstyle\sum_1^{(n)} + \sum_2^{(n)}.$$

Since

$$\lim_{n \to \infty} \left\| \textstyle\sum_1^{(n)} \right\| = 0, \qquad \left\| \textstyle\sum_2^{(n)} \right\| \leq \frac{1}{n} \sum_{k=0}^{n-1} \| U^k (f - f_\varepsilon) \| < \varepsilon,$$

we have

$$\overline{\lim_{n \to \infty}} \left\| \frac{1}{n} \sum_{k=0}^{n-1} U^k f \right\| \leq \varepsilon$$

and since ε was arbitrary, we finally obtain

$$\lim_{n \to \infty} \left\| \frac{1}{n} \sum_{k=0}^{n-1} U^k f \right\| = 0.$$

It follows from (14) that $P_U f = 0$ if $f \in G_U$, i.e., the statement of the theorem is valid for all $f \in G_U$.

Now let us show that $H = H_U^{\text{inv}} \oplus G_U$, i.e., that

$$G_U^\perp = H_U^{\text{inv}},$$

where $G_U^{\perp}$ is the orthogonal complement to G_U. It follows from (13) that $H_U^{\text{inv}} \subset G_U^{\perp}$. If $h \in G_U^{\perp}$, then for all $g \in H$ we have $(h, Ug - g) = 0$, i.e.,

$$(h, g) = (h, Ug) = (U^*h, g).$$

Therefore $(U^*h - h, g) = 0$ for all $g \in H$, therefore $U^*h - h = 0$, $h \in H_{U^*}^{\text{inv}}$. By the lemma, $h \in H_U^{\text{inv}}$, i.e., $H_U^{\text{inv}} \supset G_U^{\perp}$ and the equality (15) is proved.

Now any vector $f \in H$ can be presented in the form $f = f_1 + f_2, f_1 \in H_U^{\text{inv}}$, $f_2 \in G_U$. Then

$$\lim_{n\to\infty} \frac{1}{n} \sum_{k=0}^{n-1} U^k f = \lim_{n\to\infty} \frac{1}{n} \sum_{k=0}^{n-1} U^k f_1 + \lim_{n\to\infty} \frac{1}{n} \sum_{k=0}^{n-1} U^k f_2 = P_U f_1 + P_U f_2 = P_U f.$$

The theorem is proved. □

Proof of Lemma 4. Suppose $f \in H_U^{\text{inv}}$, i.e., $Uf = f$. Applying the operator U^* to both parts of the equality and using the fact that $U^*U = id$ for isometric operators, we get $f = U^*f$, i.e., $f \in H_{U^*}^{\text{inv}}$. Conversely, if $f \in H_{U^*}^{\text{inv}}$ then

$$\|Uf - f\|^2 = (Uf - f, Uf - f) = \|Uf\|^2 - (f, Uf) - (Uf, f) + \|f\|^2.$$

But

$$\|Uf\|^2 = \|f\|^2, \qquad (f, Uf) = (U^*f, f) = (f, f) = \|f\|^2,$$
$$(Uf, f) = (f, U^*f) = (f, f) = \|f\|^2.$$

Therefore $\|Uf - f\|^2 = 0$, i.e., $f \in H_U^{\text{inv}}$. The lemma is proved. □

von Neumann's ergodic theorem is also valid for one-parameter groups (semigroups) of isometric operators. We shall not carry out the corresponding proofs.

§8. Dynamical Systems on Compact Metric Spaces

The phase space of the dynamical system often has a supplementary structure (topological, algebraic, etc.) and the dynamical system preserves this structure. In this section we shall consider topological dynamical systems, i.e., systems generated by continuous maps T of a topological space M into itself.

To avoid pathology, we shall assume that M is a compact metric space.

We shall consider a measure μ given on the σ-algebra of Borel sets of the space M. Such measures are called *Borel*. For the σ-algebra $\mathfrak{S}$ we shall take the completion of the σ-algebra of Borel sets with respect to the measure μ.

For a continuous map $T: M \to M$ an invariant measure may not be given necessarily *a priori*, but the following important statement holds:

Theorem 1. *For any continuous map T of the compact metric space M into itself there exists a normalized Borel measure μ invariant with respect to T.*

Before we prove this theorem, recall that for every Borel measure μ on M, there is a uniquely determined positive linear function l on the space $C(M)$, namely $l(f) = \int_M f(x)\, d\mu$. To measures invariant with respect to T correspond functionals, which are invariant in the following sense

$$l(f(x)) = l(f(Tx)). \tag{1}$$

Sometimes functionals will be denoted by the same letter as the corresponding measure.

Proof of Theorem 1. Suppose $\mu^{(0)}$ is an arbitrary normalized Borel measure on M (e.g., concentrated at one point $\mu^{(0)}f = \int_M f\, d\mu^{(0)} = f(x_0)$, $x_0 \in M$). Consider the sequence of measures μ_n, $n = 1, 2, \ldots$:

$$\mu_n(f) = \frac{1}{n} \sum_{k=0}^{n-1} \mu^{(0)}(f(T^k x)), \qquad f \in C(M).$$

The set of normalized measures on the compact set M is weakly compact, hence the sequence $\{\mu_n\}$ has at least one limit point. Suppose μ is one of these limit points and $\{\mu_{n_s}\}$ is a subsequence of the sequence $\{\mu_n\}$ converging to μ. Let us show that the measure μ is invariant with respect to T. The weak convergence $\mu_{n_s} \to \mu$ means that

$$\int_M f(x)\, d\mu = \lim_{s\to\infty} \int_M f(x)\, d\mu_{n_s} \tag{2}$$

for any function $f \in C(M)$. But

$$\left| \int_M f(Tx)\, d\mu_s - \int_M f(x)\, d\mu_s \right|$$

$$= \left| \frac{1}{n_s} \sum_{k=0}^{n_s - 1} \int_M f(T^{k+1}x)\, d\mu^{(0)} - \frac{1}{n_s} \sum_{k=0}^{n_s-1} \int_M f(T^k x)\, d\mu^{(0)} \right|$$

$$= \frac{1}{n_s} \left| \int_M f(T^{n_s}x)\, d\mu^{(0)} - \int_M f(x)\, d\mu^{(0)} \right| \le \frac{2}{n_s} \max|f| \to 0,$$

when $s \to \infty$. Passing to the limit and using (2), we get

$$\int_M f(Tx)\, d\mu = \int_M f(x)\, d\mu.$$

The theorem is proved. □

Just as in the case of dynamical systems on general measure spaces, it is natural to choose a class of "nondecomposable" topological dynamical systems. We shall give the following definition for the case of discrete time, to be more precise, for homeomorphisms. Their restatement for the case of continuous time can be easily carried up.

Definition 1. The homeomorphism $T: M \to M$ is said to be *topologically transitive*, if the trajectory $\{T^n x: -\infty < n < \infty\}$ of some point $x \in M$ is dense in M.

The following important notions characterize "stronger" topological nondecomposability.

Definition 2. The homeomorphism $T: M \to M$ is said to be *minimal*, if the trajectory $\{T^n x: -\infty < n < \infty\}$ of any point is dense in M.

Definition 3. The homeomorphism $T: M \to M$ is said to be *uniquely ergodic*, if it has precisely one normalized Borel invariant measure μ.

The properties of minimality and unique ergodicity are close to each other in the sense that, in many natural examples, they appear or do not appear simultaneously (see §1 in Chap. 5). However, as will be shown in the sequel, neither one of them is implied by the other.

It is clear that a uniquely ergodic homeomorphism T is ergodic with respect to its unique normalized invariant measure μ. Indeed, if we could find a measurable invariant set E, $0 < \mu(E) < 1$, then we would be able to define the normalized invariant measure μ_1 by putting

$$\mu_1(A) = \frac{1}{\mu(E)}\, \mu(A \cap E) \quad \text{for all } A \in \mathfrak{S},$$

and $\mu_1 \neq \mu$.

Now let us show that for uniquely ergodic dynamical systems, we have a strengthened variant of the Birkhoff–Khinchin ergodic theorem.

Theorem 2. *Suppose T is a homeomorphism of the compact metric space M and μ is a normalized Borel measure invariant with respect to T. The following three statements are equivalent:*

(1) *T is uniquely ergodic;*
(2) *for any function $f \in C(M)$, we have*

$$\lim_{n\to\infty} \frac{1}{n} \sum_{k=0}^{n-1} f(T^k x) = \int_M f(x)\, d\mu \overset{\text{def}}{=} \mu(f), \tag{3}$$

at every point $x \in M$;

(3) *for any function $f \in C(M)$ the time means $(1/n) \sum_{k=0}^{n-1} f(T^k x)$ uniformly converge to $\mu(f)$.*

The proof is based on the following lemma:

Lemma 1. *Suppose T is a uniquely ergodic homeomorphism, μ is unique invariant measure to which corresponds an invariant normalized positive linear functional on $C(M)$. Then any (not necessarily positive) invariant continuous linear functional l on $C(M)$ is of the form*

$$l(f) = \lambda \cdot \mu(f), \qquad \lambda \in \mathbb{R}^1. \tag{4}$$

The proof of the lemma will be given later.

Proof of Theorem 2. Suppose T is uniquely ergodic. Consider the functions $f \in C(M)$ of the form $f(x) = g(Tx) - g(x)$, $g \in C(M)$ (they are sometimes called functions, cohomologic to zero). For such f we have

$$\frac{1}{n} \sum_{k=0}^{n-1} f(T^k x) = \frac{1}{n} \sum_{k=0}^{n-1} [g(T^{k+1}x) - g(T^k x)]$$

$$= \frac{1}{n} [g(T^n x) - g(x)] \le \frac{2}{n} \max|g(x)| \to 0 \quad \text{when} \quad n \to \infty.$$

Then

$$\mu(f) = \int_M f(x)\, d\mu = \int_M g(Tx)\, d\mu - \int_M g(x)\, d\mu = 0,$$

i.e., $\lim_{n\to\infty} (1/n) \sum_{k=0}^{n-1} f(T^k x) = \mu(f)$ and the convergence is uniform.

The set of functions cohomological to zero is a linear manifold in $C(M)$. Let us prove that the closure L of this manifold in $C(M)$ coincides with the subspace C_0 of functions $f \in C(M)$ such that $\mu(f) = 0$. Clearly $L \subset C_0$.

Assume that despite our statement we have $L \neq C_0$, i.e., there is a function f_0, such that $\mu(f_0) = 0$ but $f_0 \notin L$. It follows from the Hahn–Banach theorem that there exists a continuous linear functional l on the space $C(M)$ such that $l(f) = 0$ for $f \in L$ and $l(f_0) = 1$. Clearly l is an invariant linear functional, but it cannot be of the form (4), since $\mu(f_0) = 0$, $l(f_0) = 1$. The contradiction with Lemma 4 thus obtained shows that $L = C_0$.

Now suppose $f \in C_0$. For a given $\varepsilon > 0$ let us find a function $f_\varepsilon = g_\varepsilon(Tx) - g_\varepsilon(x)$ such that $\|f - f_\varepsilon\|_{C(M)} < \varepsilon/2$. Put

$$\left\| f(x) - \frac{1}{n}\sum_{k=0}^{n-1} f(T^k x) \right\| \leq \|f - f_\varepsilon\| + \left\| f_\varepsilon - \frac{1}{n}\sum_{k=0}^{n-1} f_\varepsilon(T^k x) \right\|$$

$$+ \left\| \frac{1}{n}\sum_{k=0}^{n-1} f_\varepsilon(T^k x) - \frac{1}{n}\sum_{k=0}^{n-1} f(T^k x) \right\| \stackrel{\text{def}}{=} \textstyle\sum_1^{(n)} + \sum_2^{(n)} + \sum_3^{(n)},$$

since

$$\left|\textstyle\sum_1^{(n)}\right| < \frac{\varepsilon}{2}, \qquad \left|\textstyle\sum_3^{(n)}\right| < \displaystyle\frac{1}{n}\sum_{k=0}^{n-1} \|f_\varepsilon(T^k x) - f(T^k x)\| < \frac{\varepsilon}{2},$$

while $\sum_2^{(n)} \to 0$ for $n \to \infty$, we get

$$\varlimsup_{n\to\infty} \left\| f(x) - \frac{1}{n}\sum_{k=0}^{n-1} f(T^k x) \right\| \leq \varepsilon.$$

Since ε was arbitrary, the implication (1) $\Rightarrow$ (3) is proved for any function $f \in C_0$ and therefore for any function $f \in C(M)$.

The implication (3) $\Rightarrow$ (2) is obvious, hence it remains to prove that (2) $\Rightarrow$ (1). Suppose ν is any invariant normalized Borel measure for T. According to the Lebesgue dominated convergence theorem, it follows from (3) that for any $f \in C(M)$ we have

$$\lim_{n\to\infty} \int_M \frac{1}{n}\sum_{k=0}^{n-1} f(T^k x)\, d\nu = \int_M \mu(f)\, d\nu = \mu(f). \tag{5}$$

On the other hand, by the invariance of ν, we have

$$\nu(f) = \nu\left(\frac{1}{n}\sum_{k=0}^{n-1} f(T^k x)\right) = \int_M \frac{1}{n}\sum_{k=0}^{n-1} f(T^k x)\, d\nu. \tag{6}$$

It now follows from (5) and (6) that $\nu(f) = \mu(f)$. The theorem is proved. ☐

Proof of Lemma 1. To any continuous linear functional l on the space $C(M)$ we can assign the positive linear functional $|l|$ (the variation of l) so that

(a) for the functions $f \in C(M), f \geq 0$ we have

$$|l|(f) = \sup_{\substack{|\varphi| \leq f \\ \varphi \in C(M)}} l(\varphi); \tag{7}$$

(b) the difference $l_1 = |l| - l$ is also a positive linear functional.

Now assume that l is a continuous linear functional invariant with respect to T. It follows from (7) that $|l|$ is also invariant. Indeed

$$|l|(f(Tx)) = \sup_{|\varphi(x)| \leq f(Tx)} l(\varphi(x)) = \sup_{|\varphi(T^{-1}x)| \leq f(x)} l(\varphi(T^{-1}x))$$

$$= \sup_{|\psi(x)| \leq f(x)} l(\psi(x)) = |l|(f(x)),$$

for $f(x) \geq 0$, and therefore for all $f \in C(M)$. Since T is uniquely ergodic, while $|l|$ is a positive linear functional, i.e., a measure, we have $|l| = c \cdot \mu$, $c \in \mathbb{R}^1$. In a similar way, $l_1 = c_1\mu, c_1 \in \mathbb{R}^1$. Therefore $l = |l| - l_1 = (c - c_1)\mu$. The lemma is proved. □

Now let us give a different proof of the implication (1) ⇒ (2) in Theorem 2, avoiding the use of the Hahn–Banach theorem.

For any point $x \in M$ denote by $\delta_x^{(n)}$ the discrete measure concentrated at the points $T^k x$, $0 \leq k < n$, $\delta_x^{(n)}(\{T^k x\}) = \frac{1}{2}$. Our statement is equivalent to saying that for any $x \in M$ the measures $\delta_x^{(n)}$ weakly converge to μ when $n \to \infty$. Since M is a compact metric space, the set of normalized measures on M is weakly compact. It suffices to prove that any limit point of the sequence of measures $\delta_x^{(n)}$ coincides with μ.

Suppose $\delta_x^{(n_i)}$ for $i \to \infty$ converges weakly to the normalized measure ν. Let us show that ν is an invariant measure for T. Indeed, for the measure $T\nu$ defined by the equality

$$T\nu(A) = \nu(T^{-1}A), \qquad A \in \mathfrak{S},$$

we have

$$T\nu = \lim_{i \to \infty} T\delta_x^{(n_i)}, \qquad T\nu - \nu = \lim_{i \to \infty} (T\delta_x^{(n_i)} - \delta_x^{(n_i)}).$$

But the definition of $\delta_x^{(n)}$ implies that $T\delta_x^{(n)}$ is a measure concentrated at the points $T^k x$, $1 \leq k < n + 1$ and that the measure of each point equals $1/n$. Therefore the difference $T\delta_x^{(n)} - \delta_x^{(n)}$ is concentrated at the points $T^n x$, x and its value at each of these points is equal to $1/n$. In other words $T\delta_x^{(n)} - \delta_x^{(n)} \to 0$ weakly for $n \to \infty$, which was to be proved.

Theorem 2 has the following consequence.

Corollary. *If the homeomorphism T is uniquely ergodic and the invariant measure μ satisfies $\mu(G) > 0$ for any nonempty open set $G \subset M$, then T is a minimal homeomorphism.*

Proof. Suppose $G \subset M$ is a nonempty open set. Since $\mu(G) = \sup \mu(F)$, where the upper bound is taken over all the closed sets $F \subset G$, we can find a closed subset $F_0 \subset G$ such that $\mu(F_0) > 0$. The sets F_0 and $M \setminus G$ are disjoint and closed, therefore there exists a function $f \in C(M)$, $f \geq 0$ such that $f(x) = 1$ for $x \in F_0$, $f(x) = 0$ for $x \in M \setminus G$. Clearly $\int_M f(x)\, d\mu = \mu(f) > 0$. For an arbitrary point $x_0 \in M$, by Theorem 2 we have

$$\lim_{n\to\infty} \frac{1}{n} \sum_{k=0}^{n-1} f(T^k x) = \mu(f) > 0.$$

Hence we can find an n such that $f(T^n x_0) > 0$, i.e., $T_n x_0 \in G$. But this means that the trajectory of the point x_0 is dense, i.e., T is a minimal homeomorphism. □

The converse statement is also valid for minimal homeomorphisms.

Theorem 3. *If the homeomorphism T is minimal, then $\mu(G) > 0$ for any normalized invariant measure μ and any nonempty open set $G \subset M$.*

Proof. If μ is an arbitrary normalized invariant measure on M, then we can find a point $x_0 \in M$ such that for any neighborhood $U = U(x_0)$ we have $\mu(U) > 0$. Indeed, if for any $x \in M$ we can find a neighborhood with zero measure, this would mean that $\mu(M) = 0$. Consider the trajectory of the point x_0. Since this trajectory is dense, for the given open set $G \subset M$, there is an n such that $T^n x_0 \in G$. But since T is a homeomorphism, the image $T^n U(x_0)$ of some sufficiently small neighborhood of this point is contained in G. From the invariance of the measure μ we get $\mu(G) \geq \mu(T^n U) = \mu(U) > 0$. The theorem is proved. □

Chapter 2

Smooth Dynamical Systems on Smooth Manifolds

§1. Invariant Measures Compatible with Differentiability

One of the most important classes of dynamical systems are those which are determined by differentiable maps of smooth manifolds. As a rule, by a manifold we shall mean an m-dimensional compact closed orientable manifold of class C^∞ ($m \geq 1$).

Recall that a Hausdorf space M is said to be a *closed m-dimensional manifold* if every point $x \in M$ possesses a neighborhood U which is homeomorphic to the open ball D of the space $\mathbb{R}^m$. If M is compact, then there exists a finite system of neighborhoods U_j, $1 \leq j \leq N$ which satisfy this condition and cover all of M. Every such system of neighborhood U_j together with fixed homeomorphisms $\phi_j: U_j \to D$, $1 \leq j \leq N$ is said to be an *atlas* on M (consisting of the *charts* (U_i, φ_i)) and is denoted by $\{U, \varphi\}$. By means of the homeomorphism φ_j one can introduce a system of coordinates in M in a natural way: to every point $x \in U_j$ corresponds the Euclidean coordinate of its image $\varphi_j(x)$. M is said to be a *manifold of class* C^r, $r = 1, 2, \ldots, \infty$, if $x \in U_i \cap U_j$, $1 \leq i \neq j \leq N$ implies that the coordinates of the point x in U_i are related to the coordinates of the point x in U_j by functions of class C^r.

Now choose some atlas $\{U, \varphi\}$ on M. Consider the homeomorphic map T of the manifold M onto itself. We shall say that T is a *diffeomorphism of class* C^r of the manifold M, if for any point $x \in M$, $x \in U_i$, the coordinates of the point $y = Tx \in U_j$ are expressed by the formulas

$$y_k = f_k(x_1, \ldots, x_m), \qquad k = 1, 2, \ldots, m,$$

where the functions f_k are r times continuously differentiable and $\det \| \partial f_k / \partial x_l \| \neq 0$ at all points $x \in M$.

It follows from the implicit function theorem that the inverse map T^{-1} is also a diffeomorphism.

By Theorem 1, §8, Chap. 1, there exists a normalized Borel measure μ on M, invariant with respect to the diffeomorphism T. Further we shall be interested in invariant measures which are compatible with the differentiable structure, in other words—smooth invariant measure. The smoothness condition means that in every atlas (U, φ) on M the measure μ is defined by its density, i.e., for any neighborhood U_i contained in the atlas there exists a positive measurable function $p(x)$, $x \in U_i$, such that for any measurable $A \subset U_i$ we have

$$\mu(A) = \int_A p(x)\, dx_1 \dots dx_m. \tag{1}$$

Let us write the invariance conditions of the measure (1) in terms of its density. The diffeomorphism T sends the point $x = (x_1, \dots, x_m)$ into the point y with coordinates

$$y_k = f_k(x_1, \dots, x_m), \qquad 1 \leq k \leq m.$$

Hence the equality $\mu(A) = \mu(T^{-1}A)$ can be written as

$$\int_A p(x)\, dx_1 \dots dx_m = \int_{T^{-1}A} p(y)\, dy_1 \dots dy_m,$$

or

$$\int_A p(x)\, dx_1 \dots dx_m = \int_A p(Tx) \frac{D(y_1, \dots, y_m)}{D(x_1, \dots, x_m)}\, dx_1 \dots dx_m,$$

hence, since A was arbitrary, we get

$$p(x) = p(Tx) \frac{D(y_1, \dots, y_m)}{D(x_1, \dots, x_m)}, \tag{2}$$

almost everywhere and if the density $p(x)$ is continuous, then the equality (2) holds everywhere. Let us apply this equality n times in the case of a continuous density at the points $x^{(0)}, Tx^{(0)}, \dots, T^{n-1}x^{(0)}$ which form a periodic trajectory. We shall obtain

$$\det \left\| \frac{\partial f_i^{(n)}}{\partial x_j} \right\| = 1 \quad \text{for } x = x^0, \tag{3}$$

where $f_i^{(n)}$ are functions defining the transformation T^n in some neighborhood of the point x_0. Clearly the last equality does not depend on our choice of the coordinate system. Hence the equality (3) is a necessary condition for the existence of a measure invariant with respect to T and possessing a continuous density.

By using this necessary condition it is easy to indicate certain cases when the existence of a smooth invariant measure is certainly impossible. For example, suppose x_0 is a fixed point of the transformation T. Then the necessary condition (3) can be written in the form

$$\det \left\| \frac{\partial f_i}{\partial x_j} \right\| = 1. \tag{4}$$

If $x^{(0)}$ is an attracting fixed point, then $\det \|\partial f_i/\partial x_j\| < 1$, and the existence of a smooth invariant measure is impossible. There is no finite invariant smooth measure with positive density also in the case when the neighborhood of some point is mapped by the transformation T into its own proper part.

Endomorphisms of Smooth Manifolds. Now suppose T is some smooth (of class C^r) nondegenerate, but now no longer bijective, map of the connected manifold M onto itself. Suppose d is the degree of this map. If $d = 1$, the map is one-to-one and we have a diffeomorphism. If $d > 1$, then the inverse map is multivalued, and by the nondegeneracy of the inverse image of every point of M consists of the same number (namely d) of points.

EXAMPLES

1. Suppose M is the unit circle S^1 with the Lebesgue measure ρ; we again view S^1 as the set $\{x: 0 \leq x < 1\}$. For any positive integer $r > 1$ define the transformation T_r by putting

$$T_r x = rx \pmod 1.$$

The measure ρ is invariant with respect to T_r, i.e., T is an endomorphism. The ergodic properties of this endomorphism are intimately related to the properties of the decomposition of numbers into a r-adic fractions.

2. Suppose that a smooth function $f(x)$ is given on the closed interval $[0, 1]$ and that it satisfies $f(0) = 0, f(1) = m$ (m is an integer greater than 1) and $f'(x) > 1$ for all x. Consider the transformation

$$Tx = f(x) \pmod 1.$$

If, for this transformation, there exists an invariant measure with density $p(x)$, then this density must satisfy the equation

$$p(x_0) = \sum_{x_i : Tx_i = x_0} \frac{p(x_i)}{f'(x_i)}, \qquad x_0 \in [0, 1].$$

Flows on Manifolds. Flows (i.e., continuous groups of diffeomorphisms) naturally arise when one considers vector fields on manifolds. As is well known, one of the main sources of ergodic theory was the study of systems of differential equations on manifolds. The use of vector fields on manifolds allows us to have an invariant notation (with respect to the choice of coordinate systems) for our systems of equations.

Now again assume that M is an m-dimensional compact orientable manifold of class C^r, $r \geq 2$. At each point $x \in M$ consider the tangent linear space T_x. The spaces T_x constitute the tangent bundle TM of the manifold M, $TM = \bigcup_{x \in M} T_x$.

A *vector field* of class C^s, $s \leq r - 1$ on M is, by definition, a section of class C^s of the tangent bundle TM. In other words, the choice of the vector field X on M means that we have assigned to each point $x \in M$ a certain vector X_x belonging to the tangent space T_x. In this case, by the smoothness of the vector field we mean the smoothness of the vector function X.

If M is supplied with an atlas (U, φ), then it naturally gives rise to a corresponding atlas (W, ψ) in the tangent bundle TM. The vector field X on M can then be written in the form of m functions

$$X_1(x_1, \ldots, x_m), \ldots, X_m(x_1, \ldots, x_m),$$

of class C^s.

The components of the vector field in two atlases (U, φ) and (U', φ') are related in the following way:

$$X'_k(x'_1, \ldots, x'_m) = \sum_{l=1}^{m} X_l(x_1, \ldots, x_m) \frac{\partial x'_k}{\partial x_l}, \qquad k = 1, \ldots, m.$$

Any vector field on M generates, on this manifold, a differential operator of the first order which (in a local coordinate system) can be written in the form

$$D_X f = \sum_{k=1}^{m} \frac{\partial f}{\partial x_k} \cdot X_k, \qquad f \in C^1(M).$$

Since X is a vector field of class C^s, $s \geq 1$, the differential equation corresponding to this differential operator satisfies the uniqueness and existence theorem for the Cauchy problem, or, which is the same thing, the uniqueness and existence theorem for the solution of the system of ordinary differential equations

$$\frac{dx_k}{dt} = X_k(x_1, \ldots, x_m) \tag{5}$$

This system of equations may be regarded as defining an infinitely small translation on M.

Assuming that for every initial point $x \in M$ the system (5) has one and only one solution, defined for all $t \in \mathbb{R}^1$, we obtain on M a one-parameter group of diffeomorphisms $\{T^t\}$ which acts as follows: if $x \in M$, then $T^t x$ is the point on the integral curve of the system (5) with origin at the point x, corresponding to the given value of t.

§2. Liouville's Theorem and the Dynamical Systems of Classical Mechanics

For many dynamical systems on smooth manifolds we can explicitly indicate the form of a smooth invariant measure. This can be done by means of Liouville's theorem, which has been mentioned previously and which shall be proved below. Concerning the general notions of classical mechanics and the theory of smooth manifolds used in this section see V. I. Arnold's book *Mathematical Methods of Classical Mechanics* (Arnold [3]).

Suppose M is an m-dimensional compact closed orientable manifold of class C^∞, and X is a vector field of class C^∞ on M. In local coordinates $(x_1, \ldots, x_m)$, the vectors of the field X are of the form

$$X(x_1, \ldots, x_m) = (X_1(x_1, \ldots, x_m), \ldots, X_m(x_1, \ldots, x_m)),$$

where $X_k(x_1, \ldots, x_m) \in C^\infty(M)$, $1 \le k \le m$. Consider, on M, the system of differential equations

$$\begin{cases} \dfrac{dx_1}{dt} = X_1(x_1, \ldots, x_m), \\ \vdots \\ \dfrac{dx_m}{dt} = X_m(x_1, \ldots, x_m). \end{cases} \tag{1}$$

The usual uniqueness and existence theorems can be applied to it. Therefore we can introduce the one-parameter group $\{T^t\}$ of diffeomorphisms of class C^∞ on the manifold M, where the transformation T^k is the translation of each point x along the solution of the system (1) which it determines in time t.

Choose a differential m-form ω of class C^∞ on M. This form determines a continuous linear functional on the space $C(M)$ in accordance to the formula

$$\omega(f) = \int_M f(x)\omega(dx), \qquad f \in C(M).$$

We shall consider only positive functionals, i.e., such that $\omega(f) > 0$ if $f > 0$. Each such form defines the measure μ_ω on M, if we put $\mu_\omega(f) = \omega(f)$.

Choose some atlas on M constituted by the charts (U_i, φ_i) for which we require that the Jacobian $\det\|\varphi_i \circ \varphi_j^{-1}\|$ be positive at any point $x \in U_i \cap U_j$. This may be done since M is an orientable manifold. Then the form ω, at every point of the domain U_i, may be given by the nonnegative density $p(x)$, which is a function of class C^∞.

It is clear that for any form ω the form ω_t defined by the relation $\omega_t(f) = \omega(f(T^t x))$ also will be positive, while the measures μ_ω and μ_{ω_t} will be equivalent.

Now consider the question of the existence of an invariant form, i.e., of form ω for which $\omega_t = \omega$ for all t. If such a form exists, then μ_ω will be an invariant measure for the group $\{T^t\}$.

Suppose the form ω is given by the density $p(x)$, i.e., the differential of the measure $\mu = \mu_\omega$ in local coordinates can be expressed as follows: $d\mu = p(x_1, \ldots, x_m)\, dx_1 \ldots dx_m$.

Theorem 1 (The Liouville Theorem on Invariant Measure). *In order for a measure μ (finite or σ-finite) be invariant with respect to $\{T^t\}$, it is necessary and sufficient to have the relation*

$$\sum_{k=1}^{m} \frac{\partial}{\partial x_k}(pX_k) = 0. \tag{2}$$

Proof. Consider the function $f \in C^\infty(M)$ concentrated in some coordinate neighborhood U_i, i.e., such that the closure of the set $\{x \in M: f(x) \neq 0\}$ is contained in U_i. We can find such a $t_0 > 0$ that for all t satisfying $|t| < t_0$ the function $f(T^t x)$ will also be concentrated in U_i. For the invariance of the measure μ_ω it is necessary and sufficient that for all such f and t_0 we have the equality

$$\int_M f p\, dx_1 \ldots dx_m = \int_M f(T^t x) \cdot p\, dx_1 \ldots dx_m,$$

for $|t| < t_0$. The right-hand side of this equation is continuously differentiable with respect to t. Therefore it is equivalent to the relation

$$\frac{d}{dt} \int_M f(T^t x) \cdot p(x)\, dx_1 \ldots dx_m \bigg|_{t=0} = 0.$$

Now taking into consideration that $f(x) = 0$ for $x \notin U_i$, we can write

$$\begin{aligned}
0 &= \frac{d}{dt} \int_M f(T^t x) \cdot p(x)\, dx_1 \ldots dx_m \bigg|_{t=0} \\
&= \frac{d}{dt} \int_{U_i} f(T^t x) p(x)\, dx_1 \ldots dx_m \bigg|_{t=0} \\
&= \int_{U_i} \sum X_k(x) \frac{\partial f}{\partial x_k} \cdot p(x)\, dx_1 \ldots dx_m \\
&= \int_{U_i} \sum \frac{\partial}{\partial x_k}(pX_k) f(x)\, dx_1 \ldots dx_m \\
&= \int_M \sum \frac{\partial}{\partial x_k}(pX_k) f(x)\, dx_1 \ldots dx_m.
\end{aligned}$$

Since the last relation holds for any function f which satisfies the conditions listed above, it implies (2). The theorem is proved. □

Corollary. *Consider, together with the system of equations* (1), *the system*

$$\begin{cases} \dfrac{dx_1}{dt} = w(x_1, \dots, x_m)X_1(x_1, \dots, x_m) \\ \vdots \\ \dfrac{dx_m}{dt} = w(x_1, \dots, x_m)X_m(x_1, \dots, x_m), \end{cases}$$

where $w(x)$ is a positive function of class C^∞ on M. This system also defines a certain one-parameter group $\{\bar{T}^t\}$ of diffeomorphisms of the manifold M. It is clear that the trajectories of $\{\bar{T}^t\}$ are the same as those of the given flow $\{T^t\}$, while the velocity of motion of every point $x \in M$ for $\{\bar{T}^t\}$ is $w(x)$ times larger than that under the action of $\{T^t\}$.

We shall say that $\{\bar{T}^t\}$ is obtained from $\{T^t\}$ by means of a change of time defined by the function $w(x)$.

Assume that the flow $\{T^t\}$ has a smooth invariant measure given by the density $p(x) \in C^\infty$, $p(x) > 0$. Then the function $\bar{p}(x) = [1/w(x)]p(x)$ is the density of the invariant measure for $\{\bar{T}^t\}$. Indeed

$$\sum_{k=1}^{m} \frac{\partial}{\partial x_k}(\bar{p} \cdot wX_k) = \sum_{k=1}^{m} \frac{\partial}{\partial x_k}(pX_k) = 0.$$

By Liouville's theorem, the measure with density $\bar{p}$ is invariant with respect to $\{\bar{T}^t\}$.

Remark 1. It is useful to look at Liouville's theorem from the functional point of view. Suppose $(M, \mathfrak{S}, \mu)$ is a measure space. Suppose that the space $L^2(M, \mathfrak{S}, \mu)$ contains a dense subset A consisting of bounded (mod 0) functions closed with respect to multiplication; suppose that on A a derivation operator D is defined, i.e., a linear map of the set A into the space of measurable functions on M such that

$$D(f_1 \cdot f_2) = (Df_1) \cdot f_2 + f_1 \cdot (Df_2)$$

for all $f_1, f_2 \in A$. Such an operator D is said to be *symmetric*, if

$$\int_M Df_1 \cdot \bar{f}_2 \, d\mu = \int_M f_1 \cdot \overline{Df_2} \, d\mu, \qquad f_1, f_2 \in A.$$

In these terms, Liouville's theorem means that the differential operator iD, where D is the derivation defined by the vector field X, is symmetric if the invariant measure μ_ω is taken for μ. In the case of C^∞ vector fields, iD also turns out to be a self-adjoint operator, which follows from the usual existence and uniqueness theorems for systems of ordinary differential equations. For infinite-dimensional dynamical systems or dynamical systems generated by a vector fields with singularities, the proof of the self-adjointness may turn out to be somewhat difficult (see in particular, §2 Chap. 8).

Remark 2. Liouville's theorem, as can be seen from its proof, is local in character: since the vector field X is continuous, it suffices to prove the invariance of the measure only in the neighborhood of each point. Further we shall often use this.

Now consider some examples of applications of Liouville's theorem.

1. *The motion of a charged particle in a stationary electromagnetic field.* The motion of the particle of mass m and charge q in an electromagnetic field given by the electric field $E(x)$ and a magnetic field $B(x)$ can be described by the Lorentz equation

$$m\frac{dv}{dt} = q(E + v \times B). \tag{3}$$

If the motion takes place in three-dimensional space $\mathbb{R}^3$, then the phase space is of six dimensions and a complete system of equations is obtained if to equation (3) we also add the equation $dx/dt = v$.

Let us show that the measure μ, where $d\mu = dx_1\,dx_2\,dx_3\,dv_1\,dv_2\,dv_3$, i.e., the measure with density $p(x) \equiv 1$ is invariant. Here $X_k = v_k$, $k = 1, 2, 3$, and $X_k = q(E_k + (v \times B)_k)$, $k = 4, 5, 6$. Therefore

$$\sum_{k=1}^{6}\frac{\partial X_k}{\partial x_k} = \sum_{k=1}^{3}\frac{\partial(v \times B)_k}{\partial v_k}.$$

It follows from the definition of the vector product that this expression vanishes. According to Liouville's theorem, we obtain the invariance of the measure μ.

2. *The problem of magnetic surfaces.* It is known that rarified plasma in a strong magnetic field moves according to a first approximation along the magnetic lines of force. Therefore the trajectory of the particles of plasma is determined by the properties of these lines. The possibility of applying ergodic theory here follows from the Maxwell equations. In case of a magnetic

field which does not depend on time and is described by the intensity vector $B = (B_1, B_2, B_3)$, the equations of magnetic lines of force are of the form

$$\frac{dx_k}{dt} = B_k, \qquad k = 1, 2, 3. \tag{4}$$

Thus a dynamical system is defined, and it corresponds to motion along magnetic lines of force. One of the Maxwell equations is div $B = 0$. This means that the dynamical system under consideration has an invariant measure μ, where $d\mu = dx_1\, dx_2\, dx_3$.

A two-dimensional surface consisting of magnetic lines of force is said to be a magnetic surface. In many cases the problem of the existence of magnetic surfaces can be stated as the ergodicity problem (more exactly, the non-ergodicity problem) of the corresponding dynamical system.

For example, assume that the magnetic field possesses a magnetic surface S which is diffeomorphic to the torus (toroidal fields). Then it is clear that the inside of S consists of magnetic lines of force. If the dynamical system (4) is ergodic inside of S, then this implies that there are no other magnetic surfaces inside of S.

3. *Hamiltonian systems.* Suppose Q is an m-dimensional manifold of class C^∞, T^*Q is the cotangent bundle over Q, i.e., the bundle of differential 1-forms on Q. Choose the coordinate neighborhood U with coordinates $q^1, \ldots, q^m$. Then every 1-form on U is given by its m components $p_1, p_2, \ldots, p_m$. The nondegenerate differential 2-form $\omega = dp \wedge dq = \sum_{i=1}^m dp_i \wedge dq^i$ determines a simplectic structure on T^*Q. For any smooth function $H(p, q)$ on T^*Q, construct the system of differential equations on T^*Q which, in the variables q, p, is of the form

$$\frac{dq^i}{dt} = \frac{\partial H}{\partial p_i}, \qquad \frac{dp_i}{dt} = -\frac{\partial H}{\partial q^i}, \qquad (i = 1, 2, \ldots, m). \tag{5}$$

The function H is said to be the *Hamiltonian*, the system of differential equations (4) the *Hamiltonian system* generated by the function H. It follows immediately from the form of this system and Liouville's theorem that the function $\rho(p, q) = \text{const}$ is the density of the invariant measure for the flow $\{T^t\}$ corresponding to the system (5). This measure is infinite. Now notice that $H(p, q)$ itself is a first integral of the system (5), i.e.,

$$\frac{dH}{dt} = -\sum_{i=1}^m \frac{\partial H}{\partial p_i} \cdot \frac{\partial H}{\partial q^i} + \sum_{i=1}^m \frac{\partial H}{\partial q^i} \cdot \frac{\partial H}{\partial p_i} \equiv 0.$$

Consider the "level surface" of the function H, i.e., the set of the form $\Gamma_c = \{(p, q): H(p, q) = c\}$. In many cases Γ_c turns out to be compact and the Liouville measure induces on it a finite invariant measure. One such example is considered in the following subsection.

4. *Geodesic flows on Riemann surfaces.* Suppose Q is a compact closed m-dimensional Riemann manifold of class C^∞ and $M' = TQ$ is the tangent bundle over Q. The Riemann structure in Q generates a Euclidian structure on every tangent plane Γ_q. In a coordinate neighborhood U with coordinates $q^1, \ldots, q^m$ the metric tensor is of the form $ds^2 = \sum g_{ij}(q)\, dq^i\, dq^j$. If we choose a basis of vectors in the tangent plane Γ_q, so that these vectors are tangent to the coordinate axes passing through $q \in U$, then for any vector $v = (v^1, \ldots, v^m) \in \Gamma_q$, its norm $\|v\|$ equals $\sqrt{\sum g_{ij}(q) v^i v^j}$. Let us denote by $\|g^{ij}(q)\|$ the matrix inverse to $\|g_{ij}(q)\|$.

Consider the cotangent bundle T^*Q. The Euclidian structure in Γ_q establishes natural isomorphism between Γ_q and Γ_q^*. For any vector $v = (v^1, \ldots, v^m) \in \Gamma_q$ introduce the coordinates $p_i = \sum_j g_{ij}(q) v^j$ and consider Γ_q^* as the set of pairs $x = (q, p)$, $p = (p_1, \ldots, p_m)$. Then

$$\|v\|^2 = \sum g_{ij} v^i v^j = \sum g^{ij} p_i p_j = \|p\|^2.$$

The form $\sum dq^i \wedge dp_i$ is the simplectic form on T^*Q considered in the previous subsection. Consider the Hamiltonian function

$$H(x) = \tfrac{1}{2} \sum g^{ij}(q) p_i p_j = \tfrac{1}{2} \|p\|^2,$$

and the corresponding Hamiltonian flow $\{T^t\}$. Since $H(x)$ is a first integral, the unit tangent bundle $M = \{x \in M' \colon \|p\| = 1\}$ is invariant with respect to the action of $\{T^t\}$.

Definition 1. The restriction of $\{T^t\}$ to M is the *geodesic flow* on the manifold Q.

The meaning of this term shall become clear somewhat later. Introduce the measure (on M) whose differential is of the form $d\mu = d\sigma(q)\, d\omega_q$, where $d\sigma$ is the differential of the measure generated by the Riemann metric on Q and $d\omega_q$ is the Lebesgue measure on the unit tangent sphere $S_q \subset \Gamma_q$. An exact definition of the measure μ may be given as follows. For any continuous function $f(p, q)$ its integral $\int f\, d\mu$ must be computed in two steps. First fix q and consider f as a function on S_q. Its integral $\int f(q, p)\, d\omega_q(p)$ is a function $f_1(q)$, $q \in Q$ and $\int f\, d\mu \overset{\text{def}}{=} \int f_1\, d\sigma$. In view of the fact that Q is compact, the measure μ is finite.

Theorem 2. *Any geodesic flow $\{T^t\}$ preserves the measure μ.*

Proof. Suppose $D \subset M$ is an open domain with a smooth boundary. Denote by D_ε the domain in M' of the form

$$D_\varepsilon = \left\{(q, p) \colon \left(q, \frac{p}{\|p\|}\right) \in D,\ 1 - \varepsilon \le \|p\| < 1 + \varepsilon\right\}.$$

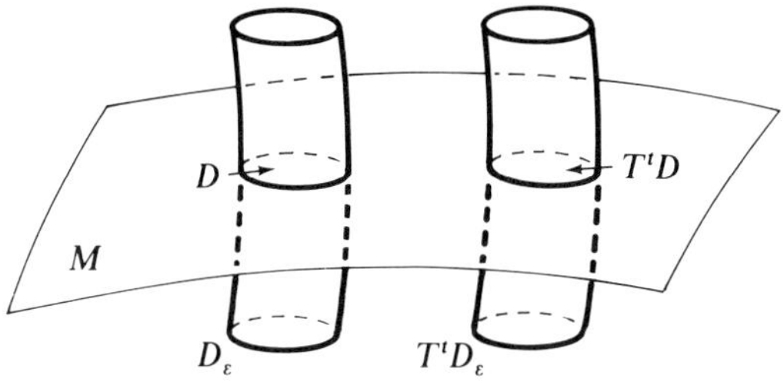

Figure 1

Then $\int_{D_\varepsilon} dq\, dp = \text{const} \cdot \varepsilon \cdot \mu(D) + o(\varepsilon)$, where const does not depend on D. Further, in accordance to Liouville's theorem, $\int_{D_\varepsilon} dq\, dp = \int_{T^tD_\varepsilon} dq\, dp$. It is easy to see (Fig. 1) that

$$\int_{T^tD_\varepsilon} dq\, dp = \text{const} \cdot \varepsilon \cdot \mu(T^tD) + o(\varepsilon).$$

Hence, if $\varepsilon \to 0$, we obtain the necessary statement. The theorem is proved. $\square$

The following theorem explains the meaning of the term "geodesic flow."

Theorem 3. *The trajectories of the geodesic flow $\{T^t\}$ are tangent vectors to geodesic lines in Q. A specific transformation T^t sends the pair (q_0, p_0) into the pair $(q_t, p_t) = T^t(q_0, p_0)$ where, in order to obtain q_t, it is necessary to construct the geodesic line, passing through q_0 in the direction of p_0, and then q_t will be located at a distance t from q_0 (along the geodesic), while the vector p_t will be tangent to this geodesic at q_t and directed in the same way as p_0.*

In the proof of this theorem, we shall use certain definitions and facts from differential geometry. By the *Christophel symbols* of an m-dimensional Riemann manifold Q with metric tensor $ds^2 = \sum g_{ij}(q)\, dq^i\, dq^j$, we mean the expressions

$$\Gamma^i_{kl} = \frac{1}{2} \sum_{j=1}^{m} g^{ij}\left(\frac{\partial g_{jl}}{\partial q^k} + \frac{\partial g_{jk}}{\partial q^l} - \frac{\partial g_{kl}}{\partial q^j}\right), \qquad 1 \le i, k, l \le m.$$

By using the Christophel symbols, we can write the differential equations for the geodesic lines on Q in the following way:

$$\frac{d^2 q^i}{dt^2} + \sum_{k,l=1}^{m} \Gamma^i_{kl} v^k v^l = 0, \qquad 1 \le i \le m. \tag{6}$$

Here $(v^1, \ldots, v^m)$ are coordinates in the tangent plane Γ_q. Moreover, in the proof of Theorem 3, we shall use the following convention often used in tensor analysis: the sign of the sum is omitted but, if some index is found twice in some expression, once as an upper index and once as a lower index, then we assume that the sum is taken over this index from 1 to m.

Proof. Let us write the Hamiltonian equations

$$\frac{dq^i}{dt} = g^{ij} p_j, \qquad \frac{dp_i}{dt} = -\frac{1}{2}\frac{\partial g^{kl}}{\partial q^i} p_k p_l.$$

Now passing to the original contravariant variables $v^i = g^{ij}p_j$, we get

$$\frac{dq^i}{dt} = v^i,$$

$$\frac{dv^i}{dt} = \frac{\partial g^{ij}}{\partial q^k} v^k p_j - \tfrac{1}{2} g^{ij} \frac{\partial g^{kl}}{\partial q^j} p_k p_l = \frac{\partial g^{ij}}{\partial q^k} \cdot g_{jl} \cdot v^k v^l - \tfrac{1}{2} g^{ij} \frac{\partial g^{kl}}{\partial q^j} g_{km} g_{ln} v^m v^n.$$

It follows from the relation $g^{kl} g_{ln} = \delta^k_n$, where δ^k_n is the Kronecker symbol, that

$$\frac{\partial g^{kl}}{\partial q^j} g_{ln} = -g^{kl} \frac{\partial g_{ln}}{\partial q^j},$$

$$g^{ij} \frac{\partial g^{kl}}{\partial q^j} g_{km} g_{ln} v^m v^n = -g^{ij} g^{kl} g_{km} \frac{\partial g_{ln}}{\partial q^j} v^m v^n = -g^{ij} \frac{\partial g_{ln}}{\partial q^j} v^l v^n.$$

Changing the notations for the indices, we obtain

$$\begin{aligned}\frac{dv^i}{dt} &= \frac{\partial g^{ij}}{\partial q^k} g_{jl} v^k v^l + \tfrac{1}{2} g^{ij} \frac{\partial g_{kl}}{\partial q^j} v^k v^l \\ &= \left(\frac{\partial g^{ij}}{\partial q^k} g_{jl} + \frac{1}{2}\frac{\partial g_{kl}}{\partial q^j}\right) v^k v^l \\ &= \left(\tfrac{1}{2} g^{ij} \frac{\partial g_{kl}}{\partial q^j} - g^{ij} \frac{\partial g_{jl}}{\partial q^k}\right) v^k v^l \\ &= -\tfrac{1}{2} g^{ij} \left(\frac{\partial g_{il}}{\partial q^k} + \frac{\partial g_{jk}}{\partial q^l} - \frac{\partial g_{kl}}{\partial q^j}\right) v^k v^l = -\Gamma^i_{kl} v^k v^l,\end{aligned}$$

where the Γ^i_{kl} are the Christophel symbols. As a result we come to the system of equations (6) for geodesic lines. The theorem is proved. □

5. *Generalized geodesic flows*

1. Suppose Q is a compact closed two-dimensional Riemann surface of class C^∞, M' is the tangent bundle over Q and M the unit tangent bundle over Q. By π let us denote the natural projection of M' onto Q; suppose $d\mu' = d\sigma(q)\, d\alpha_q$ is the measure on M', $d\mu = d\sigma(q)\, d\omega_q$ is the measure on M, where $d\sigma$ is the measure on Q induced by the Riemann metric, $d\alpha_q(d\omega_q)$ is the Lebesgue measure on the tangent plane (unit tangent bundle) at the point q. The measure μ is obviously finite.

Choose an arbitrary C^∞-function $\kappa(q)$ on Q. It is known from differential geometry that for any point $x_0 = (q_0, v_0) \in M'$ there exists a unique curve $l \subset Q$ such that

(1) at any point $q \in l$, the geodesic curvature of l equals $\kappa(q)$;[1]
(2) $q_0 \in l$, and the tangent vector to l at the point q_0 coincides with v_0.

Define a flow $\{T^t\}$ on M' such that (q_0, v_0) is mapped by any T^t into the point (q_t, v_t), where $q_t \in l$ is such that the distance from it to q_0 equals $t\|v_0\|$ (if measured along l) and v_t is the tangent vector to l at the point q_t pointing in the same direction as v_0.

Theorem 4. *The flow $\{T^t\}$ in the space M' leaves M invariant and preserves the measures μ' and μ.*

Proof. The first statement of the theorem is obvious. The second statement follows from the fact that for small t the transformation T^t is the composition of a parallel translation and the rotation by an angle $\kappa(q) + o(t)$, while each of these transformations preserves the measures μ', μ.

We shall also give an analytic proof of the invariance of the measure μ'. Consider the coordinate neighborhood U on Q with coordinates q^1, q^2. Suppose Γ^i_{kl} are the Christophel symbols of the surface Q. Then the equations of motion corresponding to the flow $\{T^t\}$ are of the form:

$$\frac{d^2q^1}{dt^2} + \sum \Gamma^1_{kl} \frac{dq^k}{dt}\frac{dq^l}{dt} - \kappa(q^1, q^2)\frac{dq^2}{dt} = 0,$$

$$\frac{d^2q^2}{dt^2} + \sum \Gamma^2_{kl} \frac{dq^k}{dt}\frac{dq^l}{dt} + \kappa(q^1, q^2)\frac{dq^1}{dt} = 0.$$

Suppose that the coordinates q^1, q^2 are orthogonal, i.e., the Riemann metric can be written in the form

$$ds^2 = a_1^2(q^1, q^2)(dq^1)^2 + a_2^2(q^1, q^2)(dq^2)^2.$$

[1] Recall the definition of geodesic curvature. Suppose l is a curve, $t_l(q)$ the unit tangent vector to l at the point q, $n_l(q)$ the unit normal vector to l at the point q, where the rotation from $t_l(q)$ to $n_l(q)$ is counter-clockwise. Choose points $q, q' \in l$ located near each other and take the parallel translation of the normal vector $n_l(q)$ to the point q'. The angle between $n_l(q')$ and the translate of the vector $n_l(q)$ equals $\kappa(q)\, dq + o(dq)$, where $dq = \|q' - q\|$ and $\kappa(q)$ is the geodesic curvature.

Then

$$d\sigma(q^1, q^2) = a_1(q^1, q^2) \cdot a_2(q^1, q^2)\, dq^1\, dq^2.$$

Introduce the variables

$$v^1 = \frac{dq^1}{dt}, \qquad v^2 = \frac{dq^2}{dt}.$$

We will show that

$$d\mu' = a_1^2(q^1, q^2) \cdot a_2^2(q^1, q^2)\, dq^1\, dq^2\, dv^1\, dv^2,$$

is the differential of the invariant measure, i.e., the density $\rho = a_1^2 a_2^2$ satisfies the conditions of Liouville's theorem. We have

$$\frac{\partial}{\partial q^1}(\rho v^1) + \frac{\partial}{\partial q^2}(\rho v^2) + \frac{\partial}{\partial v^1}[\rho(-\sum \Gamma_{kl}^1 v^k v^l + \kappa \cdot v^2)]$$

$$+ \frac{\partial}{\partial v^2}\left[\rho\left(-\sum_k \Gamma_{kl}^2 v^k v^l - \kappa v^1\right)\right]$$

$$= v^1 \frac{\partial \rho}{\partial q^1} + v^2 \frac{\partial \rho}{\partial q^2} - 2\rho[\sum \Gamma_{1l}^1 v^l + \sum \Gamma_{2l}^2 v^l].$$

Note that exactly the same expression would have been obtained if we had considered the case of an ordinary geodesic flow corresponding to $\kappa \equiv 0$. It is useful nevertheless to check directly that this last expression vanishes. The expression of the coefficients Γ_{kl}^i in terms of the metric gives us

$$\Gamma_{11}^1 = \frac{1}{a_1}\frac{\partial a_1}{\partial q^1}, \; \Gamma_{12}^1 = \frac{1}{a_1}\frac{\partial a_1}{\partial q^2}, \; \Gamma_{21}^1 = \frac{1}{a_2}\frac{\partial a_2}{\partial q^1}, \; \Gamma_{22}^2 = \frac{1}{a_2}\frac{\partial a_2}{\partial q^2}.$$

Substituting them into the last sum, we see that it vanishes.

The proof of the invariance of the measure μ, based on the invariance of the measure μ', can be carried out in the same way as in the case of an ordinary geodesic flow. The theorem is proved. □

2. Suppose Q is a compact closed m-dimensional Riemann manifold of class C^∞, M is the space of tangent m-frames on Q. This means that a point of M is a couple $x = (q, r)$ where $q \in Q$ and r is an ordered family of m pairwise orthogonal unit tangent vectors to Q at the point q, i.e., $r = (v^1, v^2, \ldots, v^m)$ $(v^i, v^j) = 0$ for $i \neq j$, $\|v^i\| = 1$. It is clear that M is a C^∞-bundle over Q with fiber R isomorphic to the group $O(n)$ of $n \times n$ orthogonal matrices. In each fiber R_q introduce the measure ω_q, which is the Haar measure, thus trans-

forming M into a measure space by setting $d\mu = d\sigma(q)\,d\omega_q$, where $d\sigma$ is the differential of the volume generated by the Riemann metric.

Consider the flow $\{T^t\}$ on M, such that every individual transformation T^t is a parallel translation by a distance t of the frame $r = (v^1, \ldots, v^m)$ along the geodesic line with origin at q in the direction of v^1. The flow $\{T^t\}$ preserves the measure μ. This follows from the fact that the parallel translation preserves angles between vectors. We will not give a full proof of this statement.

6. *The reduction of Hamiltonian systems to geodesic flows.* Suppose Q is an m-dimensional manifold of class C^∞, and $M' = TQ$ is the tangent bundle over Q. Consider the Lagrange function of the form

$$L(q, v) = \tfrac{1}{2} \sum a_{ij}(q) v^i v^j - V(q^1, \ldots, q^m).$$

Here V is the potential energy, $T = \frac{1}{2} \sum a_{ij} v^i v^j$ the kinetic energy. Suppose that the matrix $\|a_{ij}\|$ is positive definite. The Lagrange equations of motion are of the form (see Arnold [3]):

$$\frac{d}{dt}\frac{\partial L}{\partial \dot{q}^i} - \frac{\partial L}{\partial q^i} = 0, \qquad 1 \leq i \leq m.$$

Now let us pass from the Lagrange function L to the Hamiltonian function:

$$H = \tfrac{1}{2} \sum a^{ij}(q) p_i p_j + V(q),$$

where $p_i = \sum a_{ij} v^j$, $\|a^{ij}\| = \|a_{ij}\|^{-1}$. The phase space will now be the cotangent bundle T^*Q. The Maupertui–Lagrange–Jacobi variational principle (M.–L.–J.) applied to the system under consideration consists in the following: on a manifold of constant energy $H(q, p) \equiv h = \text{const}$, the trajectories of our system are the extremums of the functional

$$\int_A^B \sum p_i\, dq^i, \qquad A, B \in T^*Q.$$

Using the Hamiltonian equations and the invariance of the function H (the law of conservation of energy), this last integral may be rewritten in the form

$$\int_A^B \sum_i p_i \frac{dq^i}{dt}\, dt = \frac{1}{2} \int_A^B \sum_i p_i \sum_j a^{ij} p_j\, dt = \frac{1}{2} \int_A^B \sum_{i,j} a^{ij} p_i p_j\, dt$$

$$= \frac{1}{2} \int_A^B \sum_{i,j} a_{ij} v^i v^j\, dt = \int_A^B T\, dt = \int_A^B \sqrt{h - V} \sqrt{\sum a_{ij} \frac{dq^i}{dt} \cdot \frac{dq^j}{dt}}\, dt$$

$$= \int_A^B \sqrt{h - V} \sqrt{\sum a_{ij}\, dq^i\, dq^j}$$

Introduce the Riemann metric $ds^2 = (h - V(q)) \sum a_{ij} dq^i dq^j$ on Q and consider such manifolds of constant energy for which we have $h - V(q) > 0$. The M.-L.-J. principle shows that the trajectories of our system are geodesic lines in Q corresponding to the metric introduced above. The velocity of motion along the geodesic lines equals $h - V(q)$. This means that the phase space of our system has become a bundle, whose base is Q and whose fibers over the points q are $(m - 1)$-dimensional spheres of tangent vectors of length $h - V(q)$. Consider the map of this space into the unit tangent bundle over Q which sends each tangent vector into the unit tangent vector of the same direction. Under this map, the trajectories of the original system are projected into geodesic lines on Q, but the velocity of motion on them now equals $h - V(q)$. To obtain a geodesic flow, it remains to carry out a change of time in accordance to the function $w = (h - V(q))^{-1}$.

§3. Integrable Dynamical Systems

There are examples of smooth dynamical systems, important for the applications, in which the phase space M, $\dim M = m$, splits into invariant manifolds of smaller dimension which are diffeomorphic to the torus Tor^r, $r < m$, so that on each of these tori the motion can be represented in an appropriate coordinate system as a conditionally periodic motion (see §1, Chap. 1). If the frequencies of such a motion are rationally independent, then, as will be shown in §1 of Chap. 3, it is ergodic. Thus we have obtained the decomposition of the phase space into ergodic components (see §3, Chap. 1).

Dynamical systems possessing this property are often called integrable. This terminology is related to the fact that differential equations describing such systems turn out to be integrable in quadratures.

In this section we consider some examples of integrable systems. All of them will be Hamiltonian. The integrability of Hamiltonian systems can usually be established by using the following important Liouville theorem.

Theorem 1 (The Liouville Theorem on Integrable Systems). *Suppose $\{T^t\}$ is a Hamiltonian system generated by the Hamiltonian function H_0 and the phase space M is a simplectic manifold of dimension $2m$. Suppose we are given a family of $m - 1$ functions $H_1, H_2, \ldots, H_{m-1}$ on M such that the system of functions $H_0, H_1, \ldots, H_{m-1}$ is in involution, i.e., the Poisson brackets vanish:*

$$(H_i, H_j) = 0, \qquad 0 \le i, j \le m - 1.^1$$

[1] If $q^1, \ldots, q^m, p_1, \ldots, p_m$ are canonical coordinates on M, then the Poisson bracket of the two functions F, H is by definition

$$(F, H) = \sum_{k=1}^{m} \left(\frac{\partial F}{\partial q^k} \cdot \frac{\partial H}{\partial p_k} - \frac{\partial F}{\partial p_k} \cdot \frac{\partial H}{\partial q^k} \right).$$

Consider the set of common levels of the functions $H_0, H_1, \ldots, H_{m-1}$:

$$M_h = \{x \in M: H_k(x) = h_k, 0 \le k \le m-1\}, \qquad h = (h_0, \ldots, h_{m-1}) \in \mathbb{R}^m.$$

Assume that the gradients grad $H_k(x)$, $0 \le k \le m-1$, *are linearly independent at each point* $x \in M_h$. *Then* M_h *is invariant and diffeomorphic to the Cartesian product of the Euclidian space* $\mathbb{R}^k$, *for some* $k \le m$, *and the torus* Tor^{m-k}.

In the coordinates $(y_1, \ldots, y_k)$ *on* $\mathbb{R}^k$ *and the cyclic coordinates* $(x_1, \ldots, x_{m-k})$ *on* Tor^{m-k}, *the equations of motion on* M_h *may be written in the form*

$$\frac{dy_i}{dt} = c_i, \quad 1 \le i \le k, \qquad \frac{dx_j}{dt} = \omega_j, \quad 1 \le j \le m-k,$$

where $c_1, \ldots, c_k, \omega_1, \ldots, \omega_{m-k}$ *are constants.*

We shall not give the proof of this theorem because it can be found in many textbooks (see, for example, Arnold [3]).

Note that if the manifold M_h, which appears in the Liouville theorem, is compact, then it is diffeomorphic to the torus Tor^m and the motion on this torus is conditionally periodic.

In the general case M_h is a cylinder with k-dimensional generators and a $m-k$ dimensional torus for its base.

Integrable Geodesic Flows. The simplest examples of integrable geodesic flows can be constructed in the case of surfaces. Suppose Q is a two-dimensional compact orientable Riemann manifold of class C^∞, $M' = TQ$ is the tangent bundle over Q, M is the unit tangent bundle, which is the phase space of a geodesic flow (see §2). In the case considered dim $M = 3$, while the Hamiltonian function generated by the geodesic flow is of the form $H_0 = \|v\|^2$. Assume that the geodesic flow possesses a supplementary independent first integral H_1. The level surface $M_h = \{x = (q, v): H_0(x) = 1, H_1(x) = h\}$ is a two-dimensional surface (if the gradients grad H_0, grad H_1 are linearly independent at all points), which is invariant with respect to the geodesic flow. The vector field induced by the geodesic flow on M_h does not vanish. By Poincaré's well known theorem its Euler characteristic is zero and if M_h is orientable it is a two-dimensional torus. Thus, the phase space splits into invariant two-dimensional tori.

The function H_1 is in involution with H_0, which immediately follows from the fact that H_1 is first integral. Therefore Liouville's theorem immediately implies that the geodesic flow is nonergodic, and the motion on the invariant two-dimensional tori is conditionally periodic.

EXAMPLES

1. *The geodesic flow on a surface of rotation.* Suppose l is a smooth closed curve in $\mathbb{R}^3$ with coordinates x_1, x_2, x_3 and that the surface Q is obtained by rotating l around the x_3 axis. The distance on Q is induced by the usual metric in $\mathbb{R}^3$. The symmetry group is a one-parameter group of rotations about the axis x_3. The first integral of the geodesic flows exists here in view of Noether's theorem and is called, in differential geometry, the Clairaut integral. It follows from the above that the geodesic flow is nonergodic, while the ergodic components are two-dimensional tori on which the flow is conditionally periodic.

2. *Geodesic flows on the surface of an ellipsoid.* In the case of an ellipsoid of rotation, the phase space splits into invariant two-dimensional tori, since such an ellipsoid is the surface obtained by rotating an ellipse about its axis. In the general case, as was shown by Jacobi, a similar decomposition also exists, although it is not related to any symmetry.

3. *Point vortex systems.* Suppose $z_k = p_k + iq_k$, $1 \le k \le m$. A system of m point vortices is by definition a Hamiltonian system with a Hamiltonian of the form

$$H = H(p, q) = \sum_{k,l:\, k \neq l} \kappa_k \kappa_l \log|z_k - z_l|.$$

The coefficient κ_k is called the intensity of the vortex. For a given m we have a Hamiltonian system with phase space $\mathbb{R}^{2m}$. It has three obvious first integrals which are in involution:

$$I_1 = \sum_{k=1}^{m} q_k, \quad I_2 = \sum_{k=1}^{m} p_k, \quad I_3 = H.$$

If all the κ_k are of the same sign, the surface $H =$ const is compact. Therefore the submanifold obtained by fixing I_1, I_2, I_3 is also compact. Further the system considered has one more first integral $I_4 = \sum \kappa_k |z_k|^2$ related to the invariance of the system with respect to rotation about a fixed point. This integral is not in involution with I_1, I_2. Nevertheless for $m = 3$, we have a six-dimensional phase space and four first integrals. If we fix the values of all these integrals a two-dimensional orientable surface with a nondegenerate vector field on it will be defined. As was mentioned before, such a surface can only be a torus. However, we now cannot claim that the induced flow on this torus automatically reduces to a conditionally periodic motion. The properties of such flows will be studied further in Chap. 16.

4. *The* (L, A) *pair method.* This method for finding integrable Hamiltonian systems has appeared only recently, it consists of the following: assume that on the phase space M of the smooth dynamical system $\{T^t\}$ two functions

$L(x)$, $A(x)$ with values in the group of square complex matrices of order m are given. Suppose that we have

$$\frac{d}{dt} L = [L, A] \stackrel{\text{def}}{=} LA - AL.$$

By tr B let us denote the trace of the matrix B.

Lemma 1. *For any integer $p > 0$, the function $H_p(x) = \operatorname{tr} L^p$ is a first integral of the system $\{T^t\}$.*

Proof. We shall use the relation $\operatorname{tr} B_1 B_2 = \operatorname{tr} B_2 B_1$, valid for arbitrary matrices B_1, B_2. We have

$$\frac{d}{dt} H_p = \frac{d}{dt} \operatorname{tr} L^p = \operatorname{tr} \frac{d}{dt} L^p = p \cdot \operatorname{tr} L^{p-1} \cdot \frac{d}{dt} L$$

$$= p \cdot \operatorname{tr} L^{p-1}(LA - AL) = p(\operatorname{tr} L^p A - \operatorname{tr} L^{p-1} AL) = 0.$$

The lemma is proved. □

If L is a Hermitian matrix, i.e., $l_{ij} = \bar{l}_{ji}$, where l_{ij} are the elements of the matrix L, then tr L^p for any p is a real-valued function. Since only m functions among the functions H_p are functionally independent, it follows from the lemma that the system has m first integrals.

Consider some applications of this method. We shall consider a Hamiltonian system of m pointlike particles in $\mathbb{R}^1$, whose Hamiltonian function is of the form

$$H_0 = \sum_{i=1}^{m} \frac{p_i^2}{2} + \sum_{i,j} U(q_i - q_j),$$

where $U(q_i - q_j)$ is the potential of interaction of the ith and the jth particle and $U(-q) = U(q)$. The phase space of the system is $\mathbb{R}^{2m}$. Assume, from the purely formal point of view, that for some smooth functions $\alpha(x)$, $\beta(x)$, where $x \neq 0$, the matrices L and A have elements of the form:

$$l_{ij}(p, q) = \begin{cases} p_i & \text{for } i = j \\ \alpha(q_i - q_j) & \text{for } i \neq j, \end{cases}$$

$$a_{ij}(p, q) = \begin{cases} \sum_{k \neq i} \beta(q_i - q_k) & \text{for } i = j \\ \alpha'(q_i - q_j) & \text{for } i \neq j, \end{cases}$$

while the potential $U(q)$ is of the form $U(q) = -\alpha^2(q) + \text{const}$. Also assume that the function α is odd, while the function β is even. The equations of motion will be of the form

$$\frac{dq_i}{dt} = p_i, \qquad \frac{dp_i}{dt} = -\sum_{k \neq i} U'(q_i - q_k).$$

Then

$$\frac{dl_{ij}(p, q)}{dt} = \begin{cases} -\sum_{k \neq i} U'(q_i - q_k) & \text{for } i = j, \\ \alpha'(q_i - q_j)(p_i - p_j) & \text{for } i \neq j. \end{cases}$$

On the other hand, if $B = LA - AL$, then the elements b_{ij} of the matrix B are of the form

$$b_{ij}(p, q) = \begin{cases} \sum_{k \neq i} (\alpha(q_i - q_k)\alpha'(q_k - q_i) - \alpha'(q_i - q_k)\alpha(q_k - q_i)) \quad \text{for } i = j \\ \alpha(q_i - q_j)\left[\sum_{k \neq j} \beta(q_j - q_k) - \sum_{k \neq i} \beta(q_i - q_k)\right] + \alpha'(q_i - q_j)(p_i - p_j) \\ + \sum_{k \neq i, j} [\alpha(q_i - q_k)\alpha'(q_k - q_l) - \alpha'(q_i - q_k)\alpha(q_k - q_j)] \quad \text{for } i \neq j. \end{cases}$$

It follows immediately from the relation $U(q) = -\alpha^2(q) + \text{const}$ that $dl_{ii}/dt = b_{ii}$. The other equalities $dl_{ij}/dt = b_{ij}$, $i \neq j$, will hold if the functions α and β are related by the following functional equation

$$\alpha'(y)\alpha(z) - \alpha(y)\alpha'(z) = \alpha(y + z)[\beta(y) - \beta(z)].$$

The general solution of this functional equation can be expressed in terms of elliptic functions. We shall give two particular solutions:

(1) $\alpha(x) = \dfrac{ig}{x}, \qquad \beta(x) = -\dfrac{ig}{x^2};$

here $\qquad U(x) = \dfrac{g^2}{x^2} + \text{const}.$

(2) $\alpha(x) = \dfrac{iga}{\sinh ax}, \qquad \beta(x) = -\dfrac{iga \cosh^2 x}{\sinh^2 x};$

here $\qquad U(x) = \dfrac{g^2 a^2}{\sinh^2 ax} + \text{const}.$

Another example which can be studied by the (L, A)-pair method is the so-called *Toda chain.* The Hamiltonian of the Toda chain of m particles is of the form

$$H = \sum_{i=1}^{m} \frac{p_i^2}{2} + \sum_{i=1}^{m-1} \exp(q_{i+1} - q_i)$$

In this last expression there is a definite lack of symmetry in the potential of interaction of the particles with numbers 1 and m. We can imagine that actually there are also particles with numbers 0 and $m + 1$, one of which has been shifted to $-\infty$, while the other goes to $+\infty$. For a Toda chain a (L, A)-pair can also be constructed.

Chapter 3

Smooth Dynamical Systems on the Torus

§1. Translations on the Torus

Diffeomorphisms and flows on tori are of particular importance from various points of view. It might at first seem that this is a very special class of dynamical systems. However, this is not so: many important dynamical systems turn out to be nonergodic and their phase spaces split into invariant tori (see §3, Chap. 2).

At first we will recall the simplest type of diffeomorphisms of the torus, the so-called translations (see §1, Chap. 1).

Suppose that $\mathrm{Tor}^m = S^1 \times S^1 \times \cdots \times S^1$ (m factors) is the product of m circles. A point on the torus may be given either in the multiplicative notation as the system of complex numbers $(z_1, \ldots, z_m)$, $|z_k| = 1$, $1 \leq k \leq m$ or by putting $z_k = \exp(2\pi i x_k)$), in the additive notation, as the system of m real numbers $x_1, \ldots, x_k$ considered mod 1. In this case, we may assume that $0 \leq x_k < 1$, $1 \leq k \leq m$.

Using the additive notation, define the transformation T of the torus Tor^m in the following way: for $x = (x_1, \ldots, x_m) \in \mathrm{Tor}^m$ set

$$Tx = (x_1 + \alpha_1 \pmod 1), x_2 + \alpha_2 \pmod 1), \ldots, x_m + \alpha_m \pmod 1)),$$

where $\alpha_1, \ldots, \alpha_m$ is a fixed sequence of real numbers. The transformation T is said to be a *translation* on the torus and, in the one-dimensional case, a *rotation* of the circle. It is clear that the Lebesgue measure on Tor^m, $d\mu = \prod_{k=1}^m dx_k$, is invariant with respect to T.

Theorem 1. *For a transformation T to be ergodic it is necessary and sufficient that the numbers $1, \alpha_1, \ldots, \alpha_m$ be rationally independent, i.e., that equalities of the form $\sum_{k=1}^m s_k \alpha_k = p$, where p, s_k are integers, be possible only in the case $s_1 = s_2 = \cdots = s_k = 0$.*

Proof. First let us prove the sufficiency. To do this let us establish that every measurable function $f(x)$ invariant (mod 0) with respect to T is a constant (mod 0). Without loss of generality, we may assume that the function f is bounded. Indeed, in the converse case we can put

$$E_N(f) = \{x \in \mathrm{Tor}^m : |f(x)| \leq N\}, \tag{1}$$

and denote by χ_N the indicator of the set E_N. It follows from the invariance of f that the functions $f \cdot \chi_N$, $N = 1, 2, \ldots$, are invariant. Having proved that $f \cdot \chi_N = \text{const}$ (mod 0), we may obtain the same result for f by passing to the limit when $N \to \infty$.

The bounded measurable function $f(x)$ on Tor^m may be developed into a Fourier series which converges in the quadratic mean

$$f(x) = \sum_s c_s \exp[2\pi i(s, x)], \tag{2}$$

where $x = (x_1, \ldots, x_m)$, $s = (s_1, \ldots, s_m)$, $(s, x) = \sum_{k=1}^m s_k x_k$, and the sum $\sum_s$ is taken over all families of integers $s = (s_1, \ldots, s_m)$.

From the invariance of f we obtain, by means of (1) and (2),

$$f(Tx) = \sum_s c_s \exp[2\pi i(s, x + \alpha)] = \sum_s c_s \exp[2\pi i(s, \alpha)] \exp[2\pi i(s, x)]$$

$$= f(x) = \sum_s c_s \exp[2\pi i(s, x)] \text{ (mod 0)}.$$

In view of the unicity of the Fourier coefficient,

$$c_s = c_s \exp[2\pi i(s, \alpha)],$$

i.e., for every s either $c_s = 0$ or $\exp[2\pi i(s, \alpha)] = 1$, i.e., $\sum_{k=1}^m s_k \alpha_k = p$, where p is an integer. But, by assumption, the last equality is possible only when $s_1 = \cdots = s_m = 0$. Thus among the Fourier coefficients only c_0 does not vanish. This means that we have $f(x) = c_0 = \text{const}$ (mod 0).

Now let us prove the necessity. Suppose there exists a nonzero vector $s = (s_1, \ldots, s_m)$ with integer coordinates such that $\sum_{k=1}^m s_k \alpha_k = p$, where p is an integer. Then the function

$$f(x) = \exp\left(2\pi i \sum_{k=1}^m s_k x_k\right),$$

is not a constant (mod 0) and is invariant with respect to T:

$$f(Tx) = \exp\left[2\pi i \sum_{k=1}^m s_k(x_k + \alpha_k)\right]$$

$$= \exp\left(2\pi i \sum_{k=1}^m s_k \alpha_k\right) \cdot \exp\left(2\pi i \sum_{k=1}^m s_k x_k\right) = f(x).$$

Therefore T is not ergodic. The theorem is proved. □

Now let us give a different proof of the ergodicity of the translation. First we shall prove the following lemma.

Lemma 1. *If the numbers* $1, \alpha_1, \ldots, \alpha_m$ *are rationally independent, then the transformation* T *is minimal, i.e., the trajectory of any point* $x \in \mathrm{Tor}^m$ *is dense in the torus.*

Proof. Let us use the following Kronecker theorem: if $1, \alpha_1, \ldots, \alpha_m$ are rationally independent, then for any $\varepsilon > 0$ and any real numbers $x_1, \ldots, x_m$ there exists an integer n and a family of integers $p_1, \ldots, p_m$ such that

$$|n\alpha_k - p_k - x_k| < \varepsilon, \qquad 1 \le k \le m.$$

Since the sequence of points

$$\{(n\alpha_1 \text{ (mod 1)}, \ldots, n\alpha_m \text{ (mod 1)}): -\infty < n < \infty\}$$

is the trajectory of the point $x_0 = (0, 0, \ldots, 0) \in \mathrm{Tor}^m$, the Kronecker theorem means that this trajectory is dense on the torus. The trajectory of an arbitrary point $x \in \mathrm{Tor}^m$ is obtained from the trajectory of the point x_0 by a translation along the vector $x - x_0$ and therefore is also dense. The lemma is proved. □

Let us continue the proof of ergodicity. Suppose $A \in \mathfrak{S}$ is a set invariant with respect to T, and $\mu(A) > 0$. Let us prove that $\mu(A) = 1$.

If, conversely, we have $\mu(A) < 1$, then there is a density point $x_1 \in M$ for the set $M \setminus A$. Choose $\delta_1 > 0$ so as to have

$$\frac{\mu((M \setminus A) \cap O_\delta(x_1))}{\mu(O_\delta(x_1))} > \frac{3}{4}, \tag{3}$$

for all δ, $0 < \delta < \delta_1$, where $O_\delta(x_1)$ is the δ-neighborhood of the point x_1. Further suppose $x_2 \in M$ is a density point of the set A, and $\delta_2 > 0$ is chosen so as to have

$$\frac{\mu(A \cap O_\delta(x_2))}{\mu(O_\delta(x_2))} > \frac{3}{4},$$

or, which is the same thing,

$$\frac{\mu((M \setminus A) \cap O_\delta(x_2))}{\mu(O_\delta(x_2))} < \frac{1}{4}, \tag{4}$$

for all δ, $0 < \delta < \delta_2$. Put $\varepsilon_1 = \min(\delta_1, \delta_2)$ and choose ε_2, $0 < \varepsilon_2 < \varepsilon_1$ so as to have

$$\mu(O_{\varepsilon_2}(x)) > \tfrac{1}{2}\mu(O_{\varepsilon_1}(x)), \qquad x \in M \tag{5}$$

Note that since μ is the Lebesgue measure on Tor^m, then the measure of any spherical neighborhood $O_\varepsilon(x)$ depends on ε, but not on x.

Since the trajectory of the point x_2 is dense in M, we can find an integer n such that $O_{\varepsilon_2}(T^n x_2) \subset O_{\varepsilon_1}(x_1)$. Then (4) and (5) imply

$$\frac{\mu((M \setminus A) \cap O_{\varepsilon_1}(x_1))}{\mu(O_{\varepsilon_1}(x_1))}$$

$$= \frac{\mu[(M \setminus A) \cap O_{\varepsilon_2}(T^n x_2) + \mu[(M \setminus A) \cap (O_{\varepsilon_1}(x_1) \setminus O_{\varepsilon_2}(T^n x_2))]}{\mu(O_{\varepsilon_1}(x_1))} \leq \frac{1}{4} + \frac{1}{2} = \frac{3}{4},$$

which contradicts (3). The theorem is proved.

Theorem 2. *Suppose T is a translation on the torus* Tor^m *given by formula* (1) *and the numbers* $1, \alpha_k$, $1 \leq k \leq m$ *are rationally independent. Then T is uniquely ergodic, i.e., the Lebesgue measure μ on* Tor^m *is the only invariant normalized Borel measure.*

Proof. By Theorem 2, §8, Chap. 1, unique ergodicity implies that for any continuous function $f(x)$ on Tor^m we have

$$\lim_{n \to \infty} \frac{1}{n} \sum_{k=0}^{n-1} f(T^k x) = \int_{\text{Tor}^m} f(x)\, d\mu, \tag{6}$$

uniformly for all $x \in \text{Tor}^m$. Take an arbitrary $\varepsilon > 0$ and, using the uniform continuity of f, find a $\delta > 0$ such that $|f(x') - f(x'')| < \varepsilon/2$ for $\text{dist}(x', x'') < \delta$. Since, by Theorem 1, the translation T is ergodic, the relation (6) holds on some set $A \subset \text{Tor}^m$, $\mu(A) = 1$.

This set is dense on Tor^m and therefore we can choose a finite δ-net $x_1, \ldots, x_r \in A$. Suppose n_0 is chosen so as to have for all $n \geq n_0$

$$\left| \frac{1}{n} \sum_{k=0}^{n-1} f(T^k x_i) - \int_{\text{Tor}^m} f(x)\, d\mu \right| < \frac{\varepsilon}{2}, \qquad 1 \leq i \leq r.$$

For any point $x \in \text{Tor}^m$ for some i, $1 \leq i \leq r$, we have $\text{dist}(x, x_i) < \delta$. Clearly, for any k, we then have

$$\text{dist}(T^k x, T^k x_i) < \delta,$$

and therefore

$$\left|\frac{1}{n}\sum_{k=0}^{n-1} f(T^k x) - \int_{\mathrm{Tor}^m} f(x)\,d\mu\right| \le \left|\frac{1}{n}\sum_{k=0}^{n-1} [f(T^k x) - f(T^k x_i)]\right|$$
$$+ \left|\frac{1}{n}\sum_{k=0}^{n-1} f(T^k x_i) - \int_{\mathrm{Tor}^m} f(x)\,d\mu\right| < \frac{\varepsilon}{2} + \frac{\varepsilon}{2} = \varepsilon,$$

for $n \ge n_0$. The theorem is proved. □

Let us give a different proof of the unique ergodicity of the translation. The arguments will be similar to those which were used to prove Theorem 2, §8, Chap. 1.

The equality

$$\exp[2\pi i(s, x + \alpha)] - \exp[2\pi i(s, x)] = [\exp[2\pi i(s, \alpha)] - 1]\exp[2\pi i(s, x)],$$

where $s = (s_1, \ldots, s_m)$, $x = (x_1, \ldots, x_m)$, $(s, x) = \sum_{k=1}^{m} s_k x_k$, implies that a function of the form

$$f_s(x) = \exp[2\pi i(s, x)],$$

for all $s \ne 0$ can be written as

$$f_s(x) = g_s(Tx) - g_s(x),$$

where

$$g_s(x) = \frac{1}{\exp[2\pi i(s, \alpha)] - 1}\exp[2\pi i(s, x)].$$

Therefore, any trigonometric polynomial

$$f(x) = \sum_s c_s \exp[2\pi i(s, x)],$$

satisfying $c_0 = 0$ may be written in the form

$$f(x) = g(Tx) - g(x), \qquad g \in C(\mathrm{Tor}^m).$$

Hence as in the proof of Theorem 2, §8, Chap. 1, we obtain

$$\left|\frac{1}{n}\sum_{k=0}^{n-1} f(T^k x)\right| = \frac{1}{n}|g(T^n x) - g(x)| \le \frac{2}{n}\max|g| \to 0 \quad \text{when } n \to \infty.$$

Since in this case we have $\int_{\mathrm{Tor}^m} f(x)\, d\mu = 0$, it follows that

$$\lim_{n\to\infty} \frac{1}{n} \sum_{k=0}^{n-1} f(T^k x) = \int_{\mathrm{Tor}^m} f(x)\, d\mu, \tag{7}$$

uniformly with respect to x.

Any continuous function on Tor^m may be uniformly approximated by trigonometric polynomials in $x_1, \ldots, x_m$. Therefore, arguing again as in the proof of Theorem 2, §8, Chap. 1, we see that relation (7) is valid for any function $f \in C(\mathrm{Tor}^m)$. The theorem is proved. □

The One-parameter Groups of Translations on the Torus. What was said above concerning translations on the torus can easily be carried over to the one-parameter group of translations. Each such group $\{T^t\}$, just as an individual translation, is given by a system of m numbers $\alpha_1, \ldots, \alpha_m$: if $x = (x_1, \ldots, x_m) \in \mathrm{Tor}^m$, then

$$T^t x = (x_1 + \alpha_1 t \,(\mathrm{mod}\ 1), x_2 + \alpha_2 t \,(\mathrm{mod}\ 1), \ldots, x_m + \alpha_m t \,(\mathrm{mod}\ 1)). \tag{8}$$

The Lebesgue measure on the torus is invariant with respect to any such group of translations. The motion on the torus under the action of flow $\{T^t\}$ is obviously a conditionally periodic motion (see §1, Chap. 1). The same arguments as the ones above given for an individual translation show that the flow $\{T^t\}$ defined by equality (8) is ergodic if and only if the numbers $\alpha_1, \ldots, \alpha_m$ are rationally independent. If this independence takes place, then the corresponding flow is strictly ergodic and the Lebesgue measure on the torus is the unique measure invariant with respect to this flow. Then for any continuous function on Tor^m and any point $x \in \mathrm{Tor}^m$ we have the equality

$$\lim_{t\to\infty} \frac{1}{2t} \int_{-t}^{t} f(T^\tau x)\, d\tau = \int_{\mathrm{Tor}^m} f(x)\, d\mu.$$

§2. The Lagrange Problem

The problem studied below was stated by Lagrange in connection with certain questions of celestial mechanics.

Suppose $a_1, \ldots, a_m$ are given complex numbers, i.e., vectors on the plane and $\alpha_1, \ldots, \alpha_m$ are real numbers. Consider the curve on the z-plane given by the equation

$$z(t) = a_1 \exp(2\pi i \alpha_1 t) + a_2 \exp(2\pi i \alpha_2 t) + \cdots + a_m \exp(2\pi i \alpha_m t).$$

Assume that $z(t)$ does not vanish for any t. Then $z(t)$ may be represented in the form

$$z(t) = r(t)\exp[2\pi i\psi(t)], \tag{1}$$

where $\psi(t)$ is a continuous function.

Lagrange formulated the following question: Does the following limit

$$\omega = \lim_{t\to\infty} \frac{1}{t}\psi(t)$$

exist and if it does, how can one find it?

This problem of course can be interpreted in the following way: the vector a_1 lies in the plane, the vector a_2 is situated so that its origin coincides with the extremity of the first vector and the second vector lies in the same plane, etc. Suppose the vector a_1 rotates about its origin with angular velocity α_1, the vector a_2 rotates around its origin, i.e., about the extremity of a_1, with angular velocity α_2, etc. Then $z(t)$ is the trajectory of the extremity of the vector a_m, and Lagrange's question is to find the velocity of the rotation of the extremity of the vector a_m about the first point of the entire chain of vectors (i.e., about the origin of the vector a_1).

The answer is obvious in the case when

$$|a_2| + |a_3| + \cdots + |a_m| < |a_1|; \tag{2}$$

then we of course have $\omega = \alpha_1$. If the inequality (2) does not hold, then the problem becomes sufficiently difficult. Lagrange himself solved it only for the case of two vectors. We shall consider the general case, but will not give complete proofs, only showing the relationship between this problem and ergodic theory.

Taking logarithms of both sides of equation (1), we obtain

$$\psi(t) = \operatorname{Re}\left[\frac{1}{2\pi i}\log z(t)\right],$$

hence

$$\frac{d\psi}{dt} = \operatorname{Re}\left[\frac{1}{2\pi i}\frac{z'(t)}{z(t)}\right] = \operatorname{Re}\frac{\sum_{k=1}^{m}\alpha_k a_k \exp(2\pi i\alpha_k t)}{\sum_{k=1}^{m} a_k \exp(2\pi i\alpha_k t)}$$

$$= \operatorname{Re}\frac{\sum_{k=1}^{m}\alpha_k|a_k|\cdot\exp[2\pi i(\alpha_k t + x_k^{(0)})]}{\sum_{k=1}^{m}|a_k|\exp[2\pi i(\alpha_k t + x_k^{(0)})]}, \tag{3}$$

where $(x_1^{(0)}, x_2^{(0)}, \ldots, x_m^{(0)})$ are the angles determining the original position of the vectors $a_1, \ldots, a_m$, i.e.,

$$a_k = |a_k| \exp(2\pi i x_k^{(0)}), \qquad 1 \leq k \leq m.$$

The sequence of these angles defined mod 1 can be considered as a point $x^{(0)} = (x_1^{(0)}, \ldots, x_m^{(0)})$ on the torus.

Now on the torus Tor^m with coordinates $(x_1, \ldots, x_m)$ consider the one-parameter group of translations $\{T^t\}$, defined by the vector $(\alpha_1, \ldots, \alpha_m)$. Assume that the numbers $\alpha_1, \ldots, \alpha_m$ are rationally independent, so that the corresponding flow is ergodic.

On the torus Tor^m define the function

$$f(x) = f(x_1, \ldots, x_m) = \mathrm{Re}\,\frac{\sum_{k=1}^{m} \alpha_k |a_k| \exp(2\pi i x_k)}{\sum_{k=1}^{m} |a_k| \exp(2\pi i x_k)}. \tag{4}$$

Then the equality (3) may be rewritten in the form

$$\frac{d\psi}{dt} = f(T^t x^{(0)}),$$

hence

$$\psi(t_2) - \psi(t_1) = \int_{t_2}^{t_1} f(T^\tau x^{(0)})\, d\tau,$$

and the limit which we wish to find may now be written as

$$\lim_{t \to \infty} \frac{\psi(t)}{t} = \lim_{t \to \infty} \frac{1}{t} \int_0^t f(T^\tau x^{(0)})\, d\tau. \tag{5}$$

If the function f were continuous, then this limit would exist for all $x^{(0)} \in \mathrm{Tor}^m$ in accordance to the unique ergodicity of the flow on the torus and would be equal to

$$\int_{\mathrm{Tor}^m} f(x)\, dx_1 \ldots dx_m. \tag{6}$$

But in the expression (4) the denominator can vanish. The condition

$$\sum_{k=1}^{m} |a_k| \exp(2\pi i x_k) = 0 \tag{7}$$

is actually a system of two equations with respect to $x_1, \ldots, x_m$ (both the real and imaginary parts of this sum must equal zero). This implies that the points where equation (7) holds constitute a submanifold of the torus of codimension 2. Therefore the set of all the trajectories which intersect the submanifold is a submanifold of dimension $m - 1$ and its measure equals zero. Thus for a "randomly chosen" trajectory equation (7) does not hold with probability 1. Using these considerations, assume that the ergodic theorem is applicable and replace the integral along the trajectory in (5) by the space integral (6).

We have

$$\int_{\mathrm{Tor}^m} f(x)\, d\mu = \mathrm{Re} \int_{\mathrm{Tor}^m} \frac{\sum_{k=1}^{m} \alpha_k |a_k| \exp(2\pi i x_k)}{\sum_{k=1}^{m} |a_k| \exp(2\pi i x_k)}\, dx_1 \ldots dx_m = \sum_{k=1}^{m} \alpha_k |a_k| \cdot W_k,$$

where

$$W_k = \mathrm{Re} \int_{\mathrm{Tor}^m} \frac{\exp(2\pi i x_k)}{\sum_{l=1}^{m} |a_l| \exp(2\pi i x_l)}\, dx_1 \ldots dx_m.$$

We must compute W_k. Rewrite the integral in the form of a repeated integral, singling out integration with respect to x_k. Then

$$W_k = \int_{\mathrm{Tor}^{m-1}} \left[\int_0^1 \frac{\exp(2\pi i x_k)\, dx_k}{B(x_1, \ldots, x_{k-1}, x_{k+1}, \ldots, x_m) + |a_k| \exp(2\pi i x_k)} \right]$$
$$\times\, dx_1 \ldots dx_{k-1}\, dx_{k+1} \ldots dx_m, \qquad (8)$$

where $B(\cdot)$ is the sum of all the terms in the denominator which do not depend on x_k. The expression under the inner integral sign may be rewritten in the form

$$\frac{1}{2\pi i |a_k|} \frac{d \ln[B + |\alpha_k| \exp(2\pi i x_k)]}{dx_k}.$$

When x_k varies from 0 to 1, the point $z = B + |a_k| \exp(2\pi i x_k)$ describes a circle on the complex plane. If this circle bounds a disc containing the origin, then the corresponding integral equals 1; in the converse case it vanishes.

The circle will bound a disc containing the origin under the condition $|B| < |a_k|$. Thus in the expression (8) for W_k, the function standing under the sign of the $(m - 1)$-dimensional integral assumes only two values: 0 and $1/2\pi i |a_k|$. Therefore

$$W_k = \frac{1}{2\pi i |a_k|} \cdot P(\{|a_k| > |B|\})$$

where $P(\cdot)$ is the measure of the set of those points on Tor^{m-1} for which $|a_k| > |B|$. Since in the ergodic case the relative time when a "randomly chosen" trajectory remains in a given measurable set equals the measure of this set, the result obtained may be interpreted as follows: W_k is the quota of those moments of time when the rotation of the vector a_k contributes to the function $\psi(t)$.

Remark. In connection with the Lagrange problem (and certain other problems of the same type), the following question arises. In all our arguments an essential role was played by the ergodicity condition of the corresponding flow on the torus, consisting in the rational independence of the numbers $\alpha_1, \ldots, \alpha_m$. However, in real problems these numbers (in the case of Lagrange's problem—the angular velocities of the vectors $a_1, \ldots, a_m$) are known only approximately, so that the given condition is meaningless. Nevertheless it can be given a certain interpretation in this case as well. Theoretically, the cases of rationally dependent and independent parameters α_k are dramatically different, but practically there is a sufficiently smooth passage from one to the other: suppose $\alpha_1, \ldots, \alpha_m$ are rational but the equality $\sum_{k=1}^{m} s_k \alpha_k = p$ holds only for extremely large integral values s_k. Such a flow, of course, is not ergodic: its trajectories are periodic and any ergodic component consists of only one trajectory. However, each of these trajectories covers the torus rather densely and for sufficiently smooth functions on it the mean along the trajectory does not essentially differ from the mean on the entire torus. Thus, although ergodicity is an unstable property which can be destroyed by perturbations as small as we wish, we have a stable "approximate ergodicity."

§3. Homeomorphisms of the Circle

In the previous section, we studied the simplest dynamical systems—translations on the torus Tor^m. Now we shall consider more general dynamical systems, and interesting meaningful facts will appear already in the one-dimensional case, i.e., when the phase space M is the unit circle S^1. In the sequel S^1 will often be identified with the semi-interval $0 \le x < 1$, where x is the cyclic coordinate on S^1.

Consider a homeomorphism T preserving the orientation of the circle S^1. Every such homeomorphism may obviously be given in the form $Tx = f(x) \pmod 1$, $0 \le x < 1$, where $f(x)$ is a continuous monotonic increasing function defined for all $x \in \mathbb{R}^1$, and satisfying the condition

$$f(x + 1) = f(x) + 1. \tag{1}$$

We shall say that the function f *represents* the homeomorphism T. If T is a diffeomorphism, then the function f is smooth. We shall say that T is a diffeomorphism of class C^k, if $f \in C^k(S^1)$.

Clearly if f_1, f_2 represent the same homeomorphism then

$$f_1(x) = f_2(x) + k \tag{2}$$

for all $x \in \mathbb{R}^1$, where k is an integer.

If the function f represents the homeomorphism T, then T^n ($n = 0, \pm 1, \pm 2, \ldots$) is represented by the function $f^{(n)}$ defined by the recursive relation

$$f^{(n)}(x) = f(f^{(n-1)}(x)), \qquad f^{(0)}(x) = x, f^{(1)}(x) = f(x).$$

All the functions $f^{(n)}$ possess the same properties as f, i.e., are continuous monotonic and satisfy relation (1).

Theorem 1. *For any orientation preserving homeomorphism T of the circle S^1 and any function $f(x)$ representing T, the limit*

$$\lim_{n\to\infty} \frac{f^{(n)}(x)}{n} = \alpha \tag{3}$$

exists and does not depend on the choice of the point $x \in \mathbb{R}^1$. The number α is rational if and only if the homeomorphism T taken to some nonzero power has a fixed point.

Proof. In view of (2) we can obviously consider only one function $f(x)$ representing T: if $f_2(x) = f_1(x) + k$, then

$$\lim_{n\to\infty} \frac{f_2^{(n)}(x)}{n} = \lim_{n\to\infty} \frac{f_1^{(n)}(x)}{n} + k.$$

First let us assume that the limit (3) exists for at least one point x_0; choose an arbitrary point $x \in \mathbb{R}^1$. Now take an integer m such that

$$x_0 + m \leq x < x_0 + m + 1.$$

In view of the fact that the function $f^{(n)}$ is monotonic, we have for any n:

$$f^{(n)}(x_0 + m) \leq f^{(n)}(x) < f^{(n)}(x_0 + m + 1),$$

and by (1)

$$f^{(n)}(x_0) + m \leq f^{(n)}(x) < f^{(n)}(x_0) + m + 1.$$

This implies the relation

$$\frac{f^{(n)}(x) - f^{(n)}(x_0)}{n} \to 0 \quad \text{for } n \to \infty,$$

i.e., if the limit (3) exists for some point x_0, it also exists for all $x \in \mathbb{R}^1$ and does not depend on x.

Now let us prove the existence of the limit for some point x_0.

First consider the case when some power of the homeomorphism T, say T^k, $k \neq 0$, has a fixed point: $T^k x_0 = x_0$. Since x_0 is a fixed point for T^{-k} as well, we can assume that $k > 0$.

Clearly $f^{(k)}(x_0) = x_0 + r$, where r is an integer. Hence for any $l = 0, \pm 1, \pm 2, \ldots$, we have

$$f^{(lk)}(x_0) = x_0 + lr.$$

Any integer n can be written in the form $n = lk + s$, $0 \leq s < k$. Then

$$f^{(n)}(x_0) = f^{(lk+s)}(x_0) = f^{(s)}(f^{(lk)}(x_0)) = f^{(s)}(x_0 + lr) = f^{(s)}(x_0) + lr,$$

hence

$$\frac{f^{(n)}(x_0)}{n} = \frac{f^{(s)}(x_0)}{n} + \frac{lr}{lk+s} \to \frac{r}{k} \quad \text{when } n \to \infty,$$

i.e., in the case considered the limit (3) exists and is rational.

Now suppose no power of T possesses a fixed point. Choose an arbitrary positive integer k. The difference $f^{(k)}(x) - x$ in this case cannot assume integer values and therefore for all $x \in \mathbb{R}^1$, we have

$$x + r < f^{(k)}(x) < x + r + 1, \tag{4}$$

for some integer r.

Choose a natural number n and apply the inequality (4) to the points $x_0 = 0, f^{(k)}(x_0), f^{(2k)}(x_0), \ldots, f^{((n-1)k)}(x_0)$; then

$$f^{(lk)}(x_0) + r < f^{((l+1)k)}(x_0) < f^{(lk)}(x_0) + r + 1, \qquad l = 0, 1, \ldots, n-1.$$

Adding all these inequalities, we obtain

$$nr < f^{(nk)}(x_0) < n(r+1). \tag{5}$$

Let us write out separately the inequalities obtained for $n = 1$:

$$r < f^{(k)}(x_0) < r + 1. \tag{6}$$

The relations (5) and (6) imply that

$$\left| \frac{f^{(nk)}(x_0)}{nk} - \frac{f^{(k)}(x_0)}{k} \right| < \frac{2}{k}.$$

Since n, k are arbitrary, we can interchange them and add the two inequalities thus obtained:

$$\left|\frac{f^{(n)}(x_0)}{n} - \frac{f^{(k)}(x_0)}{k}\right| < \frac{2}{n} + \frac{2}{k}.$$

From this we see that the limit $\alpha = \lim_{n\to\infty} f^{(n)}(x_0)/n$ exists for $x_0 = 0$ and therefore for all $x \in \mathbb{R}^1$.

It remains to show that if α is rational, then some nonzero power of T has a fixed point.

First suppose $\alpha = 0$. Let us show that T has a fixed point. Assume that no such point exists. Then $f(x) - x \neq 0$ for any x, and therefore we can assume that $f(x) > x$ for all real x. In particular, $f(0) > 0$ and therefore $f^{(n)}(0) > f^{(n-1)}(0) > \cdots > 0$ in view of the fact that f is monotonic. Thus $\{f^{(n)}(0)\}$ is a monotonically increasing sequence. Moreover, $f^{(n)}(0) < 1$ for all n. Indeed, if for some n_0 we would have $f^{(n_0)}(0) \geq 1$, then $f^{(2n_0)}(0) \geq f^{(n_0)}(1) = f^{(n_0)}(0) + 1 \geq 2$ and generally $f^{(kn_0)}(0) \geq k$, hence $f^{(kn_0)}(0)/kn_0 \geq 1/n_0$, which contradicts the assumption $\alpha = 0$.

Thus the sequence $\{f^{(n)}(0)\}$ is monotonic and bounded. Suppose $x_0 = \lim_{n\to\infty} f^{(n)}(0)$. Then

$$f(x_0) = \lim_{n\to\infty} f(f^{(n)}(0)) = \lim_{n\to\infty} f^{(n+1)}(0) = x_0,$$

i.e., x_0 defines a point of the circle which is fixed with respect to T.

Now suppose α is any rational number, $\alpha = r/k$. Then the function $g(x) = f^{(k)}(x) - r$ represents the homeomorphism T^k; moreover

$$\lim_{n\to\infty} \frac{g^{(n)}(x)}{n} = \lim_{n\to\infty} \frac{f^{(kn)}(x)}{n} - r = k \cdot \lim_{n\to\infty} \frac{f^{(kn)}(x)}{kn} - r = 0.$$

Therefore, as was shown above, there exists a point which is fixed with respect to T^k. The theorem is proved. $\square$

The number α defined by formula (3) depends of course on the function f representing the homeomorphism T: $\alpha = \alpha(T, f)$. But if $f_2(x) = f_1(x) + k$, then obviously $\alpha(T, f_2) = \alpha(T, f_1) + k$. This motivates the following important definition.

Definition 1. Suppose T is an orientation-preserving homeomorphism of the circle S^1, and $f(x)$ is a function which represents it. The number

$$\alpha = \alpha(T) = \lim_{n\to\infty} \frac{f^{(n)}(x)}{n} \pmod 1, \qquad x \in \mathbb{R}^1,$$

is said to be the *rotation number* of the homeomorphism T.

It is easy to see that the rotation number is an invariant in the following sense. If T_1, T_2 are two orientation-preserving homeomorphisms of the circle S^1 and there exists a continuous map $\varphi: S^1 \to S^1$ sending T_1 into T_2, i.e., such that $\varphi(T_1 x) = T_2 \varphi(x)$ for all $x \in S^1$, then $\alpha(T_1) = \alpha(T_2)$.

Among the functions $f(x)$ representing the homeomorphism T, there is exactly one function such that $\alpha(T, f) = \alpha(T)$. In the sequel, when we speak of the function representing a homeomorphism, we will as a rule have in mind this particular function.

Since we always have $0 \leq \alpha(T) < 1$, α may be viewed as a point of the circle S^1. Also the value of $\alpha(T)$ may be considered to be a function (with values in S^1), on the compact metric space C of all orientation-preserving homeomorphisms of the circle with the metric

$$\operatorname{dist}(T_1, T_2) = \sup_{x \in S^1} \operatorname{dist}(T_1 x, T_2 x) + \sup_{x \in S^1} \operatorname{dist}(T_1^{-1} x, T_2^{-1} x)$$

Theorem 2. *The function $\alpha(T)$ is continuous at each point $T_0 \in C$.*

Proof. Suppose we are given $\varepsilon > 0$. Choose a natural number $k > 1/\varepsilon$ and an integer r such that

$$\frac{r}{k} < \alpha(T_0) < \frac{r+1}{k}. \tag{7}$$

Suppose the function $f_0(x)$ represents the homeomorphism T_0 and $\alpha(T_0) = \lim_{n \to \infty} f_0^{(n)}(x)/n$. Let us show that for all $x \in \mathbb{R}^1$ we have

$$f_0^{(k)}(x) > x + r. \tag{8}$$

Indeed, if (8) holds for some but not for all x, then we can find an $x_0 \in \mathbb{R}^1$ for which $f_0^{(k)}(x_0) = x_0 + r$. But then

$$\alpha(T_0) = \lim_{l \to \infty} \frac{f_0^{(lk)}(x_0)}{lk} = \lim_{l \to \infty} \frac{x_0 + lr}{lk} = \frac{r}{k},$$

which contradicts (7). If $f_0^{(k)}(x) < x + r$ for all x, then $\alpha(T_0) \leq r/k$. Thus (8) is proved. In a similar way we can show that $f_0^{(k)}(x) < x + r + 1$ for all x. Since $f_0^{(k)}(x)$ is a continuous periodic function, it follows from the above that for some $\eta > 0$

$$r + \eta < f_0^{(k)}(x) - x < r + 1 - \eta.$$

Since T^k varies continuously with respect to T, we can find a $\delta > 0$ such that for any homeomorphism $T \in C$ satisfying the condition $\operatorname{dist}(T, T_0) < \delta$ and for some function $f(x)$ representing T, the following inequality

$$|f^{(k)}(x) - f_0^{(k)}(x)| < \eta, \qquad x \in \mathbb{R}^1,$$

holds. For such T we have

$$r < f^{(k)}(x) - x < r + 1.$$

Hence, repeating the previous argument, we obtain

$$\frac{r}{k} < \lim_{n\to\infty} \frac{f^{(n)}(x)}{n} < \frac{r+1}{k},$$

i.e.,

$$\left| \lim_{n\to\infty} \frac{f^{(n)}(x)}{n} - \alpha(T_0) \right| < \frac{1}{k} < \varepsilon.$$

But $\alpha(T) = \lim_{n\to\infty} f^{(n)}(x)/n \pmod 1$. Therefore $\operatorname{dist}(\alpha(T), \alpha(T_0)) < \varepsilon$. The theorem is proved. □

Now let us study in greater detail the structure of the homeomorphism T with an irrational rotation number α. To do this, consider, besides T, the rotation T_α of S^1 by an angle α: $T_\alpha x = x + \alpha \pmod 1$.

Theorem 3. (1) *If α is irrational, then there exists a continuous map $\varphi: S^1 \to S^1$ sending T to the rotation T_α, i.e., $\varphi(Tx) = T_\alpha(\varphi(x))$ for all $x \in S^1$.*

(2) *The map φ is bijective (i.e., is a homeomorphism of S^1 onto itself) if and only if T is a minimal homeomorphism.*

Proof. (1) Suppose μ is any normalized Borel measure on S^1 invariant with respect to T. By Theorem 1, §8, Chap. 1, such measures exist. Choose the point $x_0 = 0$ on S^1 and, for any $x \in S^1$, put

$$\varphi(x) = \mu([x_0, x]). \tag{9}$$

Since T has no periodic points (α being irrational), the measure μ is continuous and therefore so is the map φ. Therefore, for any $x_1, x_2, x_3 \in S^1$, we have

$$\mu([x_1, x_3]) = (\mu([x_1, x_2]) + \mu([x_2, x_3]))(\text{mod } 1),$$

where in the case $b < a$ by the closed interval $[a, b]$ we mean $[a, b] = [a, 1) \cup [0, b]$.

Using this fact we can write for an arbitrary point $x \in S^1$:

$$\begin{aligned}\varphi(Tx) &= \mu([x_0, Tx]) = (\mu([x_0, Tx_0]) + \mu([Tx_0, Tx]))\,(\text{mod } 1)\\ &= (\mu([x_0, Tx_0]) + \mu([x_0, x]))\,(\text{mod } 1) = (\varphi(x) + \beta)\,(\text{mod } 1),\end{aligned}$$

where $\beta = \mu([x_0, Tx_0])$.

Let us show that $\beta = \alpha$. Suppose $f(x)$, $x \in \mathbb{R}^1$ is the function representing T and $0 \leq f(x_0) < 1$. Consider the infinite measure ν on $\mathbb{R}^1$ coinciding on every closed interval $[k, k+1]$, where k is an integer, with the measure μ carried over to $[k, k+1]$. It follows from the invariance of μ, that

$$\nu([f^{(n)}(x_0), f^{(n+1)}(x_0)]) = \nu([x_0, f(x_0)]), \qquad n = 0, \pm 1, \pm 2, \ldots.$$

Therefore

$$\begin{aligned}\beta &= \mu([x_0, Tx_0]) = \nu([x_0, f(x_0)]) \\ &= \frac{1}{n}\sum_{k=0}^{n-1} \nu([f^{(k)}(x_0), f^{(k+1)}(x_0)]) = \frac{1}{n}\nu([x_0, f^{(n)}(x_0)]).\end{aligned}$$

Suppose $m_n < f^{(n)}(x_0) < m_n + 1$, where m_n is an integer. Then obviously $m_n < \nu([x_0, f^{(n)}(x_0)]) < m_n + 1$. Hence

$$\begin{aligned}\alpha &= \lim_{n\to\infty} \frac{f^{(n)}(x_0)}{n} \pmod 1 = \lim_{n\to\infty} \frac{m_n}{n} \pmod 1 \\ &= \lim_{n\to\infty} \frac{\nu([x_0, f^{(n)}(x_0)])}{n} \pmod 1 = \beta \pmod 1.\end{aligned}$$

Since $0 \leq \alpha, \beta < 1$, the first statement of the theorem is proved.

(2) If φ is a homeomorphism, then the trajectory $\{T^n x\}$ of any point $x \in S^1$ is the inverse image with respect to φ of the trajectory $\{T_\alpha^n \varphi(x)\}$, which is dense in S^1. Hence $\{T^n x\}$ is also dense, i.e., T is minimal.

Conversely, if T is minimal, then by Theorem 3, §8, Chap. 1, the measure μ of any closed interval $[x_0, x]$, $x \neq x_0$ is positive. It then follows from formula (9) that $\varphi(x)$ is a strictly monotonic function, i.e., T is one-to-one. The theorem is proved. □

For an arbitrary point $x_0 \in S^1$ now consider the closed set $P \subset S^1$ of all limit points of the trajectory $\{T^n x_0\}$. The following theorem holds.

Theorem 4. (1) *The set P does not depend on $x_0 \in S^1$.*

(2) *P is invariant with respect to T.*

(3) *Either $P = S^1$, or P is a perfect nowhere dense subset of S^1.*

The proof is based on the following lemma.

Lemma 1. *Suppose m, n are integers, $m \neq n$; let Δ, Δ' be two arcs of the circle S^1 joining the points $T^m x_0$, $T^n x_0$ (if for example $T^n x_0 < T^m x_0$, then $\Delta = [T^n x_0, T^m x_0]$ and $\Delta' = [T^m x_0, 1) \cup [0, T^n x_0]$). Then each of these arcs has a nonempty intersection with the trajectory of any point $x \in S^1$.*

Proof of Lemma 1. Consider, for example, the arc Δ. The arcs Δ, $T^{m-n}\Delta$, $T^{2(m-n)}\Delta, \ldots$ are adjacent to each other. Besides the union

$$\Delta \cup T^{m-n}\Delta \cup \cdots \cup T^{k(m-n)}\Delta$$

for k sufficiently large covers all of the circle S^1. Indeed, otherwise the sequence of numbers $f^{(k(m-n))}(x_0)$, $k = 1, 2, \ldots$, where $f(x)$ is the function representing T, would be monotonic and bounded on $\mathbb{R}^1$, i.e., would have a limit $\tilde{x}$. But then

$$f^{(m-n)}(\tilde{x}) = \lim_{k\to\infty} f^{(m-n)}(f^{(k(m-n))}(x_0)) = \lim_{k\to\infty} f^{((k+1)(m-n))}(x_0) = \tilde{x}.$$

This implies that $\tilde{x}$ (mod 1) is a fixed point for T^{m-n}, which is impossible since α is irrational. Therefore for any $x \in S^1$, we can find an l such that $x \in T^{l(m-n)}\Delta$, i.e., $T^{-l(m-n)}(x) \in \Delta$. The lemma is proved. □

Proof of Theorem 4. (1) Suppose $x_1, x_2 \in S^1$, and P_i is the set of limit points of the trajectory $\{T^n x_i\}$, $i = 1, 2$. For any point $x \in P_1$, we can find a sequence of points of the form $x_1^{(k)} = T^{n_k}x_1 \to x$ for $k \to \infty$. By the lemma we can find a point $x_2^{(k)}$ of the form $T^{m_k}x_2$ on the shortest arc Δ_k joining the points $x_1^{(k)}$ and $x_1^{(k+1)}$. Then $x_2^{(k)} \to x$ and therefore $x \in P_2$. Thus $P_1 \subset P_2$ so that by symmetry $P_1 = P_2$.

(2) Suppose $x \in P$. Then $x = \lim_{k\to\infty} T^{n_k}x_0$, $x_0 \in S^1$. Hence $Tx = \lim_{k\to\infty} T^{n_k+1}x_0$, $T^{-1}x = \lim_{k\to\infty} T^{n_k-1}x_0$, i.e., $Tx, T^{-1}x \in P$. But this means that P is invariant.

(3) If $x \in P$, then $x = \lim_{k\to\infty} T^{n_k}x$ for some sequence $n_k \to \infty$ of integers. Since (in view of the invariance of P) we have $T^{n_k}x \in P$, it follows that x is a limit point for P, i.e., P is a perfect set.

Assume that P is not nowhere dense. Then it is dense on some arc $\Delta \subset S^1$, and we can assume that Δ is of the form $\Delta = [T^n x_0, T^m x_0]$. But as was shown in the proof of Lemma 1, a finite number of translations with respect to T^{m-n} of such an arc covers the entire circle S^1. By the invariance of P, this means that it is dense in S^1, and since P is closed we have $P = S^1$. The theorem is proved. □

This theorem makes the following definition meaningful.

Definition 2. Suppose T is a homeomorphism of the circle S^1 with an irrational rotation number. The set P of limit points of the trajectory $\{T^n x\}$ of any point $x \in S^1$ is said to be the *derived set* for T.

Clearly $P = S^1$ if and only if T is a minimal homeomorphism (for example a translation $Tx = x + \alpha$ (mod 1)).

AN EXAMPLE OF A HOMEOMORPHISM WITH A NOWHERE DENSE DERIVED SET

Let us identify, as usual, the circle S^1 with the semi-interval $0 \leq x < 1$. Suppose $P \subset S^1$ is a perfect nowhere dense set, $\{(a_n, b_n)\}$ the system of its complementary intervals. Choose an irrational number α and on the auxiliary circle $\Gamma = \{y: 0 \leq y < 1\}$ consider the translation T_α: $T_\alpha y = y + \alpha\,(\text{mod } 1)$. Both circles Γ, S^1 shall be supplied with the natural orientation. Establish a correspondence between the trajectory with respect to T_α of the point $y = 0$, i.e., between the set $\{n\alpha\}$, $n = 0, \pm 1, \pm 2, \ldots$ on the circle Γ and the set of intervals $\{(a_n, b_n)\}$, $n = 1, 2, \ldots$, on S^1. We shall order the points $n\alpha$ in the following way:

$$0, \alpha, -\alpha, 2\alpha, -2\alpha, \ldots. \tag{10}$$

To the point $y = 0$ let us assign the interval (a_1, b_1) which we shall now denote by $(a^{(0)}, b^{(0)})$; to the point $y = \alpha$ we assign the interval $(a_2, b_2) \stackrel{\text{def}}{=} (a^{(1)}, b^{(1)})$. The interval $(a^{(-1)}, b^{(-1)})$ assigned to the point $y = -\alpha$ is the interval (a_n, b_n), with the smallest number n such that the intervals $(a^{(0)}, b^{(0)})$, $(a^{(1)}, b^{(1)})$, $(a^{(-1)}, b^{(-1)})$ have the same cyclic order on the circle S^1 as the points 0, α, $-\alpha$ on Γ.

Suppose we have already assigned to the first N points in the sequence (10) the corresponding intervals, and the $(N + 1)$st point in (10) lies between the points $k_1\alpha$ and $k_2\alpha$. Then we assign to this point the free interval with the smallest number n which lies in cyclic order between $(a^{(k_1)}, b^{(k_1)})$ and $(a^{(k_2)}, b^{(k_2)})$. Since P is perfect and nowhere dense, it is possible to find such an interval. Continuing this process, we obtain the necessary correspondence.

We shall now assume that the complementary intervals to the set P are numbered in the form of a sequence $\{(a^{(n)}, b^{(n)})\}$, $-\infty < n < \infty$. Construct the map $\varphi: S^1 \to \Gamma$ in the following way. First for the points $x \in (a^{(n)}, b^{(n)})$ put $\varphi(x) = n\alpha$. Then continue φ by monotonicity to the entire circle S^1. The image will then obviously be the entire circle Γ.

This implies that φ is continuous: if a monotone map φ is discontinuous at the point x_0, then in the image of this map the interval $(\varphi(x_0 - 0), \varphi(x_0 + 0))$ will be missing.

Consider the continuous transformation T of the circle S^1 which sends each interval $(a^{(n)}, b^{(n)})$ into $(a^{(n+1)}, b^{(n+1)})$; we shall assume that inside $(a^{(n)}, b^{(n)})$ it is a linear map:

$$T(a^{(n)} + \lambda(b^{(n)} - a^{(n)})) = a^{(n+1)} + \lambda(b^{(n+1)} - a^{(n+1)}), \qquad 0 < \lambda < 1;$$

these conditions determine T entirely.

It is clear that T is one-to-one and monotonic, i.e., it preserves cyclic order, mapping S^1 onto itself. As before, this implies that T is continuous: otherwise some interval would be missing in the image of the map T. Therefore, T is an orientation preserving homeomorphism. The map φ constructed above sends T into the rotation T_α; therefore $\varphi(Tx) = T_\alpha \varphi(x)$ for all $x \in S^1$. Hence the rotation number of T is the same as the one for T_α, i.e., it equals α. Let us show that P is the derived set for T. Choose a point $x_0 \in (a^{(0)}, b^{(0)})$. Its trajectory $\{T^n x_0\}$ contains precisely one point in each interval $(a^{(n)}, b^{(n)})$. Hence P is the set of limit points of this trajectory, which was to be proved.

Theorem 5. (1) *Any homeomorphism T of the circle S^1 with irrational rotation number α is uniquely ergodic.*

(2) *The support of the invariant measure μ on T is the derived set P, i.e., $\mu(P) = 1$ and P is the smallest closed set of measure* 1.

Proof. (1) Suppose μ_1, μ_2 are two normalized Borel measures on S^1 invariant with respect to T and $\mu = \frac{1}{2}(\mu_1 + \mu_2)$. Clearly μ is also invariant. Consider the map $\varphi\colon S^1 \to S^1$ given by the formula

$$\varphi(x) = \mu([x_0, x]), \tag{11}$$

where $x_0 \in S^1$ is a fixed point, say $x_0 = 0$.

As was shown in the proof of Theorem 3, φ is continuous and maps T into the rotation T_α. Any Borel measure ν on S^1 is mapped by φ into the measure $\varphi\nu$: $(\varphi\nu)(A) \overset{\text{def}}{=} \nu(\varphi^{-1}(A))$, where $A \in \mathfrak{S}$. Since the measures which are invariant with respect to T are mapped into measures invariant with respect to T_α, while T_α is uniquely ergodic, it follows that $\varphi\mu_1 = \varphi\mu_2 = \rho$, where ρ is the Lebesgue measure on S^1.

Suppose $\mathfrak{S}_\varphi \subset \mathfrak{S}$ is the σ-algebra of subsets of S^1 which are inverse images by φ of all possible Borel sets. Then the measures μ_1 and μ_2 coincide on the sets $A \in \mathfrak{S}_\varphi$. Indeed, if $A = \varphi^{-1}(B)$, $B \in \mathfrak{S}$, then

$$\mu_1(A) = \mu_1(\varphi^{-1}(B)) = (\varphi\mu_1)(B) = \rho(B).$$

In a similar way $\mu_2(A) = \rho(B)$, i.e., $\mu_1(A) = \mu_2(A)$. Now suppose $\Delta = (a, b) \subset S^1$ is an arbitrary interval. It follows from formula (11) that $\varphi(\Delta)$ is also an interval, say (c, d) (possibly $c = d$). Further, the inverse image $\Delta' = \varphi^{-1}(\varphi(\Delta))$ is also an interval (a', b'), and $\Delta' \supset \Delta$.

The equality $\varphi(\Delta) = \varphi(\Delta')$ means that $\varphi(a) = \varphi(a') = c$ and $\varphi(b) = \varphi(b') = d$. Hence

$$\mu(\Delta' \setminus \Delta) = \mu((a', a) \cup (b, b')) = 0.$$

Since the measures μ_1, μ_2 are absolutely continuous with respect to μ, we have $\mu_1(\Delta' \setminus \Delta) = \mu_2(\Delta' \setminus \Delta) = 0$, i.e.,

$$\mu_1(\Delta') = \mu_1(\Delta), \qquad \mu_2(\Delta') = \mu_2(\Delta). \tag{12}$$

But $\Delta' \in \mathfrak{S}_\varphi$, hence

$$\mu_1(\Delta') = \mu_2(\Delta'). \tag{13}$$

It follows from (12) and (13) that $\mu_1(\Delta) = \mu_2(\Delta)$. Since the Borel measure is defined by its values on all intervals, we get $\mu_1 = \mu_2$. The first statement of the theorem is proved.

(2) If $P = S^1$, then T is a minimal homeomorphism, $\mu(G) > 0$ for any nonempty open set G and therefore S^1 is the support of μ. Suppose P is nowhere dense, $\Delta \subset S^1$ is an interval entirely contained within some complementary interval (a, b) to P. Take the interval Δ' so that $\Delta \subset \Delta' \subset (a, b)$, the inclusions being strict, and consider the continuous function $f(x)$ on S^1 such that $0 \leq f(x) \leq 1, f(x) = 1$ for $x \in \Delta$ and $f(x) = 0$ for $x \notin \Delta'$.

In view of the unique ergodicity of T, we get

$$\lim_{n\to\infty} \frac{1}{n} \sum_{k=0}^{n-1} f(T^k x) = \int_{S^1} f(x)\, d\mu,$$

for any point $x \in S^1$. But it follows from the definition of the set P that the trajectory of any point $x \in S^1$ can only intersect Δ' a finite number of times. Therefore

$$\lim_{n\to\infty} \frac{1}{n} \sum_{k=0}^{n-1} f(T^k x) = 0, \qquad \text{i.e.,} \int_{S^1} f(x)\, d\mu = 0.$$

Thus we get $\mu(\Delta) \leq \int_{S^1} f(x)\, d\mu = 0$. Since Δ was arbitrary, this implies the relation $\mu(P) = 1$. Further, if we could find a closed set $P' \subset P$ satisfying $\mu(P') = 1$, $P' \neq P$, then $P'_1 = \bigcap_{n=-\infty}^{\infty} T^n P'$ would be an invariant closed set and therefore all the limit points of any trajectory $\{T^n x\}$, $x \in P'_1$, would be contained in P'_1, which contradicts the definition of P. The theorem is proved. □

§4. The Denjoy Theorem

In this section we study orientation preserving diffeomorphisms of the circle S^1. The results of §3 are, of course, valid for these diffeomorphisms. Moreover, these results, in the case of diffeomorphisms, can be considerably sharpened, as we see in Theorem 1 (the Denjoy theorem) proved below.

First we introduce the following definitions.

Definition 1. The homeomorphisms T_1, T_2 of the topological space M onto itself are said to be *topologically equivalent* if there exists a homeomorphism $\varphi: M \to M$ such that $\varphi T_1 = T_2 \varphi$.

Theorem 1. *Suppose T is a diffeomorphism of the circle S^1 with irrational rotation number α. If the function $f(x)$ representing T, has a continuous derivative $f'(x) > 0$ with a bounded variation on $[0, 1)$, then T is topologically equivalent to the rotation T_α.*

The proof is based on the following lemma.

Lemma 1. *Suppose T is a homeomorphism of S^1 with irrational rotation number α; let $g(x)$, $x \in S^1$, be a continuous function on S^1 of bounded variation* Var(g); *μ—the unique normalized invariant measure for T; p/q is an irreducible fraction such that $|\alpha - p/q| \leq 1/q^2$.*

Then for any $x_0 \in S^1$, we have

$$\left| \sum_{k=0}^{q-1} g(T^k x_0) - q \int_{S^1} g(x)\, d\mu \right| \leq \mathrm{Var}(g).$$

The proof of the lemma will be given later.

Proof of Theorem 1. By Theorem 3, §3, it suffices to prove that the diffeomorphism T is minimal, i.e., that its derived set P is the entire circle. Assume, conversely, that $P \neq S^1$ and Δ_0 is some complementary interval to P. Since the set P is invariant with respect to T, the intervals $\Delta_k = T^k\Delta_0$, $k = 0, \pm 1, \pm 2, \ldots$ are also complementary intervals to P. If for some $k \neq 0$, we would have $\Delta_k = \Delta_0$, then, since the orientation is preserved, the end points of the interval Δ_0 would be fixed points for T^k, which is impossible when α is irrational. This implies $\Delta_{k_1} \neq \Delta_{k_2}$ for $k_1 \neq k_2$, i.e., all the Δ_k are different complementary intervals. But then $\sum_{k=-\infty}^{\infty} \rho(\Delta_k) \leq 1$, where ρ is the Lebesgue measure.

Fix a number $\varepsilon_0 > 0$ which shall be specified later. Among the intervals Δ_k there is only a finite number, say $\Delta^{(1)}, \ldots, \Delta^{(N)}$, which satisfy $\rho(\Delta^{(i)}) \geq \varepsilon_0 \rho(\Delta_0)$. Suppose $\Delta^{(i)} = T^{k_i}\Delta_0$. Then for any

$$q > \max_{1 \leq i \leq N} \{k_i\}, \tag{1}$$

we have

$$\rho(T^q\Delta_0) < \varepsilon_0 \rho(\Delta_0). \tag{2}$$

Further

$$\rho(T^q\Delta_0) = \int_{\Delta_0} \frac{df^{(q)}}{dx}\, dx, \tag{3}$$

where $f^{(q)}$ is the function representing T^q. It follows from (2) and (3) that

$$\inf_{x \in \Delta_0} \frac{df^{(q)}}{dx} < \varepsilon_0. \tag{4}$$

Using the chain rule, we obtain

$$\frac{df^{(q)}}{dx}(x) = \prod_{k=0}^{q-1} \frac{df}{dx}(f^{(k)}(x)) = \sum_{k=0}^{q-1} \frac{df}{dx}(T^k x). \tag{5}$$

The last equality follows from the fact that df/dx is of period one. Put $g(x) = \log df/dx$. The function g has a bounded variation Var g on $[0, 1)$.

Taking logarithms in (5), we obtain

$$\log \frac{df^{(q)}}{dx}(x) = \sum_{k=0}^{q-1} g(T^k x).$$

Let us find the value of $\int_{S^1} g(x)\, d\mu \overset{\text{def}}{=} a$. In view of the unique ergodicity of T, we get

$$\lim_{q \to \infty} \frac{1}{q} \log \frac{df^{(q)}}{dx}(x) = \lim_{q \to \infty} \frac{1}{q} \sum_{k=0}^{q-1} g(T^k x) = a,$$

uniformly for all $x \in S^1$. If $a > 0$, we would have $df^{(q)}/dx \to \infty$, if $a < 0$, then $df^{(q)}/dx \to 0$ and in both cases the convergence would be uniform. Each of these relations contradicts the fact that $\int_{S^1} (df^{(q)}/dx)\, dx = 1$, so that $a = 0$.

For any α we can find a natural number q as large as we wish, such that for some p, which is relatively prime to q, we have $|\alpha - p/q| \leq 1/q^2$. Choose q so as to satisfy (1). Then by Lemma 1,

$$\left| \log \frac{df^{(q)}}{dx}(x) \right| = \left| \sum_{k=0}^{q-1} g(T^k x) \right| \leq \operatorname{Var}(g). \tag{6}$$

But the inequality (4) means that

$$\sup_{x \in \Delta_0} \left| \log \frac{df^{(q)}}{dx} \right| > \ln \frac{1}{\varepsilon_0}. \tag{7}$$

But for any ε_0 satisfying $\log 1/\varepsilon_0 > \operatorname{Var}(g)$, relation (6) contradicts (7). The theorem is proved. □

Proof of Lemma 1. We can assume that $x_0 = 0$. The general case reduces to this case by the change of variables $x' = x - x_0 \pmod 1$. We shall further assume that

$$0 < \alpha - \frac{p}{q} \le \frac{1}{q^2}; \tag{8}$$

the converse inequality $0 < p/q - \alpha \le 1/q^2$ is considered in a similar way.

For $k = 1, 2, \ldots, q - 1$ choose a point $x_k \in S^1$ such that $\mu([0, x_k]) = k/q$; for $k = q$, put $x_q = 1$. Denote by Δ_k the open interval (x_k, x_{k+1}); let us show that

$$T^k x_0 \in \Delta_{kp (\mathrm{mod}\, q)}, \qquad k = 1, 2, \ldots, q - 1. \tag{9}$$

To do this, consider the map φ constructed in Theorem 3, §3, sending T into the rotation T_α. Then the point $T^k x_0$ is mapped into $T_\alpha^k x_0 = k\alpha \pmod 1$, while the interval Δ_k is mapped to the interval $\tilde{\Delta}_k = (k/q, (k+1)/q)$.

It follows from inequality (8) that

$$0 < k\alpha - \frac{kp}{q} \le \frac{k}{q^2} < \frac{1}{q},$$

i.e.,

$$T_\alpha^k x_0 \in \Delta_{kp (\mathrm{mod}\, q)}. \tag{10}$$

Since the map φ does not change the cyclic order of points on the circle (10) implies (9). The inclusion (9) is valid also for $k = 0$ if we put $\Delta_0 = [0, 1/q)$. Note that since p, q are relatively prime, the sequence $\{\Delta_{kp (\mathrm{mod}\, q)}\}$, $0 \le k < q - 1$, is simply the sequence $\Delta_0, \Delta_1, \ldots, \Delta_{q-1}$ given in a different order. Using this, we get

$$\begin{aligned}
\left| \sum_{k=0}^{q-1} g(T^k x_0) - q \int_{S^1} g(x)\, d\mu \right| &= \left| \sum_{k=0}^{q-1} g(T^k x_0) - q \sum_{k=0}^{q-1} \int_{\Delta_k} g(x)\, d\mu \right| \\
&= \left| \sum_{k=0}^{q-1} g(T^k x_0) - q \sum_{k=0}^{q-1} \int_{\Delta_{kp (\mathrm{mod}\, q)}} g(x)\, d\mu \right| \\
&\le q \cdot \sum_{k=0}^{q-1} \left| \int_{\Delta_{kp (\mathrm{mod}\, q)}} [g(x) - g(T^k x_0)]\, d\mu \right| \\
&\le \sum_{k=0}^{q-1} \sup_{x \in \Delta_{kp (\mathrm{mod}\, q)}} |g(x) - g(T^k x_0)| \le \mathrm{Var}(g).
\end{aligned}$$

The lemma is proved. $\square$

§5. Arnold's Example

According to Denjoy's theorem, any diffeomorphism T of class C^2 of the circle with irrational rotation number is topologically equivalent to a rotation. However, it is impossible to make any conclusion concerning the differentiability of the linking homeomorphism φ. As it turns out, it is not necessarily differentiable. In this section we shall discuss an appropriate example, due to V. I. Arnold [1].

The invariant measure μ on T is given by the formula $\mu([x_0, x]) = \varphi(x)$; $x_1 x_0 \in S^1$, hence if $\varphi(x)$ is a differentiable function, then the measure μ is absolutely continuous with respect to the Lebesgue measure ρ. We shall show that even in the class of diffeomorphisms given by analytic functions some are such that the invariant measure is not necessarily absolutely continuous.

The idea of the construction is as follows. Diffeomorphisms with a rational rotation number possess an invariant measure concentrated on a finite number of points—on a periodic trajectory. It is possible to construct a sequence of diffeomorphisms, converging to a transformation T with irrational rotation number, such that the invariant measure for T still is singular, although it is no longer discrete. This construction necessitates that the rotation number $\alpha(T)$, although irrational, be approximated abnormally quickly by rational numbers.

The construction of the example is based on a series of lemmas.

Lemma 1. *Suppose T is a homeomorphism of the circle S^1 and μ is a normalized invariant measure for T. Assume moreover that there exists a sequence of sets $G_n \subset S^1$ and a sequence of natural numbers N_n $(n = 1, 2, \ldots)$ such that*

(1) $\lim_{n\to\infty} \rho(G_n) = 0$, *where ρ is a normalized Lesbesgue measure*;
(2) $T^{N_n}(S^1 \setminus G_n) \subset G_n$.

Then the measure μ is not absolutely continuous with respect to ρ.

Proof. Since μ is invariant, condition (2) implies

$$\mu(S^1 \setminus G_n) \le \mu(G_n),$$

i.e., $\mu(G_n) \ge \frac{1}{2}$. Hence condition (1) implies the lemma. $\square$

Thus the problem reduces to constructing an analytic transformation of the circle with rational rotation number which possesses properties (1) and (2).

The first step consists in constructing an auxiliary transformation with rational rotation number.

First let us introduce some special definitions.

Definition 1. Suppose p is an integer and q a natural number. The homeomorphism T of the circle S^1, defined by the function $f(x)$ is called *(p, q)-stable forward* if for $x \in \mathbb{R}^1$, we have

$$f^{(q)}(x) \geq x + p,$$

and some x, for which we actually have the equality, exist.

Definition 2. The diffeomorphism T of the circle S^1 is said to be *real-analytic* in the band $|\text{Im } z| < \delta$, $\delta > 0$, if the function $f(x)$, which determines T, is analytic in this band.

Suppose T is a (p, q)-stable forward real-analytic diffeomorphism (in some neighborhood U of the real axis) of the circle S^1, and assume that the points $x_0, x_1, \ldots, x_{q-1} \in S^1$ form a cycle with respect to T, i.e., $Tx_{i-1} = x_i$, $1 \leq i \leq q - 1$, $Tx_{q-1} = x_0$.

Lemma 2. *For any $\delta > 0$, there exists a (p, q)-stable forward real-analytic diffeomorphism $\tilde{T}$ in U, such that* $\text{dist}(T, \tilde{T}) < \delta$ *and* $(x_0, \ldots, x_{q-1})$ *is the only cycle of length q for $\tilde{T}$.*

Proof. Suppose $\Delta(x)$ is an analytic function in U of period 1,

$$\begin{aligned} &\Delta(x_i) = 0, \quad 0 \leq i \leq q - 1, \\ &\Delta(x) > 0 \quad \text{for other real values of } x. \end{aligned} \tag{1}$$

A function with such property may be found in the class of trigonometric polynomials. Put

$$f_\varepsilon(x) = f(x) + \varepsilon\Delta(x), \qquad \varepsilon > 0,$$

where $f(x)$ is the function which represents T. If $\varepsilon < \min|f'(x)|/\max|\Delta'(x)|$, then $f'_\varepsilon(x) > 0$ for all $x \in \mathbb{R}^1$, hence $f_\varepsilon(x)$ represents some diffeomorphism T_ε which is analytic in U. Moreover, for sufficiently small ε, we have $\text{dist}(T, T_\varepsilon) < \delta$. Fix such an ε and put $\tilde{T} = T_\varepsilon, \tilde{f} = f_\varepsilon$. Since $\tilde{f}^{(q)}(x) \geq f^{(q)}(x)$ and

$$\tilde{f}^{(q)}(x_i) = x_i + p, \qquad 0 \leq i \leq q - 1, \tag{2}$$

we see that $\tilde{T}$ is (p, q)-stable forward.

Clearly the points $x_0, \ldots, x_{q-1}$ constitute a cycle with respect to $\tilde{T}$. Let us show that no other cycles exist. If x is a fixed point for $\tilde{T}^q$, we have

$$\tilde{f}^{(q)}(x) = x + r, \quad \text{where } r \text{ is an integer.} \tag{3}$$

It follows from (3) that $\alpha(\tilde{T}) = r/q$. On the other hand, by (2), $\alpha(\tilde{T}) = p/q$. Thus $r = p$ and by (1) the point x coincides with one of the x_i, $0 \leq i \leq q - 1$. The lemma is proved. $\square$

Lemma 3. *If the homeomorphism T is (p, q)-stable forward and has a unique cycle $(x_0, x_1, \ldots, x_{q-1})$ of length q, then for any $\varepsilon > 0$, the set G_ε which is the ε-neighborhood of this cycle satisfies the condition*

$$T^N(S^1 \backslash G_\varepsilon) \subset G_\varepsilon$$

for all sufficiently large N.

Proof. Consider one of the intervals, say (x_i, x_j), into which the points of the cycle subdivide S^1. We shall assume that $x_i < x_j$ and will choose a point x, $x_i < x < x_j$. Since $T^q x_i = x_i$, $T^q x_j = x_j$, we have $T^{lq}x \in (x_i, x_j)$, $l = 1, 2, \ldots$. If f is the function representing T, it follows from (p, q)-stability forward that $f^{(lq)}(x) > x + lp$. The inequality here is a strict one because x is not a point of the cycle. This means that the points $T^{lq}x$, $l = 1, 2, \ldots$ constitute a monotone sequence in (x_i, x_j) and, since its limit must be a fixed point for T^q, we have $\lim_{l\to\infty} T^{lq}x = x_j$. Therefore for any s, $0 \le s < q$, we obtain $\lim_{l\to\infty} T^{lq+s}x = T^s x_j$. Now take, in the role of x, successively the points $x_0 + \varepsilon, x_1 + \varepsilon, \ldots, x_{q-1} + \varepsilon$. It follows from the above that we can find $N_0, N_1, \ldots, N_{q-1}$ such that $T^{N_i}(x_i + \varepsilon) \in G_\varepsilon$. It is clear that the number $N = \max_{0 \le i \le q-1} N_i$ is the one we need. The lemma is proved. □

If the homeomorphism T satisfies the assumptions of the lemma, then, since T^N depends continuously on T, we can find, for a given $\varepsilon > 0$, a number $\delta > 0$ such that $\bar{T}^N(S^1 \backslash G_{2\varepsilon}) \subset G_{2\varepsilon}$, for any $\bar{T}$, such that $\operatorname{dist}(T, \bar{T}) < \delta$.

Lemma 4. *If T is a (p, q)-stable forward homeomorphism (p, q are arbitrary) and*

$$T_\lambda x = (Tx + \lambda) \pmod 1, \qquad \lambda > 0,$$

then, for sufficiently small λ the rotation number $\alpha(T_\lambda)$ is greater than $\alpha(T)$.

Proof. Clearly $\alpha(T) = p/q$. Suppose $f(x)$ is the function representing T. Then $f_\lambda(x) = f(x) + \lambda$ represents T_λ and since $f_\lambda^{(q)}(x) > x + p$, we see that $\alpha(T_\lambda) > p/q$. The lemma is proved. □

Now let us construct the sequence of diffeomorphisms $T_1, \ldots, T_n, \ldots$ whose limit will be the transformation we wish to construct. For T_1 take the diffeomorphisms with the following properties:

(1) T_1 is real analytic in the band $|\operatorname{Im} z| < 1$;
(2) $|f_1(z) - z| < 1$ for $|\operatorname{Im} z| < 1$, where f_1 is the function representing T_1;
(3) the rotation number $\alpha(T_1) = p/q$ is rational;
(4) T_1 is (p_1, q_1)-stable forward;
(5) T_1 has a unique cycle of length q_1.

Such a diffeomorphism may be obtained from the rotation of the circle $Tx = x + p_1/q_1 \pmod 1$, $0 < p_1 < q_1$, by applying Lemma 2.

Now let us prove the following inductive lemma.

Lemma 5. *Suppose we are given $\delta_n > 0$ and for $s = 1, 2, \ldots, n$ the diffeomorphisms T_s have been constructed so as to satisfy the following conditions:*

(1_n) *T_s is analytic in the band $|\mathrm{Im}\, z| < 1$;*
(2_n) *$|f_s^{(\pm 1)}(z) - f_{s-1}^{(\pm 1)}(z)| < 1/2^{s-1}$, for $|\mathrm{Im}\, z| < 1$, where $f_s^{(\pm 1)}$ is the function representing $T_s^{(\pm 1)}(f_0(x) \equiv x)$;*
(3_n) *the rotation numbers $\alpha(T_s) = p_s/q_s$ are rational and*

$$\left|\frac{p_s}{q_s} - \frac{p_{s-1}}{q_{s-1}}\right| < \frac{1}{2(s-1)^2 \max\limits_{1 \le l \le s} q_l^2}, \qquad s \ge 2;$$

(4_n) *T_s is (p_s, q_s)-stable forward;*
(5_n) *T_s has a unique cycle of length q_s.*

Then there exists a diffeomorphism T_{n+1} possessing the properties (1_{n+1})–(5_{n+1}) *and*

(6_{n+1}) $$\mathrm{dist}(T_n, T_{n+1}) < \delta_n.$$

Proof. Consider the family of transformations T_λ:

$$T_\lambda x = (T_n x + \lambda) \pmod 1, \qquad \lambda > 0.$$

According to Theorem 2, §3, the rotation number $\alpha(T_\lambda)$ depends on λ continuously, so that for a sufficiently small $\lambda_0 > 0$, we have

$$\alpha(T_{\lambda_0}) < \frac{p_n}{q_n} + \frac{1}{n^2 \max\limits_{1 \le l \le n} q_l^2}.$$

Assume λ_0 so small that $\mathrm{dist}(T_{\lambda_0}, T_n) < \delta_n/2$, $|f_{\lambda_0}^{(\pm 1)}(z) - f_n^{(\pm 1)}(z) < 1/2^n$ for $|\mathrm{Im}\, z| < 1$, where f_{λ_0} represents T_{λ_0}. Further, by Lemma 4, $\alpha(T_{\lambda_0}) > \alpha(T_n) = p_n/q_n$. Choose a fraction p_{n+1}/q_{n+1} satisfying the condition

$$\frac{p_n}{q_n} < \frac{p_{n+1}}{q_{n+1}} < \alpha(T_{\lambda_0}),$$

and suppose λ_1 is the largest number $\lambda < \lambda_0$ for which $\alpha(T_\lambda) = p_{n+1}/q_{n+1}$. Let us prove that T_{λ_1} is (p_{n+1}, q_{n+1})-stable forward. Indeed, if at some point $x \in \mathbb{R}^1$ we have $f_{\lambda_1}^{(q_{n+1})}(x) < p_{n+1}$, then for $\varepsilon > 0$ sufficiently small $f_{\lambda_1+\varepsilon}^{(q_{n+1})}(x) < p_{n+1}$. This last inequality cannot possibly hold for all x, since this would imply $\alpha(T_{\lambda_1+\varepsilon}) < p_{n+1}/q_{n+1}$. Hence at some point $x_0 \in \mathbb{R}^1$, we have $f_{\lambda_1+\varepsilon}^{(q_{n+1})}(x_0) =$

p_{n+1}, i.e., $\alpha(T_{\lambda_1+\varepsilon}) = p_{n+1}/q_{n+1}$, contradicting the maximality of λ_1. Now applying Lemma 2 to T_{λ_1} for a sufficiently small δ, we obtain a diffeomorphism T_{n+1} possessing properties (1_{n+1})–(6_{n+1}). The lemma is proved. □

To conclude our inductive construction, it remains to describe the choice of the numbers δ_n. On the nth step first choose an $\varepsilon_n > 0$ such that $\rho(G_n^*) < 1/2^{n+1}$, where G_n^* is an ε_n-neighborhood of the unique cycle of length q_n for T_n. Then, by Lemma 3, find an N_n such that

$$T_n^{N_n}(S^1 \backslash G_n^*) \subset G_n^*.$$

In view of the remark which follows this lemma, we can find a δ_n^* such that for any T satisfying $\operatorname{dist}(T, T_n) < 2\delta_n^*$, we have

$$T^{N_n}(S^1 \backslash G_n) \subset G_n,$$

where G_n is the $2\varepsilon_n$-neighborhood of the cycle and therefore $\rho(G_n) < 1/2^n$. Put $\delta_1 = \delta_1^*/2$, $\delta_{n+1} = \min(\delta_n/2, \delta_n^*/2)$, $n = 1, 2, \ldots$. The sequence of diffeomorphisms T_n can now be entirely determined. The sequence of functions f_n which represent them (as well as the functions $f_n^{(-1)}$) converge uniformly in the band $|\operatorname{Im} z| < 1$, so that the limit function f (as well as $f^{(-1)}$) is analytic in this band and satisfies the relation $f(x+1) = f(x) + 1$. Thus $f(x)$ determines an analytic diffeomorphism T. For any n, we have

$$|f^{(\pm 1)}(x) - f_n^{(\pm 1)}(x)| \le \sum_{k=0}^{\infty} |f_{n+k+1}^{(\pm 1)}(x) - f_{n+k}^{(\pm 1)}(x)| \le \delta_n \sum_{k=0}^{\infty} (\tfrac{1}{2})^k \le 2\delta_n,$$

i.e., $\operatorname{dist}(T, T_n) < 2\delta_n$ and therefore

$$T^{N_n}(S^1 \backslash G_n) \subset G_n.$$

According to Lemma 1, an invariant measure μ for T cannot possibly be absolutely continuous. Finally, the rotation number $\alpha(T)$ is irrational. Indeed $\alpha(T) = \lim_{n\to\infty} \alpha(T_n) = \lim_{n\to\infty}(p_n/q_n)$ and for any n

$$\left|\alpha(T) - \frac{p_n}{q_n}\right| \le \sum_{k=n}^{\infty} \left|\frac{p_{k+1}}{q_{k+1}} - \frac{p_k}{q_k}\right| < \sum_{k=n}^{\infty} \frac{1}{2k^2 \max\limits_{1 \le l \le k} q_l^2} \le \sum_{k=1}^{\infty} \frac{1}{2k^2} \frac{1}{q_n^2} < \frac{1}{q_n^2}. \tag{4}$$

But if α were an irreducible fraction $\alpha = p/q$, then for any n satisfying $q_n > q$, we would have

$$\left|\frac{p}{q} - \frac{p_n}{q_n}\right| = \left|\frac{pq_n - qp_n}{qq_n}\right| \ge \frac{1}{q_n^2},$$

which contradicts (4).

§6. The Ergodicity of Diffeomorphisms of the Circle with Respect to Lebesgue Measure

If the invariant measure μ for some homeomorphism T of the circle with irrational rotation number is equivalent to the Lebesgue measure ρ, then the ergodicity of T with respect to μ implies that any T-invariant set $A \in \mathfrak{S}$ is of Lebesgue measure 0 or 1.

It is natural to call this property *ergodicity with respect to the measure* ρ (which is not necessarily invariant). Although, as we see from the example constructed in the previous section, an invariant measure μ may be singular even in the case of a real analytic diffeomorphism, any sufficiently smooth transformation T is ergodic in the sense indicated above with respect to the Lebesgue measure.

Theorem 1. *If T is a diffeomorphism of class C^2 of the circle S^1 with irrational rotation number α, then any set $A \in \mathfrak{S}$ invariant with respect to T is of Lebesgue measure* 0 *or* 1.

The proof is based on the following lemma.

Lemma 1. *Suppose T is a minimal homeomorphism of the circle S^1 with irrational rotation number α. For any point $x_0 \in S^1$ and $\delta > 0$ we can find an interval $\Delta_0 = (x_0 - \delta, x_0 + \delta_0)$, $\delta_0 < \delta$, and a natural number N such that*

(1) $\bigcup_{k=0}^{N-1} T^k\Delta_0 = S^1$;
(2) *every point $x \in S^1$ belongs to no more than five intervals $T^k\Delta_0$, $0 \leq k < N$.*

The proof of the lemma will be given below.

Proof of Theorem 1. Suppose $\rho(A) > 0$. Then A has a density point $x_0 \in S^1$. Choose an $\varepsilon > 0$. By definition of density points, we can find a $\delta > 0$ such that for any interval Δ satisfying the conditions $x_0 \in \Delta$, $\Delta \subset (x_0 - \delta, x_0 \in \delta)$, we have $\rho(A \cap \Delta) \geq (1 - \varepsilon)\rho(\Delta)$, or, in other words, $\rho(B \cap \Delta) < \varepsilon\rho(\Delta)$, where $B = S^1 \setminus A$. Since, by Denjoy's theorem, T is minimal, Lemma 1 shows that we can find an interval Δ_0, $\Delta_0 = (x_0 - \delta_0, x_0 + \delta_0)$, $\delta_0 < \delta$, and a number N such that

(1) $\bigcup_{k=0}^{N-1} T^k\Delta_0 = S^1$;
(2) each point $x \in S^1$ belongs to no more than five intervals $\Delta_k = T^k\Delta_0$, $0 \leq k < N$.

Note that property (2) implies $\sum_{k=0}^{N-1} \rho(\Delta_k) \leq 5$. Choose any pair of points $x_1, x_2 \in \Delta_0$. Suppose the function $f^{(k)}$ represents the diffeomorphism T^k. By the chain rule, for any k, $0 \leq k \leq N - 1$, we have

$$\frac{df^{(k)}}{dx}(x_i) = \prod_{s=0}^{k-1} \frac{df}{dx}(f^{(s)}(x_i)) = \prod_{s=0}^{k-1} \frac{df}{dx}(T^s x_i), \qquad i = 1, 2, \ldots.$$

The last relation follows from the fact that df/dx is of period 1. Further

$$\frac{(df^{(k)}/dx)(x_1)}{(df^{(k)}/dx)(x_2)} = \prod_{s=0}^{k-1} \frac{(df/dx)(T^s x_1)}{(df/dx)(T^s x_2)}$$

$$= \prod_{s=0}^{k-1} \left(1 + \frac{(df/dx)(T^s x_1) - (df/dx)(T^s x_2)}{(df/dx)(T^s x_2)}\right)$$

$$= \exp\left(\sum_{s=0}^{k-1} \log\left(1 + \frac{(df/dx)(T^s x_1) - (df/dx)(T^s x_2)}{(df/dx)(T^s x_2)}\right)\right)$$

$$\leq \exp\left(\sum_{s=0}^{k-1} \frac{|(df/dx)(T^s x_1) - (df/dx)(T^s x_2)|}{|(df/dx)(T^s x_2)|}\right)$$

$$\leq \exp\left(\frac{\max\limits_{x \in S^1} |d^2 f/dx^2|}{\min\limits_{x \in S^1} |df/dx|}\right) \cdot \exp\left(\sum_{s=0}^{k-1} |T^s x_1 - T^s x_2|\right)$$

$$\leq \text{const} \cdot \exp \sum_{s=0}^{k-1} \rho(\Delta_s) \leq e^5 \cdot \text{const} = \text{const}. \tag{1}$$

On the other hand

$$\rho(\Delta_k) = \int_{\Delta_0} \frac{df^{(k)}}{dx}\, dx,$$

so that we can find a point $\tilde{x} \in \Delta_0$ satisfying

$$\frac{df^{(k)}}{dx}(\tilde{x}) = \frac{\rho(\Delta_k)}{\rho(\Delta_0)}.$$

Setting $x_1 = x$, $x_2 = \tilde{x}$ in (1), we see that for any $x \in \Delta_0$

$$\frac{df^{(k)}}{dx}(x) \leq \text{const} \cdot \frac{\rho(\Delta_k)}{\rho(\Delta_0)}, \qquad 0 \leq k \leq N - 1.$$

Hence

$$\rho(B \cap \Delta_k) = \rho(T^k B \cap T^k \Delta_0) = \rho(T^k(B \cap \Delta_0))$$

$$= \int_{B \cap \Delta_0} \frac{df^{(k)}}{dx}\, dx \leq \text{const} \cdot \frac{\rho(\Delta_k)}{\rho(\Delta_0)} \cdot \rho(B \cap \Delta_0) < \text{const} \cdot \varepsilon \cdot \rho(\Delta_k).$$

Taking the sum over k from 0 to $N - 1$:

$$\rho(B) = \sum_{k=0}^{N-1} \rho(B \cap \Delta_k) \leq \text{const} \cdot \varepsilon \sum_{k=0}^{N-1} \rho(\Delta_k) = \text{const} \cdot \varepsilon.$$

Since ε was arbitrary, $\rho(B) = 0$. The theorem is proved. □

Proof of Lemma 1. It suffices to consider the case where $T = T_\alpha$. Indeed, in the general case, by Theorem 3, §3, there exists a homeomorphism $\varphi: S^1 \to S^1$ such that $\varphi T = T_\alpha \varphi$. Choose a point $x_0' = \varphi^{-1}(x_0)$ and a number δ' such that $\varphi((x_0' - \delta', x_0' + \delta')) \subset (x_0 - \delta, x_0 + \delta)$. Suppose that we have shown, for the rotation T_α, that there is an interval $\Delta_0' = (x_0' - \delta_0', x_0 + \delta_0')$, $\delta_0' < \delta'$ and a number N' such that conditions (1) and (2) hold for $T = T_\alpha$, $x_0 = x_0'$, $\delta = \delta'$. Since φ preserves the cyclic order of points on the circle, we can put $\Delta_0 = \varphi(\Delta_0')$, $N = N'$.

Thus we shall assume that $T = T_\alpha$. We can find such an irreducible fraction p/q that

$$\left|\alpha - \frac{p}{q}\right| \leq \frac{1}{q^2}, \qquad \frac{1}{q} < \frac{\delta}{2}. \tag{2}$$

Let us prove that statements (1) and (2) of the lemma are satisfied if we choose

$$\Delta_0 = \left(x_0 - \frac{2}{q}, x_0 + \frac{2}{q}\right), \qquad N = q.$$

Now denote by $\tilde{T}$ the rotation by the angle p/q:

$$\tilde{T}x \overset{\text{def}}{=} x + \frac{p}{q} \pmod 1, \qquad \tilde{\Delta}_0 \overset{\text{def}}{=} \left[x_0 - \frac{1}{q}, x_0 + \frac{1}{q}\right].$$

Since p, q are relatively prime, $\bigcup_{k=0}^{q-1} \tilde{T}^k \tilde{\Delta}_0 = S^1$. Further (2) implies that

$$\left|k\alpha - k\frac{p}{q}\right| \leq \frac{k}{q^2} < \frac{1}{q},$$

and this means

$$\operatorname{dist}(T^k x_0, \tilde{T}^k x_0) < \frac{1}{q}. \tag{3}$$

Therefore

$$T^k x_0 - \frac{2}{q} < \tilde{T}^k x_0 - \frac{1}{q}, \qquad T^k x_0 + \frac{2}{q} > \tilde{T}^k x_0 + \frac{1}{q},$$

i.e., $T^k\Delta_0 \supset \tilde{T}^k\tilde{\Delta}_0$. Hence

$$\bigcup_{k=0}^{q-1} T^k\Delta_0 \supset \bigcup_{k=0}^{q-1} \tilde{T}^k\tilde{\Delta}_0 = S^1.$$

Now choose a point $x \in S^1$ such that

$$x \in T^{k_1}\Delta_0,\ x \in T^{k_2}\Delta_0, \qquad 0 \leq k_1, k_2 < q.$$

Then $\operatorname{dist}(x, T^{k_1}x_0) < 2/q$, $\operatorname{dist}(x, T^{k_2}\Delta_0) < 2/q$. Taking (3) into consideration, we get

$$\operatorname{dist}(x, \tilde{T}^{k_1}x_0) < \frac{3}{q}, \qquad \operatorname{dist}(x, \tilde{T}^{k_2}x_0) < \frac{3}{q},$$

i.e.,

$$\operatorname{dist}(\tilde{T}^{k_1}x_0, \tilde{T}^{k_2}x_0) < \frac{6}{q}. \tag{4}$$

Now assume, in contradiction to our statement, that x belongs to at least six sets $T^k\Delta$. Consider the points $\tilde{T}^k x_0$ for those k which satisfy $x \in T^k\Delta$. They are all vertices of a regular polygon with q sides, inscribed in S^1, and since there are at least six of them, we can find two for which inequality (4) does not hold. The lemma is proved. □

Chapter 4

Dynamical Systems of Algebraic Origin

In this chapter we shall consider certain classes of dynamical systems of algebraic origin. In these systems the phase space possesses some sort of symmetry, and the action of the dynamical system preserves this symmetry. We shall make use of the theory of characters of commutative compact groups. This theory is developed, for example, in the book by Pontrjagin [1].

§1. Translations on Compact Topological Groups

Suppose the measure space M (also denoted by G)[1] is a compact topological group (not necessarily commutative) with a left invariant Haar measure μ defined on a Borel σ-algebra $\mathfrak{S}$. For any element $g_0 \in G$ consider the transformation $T_{g_0}: M \to M$ given by the relation $T_{g_0}x = g_0 x$. Here the multiplication is the group operation in G.

Definition 1. The transformation T_{g_0} is called a *left group translation* on G. (The word "left" will usually be omitted.)

It is clear that T_{g_0} preserves the measure μ.

In the previous chapter we have already considered a particular case of group translations—the translations on the torus Tor^m.

Lemma 1. *If the translation T_{g_0} is ergodic, the group G is commutative.*

Proof. Since for any nonempty open set $U \subset G$ the Haar measure satisfies $\mu(U) > 0$, the ergodicity of T_{g_0} implies that there exists a point $x_0 \in M$ whose trajectory $\{g_0^n x, -\infty < n < \infty\}$ is dense in M. Since the map $\varphi: M \to M$, $\varphi(x) = x \cdot x_0^{-1}$ is a homeomorphism, the trajectory of the point $x_0 = e$ (where e is the unit of the group) i.e., the subgroup $\{g_0^n, -\infty < n < \infty\}$ itself is dense in M.

[1] As a rule we shall use the letter G when we are concerned with group properties, and the letter M when we are dealing with the properties of measure spaces.

If $y_1, y_2 \in M$, $y_1 = \lim_{i\to\infty} g_0^{n_i}$, $y_2 = \lim_{j\to\infty} g_0^{m_j}$, then

$$\begin{aligned} y_1 y_2 &= \lim_{i\to\infty} g_0^{n_i} \cdot \lim_{j\to\infty} g_0^{m_j} = \lim_{i\to\infty,\, j\to\infty} g_0^{n_i+m_j} \\ &= \lim_{i\to\infty,\, j\to\infty} g_0^{m_j+n_i} = \lim_{j\to\infty,\, i\to\infty} g_0^{m_j} \cdot g_0^{n_i} = y_2 \cdot y_1. \end{aligned}$$

The lemma is proved. □

Further we shall only consider group translations on commutative groups. Note that the phase space of M is a compact metric space and its translations T_{g_0} are homeomorphisms.

Theorem 1. *The following conditions are equivalent:*

(i) *the translation T_{g_0} is ergodic;*
(ii) *the homeomorphism T_{g_0} is minimal;*
(iii) *the homeomorphism T_{g_0} is uniquely ergodic;*
(iv) *$\chi(g_0) \neq 1$ for any nontrivial continuous character of the group G.*

Proof. (1) The implication (i) ⇒ (ii) was essentially proved in the proof of Lemma 1.

(2) Let us prove (i) ⇒ (iii). In view of Theorem 2, §8, Chap. 1, unique ergodicity means that for any function $f \in C(M)$ we have

$$\lim_{n\to\infty} \frac{1}{n} \sum_{k=0}^{n-1} f(T_{g_0}^k x) = \int_M f(x)\, d\mu,$$

for all $x \in M$. Take an arbitrary function $f \in C(M)$. By the Birkhoff–Khinchin ergodic theorem, we can find a point $x_0 \in M$, for which

$$\lim_{n\to\infty} \frac{1}{n} \sum_{k=0}^{n-1} f(T_{g_0}^k x_0) = \int_M f(x)\, d\mu.$$

Let us choose $\varepsilon > 0$ and show that for any point $x \in M$ we have

$$\left| \frac{1}{n} \sum_{k=0}^{n-1} f(T_{g_0}^k x) - \frac{1}{n} \sum_{k=0}^{n-1} f(T_{g_0}^k x_0) \right| < \varepsilon, \tag{1}$$

as soon as n is sufficiently large. Since f is uniformly continuous, we can choose a neighborhood V of the unit $e \in G$, such that $y_1, y_2 \in G$, $y_2 \in V y_1 = \{z y_1 : z \in V\}$ implies $|f(y_2) - f(y_1)| < \varepsilon/2$. According to the implication (i) ⇒ (ii) proved above, the trajectory of the point x is dense in $M = G$ so that for some $n_0 > 0$ the point $T_{g_0}^{n_0}x$ belongs to $Vx_0 = \{zx_0 : z \in V\}$. But then for all k, $0 \le k \le n - n_0 - 1$, we have

$$T_{g_0}^{n_0+k}x \in T_{g_0}^k(Vx_0) = VT_{g_0}^k x_0.$$

Here we have used the commutativity of the group G. Thus

$$|f(T_{g_0}^{n_0+k}x) - f(T_{g_0}^k x_0)| < \frac{\varepsilon}{2}.$$

For an arbitrary $n > n_0$, we get

$$\left| \frac{1}{n}\sum_{k=0}^{n-1} f(T_{g_0}^k x) - \frac{1}{n}\sum_{k=0}^{n-1} f(T_{g_0}^k x_0) \right| \le \frac{1}{n}\sum_{k=0}^{n-n_0-1} |f(T_{g_0}^{n_0+k}x) - f(T_{g_0}^k x_0)|$$
$$+ \frac{1}{n}\left| \sum_{k=0}^{n_0-1} f(T_{g_0}^k x) \right| + \frac{1}{n}\sum_{k=n-n_0}^{n-1} |f(T_{g_0}^k x_0)| < \frac{n-n_0}{n}\cdot\frac{\varepsilon}{2}$$
$$+ 2\max|f|\cdot\frac{n_0}{n}.$$

Hence, for n sufficiently large, we obtain the inequality (1). Since ε was an arbitrary number, the implication (i) ⇒ (iii) is proved.

(3) Let us prove (iii) ⇒ (ii). This follows from Theorem 2, §8, Chap. 1 (also see the corollary to this theorem) and from the fact that the Haar measure $\mu(U)$ is positive for any nonempty open set $U \subset G$.

(4) Let us prove (ii) ⇒ (iv). Assume that our statement is false, i.e., $\chi(g_0) = 1$ for some nontrivial continuous character χ. Then $\chi(g_0^n) = [\chi(g_0)]^n = 1$ for all n, i.e.,

$$\{g_0^n, -\infty < n < \infty\} \subset \{g \in G : \chi(g) = 1\}.$$

Since the character χ is nontrivial, this last set is a closed subgroup which does not coincide with the entire group G. Therefore $\{g_0^n, -\infty < n < \infty\}$ cannot be dense. The contradiction thus obtained proves (ii) ⇒ (iii).

(5) Let us prove (iv) ⇒ (i). We shall use the theory of Fourier series on the group $G = M$. For any function $f \in L^2(M, \mathfrak{S}, \mu)$ we can determine its Fourier coefficients $\hat{f}_\chi = \int_M f \cdot \bar{\chi}\, d\mu$, $\chi \in \hat{G}$, where $\hat{G}$ is the character group of the group G. Since G is compact, the group $\hat{G}$ is countable and $f(x) = \sum_{\chi \in \hat{G}} \hat{f}_\chi \cdot \chi(x)$, where the series in the right-hand side converges with respect to the norm in $L^2(M, \mathfrak{S}, \mu)$. Consider the unitary operator U_{g_0} adjoint to the translation T_{g_0}. Then

$$(U_{g_0}f)(x) = \sum \hat{f}_\chi \cdot U_{g_0}\chi(x) = \sum \hat{f}_\chi \cdot \chi(g_0 x) = \sum (\hat{f}_\chi \cdot \chi(g_0)) \cdot \chi(x),$$

i.e., the action of the operator U_{g_0} is the multiplication of the Fourier coefficient $\hat{f}_\chi$ by $\chi(g_0)$. Therefore, for a invariant (mod 0) function f (i.e., a function such that $U_{g_0} f = f$) we get $\hat{f}_\chi = \hat{f}_\chi \cdot \chi(g_0)$ for all $\chi \in \hat{G}$. By assumption $\chi(g_0) \neq 1$ if $\chi \not\equiv 1$. Therefore, $\hat{f}_\chi = 0$ for $\chi \not\equiv 1$ and $f(x) = \text{const(mod 0)}$. The theorem is proved. □

Remark. It is easy to verify that no group translation possesses a mixing. Indeed, for any nontrivial character $\chi \not\equiv 1$ we have

$$(U^n_{g_0}\chi, \chi) = \int \chi(g_0^n x)\overline{\chi(x)}\, d\mu = \chi(g_0^n) \int |\chi(x)|^2\, d\mu = \chi(g_0^n) = (\chi(g_0))^n,$$

and this scalar product does not vanish when $n \to \infty$, since $|\chi(g_0)| = 1$. At the same time $\int \chi(x)\, d\mu = (\chi, 1) = 0$.

Now assume $\{g_t, -\infty < t < \infty\}$ is a continuous one-parameter subgroup of the commutative compact group $G = M$. Such a subgroup generates a flow $\{T^t\}$ on M in accordance to the formula

$$T^t x = g_t x, \qquad x \in M.$$

It is obvious that this flow preserves the Haar measure μ. The properties of such flows are similar to those of individual group translations.

Theorem 1′. *The following conditions are equivalent:*

(i) *the flow $\{T^t\}$ is ergodic;*
(ii) *the one-parameter group of homeomorphisms $\{T^t\}$ is minimal;*
(iii) *the one-parameter group of homeomorphisms $\{T^k\}$ is uniquely ergodic;*
(iv) *for any nontrivial character $\chi \in \hat{G}$, the derivative $(d/dt)\chi(g_t)|_{t=0}$ does not vanish.*

Moreover the flow $\{T^t\}$ does not possess a mixing.

The proof differs from that of Theorem 1 only by unimportant details and hence will not be provided.

EXAMPLES

1. Denote by S^1 the unit circle with normalized Lebesgue measure viewed as a commutative group; $\{S_i^1\}$ is an infinite number of copies of such groups. Put $M_1 = \prod_{i=0}^{\infty} S_i^1, M_2 = \prod_{i=-\infty}^{\infty} S_i^1$. It is clear that M_1 and M_2 are compact topological groups in the Cartesian product topology. Any character χ is of the form $\chi = \exp(2\pi i \sum n_k x_k)$ where x_k is the cyclic coordinate on S_k^1, $0 \leq x_k < 1$, n_k are integers, $-\infty < n_k < \infty$ and the sum $\sum n_k x_k$ consists of

only a finite number of nonzero summands (i.e., only finite number of the n_k do not vanish), and in the first case we have $k \geq 0$, in the second $-\infty < k < \infty$.

For any real σ construct a flow on M_1, M_2 defined by the formula

$$\{T^t x\}_k = (x_k + \sigma^k t)(\text{mod } 1).$$

Such a flow will be ergodic if and only if the number σ is transcendental. Indeed for the character $\chi = \exp(2\pi_i \sum n_k x_k)$, we have

$$\frac{d}{dt}\chi(g_t)|_{t=0} = 2\pi i \sum_k n_k \sigma^k.$$

The ergodicity condition $\sum_k n_k \sigma^k \neq 0$ for any integers n_k is the condition for the number σ to be transcendental.

2. Suppose $\mathbb{Z}_p$, $1 < p < \infty$ is the commutative group consisting of p elements. The group operation on $\mathbb{Z}_p$ will be written additively. Suppose further $\mathbb{Z}_p^{(i)}$, $i = 1, 2, 3, \ldots$ is an infinite number of copies of such groups and $G = M = \prod_i \mathbb{Z}_p^{(i)}$. Consider an arbitrary element $g \in G$, $g = \{a_i\}$, $a_i \in \mathbb{Z}_p^{(i)}$. Let us show that in this case no translation T_g is ergodic. Indeed there exist numbers $i_1, i_2, i_1 \neq i_2$ such that $a_{i_1} = a_{i_2}$. In that case the set $\{x: x_{i_1} = x_{i_2}\}$ will be a nontrivial invariant subset of M, since $T_g x = \{x_i + a_i\}$.

§2. Skew Translations and Compound Skew Translations on Commutative Compact Groups

Assume that the measure space M is the direct product $M = M_1 \times M_2$ where M_1, M_2 are commutative compact groups. Then M is also a commutative compact group, which is the direct product of the groups M_1, M_2. The measure μ on M is the normalized Haar measure $\mu = \mu_1 \times \mu_2$, where μ_i is the Haar measure on M_i, $i = 1, 2$. The group operation in this section will be written additively.

For any measurable function $\varphi: M_1 \to M_2$ in the sense of Borel and any $g_0 \in M_1$, consider the transformation T of the space M given by

$$T(x_1, x_2) = (x_1 + g_0, x_2 + \varphi(x_1)). \tag{1}$$

Definition 1. The transformation T is said to be a *skew translation* on the group.

It is clear that a skew translation preserves the Haar measure μ.

EXAMPLE

1. $M_1 = M_2 = S^1$. Then M is the two-dimensional torus Tor^2. An example of the skew translation on it is the transformation of the form

$$T(x_1, x_2) = (x_1 + \alpha(\mathrm{mod}\ 1), x_2 + kx_1(\mathrm{mod}\ 1)), \tag{2}$$

where k is an integer, $\alpha \in \mathbb{R}^1$, x_1, x_2 are cyclic coordinates on M_1, M_2.

The skew translation used in formula (1) is not necessarily ergodic even if the translation T_{g_0} on M is ergodic. For example, if $M_1 = M_2$, $\varphi(x_1) \equiv g_0$, then T is not ergodic for any g_0. Let us show, however, that the skew translation (2) is ergodic for any irrational α and any $k \neq 0$.

Indeed, if $s = (s_1, s_2), f_s(x) = f_{(s_1, s_2)}(x_1, x_2) = \exp[2\pi i(s_1 x_1 + s_2 x_2)]$, then $f_s(Tx) = \exp[2\pi i s_1 \alpha \cdot f_{\theta(s)}(x)]$, where $\theta(s) = (s_1 + ks_2, s_2)$. This means that if $g(x)$ is a (mod 0)-invariant function, $g \in L^2(M, \sigma, \mu)$, $g(x) = \sum_s c_s f_s(x)$ is its Fourier series, then $|c_s| = |c_{\theta^{-1}(s)}|$. The condition $\sum |c_s|^2 < \infty$ implies $c_s = 0$ if $s_2 \neq 0$. In the case $s_2 = 0$, for $s = (s_1, 0)$ we get the equality $c_s = c_s \exp(2\pi i s_1 \alpha)$. Since α is irrational, $c_s = 0$ for all s, except $s = (0, 0)$. Hence $g(x) \equiv \mathrm{const}(\mathrm{mod}\ 0)$.

In a number of problems, transformations more complicated than skew translations appear. Indeed, suppose $M = M_1 \times \cdots \times M_r$ $(r \geq 1)$ is the direct product of r commutative compact groups. Assume further that $g_0 \in M_1$, $\varphi_s: M_1 \times \cdots \times M_s \to M_{s+1}$ are functions measurable in the sense of Borel, $s = 1, 2, \ldots, r - 1$. Consider the transformation T of the space M given by the formula

$$\begin{aligned} Tx &= T(x_1, \ldots, x_r) \\ &= (x_1 + g_0, x_2 + \varphi_1(x_1), x_3 + \varphi_2(x_1, x_2), \ldots, x_r + \varphi_{r-1}(x_1, \ldots, x_{r-1}). \end{aligned} \tag{3}$$

Definition 2. The transformation T is said to be a *compound skew translation* on the group M.

It is clear that any compound skew translation preserves the Haar measure μ on M. In many cases it is possible to establish the ergodicity of a compound skew translation (as in the example above) by using the Fourier series.

Now let us prove the main theorem on compound skew translations. Note preliminarily, that the transformation (3) is a homeomorphism of the space M viewed as a compact metric space.

Theorem 1. *If the compound skew translation* (3) *is ergodic with respect to the measure* μ, *then it is uniquely ergodic.*

The proof shall be carried out by induction on r. For $r = 1$, we have an ordinary group translation and its unique ergodicity follows from Theorem 1, §1. Now take $r > 1$ and assume that the theorem has been proved for dimensions less than r. Denote $\overline{M}_1 = M_1 \times \cdots \times M_{r-1}$, $\overline{M}_2 = M_r$; μ_1, μ_2 are the normed Haar measures on $\overline{M}_1, \overline{M}_2$. Consider the automorphism T_1 of the space $\overline{M}_1$:

$$\begin{aligned} T_1 y_1 &= T_1(x_1, \ldots, x_{r-1}) \\ &= (x_1 + g_0, x_2 + \varphi_1(x_1), \ldots, x_{r-1} + \varphi_{r-2}(x_1, \ldots, x_{r-2})). \end{aligned}$$

Then $M = \overline{M}_1 \times \overline{M}_2$ and the automorphism T may be written in the form

$$T(y_1, y_2) = (T_1 y_1, y_2 + \varphi(y_1)), \tag{4}$$

where $\varphi(y_1) = \varphi_r(x_1, \ldots, x_{r-1})$, $y_i \in \overline{M}_i$ $(i = 1, 2)$. The ergodicity of T implies that of T_1 so that by the induction hypothesis, T_1 is uniquely ergodic.

Note that under the natural projection $\pi\colon M \to \overline{M}_1$, $\pi(y_1, y_2) = y_1$, any normalized invariant measure ν for T is mapped into the unique normalized invariant measure μ_1 for T_1. Therefore the statement of the theorem will follow from the lemma below (in which we write M_1, M_2 instead of $\overline{M}_1, \overline{M}_2$, and denote the corresponding σ-algebras by $\mathfrak{S}_1, \mathfrak{S}_2$).

Lemma 1. *Suppose an ergodic automorphism of the space $M = M_1 \times M_2$ is given by formula* (4). *Then any normalized invariant Borel measure ν (for T) which projects under the map $\pi\colon M \to M_1$ into the measure μ_1 coincides with the measure $\mu = \mu_1 \times \mu_2$.*

Proof. The proof of the lemma will be split up into several steps.

1. The space $M = M_1 \times M_2$ may be decomposed into fibers $M_{x_1} = \{x_1\} \times M_2$ where $x_1 \in M_1$. On each fiber the measure ν induces the conditional measure $\nu(\cdot|x_1)$. For any $z \in M_2$ denote by $\nu_z(\cdot|x_1)$ the measure defined on M_{x_1} by means of the formula

$$\nu_z(E|x_1) = \nu(E - z|x_1), \qquad E \in \mathfrak{S}_2,$$

(here $E - z = \{x_2 - z : x_2 \in E\}$). Introduce the measure ν_z on $\mathfrak{S}$ for which

$$\nu_z(A_1 \times A_2) = \int_{A_1} \nu_z(A_2|x_1)\, d\mu_1(x_1), \qquad A_1 \in \mathfrak{S}_1, A_2 \in \mathfrak{S}_2.$$

Clearly ν_z, as well as the measures μ, ν, are projected into the measure μ_1.

2. Let us prove that for any $z \in M_2$ the measure ν_z is invariant with respect to T. Suppose

$$A = A_1 \times A_2, A_1 \in \mathfrak{S}_1, A_2 \in \mathfrak{S}_2 \quad \text{and} \quad \chi_A(x) = \chi_{A_1}(x_1) \cdot \chi_{A_2}(x)$$

is the indicator of the set A. Then

$$\begin{aligned}
\nu_z(T^{-1}A) &= \int_M \chi_{T^{-1}A}(x)\, d\nu_z = \int_M \chi_A(Tx)\, d\nu_z = \int_M \chi_A(T_1 x_1, x_2 + \varphi(x_1))\, d\nu_z \\
&= \int_M \chi_{A_1}(T_1 x_1)\chi_{A_2}(x_2 + \varphi(x_1))\, d\nu_z \\
&= \int_{M_1} \chi_{A_1}(T_1 x_1) \int_{M_2} \chi_{A_2}(x_2 + \varphi(x_1))\, d(\nu_z | x_1)\, d\mu_1 \\
&= \int_{M_1} \chi_{A_1}(T_1 x_1) \int_{M_2} \chi_{A_2}(x_2 + \varphi(x_1) - z)\, d(\nu | x_1)\, d\mu_1 \\
&= \int_M \chi_{A_1}(T_1 x_1)\chi_{A_2}(x_2 + \varphi(x_1) - z)\, d\nu = \nu(T^{-1}(A_1 \times (A_2 - z)) \\
&= \nu(A_1 \times (A_2 - z)) = \nu_z(A).
\end{aligned}$$

This implies the invariance of the measure ν_z.

Further, for any subset $C \in \mathfrak{S}_2$, $\mu_2(C) > 0$, introduce the measure ν_C on $\mathfrak{S}$ by putting

$$\nu_C(A) = \frac{1}{\mu_2(C)} \int_C \nu_z(A)\, d\mu_2(z), \qquad A \in \mathfrak{S}.$$

The invariance of the measures ν_z implies that ν_C is also invariant with respect to T. Moreover, ν_C, as well as ν_z, is projected into the measure μ_1.

3. Let us show that ν_C is absolutely continuous with respect to μ. This will imply the relation $\nu_C = \mu$. Indeed, since the measure μ is ergodic, any normalized invariant measure which is absolutely continuous with respect to μ coincides with μ (see §2, Chap. 1).

Using the Fubini theorem for the set $A = A_1 \times A_2 \in \mathfrak{S}$, we can write

$$\begin{aligned}
\nu_C(A) &= \frac{1}{\mu_2(C)} \int_C \nu_z(A)\, d\mu_2(z) = \frac{1}{\mu_2(C)} \int_C d\mu_2(z) \int_{A_1} \nu_z(A_2 | x_1)\, d\mu_1(x_1) \\
&= \int_{A_1} d\mu_1(x_1) \cdot \frac{1}{\mu_2(C)} \int_C \nu_z(A_2 | x_1)\, d\mu_2(z).
\end{aligned}$$

This means that the conditional measure $\nu_C(\cdot|x_1)$ is of the form

$$\nu_C(A_2|x_1) = \frac{1}{\mu_2(C)} \int_C \nu_z(A_2|x_1)\, d\mu_2(z), \tag{5}$$

and to prove the absolute continuity of ν_C it suffices to establish that the measure $\nu_C(\cdot|x_1)$ is absolutely continuous on any fiber M_{x_1} with respect to the Haar measure μ_2 (we are using the natural identification of the fiber M_{x_1} with the space M_2).

To do this consider the direct product $M_2 \times M_2 = \{(z_1, z_2): z_1, z_2 \in M_2\}$ and introduce the measure $d\lambda(z_1, z_2) = d\mu_2(z_1)\, d(\nu_{z_1}|x)(z_2)$ on this product. For any set $A_2 \subset M_2$, $\mu_2(A_2) = 0$ we will obviously have $\lambda(D) = 0$ where $D = A_1 \times A_2$. Using Fubini's theorem again, we see that for almost all (with respect to the measure μ_1) points z_1 the relation $\nu_{z_1}(A_2|x_1) = 0$ holds. Now (5) implies $\nu_C(A_2|x_1) = 0$. Thus the absolute continuity of ν_C is proved. In view of the remark at the beginning of this subsection, we see that $\nu_C - \mu$ for any $C \in \mathfrak{S}_2$, $\mu_2(C) > 0$.

4. Let us deduce from the above that $\nu = \mu$. If we have $\nu \neq \mu$, then for a set of positive μ_1-measure of conditions x_1 we would have $\nu(\cdot|x_1) \neq \mu_2$ (we again identify M_{x_1} with M_2). Since the measures $\nu(\cdot|x_1)$ are Borel, there exists a countable system of Borel sets $B = \{B_i\}$, $B_i \in M_2$ such that all these measures are entirely determined by their values on the sets B_i. Hence we can find a number i for which $\nu(B_i|x_1) \neq \mu_2(B_i)$ for a set of conditions x_1 of positive μ_1-measure, and we can assume, for example, that $\nu(B_i|x_1) < \mu_2(B_i)$. It now follows from the uniqueness of the Haar measure that we can find a $C \in \mathfrak{S}_2$, $\mu(C) > 0$ satisfying $\nu_z(B_i|x_1) < \mu_2(B_i)$ for all $z \in C$ and for a set of conditions x_1 of positive μ_1-measure. But then, by (5), $\nu_C(B_i|x_1) < \mu_2(B_i)$. On the other hand, the equality $\nu_C = \mu$ implies that for almost all (with respect to the measure μ_1) conditions x_1 we must have $\nu_C(\cdot|x_1) = \mu_2$. The contradiction thus obtained shows $\nu = \mu$. The lemma is proved, thus concluding the proof of Theorem 1. □

§3. Endomorphisms and Automorphisms of Commutative Compact Groups

In this section the measure space will again be a commutative compact group $M = G$ with normalized Haar measure μ given on the Borel σ-algebra $\mathfrak{S}$. An important role will be played by the character group $\hat{G}$. According to Pontrjagin's duality theorem, the group $\hat{G}$ is countable in the case considered.

Definition 1. A *group endomorphism* of the group G is a continuous map of G onto itself which commutes with the group operation. If this map is one-to-one, then it is called a *group automorphism*.

In other words, T is a group endomorphism if T maps G onto G continuously and $T(x \pm y) = Tx \pm Ty$ for all $x, y \in G$ and $Te = e$, where e is the unit of the group G. If T is a group automorphism, then T^{-1} is also a group automorphism.

For a group endomorphism the inverse image of the unit is a subgroup of the group G. In the case of a group automorphism, this subgroup consists only of the unit of the group.

For any group endomorphism T introduce the adjoint transformation T^* which acts in the group $\hat{G}$ according to the formula

$$(T^*\chi)(x) = \chi(Tx), \qquad \chi \in \hat{G}. \tag{1}$$

It follows in particular from this formula that $T^*\chi \in \hat{G}$. The group $\hat{G}$ may be viewed as a subset of the space $L^2(M, \mathfrak{S}, \mu)$ and the formula (1) shows that T^* is the restriction of the unitary operator U_T adjoint to T to an invariant subset $\hat{G}$. Further for all $\chi_1, \chi_2 \in \hat{G}$, $x \in G$, we have

$$T^*(\chi_1 \pm \chi_2)(x) = (\chi_1 \pm \chi_2)(Tx) = \chi_1(Tx) \pm \chi_2(Tx) = T^*\chi_1(x) \pm T^*\chi_2(x),$$

i.e., $T^*(\chi_1 \pm \chi_2) = T^*\chi_1 \pm T^*\chi_2$. This means that T^* also commutes with the group operation in the group $\hat{G}$. However, T^* is not necessarily a group endomorphism of the group $\hat{G}$ since it may not be surjective. If T is an automorphism, then $(T^{-1})^* = (T^*)^{-1}$, i.e., T^* is invertible and then T^* is a group automorphism.

Now let us show that in the case of an arbitrary endomorphism T the kernel of the homomorphism T^* is trivial. Suppose $\chi \in \hat{G}$ satisfies $T^*\chi = \hat{e}$, where $\hat{e}$ is the unit of the group $\hat{G}$. Then

$$\chi(Tx) = (T^*\chi)(x) = \hat{e}(x) = 1,$$

for any $x \in G$. Since T maps G onto G, Tx ranges over the entire group and therefore $\chi = \hat{e}$.

Lemma 1. *Suppose T^* is a homomorphism of the group $\hat{G}$ with trivial kernel. Then the formula* (1) *determines a group endomorphism T of the group G.*

Proof. The formula $\chi(Tx) = (T^*\chi)(x)$, according to Pontrjagin's duality theorem, determines a continuous homomorphism of G into G. Let us show that this homomorphism maps G *onto* G. According to the same Pontrjagin duality theorem, the quotient group G/TG is the kernel of the homomorphism T^* which is trivial by assumption. Hence $TG = G$. The lemma is proved. □

Lemma 2. *Formula* (1) *establishes a one-to-one correspondence between the group endomorphisms of the group G and the homomorphisms of the group $\hat{G}$ into itself with trivial kernels.*

Proof. Let us show that $T_1 \neq T_2$ implies $T_1^* \neq T_2^*$. Indeed, we can find an $x \in G$ such that $T_1 x \neq T_2 x$. But then it is possible to find a $\chi \in \hat{G}$, for which $\chi(T_1 x) \neq \chi(T_2 x)$, i.e., $T_1^* \chi \neq T_2^* \chi$. In the same way, we can show that $T_1^* \neq T_2^*$ implies that the corresponding group endomorphisms T_1 and T_2 do not coincide. The lemma is proved. □

Lemma 3. *Any group endomorphism T preserves the Haar measure μ.*

Proof. Suppose $f(x) = \sum_{\chi \in \hat{G}} c_\chi \cdot \chi(x)$, where only a finite number of coefficients c_χ do not vanish. Then

$$\int f(Tx)\, d\mu(x) = \int \sum c_\chi \chi(Tx)\, d\mu(x)$$

$$= \int \sum c_\chi \cdot T^*\chi(x)\, d\mu(x) = \sum c_\chi \int T^*\chi(x)\, d\mu(x).$$

The last integral does not vanish only when $\chi = \chi_0$. But then

$$\int f(Tx)\, d\mu(x) = c_{\chi_0} = \int f(x)\, d\mu(x).$$

Since the functions under consideration are dense in $C(M)$, we have, for any continuous function $f(x)$,

$$\int f(Tx)\, d\mu(x) = \int f(x)\, d\mu(x),$$

i.e., the measure μ is invariant. The lemma is proved. □

By using Lemmas 1 and 2, we shall now construct several examples of group endomorphisms and automorphisms of commutative groups.

EXAMPLES

1. G is the m-dimensional torus $\mathrm{Tor}^m = S^1 \times \cdots \times S^1$ (m terms). Then the character group of the group G is the additive group $\mathbb{Z}^m$ of points of the space $\mathbb{R}^m$ with integer coordinates. Any homomorphism T^* of the group $\mathbb{Z}^m$ into itself is given by a matrix $A^* = \|a_{ij}^*\|_{i,j=1}^m$ with integer elements. This homomorphism has a trivial kernel if and only if $\det A^* \neq 0$. T^* will be an automorphism of the group $\mathbb{Z}^m$ if and only if $\det A^* = \pm 1$. According to Lemma 1, the homomorphism T^* corresponds to the group endomorphism

T of the torus Tor^m whose action in the cyclic coordinates $x_1, \ldots, x_m$ may be written as follows

$$Tx = x', \quad \text{where} \quad x_i' = \sum_{j=1}^{m} a_{ij} x_j (\bmod 1), \qquad 1 \le i \le m.$$

The matrix $A = \|a_{ij}\|$ is the transposed matrix with respect to A^*, i.e., $a_{ij} = a_{ji}^*$. If $\det A^* = \pm 1$, then $\det A = \pm 1$ and T is a group automorphism.

2. $\hat{G}$ is a countable subgroup of the additive group of real numbers. For example, $\hat{G}$ is the additive group of rational numbers, or the group of binary rational numbers, i.e., number of the form $p/2^k$, where p, k are integers. Consider a real number α such that $\alpha\hat{G} \subset \hat{G}$. It is clear that the transformation T^* which acts according to the formula $T^*\chi = \alpha\chi$, $\chi \in \hat{G}$ is a homomorphism of the group $\hat{G}$ with trivial kernel. By Lemma 1 it generates a group endomorphism of the character group G of the group $\hat{G}$. If $\alpha^{-1} \in \hat{G}$, this endomorphism will be an automorphism.

Lemma 3 implies that an arbitrary group endomorphism preserves the Haar measure. Now we shall obtain a necessary and sufficient condition for the ergodicity and the mixing property of such an endomorphism.

Suppose T^* is the homomorphism adjoint to the group endomorphism T of the character group $\hat{G}$. Consider the sequence of subgroups $\{\hat{G}_n = (T^*)^n\hat{G}\}$, $n = 0, 1, \ldots$. Clearly, $\hat{G}_0 = \hat{G}$, $\hat{G}_{n+1} \subset \hat{G}_n$, $n = 0, 1, \ldots$ and the subgroup $\hat{G}_n$ consists of the elements for which $(T^*)^{-n}$ is defined. Put $\hat{G}_\infty = \bigcap_n \hat{G}_n$. Then $\hat{G}_\infty$ is a subgroup of the group $\hat{G}$ and the restriction of T^* to $\hat{G}_\infty$ is an automorphism, which we shall also denote by T^*. If T is a group automorphism, then $\hat{G} = \hat{G}_\infty = \hat{G}_n$ for all n.

First assume that T is a group automorphism and therefore T^* is also a group automorphism.

Theorem 1. *For the ergodicity of T it is necessary and sufficient that the trajectory of any nonzero element of the group $\hat{G}$ under the action of T^* be infinite. If this condition holds, then T also is mixing.*

Proof. Necessity. Suppose T is ergodic. If we can find a nontrivial character χ for which $(T^*)^n\chi = \chi$ for some n, then by the orthogonality of distinct characters the function $(1/n)(\chi + T^*\chi + \cdots + (T^*)^{n-1}\chi)$ is invariant and is not a constant (mod 0). Therefore T is nonergodic, proving the necessity of our condition.

Sufficiency. We shall prove at once that if the conditions of the theorem hold, then T is mixing. Suppose

$$f_1(x) = \sum_{\chi \ne \hat{e}} c_\chi^{(1)} \cdot \chi(x), \qquad f_2(x) = \sum_{\chi \ne \hat{e}} c_\chi^{(2)} \cdot \chi(x),$$

where in the sums only a finite number of terms do not vanish. Then

$$(U_T^n f_1, f_2) = \sum_{\chi_1, \chi_2} c_{\chi_1}^{(1)} \cdot \bar{c}_{\chi_2}^{(2)} \cdot ((T^*)^n \chi_1, \chi_2).$$

The scalar product $((T^*)^n \chi_1, \chi_2)$ does not vanish only if $(T^*)^n \chi_1 = \chi_2$. But since the trajectory of each nontrivial character is infinite, this last equality is impossible for sufficiently large n. Hence $(U_T^n f_1, f_2) = 0$ for n large enough. Since the functions of the type considered are dense in $L_0^2(M, \mathfrak{S}, \mu)$, we have $\lim_{n\to\infty}(U_T^n f_1, f_2) = 0$ for all $f_1, f_2 \in L_0^2(M, \mathfrak{S}, \mu)$. This obviously implies that T is mixing. The theorem is proved. □

Theorem 2. *A group endomorphism T is ergodic, if and only if, the automorphism T^* of the group $\hat{G}_\infty$ satisfies the assumptions of Theorem 1. In this case T also is mixing.*

Proof. Necessity can be proved just as in Theorem 1.

Sufficiency. Put $\hat{G}^{(m)} = \hat{G}_m \setminus \hat{G}_{m+1}$, $m = 0, 1, \ldots$. Then the $\hat{G}^{(m)}$ are pairwise disjoint and

$$(T^*)^n \hat{G}^{(m)} = \hat{G}^{(n+m)}, \tag{2}$$

for all $m, n \geq 0$. Suppose the functions f_1, f_2 are the same as in the proof of Theorem 1. Present f_2 in the form

$$f_2 = \sum_{m=0}^{\infty} f_2^{(m)} + f_2^{(\infty)}, \tag{3}$$

where

$$f_2^{(m)} = \sum_{\chi \in \hat{G}^{(m)}} c_\chi^{(2)} \cdot \chi, \qquad f_2^{(\infty)} = \sum_{\chi \in \hat{G}^{(\infty)}} c_\chi^{(2)} \cdot \chi.$$

Note that in the sums above only a finite number of terms do not vanish. In accordance to (3), represent the scalar product $(U_T^n f_1, f_2)$ in the form

$$(U_T^n f_1, f_2) = \sum_{m=0}^{\infty} (U_T^n f_1, f_2^{(m)}) + (U_T^n f_1, f_2^{(\infty)}). \tag{4}$$

For a fixed m each term of the last sum (over m) vanishes if n is sufficiently large. This immediately follows from (2) and the fact that distinct characters are orthogonal. To the last summand in (4) we can apply the arguments used in proving the sufficiency in the previous theorem and it also vanishes for sufficiently large n. Since the sum over m in (4) contains only a finite number of nonzero terms, we see that $(U_T^n f_1, f_2) = 0$ for n sufficiently large. The theorem is proved. □

EXAMPLES

1. $G = \mathrm{Tor}^2$, T is the group automorphism given by the matrix

$$T = \begin{Vmatrix} a & b \\ c & d \end{Vmatrix}$$

where a, b, c, d are integers and $ad - bc = \pm 1$. Then the automorphism T^* is given by the matrix

$$T^* = \begin{Vmatrix} a & c \\ b & d \end{Vmatrix}.$$

The conditions of Theorem 1 reduce to the fact that the matrix T^*, and therefore the matrix T as well, has no eigen-values which are roots of 1. In the two-dimensional case this is equivalent to the fact that T has two real eigen-values whose absolute value differs from 1.

2. $\hat{G}$ is a countable subgroup of $\mathbb{R}^1$, the transformation T^* is the multiplication by a fixed element $r \in \hat{G}$. Then the automorphism T of the group G corresponding to the transformation T^* is ergodic for $r \neq 1$.

The theory of characters enables us to study in detail the structure of periodic points of group automorphisms of commutative compact groups. We shall assume that the group automorphism T is ergodic and therefore satisfies the assumptions of Theorem 1. The fact that each orbit $\{(T^*)^n\chi\}$ is infinite for any nontrivial χ means that $(T^*)^n\chi \neq \chi$ for any $n \neq 0$. In other words, the kernel of $(T^*)^n - I$ is trivial for any $n \neq 0$, where I is the identical transformation. Therefore Lemma 1 implies that $T^n - I$ is a group endomorphism for $n \neq 0$ and hence maps the group G onto itself. Put $N_n = \{x \in G: T^n x = x\}$. Clearly $N_n = \{x \in G: (T^n - I)x = e\}$ and therefore N_n is a subgroup of the group G. It is easy to check that $\mathrm{card}(N_n)$ equals the number of cosets in $\hat{G}/((T^*)^n - I)\hat{G}$.

Consider the sequence of subgroups $((T^*)^n - I)\hat{G}$, $n = 0, 1, \ldots$. We shall say that it *vanishes to infinity* if for any nontrivial character $\chi \in \hat{G}$ the inclusion $\chi \in ((T^*)^n - I)\hat{G}$ holds only for a finite number of values of n.

Theorem 3. *If the automorphism T is ergodic and the sequence $((T^*)^n - I)\hat{G}$ vanishes to infinity, then for any function $f \in C(G)$ we have*

$$\lim_{n\to\infty} \frac{1}{\mathrm{card}(N_n)} \sum_{z \in N_n} f(z) = \int_G f\, d\mu.$$

Proof. It suffices to prove the statement of the theorem for functions of the form $f = \chi_0$, where χ_0 is a nontrivial character of the group G, i.e., to show that

$$\lim_{n\to\infty} \frac{1}{\operatorname{card}(N_n)} \sum_{z\in N_n} \chi_0(z) = 0. \tag{5}$$

Assume n satisfies

$$\chi_0 \notin [(T^*)^n - I]\hat{G}. \tag{6}$$

For any character $\chi \in \hat{G}$ let us put

$$\chi^{(n)}(x) \stackrel{\text{def}}{=} \frac{\chi(T^n x)}{\chi(x)} = [(T^*)^n - I]\chi(x).$$

Then $\chi^{(n)}$ is also a character. It is easy to check that $\chi^{(n)}$ is constant on cosets with respect to the subgroup N_n and therefore may be viewed as a function on the quotient group G/N_n. The family of characters $\chi^{(n)}$ corresponding to all $\chi \in G$ constitutes a complete orthogonal system of functions in the space $L^2(M/N_n, \mu|N_n)$ where $\mu|N_n$ is the Haar measure on G/N_n. It follows from (6) that the character χ_0 is orthogonal to any character of the form $\chi^{(n)}$ and therefore $\int \chi(x_0)f(x)\,d\mu = 0$ for any function $f \in L^2(M, \mathfrak{S}, \mu)$ constant on the cosets with respect to N_n. Represent the last integral in the form of a repeated integral: first integrate (take the mean) with respect to the coset and then integrate over the quotient group G/N_n. We then see, since f was arbitrary, that

$$\sum_{z\in M_n} \chi_0(xz) = \chi_0(x) \sum_{z\in N_n} \chi_0(z) = 0,$$

for almost all $x \in M$. This implies (5). The theorem is proved. □

Corollaries and Examples

1. Under the conditions of Theorem 3, the set of all periodic points of the automorphism T is dense. This means that T has many invariant measures concentrated on periodic trajectories. Often group automorphisms also possess many continuous invariant measures, but the corresponding theorems are not easy.

2. Suppose T is an ergodic automorphism of the character group of the rational numbers, adjoint to the multiplication by the rational number r. Then $(T^*)^n - I$ is the multiplication by $r^n - 1$ and is therefore an automorphism of the group of rational numbers. Thus card $(N_n) = 1$ for any n and the statement of Theorem 3 on the uniform distribution of periodic points breaks down.

3. Suppose M is the direct product $\bigotimes_{-\infty}^{\infty} \mathbb{Z}_2$ of copies of the group $\mathbb{Z}_2$, X is the direct sum of such groups and T is the shift (group automorphism). Here, as can be seen easily, we have a uniform distribution of periodic points, but the statement that the sequence of subgroups $((T^*)^n - I)X$ vanishes to infinity is not easy to prove.

4. *Hopf's geometric method for proving the ergodicity of the group automorphisms of the torus.* Suppose $M = \mathrm{Tor}^m$ is an m-dimensional torus, T is its group automorphism given by an integer matrix $A = \|a_{ij}\|$, $\det A = \pm 1$. Assume that none of the eigen-values of the matrix A have absolute values equal to one.

We shall use the following algebraic statement, which immediately follows from the fact that the matrix A can be written in Jordan normal form. The space $\mathbb{R}^m$ may be presented as the direct sum of two subspaces $E^{(u)}, E^{(s)}$: $\mathbb{R}^m = E^u \oplus E^s$ such that

(1) $AE^{(u)} = A^{-1}E^{(u)} = E^{(u)}$.
(2) $AE^{(s)} = A^{-1}E^{(s)} = E^{(s)}$.
(3) There exists a constant λ, $0 < \lambda < 1$, such that

$$\|Az\| \le \lambda\|z\| \quad \text{for any } z \in E^{(s)},$$

$$\|A^{-1}z\| \le \lambda\|z\| \quad \text{for any } z \in E^{(u)}.$$

Suppose $m_u = \dim E^{(u)}$, $m_s = \dim E^{(s)}$. Then $m = m_u + m_s$ and we can introduce, on the torus Tor^m, an m_u-dimensional subgroup $\mathscr{E}^{(u)}$ of the translation group by vectors from the space $E^{(u)}$ as well as a similar subgroup $\mathscr{E}^{(s)}$.

An open set $U \subset \mathrm{Tor}^m$ which can be decomposed into connected open sets $U^{(u)}$ of orbits of the group $\mathscr{E}^{(u)}$ and into connected open sets $U^{(s)}$ or orbits of the group $\mathscr{E}^{(s)}$, where any two sets $U_1^{(u)}$, $U_2^{(s)}$ in these decompositions intersect in precisely one point, will be called a *parallelogram.* The sets $U^{(u)}$, $(U^{(s)})$ will be called *expanding* (*contracting*) *fibres* of the parallelogram U. Denote by $U^{(u)}(x)(U^{(s)}(x))$ the expanding (contracting) fibre containing the point $x \in U$. Then for any $x \in U$ the parallelogram U may be presented in the form

$$U = \bigcup_{y \in U^{(u)}(x)} U^{(s)}(y) = \bigcup_{y \in U^{(s)}(x)} U^{(u)}(y).$$

It follows from (3) that for any y_1, y_2 belonging to the same $U^{(s)}(U^{(u)})$ we have the relation $\mathrm{dist}(T^n y_1, T^n y_2) \to 0$ when $n \to \infty$ $(n \to -\infty)$. This in turn implies that if f is a continuous function on Tor^m and

$$\lim_{n\to\infty} \frac{1}{n} \sum_{k=0}^{n-1} f(T^k y_1) \qquad \left(\lim_{n\to\infty} \frac{1}{n} \sum_{k=0}^{n-1} f(T^{-k} y_1)\right)$$

exists, then we can find the same limit for the point z_2 and these limits coincide. Note that the restriction of the Haar measure μ to U may be written in the form $d\mu = d\mu^{(u)} \cdot d\mu^{(s)}$. This means that for any bounded measurable function f and any expanding fibre $U_0^{(u)}$ (contracting fibre $U_0^{(s)}$) we have the following relation

$$\int_U f\,d\mu = \int_{U_0^{(u)}} d\mu^{(u)}(x) \int_{U^{(s)}(x)} f\,d\mu^{(s)} = \int_{U_0^{(s)}} d\mu^{(s)}(y) \int_{U^{(u)}(y)} f\,d\mu^{(u)}.$$

Now suppose f is a continuous function on Tor^m. According to the Birkhoff–Khinchin ergodic theorem the following limits

$$\bar{f}^+(x) = \lim_{n\to\infty} \frac{1}{n} \sum_{k=0}^{n-1} f(T^k x), \qquad \bar{f}^-(x) = \lim_{n\to\infty} \frac{1}{n} \sum_{k=0}^{n-1} f(T^{-k} x)$$

exist and are equal to each other for almost all x. We shall show that the function $\bar{f}^+(x)$ on the parallelogram U is constant almost everywhere. Take two points $x_1, x_2 \in U$. Let us find points y_1, y_2 located in the same expanding fibre: $U^{(u)}(y_1) = U^{(u)}(y_2) = U_0^{(u)}$ such that $y_1 = U_0^{(u)} \cap U^{(s)}(x_1)$, $y_2 = U_0^{(u)} \cap U^{(s)}(x_2)$. Assume that the values $\bar{f}^+(y_1), \bar{f}^-(y_1), \bar{f}^+(y_2), \bar{f}^-(y_2)$ exist and $\bar{f}^+(y_1) = \bar{f}^-(y_1)$, $\bar{f}^+(y_2) = \bar{f}^-(y_2)$. Then

$$\bar{f}^+(x_1) = \bar{f}^+(y_1) = \bar{f}^-(y_1) = \bar{f}^-(y_2) = \bar{f}^+(y_2) = \bar{f}^+(x_2).$$

It follows from Fubini's theorem that there exist subsets $U' \subset U$, $\mu(U') = \mu(U)$ such that for any two points $x_1, x_2 \in U'$ there exist points y_1, y_2 for which our assumptions hold. Therefore $\bar{f}^+(x)$ is constant almost everywhere on U.

Now note that the torus Tor^m can be covered by parallelograms (hence by finite number of them). This means that $\bar{f}^+(x)$ is constant almost everywhere on the torus. Since continuous functions are dense in $L^2(M, \mathfrak{S}, \mu)$, we see that any function invariant with respect to T may be approximated by functions of the form $\bar{f}^+$; it is therefore constant almost everywhere. Thus the ergodicity of T is proved.

§4. Dynamical Systems on Homogeneous Spaces of the Group SL(2, ℝ)

The group SL(2, ℝ) of matrices of the second order with real elements and determinant equal to 1 is the simplest noncommutative group which is related to interesting examples of dynamical systems with various ergodic properties.

The space of the group SL(2, ℝ) may be realized as a hyperboloid in $\mathbb{R}^4$: if

$$g = \begin{Vmatrix} a & b \\ c & d \end{Vmatrix} \in \mathrm{SL}(2, \mathbb{R})$$

then $ad - bc = 1$ is the equation of this hyperboloid: in this case a, b, c, d are viewed as coordinates in $\mathbb{R}^4$). SL(2, $\mathbb{R}$) is a locally compact topological group and even a Lie group. Outside of the hyperplane $d = 0$ we can choose the numbers a, b, c to be the coordinates on SL(2, $\mathbb{R}$). There exists a Haar measure, left and right invariant, on SL(2, $\mathbb{R}$) which in the coordinates a, b, c may be written in the form $d\mu(g) = (1/|d|)\, da\, db\, dc$. This measure is infinite: $\mu(\mathrm{SL}(2, \mathbb{R})) = \infty$, but σ-finite.

Consider a discrete subgroup $\Gamma \subset \mathrm{SL}(2, \mathbb{R})$, i.e., such a subgroup that for some neighborhood U of the unit e of the group SL(2, $\mathbb{R}$) we have $\Gamma \cap U = \{e\}$. Assume that the quotient space $M \overset{\text{def}}{=} \Gamma \backslash \mathrm{SL}(2, \mathbb{R})$, i.e., the space of left cosets is of finite volume. The condition of finiteness of the volume of M means that there exists a Borel subset $D \subset \mathrm{SL}(2, \mathbb{R})$ (a fundamental domain) for which $\mu(D) < \infty$, $\gamma_1 D \cap \gamma_2 D = \varnothing$ when $\gamma_1, \gamma_2 \in \Gamma$, $\gamma_1 \neq \gamma_2$ and $\bigcup_{\gamma \in \Gamma} \gamma D = \mathrm{SL}(2, \mathbb{R})$.

As an example of a subgroup Γ with this property, we can indicate the subgroup of matrices with integer elements a, b, c, d. In this case D is non-compact although $\mu(D) < \infty$. There exist examples of subgroups Γ for which D is compact.

The Haar measure μ on SL(2, $\mathbb{R}$) is defined up to a factor which we shall choose so as to have $\mu(D) = 1$. Identifying D with M, we transform M into a space with measure which we shall continue to denote by μ.

Any one-parameter subgroup $\{g_t, -\infty < t < \infty\} \subset \mathrm{SL}(2, \mathbb{R})$ generates a dynamical system $\{T^t\}$ on M: $T^t x = x \cdot g_t$, $x \in M$. This equality is meaningful, since the left cosets with respect to the subgroup Γ are mapped by the right translation g_t into left cosets. It follows from the right invariance of the Haar measure that $\{T^t\}$ preserves the measure μ and is therefore a flow on M. Further we shall denote dynamical systems generated by one-parameter subgroups in the same way as the subgroups themselves.

Let us consider some important examples of one-parameter subgroups specifying the names of the corresponding flows. The meaning of these names will become clear later when we shall explain the relationship between the dynamical systems considered and the geometry of Lobachevsky.

EXAMPLES

1. $g_t = \begin{Vmatrix} e^t & 0 \\ 0 & e^{-t} \end{Vmatrix}$—geodesic flow.

2. $O_t^+ = \begin{Vmatrix} 1 & t \\ 0 & 1 \end{Vmatrix}, O_t^- = \begin{Vmatrix} 1 & 0 \\ t & 1 \end{Vmatrix}$—horocyclic (positive and negative) flows.

3. $a_t = \begin{Vmatrix} \cos t & \sin t \\ -\sin t & \cos t \end{Vmatrix}$—cyclic flow.

Theorem 1. *The geodesic flow $\{g_t\}$ is ergodic and mixing.*

Proof. It can be checked directly that

$$O_s^+ g_t = g_t O_{e^{-2t}s}^+, \qquad O_s^- g_t = g_t O_{e^{2t}s}^-, \tag{1}$$

for any s, t. Note further that the space $M = \Gamma \backslash \mathrm{SL}(2, \mathbb{R})$ is a manifold of class C^∞, and the vector fields on it corresponding to the flows $\{g_t\}$, $\{O_t^+\}$, $\{O_t^-\}$ are also of class C^∞ and are linearly independent at each point $x \in M$. The arguments we shall use below are similar to the proof (due to Hopf) of the ergodicity of the group automorphism of the torus given above in §3. We shall divide the argument into several steps.

1. Suppose $f(x)$ is a continuous function with compact support on M, i.e., $f(x) = 0$ outside of some compact set. By the Birkhoff–Khinchin ergodic theorem, the time means

$$\bar{f}^+(x) = \lim_{t \to \infty} \frac{1}{t} \int_0^t f(x \cdot g_u)\, du, \qquad \bar{f}^-(x) = \lim_{t \to \infty} \frac{1}{t} \int_0^t f(x \cdot g_{-u})\, du,$$

exist for almost all $x \in M$ and for almost all x we have the equality $\bar{f}^+(x) = \bar{f}^-(x)$. Let us show that if the limit $\bar{f}^+(x)(\bar{f}^-(x))$ exists for some x, then for any s we have a similar limit for the point xO_s^+ and $\bar{f}^+(xO_s^+) = \bar{f}^+(x)$ $(\bar{f}^-(xO_s^-) = \bar{f}^-(x)$ respectively). Indeed, it follows from (1) that for $u > 0$, we have

$$f(xO_s^+ g_u) = f(xg_u O_{e^{-2u}s}^+)$$

and for $u \leq 0$:

$$f(xO_s^- g_u) = f(xg_u O_{e^{2u}s}^-).$$

Therefore in view of the uniform continuity of f

$$\lim_{u \to +\infty} [f(xO_s^+ g_u) - f(xg_u)] = \lim_{u \to +\infty} [f(xg_u O_{e^{-2u}s}^+) - f(xg_u)] = 0,$$

$$\lim_{u \to -\infty} [f(xO_s^- g_u) - f(xg_u)] = \lim_{u \to -\infty} [f(xg_u O_{e^{2u}s}^-) - f(xg_u)] = 0,$$

and the limits are uniform with respect to x. Therefore

$$\lim_{t \to \infty} \frac{1}{t} \int_0^t [f(xg_u) - f(xO_s^+ g_u)] = \lim_{t \to \infty} \frac{1}{t} \int_0^t [f(xg_{-u}) - f(xO_s^- g_{-u})]\, du = 0. \tag{2}$$

Fixing $x_0 \in M$ let us construct, for $\delta_1 > 0, \delta_2 > 0, \delta_3 > 0$ first a smooth curve $\gamma_{\delta_1}(x_0) = \{x_0 O_r^- : |r| < \delta_1\}$ and then an open smooth two-dimensional surface

$$\sigma_{\delta_1, \delta_2}(x_0) = \{x_0 O_r^- g_u : |r| < \delta_1, |u| < \delta_2\} = \bigcup_{|u| < \delta_2} (\gamma_{\delta_1}(x_0)) g_u,$$

and finally an open neighborhood $U_{\delta_1, \delta_2, \delta_3}(x_0)$ of the point x_0:

$$U_{\delta_1, \delta_2, \delta_3}(x_0) = \bigcup_{|s| < \delta_3} (\sigma_{\delta_1, \delta_2}(x_0)) O_s^+.$$

It follows from the smoothness of the corresponding vector fields that for sufficiently small $\delta_1, \delta_2, \delta_3$ the surfaces $(\sigma_{\delta_1, \delta_2}(x_0)) O_s^+$ are disjoint for distinct s, $|s| < \delta_3$ and for the point $x = x_0 O_r^- g_u O_s^+ \in U_{\delta_1, \delta_2, \delta_3}(x_0)$ the numbers r, u, s are smooth coordinates in $U_{\delta_1, \delta_2, \delta_3}(x_0)$. Also, when x_0 ranges over a compact part of M, the numbers $\delta_1, \delta_2, \delta_3$ may be chosen so as to satisfy this last condition independently of x_0. We shall assume that such a choice has already been carried out.

Note that the measure μ induces conditional measures (for which no special notation will be introduced) on the surfaces $(\sigma_{\delta_1, \delta_2}(x_0)) O_s^+$.

Lemma 1. *For almost every (with respect to the measure μ) point $x_0 \in M$ the surface $\sigma_{\delta_1, \delta_2}(x_0)$ has the following property: for almost all (with respect the conditional measure) points $y \in \sigma_{\delta_1, \delta_2}(x_0)$ we have $f^+(y) = \bar{f}^-(y)$.*

Proof. Since the set of all $x \in M$ for which $\bar{f}^+(x)$ does not equal $\bar{f}^-(x)$, or which one of these means does not exist, is of vanishing measure, the statement of the lemma immediately follows from Fubini's theorem. □

Using this lemma let us conclude the proof of the ergodicity of the flow $\{g_t\}$. For this it suffices to show that $\bar{f}^+(x)$ is constant for almost all $x \in U_{\delta_1, \delta_2, \delta_3}(x_0)$.

Denote by $\tilde{U}$ the subset of $U_{\delta_1, \delta_2, \delta_3}(x_0)$ consisting of all points $x = x_0 O_r^- g_u O_s^+$ satisfying the following conditions:

(1) $\bar{f}^+(x)$ exists;
(2) for $y = x_0 O_r^- g_u \in \sigma_{\delta_1, \delta_2}(x_0)$ we have

$$\bar{f}^+(y) = \bar{f}^-(y). \tag{3}$$

Using the smoothness of the vector fields and Fubini's theorem, it is easy to show that $\tilde{U}$ is a subset of $U_{\delta_1, \delta_2, \delta_3}(x_0)$ of full measure. For any two points $x_1, x_2 \in \tilde{U}$, $x_1 = x_0 O_{r_1}^- g_{u_1} O_{s_1}^+$, $x_2 = x_0 O_{r_2}^- g_{u_2} O_{s_2}^+$, put

$$y_1 = x_0 O_{r_1}^- g_{u_1}, \qquad z_1 = x_0 O_{r_1}^-,$$

$$y_2 = x_0 O_{r_2}^- g_{u_2}, \qquad z_2 = x_0 O_{r_2}^-.$$

We then obtain

$$\bar{f}^+(x_1) = \bar{f}^+(y_1) = \bar{f}^-(y_1) = \bar{f}^-(z_1) = \bar{f}^-(z_2) = \bar{f}^-(y_2) = \bar{f}^+(y_2) = \bar{f}^+(x_2).$$

The relations $\bar{f}^+(x_1) = \bar{f}^+(y_1)$, $\bar{f}^+(x_2) = \bar{f}^+(y_2)$, $\bar{f}^-(z_1) = \bar{f}^-(z_2)$ follow from (2), the relations $\bar{f}^+(y_1) = \bar{f}^-(y_1)$, $\bar{f}^+(y_2) = \bar{f}^-(y_2)$ from (3), and the relations $\bar{f}^-(y_1) = \bar{f}^-(z_1)$, $\bar{f}^-(y_2) = \bar{f}^-(z_2)$ follow from the fact that the time means for any two points on the same trajectory of the flow $\{g_t\}$ exist or do not exist simultaneously and coincide if they exist.

Thus $\bar{f}^+(x) = \text{const}(\text{mod } 0)$ on $U_{\delta_1, \delta_2, \delta_3}(x_0)$. As we mentioned above this implies the ergodicity of $\{g_t\}$.

2. Now let us prove that the flow $\{g_t\}$ is mixing. Consider the Hilbert space $L^2(M, \mathfrak{S}, \mu)$ and introduce in it the one-parameter groups of unitary operators $\{U^t\}$, $\{V^t_+\}$, $\{V^t_-\}$ adjoint to the flow $\{g_t\}$, $\{O_t^+\}$, $\{O_t^-\}$ respectively

$$(U^t f)(x) = f(xg_t), \qquad (V^t_+ f)(x) = f(xO_t^+), \qquad (V^t_- f)(x) = f(xO_t^-).$$

The following commutation relations follow from (1)

$$V^s_+ U^t = U^t V_+^{-e^{2t}s}, \qquad V^s_- U^t = U^t V_-^{e^{2t}s}. \tag{4}$$

Introduce the subspaces

$$H_+ = \{h \in L^2(M, \mathfrak{S}, \mu) \colon V^t_+ h = h \text{ for all } t\}.$$

$$H_- = \{h \in L^2(M, \mathfrak{S}, \mu) \colon V^t_- h = h \text{ for all } t\}.$$

Lemma 2. (1) *If the vectors $f_1, f_2 \in L^2(M, \mathfrak{S}, \mu)$ are orthogonal to the subspace H_+, then*

$$\lim_{t \to +\infty} (U^t f_1, f_2) = 0.$$

(2) *If $f_1, f_2 \in L^2(M, \mathfrak{S}, \mu)$ are orthogonal to H_- then*

$$\lim_{t \to -\infty} (U^t f_1, f_2) = 0.$$

Proof. We shall only prove the first statement of the lemma, since the second one is proved in a similar way. It follows from von Neumann's ergodicity theorem that for any $\varepsilon > 0$ there exists a $N = N(\varepsilon)$ such that

$$\left\| \frac{1}{s} \int_0^s V^u_+ f_1 \, du \right\| < \varepsilon,$$

for all $s > N$. Then since U^t is unitary

$$\left|\left(U^t\left(\frac{1}{s}\int_0^s V_+^u f_1\,du\right), f_2\right)\right| < \varepsilon \cdot \|f_2\|. \tag{5}$$

But (4) implies:

$$\left(U^t\left(\frac{1}{s}\int_0^s V_+^u f_1\,du\right), f_2\right) = \frac{1}{s}\int_0^s du(U^t V_+^u f_1, f_2)$$
$$= \frac{1}{s}\int_0^s du(V_+^{e^{2t}u} U^t f_1, f_2) = \frac{1}{s}\int_0^s du(U^t f_1, V_+^{-e^{2t}u} f_2).$$

When $t \to \infty$ for a fixed s for all u, $0 \le u \le s$,

$$\lim_{t\to\infty} \|V_+^{-e^{-2t}u} f_2 - f_2\| = 0,$$

uniformly with respect to u. Therefore

$$\lim_{t\to\infty}\left[\frac{1}{s}\int_0^s du(U^t f_1, V_+^{-e^{-2t}u} f_2) - (U^t f_1, f_2)\right] = 0. \tag{6}$$

Comparing (5) and (6) we obtain

$$\overline{\lim_{t\to\infty}} |(U^t f_1, f_2)| \le \varepsilon \|f_2\|.$$

Since ε was arbitrary, the lemma is proved. □

In view of the fact that U^t is unitary we have the relation

$$(U^{-t} f_1, f_2) = (f_1, U^t f_2) = \overline{(U^t f_2, f_1)}.$$

From this remark and from Lemma 2 we see that $H = H_+ \oplus H_-$ implies $\lim_{t\to\pm\infty}(U^t f_1, f_2) = 0$ for all f_1, f_2 orthogonal to H.

3. Let us show that each of the subspaces H_+, H_- is the subspace of constants. This will imply that $\{g_t\}$ is mixing. We shall carry out the proof for H_+ only, since the argument for H_- is similar.

Lemma 3. *Assume*

$$a_\varphi = \begin{Vmatrix} \cos\varphi & \sin\varphi \\ -\sin\varphi & \cos\varphi \end{Vmatrix} \in \mathrm{SL}(2, \mathbb{R}).$$

Then for all $t > 0$, we can find $\varphi_1 = \varphi_1(t)$, $t_1 = t_1(t)$, $\varphi_2 = \varphi_2(t)$, for which $O_t^+ = a_{\varphi_1} g_{t_1} a_{\varphi_2}$. Moreover

$$\lim_{t \to \infty} \varphi_1(t) = 0, \qquad \lim_{t \to \infty} \varphi_2(t) = -\frac{\pi}{2}$$

where $t_1(t)$ is a monotone function in t for t sufficiently large and $\lim_{t \to \infty} t_1(t) = \infty$.

The proof of Lemma 3 will be given later; now we shall use it to prove the necessary statement.

Suppose $\{A^\varphi\}$ is the one-parameter group of unitary operators adjoint to the flow $\{a_\varphi\}$: $A_\varphi f(x) = f(x \cdot a_\varphi)$, $f \in L^2(M, \mu)$, $-\infty < \varphi < \infty$. Then the equality $V_+^t h = h$ implies $A^{\varphi_1} U^{t_1} A^{\varphi_2} h = h$, i.e.,

$$U^{t_1} A^{\varphi_2} h = A^{-\varphi_1} h. \tag{7}$$

Since $\varphi_2(t) \to -\pi/2$ when $t \to \infty$, we have $\lim_{t \to \infty} \|A^{\varphi_2} h - A^{-\pi/2} h\| = 0$. Therefore

$$\lim_{t \to \infty} \|U^{t_1} A^{\varphi_2} h - U^{t_1} A^{-\pi/2} h\| = \lim_{t \to \infty} \|A^{\varphi_2} h - A^{-\pi/2} h\|. \tag{8}$$

Further the relation $\varphi_1(t) \to 0$ for $t \to \infty$ implies

$$\lim_{t \to \infty} \|A^{-\varphi_1} h - h\| = 0. \tag{9}$$

Comparing (7), (8), and (9), we obtain

$$\lim_{t \to \infty} \|U^{t_1} A^{-\pi/2} h - h\| = 0. \tag{10}$$

By von Neumann's ergodic theorem, using the ergodicity of $\{g_t\}$, we get

$$\lim_{T \to \infty} \frac{1}{T} \int_0^T U^{t_1} A^{-\pi/2} h \, dt_1 = \int_M A^{-\pi/2} h \, d\mu = \text{const}, \tag{11}$$

where the convergence is with respect to the norm in $L^2(M, \mathfrak{S}, \mu)$. It follows from (10) and (11) that

$$h = \int_M A^{-\pi/2} h \, d\mu.$$

As we pointed out above, this implies the existence of a mixing for the flow $\{g_t\}$. The theorem is proved. □

Proof of Lemma 3. Consider the following matrix d_t:

$$d_t = O_t^+(O_t^+)^* = \begin{Vmatrix} 1 + t^2 & t \\ t & 1 \end{Vmatrix}.$$

This matrix is self-adjoint and unimodular. Its eigen-values satisfy $\lambda_1(t) > 1$, $\lambda_2(t) = \lambda_1^{-1}(t) < 1$ and $\lambda_1(t) \to \infty$ when $t \to \infty$. Thus we can write $d_t = a_{\varphi_1}\Lambda_t a_{\varphi_1}^{-1}$, where

$$\Lambda_t = \begin{Vmatrix} \lambda_1(t) & 0 \\ 0 & \lambda_2(t) \end{Vmatrix}, \qquad \varphi_1 = \varphi_1(t);$$

it is easy to check that $\varphi_1(t) = O(1/t)$ when $t \to \infty$. Now put

$$O_t^+ = a_{\varphi_1}\sqrt{\Lambda_t}\, b_t. \tag{12}$$

This last relation may be regarded as the definition of the matrix b_t. Let us show that $b_t = a_{\varphi_2}$ for some $\varphi_2 = \varphi_2(t)$. We have

$$d_t = O_t^+(O_t^+)^* = a_{\varphi_1}\sqrt{\Lambda_t}\, b_t b_t^* \sqrt{\Lambda_t} \cdot a_{\varphi_1}^{-1} = a_{\varphi_1}\Lambda_t a_{\varphi_1}^{-1},$$

hence $\sqrt{\Lambda_t}\, b_t b_t^* \sqrt{\Lambda_t} = \Lambda_t$ and therefore $b_t b_t^* = e$. This means that the matrix b_t is orthogonal, i.e., $b_t = a_{\varphi_2}$. Direct computations, which we omit, show that $\varphi_2(t) \to -\pi/2$ for $t \to \infty$. Since $\Lambda_t = g_{t_1}$ where $t_1 = t_1(t) = \ln \lambda_1(t)$, we see that (12) gives the necessary representation for the matrix O_t^+. The lemma is proved. □

Remark. The third part of the proof of Theorem 1 essentially contains the proof of the following statement:

The horocyclic flows $\{O_t^+\}$, $\{O_t^-\}$ are ergodic.

Now let us clarify the relationship between the dynamical systems considered and Lobachevsky's geometry. The Lobachevsky plane L will be realized in the form of the upper half plane $\operatorname{Im} z > 0$ of the complex z-plane with the metric $ds^2 = (1/y^2)(dx^2 + dy^2)$ where $z = x + iy$.

The group SL(2, ℝ) is intimately connected with the group of transformations of the form $z \to (az + b)/(cz + d)$, with real coefficients a, b, c, d. Indeed, any transformation of the complex plane of the form above which sends the upper half plane into itself satisfies the condition

$$\det \begin{Vmatrix} a & b \\ c & d \end{Vmatrix} > 0,$$

where a, b, c, d are real. We can always assume that

$$\det\begin{Vmatrix} a & b \\ c & d \end{Vmatrix} = 1,$$

i.e., that

$$\begin{Vmatrix} a & b \\ c & d \end{Vmatrix} \in \mathrm{SL}(2, \mathbb{R}).$$

Further two distinct matrices $g', g'' \in \mathrm{SL}(2, \mathbb{R})$ generate the same transformation if and only if

$$g' = g''\begin{Vmatrix} -1 & 0 \\ 0 & -1 \end{Vmatrix}.$$

This implies that the isometry group of the Lobachevsky plane is isomorphic to $\mathbb{Z}_2\backslash\mathrm{SL}(2, \mathbb{R})$, where $\mathbb{Z}_2$ is the subgroup constituted by the two matrices

$$\begin{Vmatrix} 1 & 0 \\ 0 & 1 \end{Vmatrix}, \begin{Vmatrix} -1 & 0 \\ 0 & -1 \end{Vmatrix}.$$

On the other hand, the isometry group of the Lobachevsky plane is naturally isomorphic to the set of unit tangent vectors to the Lobachevsky plane. We get this isomorphism because there exists a unique isometry of the Lobachevsky plane sending a fixed unit tangent vector h_0 into some other given unit tangent vector h. For the vector h_0 it is convenient to choose the unit vector with origin at the point i directed vertically upwards.

It is well known that any surface of constant negative curvature is the quotient space of the Lobachevsky plane L by some discrete subgroup Γ of its isometry group. It follows from this that the quotient space of the entire isometry group by its discrete subgroup is the space of unit tangent vectors of this surface. Now the relationship between the space M considered in this section and the space of unit tangent vectors to a surface of constant negative curvature becomes clear. Because we took the quotient with respect to the subgroup $\mathbb{Z}_2$, any such space possesses a two-sheeted covering by the space M (for a corresponding subgroup Γ).

Now let us clarify the names of the subgroups of the group $\mathrm{SL}(2, \mathbb{R})$ considered above. The subgroup $\{g_t\}$ corresponds to the geodesic flow on the space of unit tangent vectors on a surface of constant negative curvature. To verify this, it suffices to consider the motion of the vector h_0 mentioned above with unit velocity along the geodesic line which it determines, i.e., along the imaginary axis. During time t the origin of this vector will move from the point i to the point ie^t. This transformation is precisely the one given by the matrix g_t.

The cyclic subgroup corresponds to the motion when the unit tangent vector rotates with constant angular velocity about its origin. The horocyclic flows are related to remarkable curves on the Lobachevsky plane —the horocycles. In the Lobachevsky plane the directed geodesic lines can be decomposed into sheaves of "parallel" geodesic lines. Within one such sheaf the geodesic lines approach each other with exponential velocity. In fact Hopf's proof of the ergodicity of the geodesic flow used the existence of such a sheaf. The horocycles are orthogonal trajectories of such a sheaf. The symbols $+$, $-$ indicate whether the geodesic lines approach each other for $t \to -\infty$ or for $t \to +\infty$. The action of the horocyclic flow is the following: the unit tangent vector moves so that its origin lies on the horocycle and its direction is perpendicular to the horocycle.

The dynamical systems considered in this section have the following natural generalization. Suppose G is a locally compact Lie group, Γ is its discrete subgroup, K is a compact subgroup of the group G. Consider the space of two-sided cosets $M = \Gamma\backslash G/K$. If μ is a right invariant Haar measure on G, then it induces a measure on M. Any one-parameter subgroup $\{g_t\}$ which commutes with K, i.e., $g_{-t}Kg_t = K$, generates a flow on M which acts according to the formula

$$S_t(DgK) = Dgg_tK$$

and preserves the measure μ. The study of the ergodic properties of such flows may be carried out in sufficient detail by using the theory of representations of Lie groups.

Chapter 5

Interval Exchange Transformations

§1. Definition of Interval Exchange Transformations

Suppose the space M is the semi-interval $[0, 1)$, $\xi = (\Delta_1, \ldots, \Delta_r)$ is a partition of M into $r \geq 2$ disjoint semi-intervals, numbered from left to right, and let $\pi = (\pi_1, \ldots, \pi_r)$ be a permutation of the number $(1, 2, \ldots, r)$.

Definition 1. Suppose the transformation $T: M \to M$ is a translation $T_{\alpha_i} x = x + \alpha_i \pmod 1$ on each of the semi-intervals Δ_i (the number α_i depends on i) and "exchanges" the semi-intervals according to the permutation π, i.e., the semi-intervals $T\Delta_i = T_{\alpha_i}\Delta_i = \Delta_i'$ adhere to each other in the order $\Delta_{\pi_1}', \ldots, \Delta_{\pi_r}'$; then T is said to be the *interval exchange transformation* corresponding to the partition ξ and the permutation π.

It is clear that interval exchanges are invertible transformations of M preserving the Lebesgue measure ρ and the numbers $\alpha_1, \ldots, \alpha_r$ are well defined (mod 1) by the pair (ξ, π). If the translations T_{α_i} and $T_{\alpha_{i+1}}$ of neighboring semi-intervals Δ_i, Δ_{i+1} (or on the semi-intervals Δ_r, Δ_1) are distinct, i.e., $\alpha_i \neq \alpha_{i+1} \pmod 1$, then we say that T is an interval exchange of precisely r segments.

Example

1. Under the natural identification of the semi-interval $M = [0, 1)$ with the circle S^1, the exchange of two segments corresponds to rotations (translations) $S^1 \to S^1$. Thus interval exchange transformations may be regarded as generalizations of group translations of the circle.

Suppose T is an interval exchange transformation of the intervals Δ_i, then the map T^{-1} will be an interval exchange transformation of the intervals $T\Delta_i = \Delta_i'$, while the powers T^n for $n \geq 2$ are interval exchange transformations of intervals of the form

$$\Delta_{i_0} \cap T^{-1}\Delta_{i_1} \cap T^{-2}\Delta_{i_2} \cap \cdots \cap T^{-n+1}\Delta_{i_{n-1}}.$$

Each such intersection is either empty or is a semi-interval—this can easily be proved by induction. Indeed

$$\Delta_i \cap T^{-1}\Delta_j = T^{-1}(\Delta_j \cap T\Delta_i) = T^{-1}(\Delta_i \cap \Delta_j')$$

is a semi-interval or the empty set, while

$$\begin{aligned}&\Delta_{i_0} \cap T^{-1}\Delta_{i_1} \cap \cdots \cap T^{-n+1}\Delta_{i_{n-1}}\\ &= \Delta_{i_0} \cap T^{-1}(\Delta_{i_1} \cap T^{-1}(\Delta_{i_2} \cap T^{-1}(\Delta_{i_3} \cap \cdots \cap T^{-1}(\Delta_{i_{n-2}} \cap T^{-1}\Delta_{i_{n-1}}) \cdots).\end{aligned}$$

The set of left end points of the intervals rearranged by T shall be denoted by $\mathscr{L}(T)$. Clearly $\mathscr{L}(T^n) = \bigcup_{k=0}^{n-1} T^{-k}\mathscr{L}(T)$.

In a similar way, for $n \geq 2$ the transformation T^{-n} is an interval exchange transformation of intervals of the form

$$T(\Delta_{i_0} \cap T\Delta_{i_1} \cap \cdots \cap T^{n-1}\Delta_{i_n}), \qquad \mathscr{L}(T^{-n}) = T^n\mathscr{L}(T^n) = \bigcup_{k=1}^{n} T^k\mathscr{L}(T).$$

Any interval exchange transformation is a piecewise isometric map, which is continuous from the right at every point $x \in [0, 1)$. Therefore, if the point x_0 is a fixed point of an interval exchange transformation, then a right semi-neighborhood and, moreover, the entire exchanged interval containing x_0, consists entirely of fixed points. Applying this argument to the interval exchange transformations T^n, $n \neq 0$, whose fixed points are periodic for T, we obtain the following: either the transformation T has no periodic points, or there exists an interval entirely made up of fixed points of some power T^n of the transformation T. It is obvious that in the second case the transformation T cannot be ergodic with respect to the Lebesgue measure on $[0, 1)$, therefore we shall essentially be interested in the study of interval exchange transformations without periodic points; it is natural to call them *aperiodic*.

Theorem 1. *The following statements are equivalent:*

(a) *the interval exchange transformation T of the intervals Δ_i is aperiodic;*
(b) $\max_{i_0, \ldots, i_n} \operatorname{diam}(\Delta_{i_0} \cap T\Delta_{i_1} \cap \cdots \cap T^n\Delta_{i_n}) \to 0$ *when* $n \to \infty$;
(c) *the union of the positive semi-trajectories of the left end points d_i of the intervals Δ_i, i.e., the set*

$$\mathscr{L}^\infty(T) \stackrel{\text{def}}{=} \bigcup_i \{T^k d_i : k \geq 0\} = \bigcup_{k=0}^{\infty} T^k\mathscr{L}(T),$$

is dense in M.

Proof. If the interval exchange T were not aperiodic, then the conditions (b) and (c) would not be satisfied, since in this case there is an interval consisting of periodic points of the same period. Thus (b) or (c) imply (a). Further, since the semi-intervals $\Delta_{i_0} \cap T\Delta_{i_1} \cap \cdots \cap T^n\Delta_{i_n}$ for each n form a partition ξ_n of the set M, where ξ_{n+1} is a refinement of ξ_n, while the union of the left end points of all these semi-intervals coincides with the set $\mathscr{L}^\infty(T)$, the conditions (b) and (c) are equivalent. Finally, if condition (c) does not hold, then the open set $M \setminus \mathrm{Cl}(\mathscr{L}^\infty(T))$ (where Cl A is the closure of A) is invariant with respect to T and is the union of certain intervals I_α; it follows from the definition of $\mathscr{L}^\infty(T)$ that T exchanges these intervals in some way (all the degrees of T are continuous on I_α). Since there is a finite number of intervals of fixed length, all the intervals I_α consist of periodic points and T is not an aperiodic transformation. Thus (a) implies (b) and we have shown (a) $\Rightarrow$ (b) $\Leftrightarrow$ (c) $\Leftrightarrow$ (a). The theorem is proved. □

§2. An Estimate of the Number of Invariant Measures

In this section we shall prove the following theorem.

Theorem 1. *Suppose T is an aperiodic interval exchange transformation of r intervals. Then*

(1) *for any measure μ invariant with respect to T the semi-interval $M = [0, 1)$ may be subdivided into no more than r subsets of positive measure invariant with respect to T;*
(2) *there exist no more than r ergodic normed invariant measures with respect to T.**

Proof. First let us show that the second statement is a consequence of the first. Indeed, if $\mu_1, \ldots, \mu_p$, $p > r$, are distinct ergodic measures, then, by Theorem 2, §2, Chap. 1, we can find p invariant sets $A_1, \ldots, A_p$ such that

$$\mu_i(A_j) = \begin{cases} 1 & \text{for } i = j, \\ 0 & \text{for } i \neq j. \end{cases}$$

Take $\mu = (1/p)\sum_{i=1}^{p} \mu_i$. The measure μ is invariant and $\mu(A_i) = 1/p > 0$, $1 \leq i \leq p$.

Now let us prove the first statement. First note that the measure μ is necessarily continuous since T is aperiodic. Suppose U_T is the unitary operator in the Hilbert space $H = L^2(M, \mu)$ adjoint to T: $U_T f(x) = f(Tx)$. Consider the invariant subspace of this operator

$$H^{\mathrm{inv}} = \{f \in H : U_T f = f\}$$

* We say that the measure μ is ergodic with respect to T, if T is ergodic with respect to μ.

If M is subdivided into k subsets of positive measure, invariant with respect to T, then the subspace H_k of functions, constant mod 0 on these subsets is contained in H^{inv}, so that $k = \dim H_k \le \dim H^{\text{inv}}$. To prove the theorem, it now suffices to show that

$$\dim H^{\text{inv}} \le r. \tag{1}$$

The argument will be in several steps.

1. For $h \in H$ denote by $H(h)$ the closed subspace of the space H spanning the functions $U_T^k h$, $-\infty < k < \infty$. Suppose h^{inv} is the orthogonal projection of h onto H^{inv}, $h^{\perp} = h - h^{\text{inv}}$ and $\{h^{\text{inv}}\} = H(h^{\text{inv}})$ is the one-dimensional or zero-dimensional subspace spanning h^{inv}. According to von Neumann's ergodic theorem.

$$\frac{1}{n}\sum_{k=0}^{n-1} U_T^k h \to h^{\text{inv}} \quad \text{when } n \to \infty,$$

with respect to the norm in the space H, so that $h^{\text{inv}} \in H(h)$. Therefore $h^{\perp} = h - h^{\text{inv}} \in H(h)$ and $H(h^{\perp}) \subset H(h)$. Since U_T is a unitary operator and $h^{\perp} \perp H^{\text{inv}}$, we have $H(h^{\perp}) \perp H^{\text{inv}}$, $H(h^{\perp}) \perp \{h^{\text{inv}}\}$. It follows from the previous remarks that

$$H(h) = H(h^{\perp}) \oplus \{h^{\text{inv}}\},$$

(the sum being orthogonal) since

$$h = h^{\perp} + h^{\text{inv}} \qquad \text{and} \qquad H(h) \subset H(h^{\perp}) + H(h^{\text{inv}}).$$

2. Now assume that we have found p functions $h_1, \ldots, h_p \in H$ such that $H = H(h_1) + \cdots + H(h_p)$ (the sum is not necessarily orthogonal). Since $H(h_i) = H(h_i^{\perp}) + \{h_i^{\text{inv}}\}$, where $H(h_i^{\perp}) \perp H^{\text{inv}}$, we have

$$H^{\text{inv}} = \{h_1^{\text{inv}}\} + \cdots + \{h_p^{\text{inv}}\}$$

and $\dim H^{\text{inv}} \le p$. Therefore, in order to prove relation (1), it suffices to find r functions h_i which possess the property just indicated; this is what we shall do now.

3. Suppose $h_i = \chi_{\Delta_i}$ are the indicators of the exchanged intervals. It follows from Lemma 1 in §1 and from the continuity of the measure μ that the linear combinations of the indicators $\chi_{i_0, i_1, \ldots, i_m}$ of the sets

$$\Delta_{i_0, \ldots, i_m} \overset{\text{def}}{=} \Delta_{i_0} \cap T\Delta_{i_1} \cap \cdots \cap T^m \Delta_{i_m} \qquad (m = 1, 2, \ldots; i_k = 1, \ldots, r),$$

for aperiodic interval exchange transformations are dense in H. It is therefore sufficient to prove that all the functions $\chi_{i_0,\dots,i_m}$ may be represented in the form

$$\sum_{i=1}^{r}\sum_{k=0}^{m} c_{ik} U_T^{-k}\chi_{\Delta_i},$$

i.e., are contained in $H(\chi_{\Delta_1}) \oplus \cdots \oplus H(\chi_{\Delta_r})$. Let us prove this statement by induction on m. For $m = 0$ the statement is trivial; let us describe the passage from m to $m + 1$. We have

$$\chi_{i_0, i_1, \dots, i_{m+1}} = \chi_{\Delta_{i_0}} \cdot \chi_{T\Delta_{i_1} \cap \cdots \cap T^{m+1}\Delta_{i_{m+1}}} = \chi_{i_0} \cdot U_T^{-1}\chi_{i_1, \dots, i_{m+1}}.$$

According to the induction hypothesis, the functions $\chi_{i_1,\dots,i_{m+1}}$ may be represented in the necessary form. Now let us consider the intervals Δ_{i_0} in their natural order on the interval $M = [0, 1)$ from left to right: $i_0 = 1, 2, \dots, r$. Note that $\Delta_{i_1} \cap T\Delta_{i_2} \cap \cdots \cap T^m\Delta_{i_{m+1}} \subset \Delta_{i_1}$ so that the sets $T\Delta_{i_1} \cap T^2\Delta_{i_2} \cap \cdots \cap T^{m+1}\Delta_{i_{m+1}} = T_{\Delta_{i_1},\dots,i_{m+1}}$ are semi-intervals; they shall also be considered in their natural order from left to right.

First suppose $i_0 = 1$. The following two cases are possible:

(a$_1$) $T_{\Delta_{i_1},\dots,i_{m+1}} \subset \Delta_1$;
(b$_1$) $T_{\Delta_{i_1},\dots,i_{m+1}} \not\subset \Delta_1$.

In the case (a$_1$) we have $\chi_{1, i_1, \dots, i_{m+1}} = U_T^{-1}\chi_{i_1,\dots,i_{m+1}}$ and our statement follows from the inductive hypothesis.

In the case (b$_1$) we have the relation

$$\chi_{1, i_1, \dots, i_{m+1}} = \chi_1 - \sum \chi_{1, i'_1, \dots, i'_{m+1}},$$

where the sum is taken over all sequences $(i'_1, \dots, i'_{m+1})$ such that

$$T\Delta_{i'_1,\dots,i'_{m+1}} \subset \Delta_1.$$

By the case (a$_1$) already considered, each summand in the last equality can be represented in the necessary form, so that our statement is proved again.

Now assume that $i_0 = 2$. Here we shall also consider two cases

(a$_2$) $T\Delta_{i_1,\dots,i_{m+1}} \subset \Delta_1 \cup \Delta_2$;
(b$_2$) $T\Delta_{i_1,\dots,i_{m+1}} \not\subset \Delta_1 \cup \Delta_2$.

In the case (a$_2$) we have

$$\chi_{2, i_1, \dots, i_{m+1}} = U_T^{-1}\chi_{i_1,\dots,i_{m+1}} - \chi_{1, i_1, \dots, i_{m+1}},$$

so that the statements proved above and the induction hypothesis imply our statement.

In the case (b_2), either $T\Delta_{i_1,\ldots,i_{m+1}} \cap (\Delta_1 \cup \Delta_2) = \varnothing$ and then

$$\chi_{2,i_1,\ldots,i_{m+1}} = 0,$$

or $\chi_{2,i_1,\ldots,i_{m+1}}$ can be written in the form

$$\chi_{2,i_1,\ldots,i_{m+1}} = U_T^{-1}\chi_{i_1,\ldots,i_{m+1}} - \sum \chi_{1,i'_1,\ldots,i'_{m+1}} - \sum \chi_{2,i''_1,\ldots,i''_{m+1}},$$

where each summand corresponds to one of the cases already considered.

Similar arguments may be carried out for $i_0 = 3, 4, \ldots, r$.

Thus we have indicated r functions which satisfy the conditions of step 2. Thus the inequality (1) is proved, concluding the proof of Theorem 1. □

Example

1. Subdivide M into two semi-intervals and on each of these consider the exchange of two intervals corresponding to an ergodic translation of the circle (see Example 1, §1). The transformation T thus obtained is an aperiodic exchange of four segments which is not ergodic with respect to Lebesgue measure.

§3. Absence of Mixing

The theorem which shall be proved in this section shows that interval exchange transformations possess fairly weak statistical properties.

Theorem 1. *Suppose T is an interval exchange transformation and μ is any invariant Borel measure for T. Then T possesses no mixing with respect to the measure μ.*

The proof is based on two lemmas.

Lemma 1. *Suppose that in the space M with measure μ, possessing a transformation T which preserves this measure, a sequence of partitions $\xi_i = \{A_1^{(i)}, \ldots, A_{s_i}^{(i)}\}$, $i = 1, 2, \ldots$ has been constructed (where the $A_j^{(i)}$ are measurable subsets of M such that $\mu(A_j^{(i)} \cap A_l^{(i)}) = 0$ for $j \neq l$ and $\mu(M \setminus \bigcup_{j=1}^{s_i} A_j^{(i)}) = 0$), as well as a sequence of families of numbers $r_j^{(i)}$, $1 \leq j \leq s_i$, so that the following conditions hold:*

(i) $s_i \leq s = \text{const}$;
(ii) $\min_j r_j^{(i)} \to \infty$, *when* $i \to \infty$;
(iii) *for any measurable set* $C \subset M$

$$\lim_{i\to\infty} \mu\left(C \triangle \left(\bigcup_{j=1}^{s_i} T^{r_j^{(i)}}(A_j^{(i)} \cap C)\right)\right) = 0. \tag{1}$$

Then T is not mixing.

Proof. The relation (1) may be rewritten in the form

$$\lim_{i\to\infty} \mu\left(C \cap \left(\bigcup_{j=1}^{s_i} T^{r_j^{(i)}}(A_j^{(i)} \cap C)\right)\right) = \mu(C). \tag{2}$$

Further:

$$\mu\left(C \cap \left(\bigcup_{j=1}^{s_i} T^{r_j^{(i)}}(A_j^{(i)} \cap C)\right)\right) = \mu\left(\bigcup_{j=1}^{s_i} (C \cap T^{r_j^{(i)}}(A_j^{(i)} \cap C))\right)$$

$$\leq \sum_{j=1}^{s_i} \mu(C \cap T^{r_j^{(i)}}(A_j^{(i)} \cap C)) \leq \sum_{j=1}^{s_i} \mu(C \cap T^{r_j^{(i)}} C).$$

If T has a mixing, then (i) and (ii) imply that the upper limit of the last expression when $i \to \infty$ is no greater than $s \cdot \mu^2(C)$. In the case $s \cdot \mu(C) < 1$ this contradicts (2). The lemma is proved. □

Suppose T is an exchange of r intervals, and $M_1 = [a_0, a_1) \subset M$ is an arbitrary semi-interval contained in $[0, 1)$. Consider the induced automorphism T_1 constructed from the automorphism T and the set M_1 (see §5, Chap. 1).

Lemma 2. *The transformation T_1 is an interval exchange transformation and the number r_1 of exchanged segments satisfies $r_1 \leq r + 2$.*

Proof. For each of the $r + 1$ points $y = a_0, a_1, d_2, \ldots, d_r$ (where $d_2, d_3, \ldots, d_r$ are the discontinuity points of the transformation T, i.e., the left end points of the intervals Δ_i, except for the point $d_1 = 0$), introduce the number $s(y)$ defined as the smallest number $s \geq 0$ such that $T^{-s(y)}y \in [a_0, a_1)$ (if such s actually exist). The points $T^{-s(y)}y$ divide the semi-interval $[a_0, a_1)$ into r_1 semi-intervals $\Delta_1', \ldots, \Delta_{r_1}'$, where $r_1 \leq r + 2$. For each semi-interval Δ_i' consider the number k_i—the smallest number $k \geq 1$ such that $T^k\Delta_i' \cap M_1 \neq \varnothing$. Then for $1 \leq p \leq k_i$ the transformations T^p are continuous on Δ_i' and we have $T^{k_i}\Delta_i' \subset M_1$, since in the converse case for some p, $1 \leq p \leq k_i - 1$, the semi-interval $T^p\Delta_i'$ would contain one of the points $y = a_0, a_1, d_2, \ldots, d_r$ and then $s(y) = p$, so that the point $T^{-p}y = T^{-s(y)}y$ would lie within Δ_i', contradicting the definition of the semi-intervals Δ_i'. Recalling the definition of an induced automorphism $T_{M_1} = T_1$ and the return function $k_{M_1}(x) = k(x)$, we see that $k(x) = k_i$ for $x \in \Delta_i'$, while the transformation T_1 which coincides with T^{k_i} on the semi-interval Δ_i' is an interval exchange of the intervals Δ_i', $i = 1, 2, \ldots, r_1$. The lemma is proved. □

Note that the sets $T^p\Delta_i'$ for $0 \le p \le k_i - 1$ are pairwise disjoint semi-intervals. For a better understanding of what will follow, it is useful to view the $T^p\Delta_i'$ for $1 \le p \le k_i - 1$ in this situation as "steps" ("storeys") over $\Delta_i' \subset M_1$.

Proof of Theorem 1. First note that the invariant measure μ may be assumed continuous, and the transformation T assumed aperiodic and ergodic with respect to the measure μ. Our argument is split up into several steps.

1. In this step we shall describe an auxiliary construction, used later on to find a sequence of partitions ξ_i and of numbers $r_j^{(i)}$ satisfying the assumptions of Lemma 1.

Suppose $T = T_0$ is an aperiodic exchange on the semi-interval $M = [0, 1) = \Delta^{(0)}$; by $\Delta_j^{(0)}$, $1 \le j \le r = r^{(0)}$ denote the semi-intervals exchanged by the transformation T_0. Consider the sequence of nested semi-intervals

$$\Delta^{(0)} \supset \Delta^{(1)} \supset \cdots \supset \Delta^{(r)} \supset \cdots$$

(arbitrary at first). The transformation of the semi-interval $\Delta^{(i)}$ induced by the original interval exchange $T_0 = T$ will be denoted by T_i, and the return function corresponding to it by $k^{(i)}(x)$. According to Lemma 2, the transformation $T_i\colon \Delta^{(i)} \to \Delta^{(i)}$ is a exchange of certain semi-intervals $\Delta_j^{(i)}$, $1 \le j \le r^{(i)}$. Suppose that for all $x \in \Delta_j^{(i)}$ the function $k^{(i)}(x)$ equals $k_j^{(i)}$; note that since the original transformation was aperiodic, we have $r^{(i)} \ge 2$ (the function $k^{(i)}(x)$ is not constant); put $k_{j_0(i)}^{(i)} = \max\{k_j^{(i)}\colon 1 \le j \le r^{(i)}\}$. Now we can uniquely determine the sequence $\{\Delta^{(i)}, T_i\}$ by taking $\Delta^{(i+1)} = \Delta_{j_0(i)}^{(i)}$ for each $i \ge 0$.

Let us list some properties of the construction described above:

(a) For any $i \ge 0$ and any $x \in \Delta^{(i+1)}$, we have the inequality $k^{(i+1)}(x) \ge k_{j_0(i)}^{(i)}$. This follows from the definition of the function $k^{(i+1)}(x)$.

(b) We have the strict inequality $k_{j_0(i+1)}^{(i+1)} > k_{j_0(i)}^{(i)}$. This follows from the fact that $k^{(i)}(x) \ne \text{const}$.

(c) $m_i \overset{\text{def}}{=} \min\{k_j^{(i)}\colon 1 \le j \le r^{(i)}\} \to \infty$, when $i \to \infty$. This follows from (a) and (b).

(d) $\operatorname{diam} \Delta^{(i)} \to 0$ for $i \to \infty$. Indeed, by the Kac lemma proved in §5, Chap. 1, we have $m_i \operatorname{diam} \Delta^{(i)} \le \operatorname{diam} \Delta^{(0)} = 1$. Hence (d) follows from (c).

We now go on to the definition of the partitions ξ_i and the numbers $r_j^{(i)}$ satisfying the assumptions of Lemma 1. For any $i \ge 0$ and $j = 1, 2, \ldots, r^{(i)}$, consider the transformation $T_{i,j}$ of the interval $\Delta_j^{(i)}$ induced by the original exchange transformation $T = T_0$. Each $T_{i,j}$ is an exchange of certain intervals $\Delta_{jl}^{(i)} \subset \Delta_j^{(i)}$, where $1 \le l \le r_j^{(i)}$.

Note that according to Lemma 2 we have the following inequalities, concerning the number of intervals exchanged by the transformations T_i and $T_{i,j}$:

(e) $r^{(i)} \leq r^{(0)} + 2$, $r_j^{(i)} \leq r^{(0)} + 2$. Obviously for any $x \in \Delta_j^{(i)}$ the return function $k_j^{(i)}(x)$ corresponding to the transformation $T_{i,j}$ is not less than $k^{(i)}(x)$, so that we have

(f) if the function $k_j^{(i)}(x)$ equals $k_{j,l}^{(i)}$ for $x \in \Delta_{j,l}^{(i)}$, then $k_{j,l}^{(i)} \geq m^{(i)}$ for all l, $1 \leq l \leq r_j^{(i)}$.

Finally, for $i \geq 0$, $1 \leq j \leq r^{(i)}$, $1 \leq l \leq r_j^{(i)}$, put

$$A_{jl}^{(i)} = \bigcup_{p=0}^{k_j^{(i)}} T^p \Delta_{jl}^{(i)}, \qquad r_{jl}^{(i)} = k_{jl}^{(i)}.$$

Since the set $M_1 = \bigcup_{j,l} A_{jl}^{(i)}$ is obviously invariant with respect to the transformation T, we have, for any measure μ invariant with respect to T, either $\mu(M \setminus M_1) = 0$ or $\mu(M_1) = 0$. In the first case

$$\xi_i = \{A_{jl}^{(i)} : 1 \leq j \leq r^{(i)}, 1 \leq l \leq r_j^{(i)}\},$$

is a partition of the measure space (M, μ); in the second case, we must carry out the construction described above, starting with $M \setminus M_i$ in the role of $\Delta^{(0)}$ (it is the union of a finite number of semi-intervals $\Delta_j^{(0)}$ which is invariant with respect to T). Since $T = T_0$ is an exchange of a finite number of semi-intervals we can, by repeating our construction several times (if necessary), obtain the necessary partition $\xi_i = \{A_{j,l}^{(i)}\}$, so that the auxiliary transformations T_i and $T_{i,j}$ will continue to possess the properties (a)–(e).

2. In carrying out this step, we will show that the partitions ξ_i and the numbers $r_{j,l}^{(i)}$ satisfy all the assumptions of Lemma 1, so that the statement of Theorem 1 will be proved.

Firstly, it follows from the inequalities (e) that the number of elements $A_{j,l}^{(i)}$ of the partitions ξ_i is not greater than the constants $s = (r^{(0)} + 2)^2$. Secondly, it follows from the definition of the numbers $r_{j,l}^{(i)}$, the inequality (f) and property (c) that $\min_{j,l} r_{j,l}^{(i)} \to \infty$, when $i \to \infty$.

Finally let us verify condition (iii) of lemma 1. Suppose $C \subset M$ is an arbitrary measurable set. For any $\delta > 0$ consider the following sets:

$$C_{\delta,i}^{+} = \{x : x \in T^p \Delta_j^{(i)}, \mu(C \cap T^p \Delta_j^{(i)}) \geq (1 - \delta)\mu(\Delta_j^{(i)})\},$$

$$C_{\delta,i}^{-} = \{x : x \in T^p \Delta_j^{(i)}, \mu(C \cap T^p \Delta_j^{(i)}) \leq \delta \cdot \mu(\Delta_j^{(i)})\},$$

$$C_{\delta,i}^{0} = \{x : x \in T^p \Delta_j^{(i)}, \delta \cdot \mu(\Delta_j^{(i)}) < \mu(C \cap T^p \Delta_j^{(i)}) < (1 - \delta)\mu(\Delta_j^{(i)})\},$$

(here j and p range over the values $1 \leq j \leq r^{(i)}$, $0 \leq p \leq k_j^{(i)}$). According to the theorem on density points, it follows from the continuity of the measure μ and from relation (d) that

$$\mu(C_{\delta,i}^{0}) \to 0, \qquad \mu(C \triangle C_{\delta,i}^{+}) \to 0, \qquad \mu((M \setminus C) \triangle C_{\delta,i}^{-}) \to 0, \tag{3}$$

when $i \to \infty$. Further, since $T^{r_{jl}^{(i)}} A_{jl}^{(i)} \subset \bigcup_{p=0}^{k_j^{(i)}} T^p \Delta_j^{(i)}$ and these unions are pairwise disjoint for distinct j, we have

$$\begin{aligned}\mu&\left(C \triangle \left(\bigcup_{j,l} T^{r_{jl}^{(i)}} (A_{jl}^{(i)} \cap C)\right)\right)\\ &= \sum_{j,p} \mu(T^p \Delta_j^{(i)}) \cdot \mu\left(C \triangle \bigcup_l T^{r_{jl}^{(i)}} (A_{jl}^{(i)} \cap C) \,|\, T^p \Delta_j^{(i)}\right)\\ &= \Sigma_{\delta,i}^+ + \Sigma_{\delta,i}^- + \Sigma_{\delta,i}^0,\end{aligned}$$

where the sums are taken over those indices j, p for which $T^p \Delta_j^{(i)}$ is contained in the sets $C_{\delta,i}^+$, $C_{\delta,i}^-$, $C_{\delta,i}^0$ respectively. Since $\Sigma_{\delta,i}^0 \le \mu(C_{\delta,i}^0)$, it follows from (3) that

$$\Sigma_{\delta,i}^0 \to 0 \quad \text{when } i \to \infty. \tag{4}$$

Let us estimate the first two sums. Note that

$$T^{r_{jl}^{(i)}} A_{jl}^{(i)} \cap T^p \Delta_j^{(i)} = T^{r_{jl}^{(i)}} (T^p \Delta_{jl}^{(i)}),$$

hence $T^p \Delta_j^{(i)} \subset C_{\delta,i}^+$ implies

$$\begin{aligned}\mu&\left(C \triangle \left(\bigcup_l T^{r_{jl}^{(i)}} (A_{jl}^{(i)} \cap C)\right) \middle| T^p \Delta_j^{(i)}\right)\\ &\le \mu\left(T^p \Delta_j^{(i)} \setminus \bigcup_l T^{r_{jl}^{(i)}} (A_{jl}^{(i)} \cap C) \,|\, T^p \Delta_j^{(i)}\right) + \delta\\ &= \mu\left(T^p \Delta_j^{(i)} \setminus \bigcup_l T^{r_{jl}^{(i)}} (A_{jl}^{(i)} \cap C) \cap T^p \Delta_j^{(i)} \,|\, T^p \Delta_j^{(i)}\right) + \delta\\ &= \mu\left(T^p \Delta_j^{(i)} \setminus \bigcup_l T^{r_{jl}^{(i)}} (T^p \Delta_{jl}^{(i)} \cap C) \,|\, T^p \Delta_j^{(i)}\right) + \delta.\end{aligned}$$

Since $\bigcup_l T^{r_{jl}^{(i)}+p} \Delta_{jl}^{(i)} = T^p \Delta_j^{(i)}$, where the sets $T^{r_{jl}^{(i)}+p} \Delta_{jl}^{(i)}$ are pairwise disjoint for distinct l, while the set $C \cap T^p \Delta_j^{(i)}$ coincides with $T^p \Delta_j^{(i)}$ up to δ, we have

$$\bigcup_l T^{r_{jl}^{(i)}} (T^p \Delta_{jl}^{(i)} \cap C) \subset T^p \Delta_j^{(i)},$$

$$\begin{aligned}\mu\left(\bigcup_l T^{r_{jl}^{(i)}} (T^p \Delta_{jl}^{(i)} \cap C)\right) &= \sum_l \mu(T^{r_{jl}^{(i)}} (T^p \Delta_{jl}^{(i)} \cap C))\\ &= \sum_l \mu(T^p \Delta_{jl}^{(i)} \cap C) = \mu(T^p \Delta_j^{(i)} \cap C) \ge (1-\delta)\mu(T^p \Delta_j^{(i)});\end{aligned}$$

therefore

$$\mu\left(T^p \Delta_j^{(i)} \setminus \bigcup_l T^{r_{jl}^{(i)}} (T^p \Delta_{jl}^{(i)} \cap C) \,|\, T^p \Delta_j^{(i)}\right) \le \delta,$$

and, taking into consideration the previous estimate, we get

$$\mu\left(C \triangle \left(\bigcup_l T^{r_{jl}^{(i)}}(A_{jl}^{(i)} \cap C)\right) \middle| T^p \Delta_j^{(i)}\right) < 2\delta,$$

so that

$$\Sigma_{\delta,i}^{+} < 2\delta. \tag{5}$$

Now let us estimate the $\Sigma_{\delta,i}^{-}$. To do this, rewrite it in the form

$$\begin{aligned}\Sigma_{\delta,i}^{-} &= \mu\left(\left[C \triangle \left(\bigcup_{j,l} T^{r_{jl}^{(i)}}(A_{jl}^{(i)} \cap C\right)\right] \cap C_{\delta,i}^{-}\right) \\ &= \mu(C \cap C_{\delta,i}^{-}) + \sum_{j,p}^{-} \mu(T^p \Delta_j^{(i)}) \cdot \mu\left(\bigcup_l T^{r_{jl}^{(i)}}(A_{jl}^{(i)} \cap C) | T^p \Delta_j^{(i)}\right).\end{aligned}$$

Note that $C \cap C_{\delta}^{-} \subset (M \backslash C) \triangle C_{\delta}^{-}$, so that (3) implies

$$\mu(C \cap C_{\delta,i}^{-}) \to 0 \quad \text{when } i \to \infty. \tag{6}$$

Arguing as above, we further see that

$$\mu\left(\bigcup_l T^{r_{jl}^{(i)}}(A_{jl}^{(i)} \cap C) | T^p \Delta_j^{(i)}\right) = \mu\left(\bigcup_l T^{r_{jl}^{(i)}}(T^p \Delta_{jl}^{(i)} \cap C) | T^p \Delta_j^{(i)}\right),$$

$$\bigcup_l T^{r_{jl}^{(i)}}(T^p \Delta_{jl}^{(i)} \cap C) \subset T^p \Delta_j^{(i)},$$

$$\mu\left(\bigcup_l T^{r_{jl}^{(i)}}(T^p \Delta_{jl}^{(i)} \cap C)\right) = \mu(T^p \Delta_j^{(i)} \cap C) \leq \delta \cdot \mu(T^p \Delta_j^{(i)}),$$

so that the second sum in the formula for $\Sigma_{\delta,i}^{-}$ is no greater than δ. Since $\delta > 0$ was arbitrary, the relations (4), (5), and (6) give us

$$\lim_{i \to \infty} \mu\left(C \triangle \bigcup_{j,l} T^{r_{jl}^{(i)}}(A_{jl}^{(i)} \cap C)\right) = 0.$$

The theorem is proved. □

§4. An Example of a Minimal but not Uniquely Ergodic Interval Exchange Transformation

Interval exchange transformations are not, of course, homeomorphisms of the space $M = [0, 1)$, nevertheless it is possible to introduce the notions of minimality and unique ergodicity for them.

Definition 1. The interval exchange transformation $T\colon M \to M$ is said to be *minimal* if the trajectory $\Omega(x) = \{T^n x\colon -\infty < n < \infty\}$ of an arbitrary point $x \in M$ is dense in M.

Definition 2. The interval exchange transformation $T\colon M \to M$ is said to be *uniquely ergodic* if the Lebesgue measure ρ is the only normalized Borel invariant measure.

We now give the following minimality criterion for an interval exchange transformation.

Theorem 1. *If T is an exchange of the intervals $\Delta_1, \ldots, \Delta_r$, $r \geq 2$, where for $i = 2, \ldots, r$ the trajectory $\Omega(d_i)$ of any of the left end points d_i of the segments Δ_i (i.e., of every d_i except the point $d_1 = 0$) is infinite and these trajectories do not intersect for distinct i, then the interval exchange T is minimal.*

Proof. 1. First let us show that T is aperiodic. Indeed, in the converse case some degree T^n of T, being an exchange of the intervals $\Delta_j^{(n)}$, must have an entire semi-interval of fixed points. For the left end point d_0 of this interval, we have $T^n d_0 = d_0$, where $d_0 = T^k d_i$ for some i, $1 \leq i \leq r$ and some k. In the case $i \geq 2$, we immediately obtain a contradiction to the assumption of the theorem requiring the trajectories to be infinite.

If $d_0 = T^k d_1$, then two cases are possible: (1) $T^{-1} d_1 = d_i$ for $i \geq 2$ and then, as above, $d_0 = T^{k-1} d_i$, $i \geq 2$ and $d_i = T^n d_i$, which contradicts the fact that the trajectory $\Omega(d_i)$ is infinite; (2) $T^{-1} d_1 = d_1$; in this case the interval Δ_1 is fixed under the transformation T, hence one of the points d_i, $i \geq 2$ is mapped into d_2 by T, which again contradicts the assumption of the theorem.

2. Now let us prove the minimality of T. Suppose $x_0 \in M$; assume that the closure $\overline{\Omega(x_0)}$ of the trajectory of x_0 does not coincide with M. Then we can find a semi-interval $\Delta^{(0)} = [a, b) \subset M \backslash \overline{\Omega(x_0)}$. Consider the interval exchange T_1 induced by the interval exchange T on the semi-interval $\Delta^{(0)}$ (see Lemma 1 in §3). Suppose $\Delta_j^{(0)}$ are the intervals exchanged by the transformation T_1, $k(x)$ is the return function and $k(x) = k_j$ for $x \in \Delta_j^{(0)}$. Put

$$F = \bigcup_j \bigcup_{k=0}^{k_j - 1} T^k \Delta_j^{(0)}.$$

The set F is the union of a finite number of nonintersecting semi-intervals, hence its connected components are also semi-intervals F_s, and their number is finite. Denote by G the union of the left end points of all the semi-intervals F_s. By definition, the set F is invariant with respect to T and therefore for $x \in G$ we either have $Tx \in G$, or the point x is a point where the transformation T is disconnected, or $x = 0$, i.e., $x = d_i$ for some i, $1 \leq i \leq r$. Since the set G is finite, while the transformation T is aperiodic, for $x \in G$ and some $n \geq 0$ we have $T^n x = d_i$, $1 \leq i \leq r$. In a similar way, for an arbitrary point $x \in G$, we have either $T^{-1}x \in G$, or $T^{-1}x$ is a point where the transformation T is disconnected, i.e., $T^{-1}x = d_j$, $j \geq 2$; hence for some $m > 0$ the point $T^{-m}x$ satisfies $T^{-m}x = d_j, j \geq 2$. Hence for $x \in G$ we have

$$x = T^m d_j, \qquad T^n x = T^{m+n} d_j = d_i,$$

where $n \geq 0, m > 0$ and $i \geq 1, j \geq 2$. Hence by the conditions of the theorem, for $i, j \geq 2$ the trajectories of the points d_i and d_j do not intersect; the last relation may hold only the case when $i = 1$, i.e., if $d_i = d_1$. Then $d_j = T^{-1} d_1$ and $m = 1$, $n = 0$. Therefore $x = T d_j = d_1 = 0$ and $G = \{0\}$, i.e., $F = M$. However, by construction, $F \subset M \backslash \Omega(x_0)$, so that $F \neq M$. The contradiction thus obtained shows that we have $\overline{\Omega(x_0)} = M$, i.e., T is a minimal interval exchange transformation. The theorem is proved. □

The following theorem shows that there exist minimal interval exchange transformations which are not uniquely ergodic.

Theorem 2. *For any natural m there exists a minimal interval exchange transformation possessing m ergodic normalized invariant measures.*

Proof. We shall use the construction of a skew translation over a rotation of the circle (see §1 in Chap. 4). Suppose $S^1 = [0, 1)$ is the unit circle with Lebesgue measure ρ and T_α is the rotation automorphism by an angle α;

$$T_\alpha x = x + \alpha \ (\mathrm{mod}\ 1).$$

Consider the group $\mathbb{Z}_m = \{0, 1, \ldots, m - 1\}$ of residues modulo m with Haar measure ν (the uniform distribution) and an arbitrary measurable function $\sigma\colon S^1 \to \mathbb{Z}_m$.

Consider the automorphism T of the space $M = S^1 \times \mathbb{Z}_m$ with measure $\mu = \rho \times \nu$ defined by the formula

$$T(x, k) = (T_\alpha x, k \oplus \sigma(x)), \qquad x \in S^1, k \in \mathbb{Z}_m. \tag{1}$$

Here $\oplus$ is the group operation in $\mathbb{Z}_m$.

If $\sigma(x)$ assumes each of its values on the union of a finite number of semi-intervals of the form $[\beta, \gamma) \subset S^1$, then, identifying M with the semi-interval $[0, m)$ by pasting together the points $(x, k) \in M$ with the points $x + k \in [0, m)$, we shall obtain an interval exchange of a finite number of intervals corresponding to the transformation T. We shall need the following lemma.

Lemma 1. *If the transformation T of the form* (1) *constructed from the function $\sigma(x)$ and the irrational number α is such that there exists a measurable function $\tau: S^1 \to \mathbb{Z}_m$ such that for any k and any interval $[\beta, \gamma] \subset S^1$ the set $\{x \in [\beta, \delta], \tau(x) = k\}$ is of positive measure and*

$$\tau(T_\alpha x) = \tau(x) \oplus \sigma(x) \tag{2}$$

almost everywhere, then the interval exchange transformation corresponding to T is minimal, and T possesses m ergodic normalized invariant measures.

The proof of the lemma is carried out below; now we shall determine the number α and the functions $\sigma(x)$, $\tau(x)$ which satisfy its conditions.

Choose any $\varepsilon > 0$. Suppose the irrational number α is such that for some sequence of irreducible fractions p_s/q_s we have the inequality

$$0 < \alpha - \frac{p_s}{q_s} < \frac{1}{q_s^{2+\varepsilon}}. \tag{3}$$

It is clear that such an α exists. Choosing an appropriate subsequence of the sequence p_s/q_s, we may assume that

$$\sum_{s=1}^{\infty} \frac{1}{q_s^{q+\varepsilon}} < \frac{1}{2mq_1}, \tag{4}$$

$$\sum_{p=s+1}^{\infty} \frac{1}{q_p^{\varepsilon}} < \frac{1}{2}\{q_s\alpha\}, \qquad s = 1, 2, \tag{5}$$

where the figure brackets denote the fractional part of the number in them. Since the fraction p_s/q_s is irreducible, the points $jp_s/q_s \pmod 1$, $0 \le j < q_s$ divide the circle S^1 into q_s equal semi-intervals

$$\Delta_j = \left[j\frac{p_s}{q_s} \pmod 1, (j+1)\frac{p_s}{q_s} \pmod 1 \right).$$

Suppose $\Delta_j' = [j\alpha \pmod 1, j\alpha + mq_s\alpha \pmod 1))$. It follows from the inequalities (3) and (4) that $\Delta_j' \subset \Delta_j$, hence the semi-intervals Δ_j' are disjoint. For any s we now put

$$\sigma_s(x) = \begin{cases} 1 & \text{for } x \in [0, mq_s\alpha \pmod 1)); \\ 0 & \text{for all other } x; \end{cases}$$

$$\tau_s(x) = \begin{cases} k & \text{for } x \in [kq_s\alpha \pmod 1), (k+1)q_s\alpha \pmod 1)), 0 \le k < m \\ k \oplus 1 & \text{for } x \in [j\alpha + kq_s\alpha \pmod 1), \\ & \quad j\alpha + (k+1)q_s\alpha \pmod 1)), 1 \le j < q_s, 0 \le k < m. \end{cases}$$

Let us show that for all $x \in S^1$ we have the relation

$$\tau_s((x + \alpha) \bmod 1) = \tau_s(x) \oplus \sigma_s(x). \tag{6}$$

To do this, it suffices to notice that only the following five cases are possible:

(1) $x \in \bigcup_{j=0}^{q_s-2} \Delta'_j, \qquad T_\alpha x \in \bigcup_{j=1}^{q_s-1} \Delta'_j;$

(2) $x \in \Delta'_{q_s-1}, \qquad T_\alpha x \in \Delta'_0;$

(3) $x \in \Delta'_{q_s-1}, \qquad T_\alpha x \in M \setminus \bigcup_{j=0}^{q_s-1} \Delta'_j;$

(4) $x \in M \setminus \bigcup_{j=0}^{q_s-1} \Delta'_j, \qquad T_\alpha x \in \Delta'_0;$

(5) $x \in M \setminus \bigcup_{j=0}^{q_s-1} \Delta'_j, \qquad T_\alpha x \in M \setminus \bigcup_{j=0}^{q_s-1} \Delta'_j.$

In each of these cases the verification of (6) is straightforward.

Further consider the sequence of "shifted" functions $\tilde{\sigma}_1(x) = \sigma_1(x)$, $\tilde{\sigma}_2(x) = \sigma_2((x - mq_1\alpha) \bmod 1)$, $\tilde{\sigma}_3(x) = \sigma_3((x - mq_1\alpha - mq_2\alpha) \bmod 1), \ldots,$ and similarly

$$\tilde{\tau}_s = \tau_s\left(\left(x - \sum_{p=1}^{s-1} mq_p\alpha\right) \bmod 1\right), \qquad s = 1, 2, \ldots .$$

It follows from the definition of $\sigma_s(x)$ that the functions $\tilde{\sigma}_s(x)$ do not vanish (and are equal to 1) on adjacent semi-intervals whose union is the semi-interval $[0, \delta)$ where $\delta = \sum_{s=1}^{\infty} \{mq_s\alpha\}$. Put

$$\sigma(x) = \sum_{s=1}^{\infty} \tilde{\sigma}_s(x) = \begin{cases} 1 & \text{for } x \in [0, \delta), \\ 0 & \text{for all other } x. \end{cases}$$

Any function $\tau_s(x)$ and hence any function $\tilde{\tau}_s(x)$ is nonzero on a set of measure no greater than $q_s\{mq_s\alpha\} < m/q_s^{\varepsilon}$. Hence all the sums of the form

$$\sum_{p=s+1}^{s+N} \tilde{\tau}_p(x)$$

are nonzero on a set of measure no greater than

$$\sum_{p=s+1}^{\infty} \frac{m}{q_p^{\varepsilon}} < \frac{m}{2}\{q_s\alpha\} < \frac{m}{2} q_s^{-1-\varepsilon}.$$

Since $\sum_{s=1}^{\infty} (1/q_s^{1+\varepsilon}) < \infty$, it follows from the Borel–Cantelli lemma that the series $\sum_{s=1}^{\infty} \tilde{\tau}_s(x)$ converges almost everywhere to some measurable function.

Put $\tau(x) = \sum_{s=1}^{\infty} \tilde{\tau}_s(x)$. It is clear that for our choice of $\sigma(x)$ and $\tau(x)$ that relation (2) holds. Now let us check the second assumption of Lemma 1 for the function $\tau(x)$. The points $j\alpha \pmod 1$, $0 \leq j < q_s$ divide the circle into closed intervals of the form $[j_1\alpha, j_2\alpha]$. On each of them the function $\tau_s(x)$ assumes each value $k \in \mathbb{Z}_m$ on a set of measure no less than $\{q_s\alpha\}$. The same is true of the shifted functions $\tilde{\tau}_s(x)$ on the corresponding shifted closed intervals. Since for $p < s$ the end points of the intervals appearing in the definition of $\tau_p(x)$, as well as of the shifted intervals for the functions $\tilde{\tau}_p(x)$, are contained in the set of end points of the closed intervals indicated above, the functions $\tilde{\tau}_p$, $p < s$ are constant on each of these closed intervals. Therefore the sum $\sum_{p=1}^{s} \tilde{\tau}_p(x)$ assumes, on each of them, any of the values $k \in \mathbb{Z}_m$ on a set of measure no less than $\{q_s\alpha\}$. Since $\tau(x) = \sum_{p=1}^{s} \tilde{\tau}_p(x) + \sum_{p=s+1}^{\infty} \tilde{\tau}_p(x)$, while the second summand, by (5), differs from zero on a set of measure less than $\frac{1}{2}\{q_s\alpha\}$, we see that $\tau(x)$ assumes each value on a set of measure greater than or equal to $\frac{1}{2}\{q_s\alpha\}$. Since α is irrational, we can approximate any segment $[\beta, \gamma] \subset S^1$ by closed intervals of the form $[k_1\alpha, k_2\alpha]$. The theorem is proved. □

Proof of Lemma 1. Consider the sets

$$S_k = \{(x, \tau(x) \oplus k): x \in S^1\}, \qquad 0 \leq k \leq m - 1.$$

Since (2) implies that we have almost everywhere

$$T(x, \tau(x) + k) = (T_\alpha x, \tau(T_\alpha x) \oplus k),$$

we see that all the S_k are invariant mod 0 with respect to T. By the identification of S_k with S_1 which pastes together the points $(x, \tau(x) \oplus k)$ and $x \in S^1$, the rotation T_α of the circle S^1 will correspond to the restriction of the map T to S^k. The restrictions of the Lebesgue measure ρ to each of the subsets S_k give us the m required measures.

Now let us prove the minimality of T. Assume the converse. Then, as was shown in part 2 of the proof of Theorem 1, there is an invariant set $F \neq M$ consisting of a finite number of semi-intervals. Then, for at least one k, we will have $\rho(S_k \cap F) > 0$. Fixing such a k, we notice that $\rho(S_k \cap (M \setminus F)) > 0$. Indeed, the conditions of the lemma imply that S_k intersects each semi-interval $[\beta, \gamma) \times \{l\} \subset M$ in a set of positive measure, while $(M \setminus F)$ contains some semi-interval. The set $A = S_k \cap F$ is invariant (mod 0) with respect to T and, under the identification of S_k with S_1, it is transformed into some set B invariant with respect to T_α and satisfying $0 < \rho(B) < 1$. This contradicts the irrationality of α. The lemma is proved. □

Chapter 6

Billiards

In this chapter we consider dynamical systems of the billiards type, i.e., dynamical systems corresponding to the inertial motion of a point mass inside a domain with a piece-wise smooth boundary. Upon reaching the boundary, the point bounces off in accordance to the usual rule: "the angle of incidence is equal to the angle of reflection." Besides the intrinsic interest of the problem, systems of billiards are remarkable in view of the fact that they naturally appear in many important problems of physics.

§1. The Construction of Dynamical Systems of the Billiards Type

The rigorous construction of a dynamical system of the billiards type is not quite simple. This is because of the fact that, in the natural cases, the boundary of the domain within which the point mass moves has singularities and the continuation of the trajectory after the point reaches a singular point of the boundary is not defined in general, or rather, to be more precise, can be defined in many ways. We shall now give the general definition of billiards in Riemann spaces with piecewise smooth boundaries. Billiards may be viewed as generalization of geodesic flows.

1. *The phase space of billiards.* Suppose Q_0 is a closed Riemann manifold of class C^∞, possibly noncompact. Suppose that r functions $f_1, f_2, \ldots, f_r$ of class C^∞ are given on Q_0. The set

$$Q = \{q \in Q_0 \colon f_i(q) \geq 0, 1 \leq i \leq r\},$$

is said to be a *compact Riemann manifold with piecewise smooth boundary* if

(1) Q is compact;
(2) the set $f_i^{-1}(0)$ does not contain any critical points of the function f_i and is therefore a C^∞-submanifold of codimension 1, $1 \leq i \leq r$;
(3) the gradients grad f_i, grad f_j are linearly independent at the intersection points $q \in f_i^{-1}(0) \cap f_j^{-1}(0)$.

EXAMPLES

1. Q_0 is the Euclidian space $\mathbb{R}^d$ with coordinates $q_1, \ldots, q_d$; $Q = \{q \in Q_0: 0 \leq q_1 \leq \cdots \leq q_d \leq 1\}$;

2. Q_0 is the dr-dimensional torus with cyclic coordinates

$$f_{i_1 i_2}(q) = (2\rho)^2 - \sum_{j=1}^{d} (q_j^{(i_1)} - q_j^{(i_2)})^2,$$

where ρ is a parameter and $1 \leq i_1, i_2 \leq r$, $i_1 \neq i_2$.

Put $\partial Q_i = f_i^{-1}(0) \cap Q$, $1 \leq i \leq r$. Then the boundary satisfies $\partial Q = \bigcup_{i=1}^{r} \partial Q_i$. The sets $\partial \tilde{Q}_i = \partial Q_i \setminus \bigcup_{k \neq i} \partial Q_k$ shall be called *regular components* of the boundary. Each regular component is an open C^∞-submanifold of codimension 1. The points of the boundary $q \in \bigcup_i \partial \tilde{Q}_i$ will be called *regular points* in contrast with the other points of the boundary, which shall be called *singular*.

Suppose Γ_q for $q \in f_i^{-1}(0)$ is the tangent space to $f_i^{-1}(0)$ at the point q. By $n(q)$ denote the unit normal vector to Γ_q directed inside Q. If q is a regular point of the boundary, then $n(q)$ is a unique vector. At singular points there may be several vectors $n(q)$.

Denote by M_0 the unit tangent bundle over Q_0. The points of M_0 are of the form $x = (q, v)$, where $q \in Q_0$, $v \in S^{d-1}$, $d = \dim Q$. Suppose $\pi: M_0 \to Q_0$ is the natural projection, i.e., $\pi(q, v) = q$ for $x = (q, v) \in M_0$. Put $M = \pi^{-1}(Q)$. Clearly, if Q is a compact Riemann manifold with piece-wise smooth boundary, then M is a manifold of the same type. The boundary satisfies $\partial M = \pi^{-1}(\partial Q)$; clearly $\partial \tilde{M}_i \overset{\text{def}}{=} \pi^{-1}(\partial \tilde{Q}_i)$ is a regular component of the boundary; put $\partial \tilde{M} = \bigcup_i \partial \tilde{M}_i$. If $\dim Q = d$, then $\dim M = 2d - 1$.

For $x \in M$, the point $q = \pi(x)$ is said to be the *carrier* of x. M possesses a natural involution sending each point $x \in M$ into the point $x' \in M$ with the same carrier and opposite unit vector.

Define the measure μ on M_0 by putting $d\mu = d\rho(q)\, d\omega_q$, where $d\rho(q)$ is the volume element in Q_0 generated by the Riemann metric, ω_q is the Lebesgue measure on the $(d-1)$-dimensional sphere $S^{d-1}(q) = \pi^{-1}(x)$. This formula means that for any Borel set $A \subset M$ we have

$$\mu(A) = \int_{Q_0} d\rho(q) \int_{A \cap S^{d-1}(q)} d\omega_q(x).$$

The same letter μ will denote the restriction of this measure to M. By introducing a constant multiplier into $d\omega_q$, we may assume that the measure μ is normalized on M. Further we shall also need the measure μ_1 on $\partial \tilde{M}$, where $d\mu_1(x) = d\rho_i(q)\, d\omega_q |(n(q), x)|$, $x \in \partial \tilde{M}_i$ and $d\rho_i(q)$ is the volume element induced by the Riemann metric on $\partial \tilde{Q}_i$.

2. *The construction of dynamical systems of the billiards type.* Consider a geodesic flow on the space M_0 and the corresponding vector field $X = \{X(x), x \in M_0\}$. Here $X(x)$ is the tangent vector to M_0 at the point x. The same letter X will denote the restriction of the vector field to M. Then X determines the motion of our point with unit velocity along geodesic lines in M. Suppose $N_{i,j}$ is the set of all interior points $x \in M$ such that the segment of the geodesic line constructed in the direction x intersects ∂Q on $\partial Q_i \cap \partial Q_j$. It follows from the definition of Riemann manifolds with piecewise smooth boundary that $N_{i,j}$ is a closed submanifold of codimension 1, so that $\mu(\bigcup_{i \neq j} N_{i,j}) = 0$. Choose $x \in \operatorname{Int} M \setminus \bigcup_{i \neq j} N_{i,j}$ (here Int M is the set of interior points of M). The following cases may arise:

(i) the geodesic line constructed in the direction x does not intersect the boundary ∂Q;
(ii) the end point of a geodesic segment of some finite length s is located at a regular point of thc boundary ∂Q.

In the case (i), consider the motion of x along the geodesic half-line, i.e., the same motion as in the case of a geodesic flow. In the case (ii), denote by y the tangent vector obtained from x by parallel translation along the geodesic to the end point of the segment of length s (we assume that s is the smallest positive number for which (ii) is satisfied). Reflect y at the point $q = \pi(y)$ according to the "incidence angle equals reflection angle" rule, i.e., construct the new tangent vector $y' = y - 2(n(q), y) \cdot n(q)$. If $n(q), y) \neq 0$, then y' points inside Q. Assume that $y' \notin \bigcup_{i \neq j} N_{ij}$. Then for y' we have either (i) or (ii). If we have (i), then y' moves further as in the case of a geodesic flow. In the case (ii), we shall consider the new geodesic segment in the direction of y', its end point being the next intersection with the boundary, etc. A similar construction may be carried out in the opposite direction.

Denote by $N^{(1)}$ the set of all points $x \in M$ which will be contained in $\bigcup_{i \neq j} N_{i,j}$ at some step of this construction. A bit later we shall show that $\mu(N^{(1)}) = 0$. Consider $M \setminus N^{(1)}$. The set must contain some x for which our process will lead to an infinite number of reflections in finite time. Denote by $N^{(2)}$ the set of all x possessing this property. Further it will be shown that $\mu(N^{(2)}) = 0$. Put $M' = M \setminus (N^{(1)} \cup N^{(2)})$. Define a one-parameter group of transformations $\{T^t\}$ on M by setting (for any $x \in M'$ and any t, $-\infty < t < \infty$) $T^t x$ equal to the tangent vector obtained by a translation of x along the trajectory which it determines by a distance t. In the case when t is the moment when the boundary is reached, we put $T^t x \overset{\text{def}}{=} \lim_{t' \to t+0} T^{t'} x$. If y is a point of the boundary ∂M and q is its carrier, then it is convenient to suppose y identified with the point $y' = y - 2(n(q), y) \cdot n(q)$. Then the points $\lim_{t' \to t+0} T^{t'} x$, $\lim_{t' \to t-0} T^{t'} x$ will always be identified. The set obtained from M' as the result of such an identification will still be denoted by M'.

Definition 1. The transformation group $\{T^t\}$ is refered to as *billiards* on Q.

Definition 2. If for almost every (in the sense of the measure μ) $x \in M'$ we have (ii), the billiards are said to be *proper*.

In the sequel we consider only proper billiards and the adjective "proper" is always omitted.

The space Q is sometimes said to be the *configuration space* of the billiards and sets of the form $\pi(\{T^t x: -\infty < t < \infty\}), x \in M'$ are called *configurational trajectories* of the billiards.

The billiards $\{T^t\}$ are directly related to the transformation T_1 of the set

$$M_1 \stackrel{\text{def}}{=} \{x \in \partial\tilde{M}: (n(q), x) > 0, q = \pi(x)\}$$

defined in the following way: consider the geodesic segment in the direction of x with origin $q = \pi(x)$ and end point at the first intersection with the boundary and reflect the tangent vector from the boundary at the end of the segment. The vector y thus obtained will be put equal to $T_1 x$. Clearly if x is in the set where $\{T^t\}$ is defined, then $T_1 x = T^{f(x)}x$, where $f(x)$ is the length of the geodesic segment. Assume that $T_1 x \in M_1$. Then T_1 is defined and continuous in some neighborhood $O \subset \partial\tilde{M}$ of the point x.

Lemma 1. *The restriction of the transformation T_1 to O preserves the measure μ_1.*

Proof. By Liouville's theorem, the geodesic flow preserves the measure $\mu = d\rho\, d\omega$ (see Chap. 2, §2). The neighborhood O is contained in the $(2d - 2)$-dimensional submanifold $\pi^{-1}(f^{-1}(0))$. For any $C \subset O$ the integral $\int_C d\mu_1$ is the flux of the vector field X through C. Consider the set

$$M_1^- = \{x \in \partial\tilde{M}: (n(q), x) < 0\}$$

and the transformation T_1^- sending $x \in M_1$ into the point

$$T_1^- x \stackrel{\text{def}}{=} \lim_{\varepsilon \to +0} T^{f(x)-\varepsilon} x$$

belonging to the closure of the set M_1^-. The restriction of the measure μ_1 to M_1^- will be denoted by μ_1^-. Also introduce the map $\sigma: M_1^- \to M_1$ which acts according to the formula

$$\sigma x = x - 2(n(q), x)n(q), \qquad q = \pi(x).$$

For fixed q, the map σ is isometric. It sends the measure μ_1^- into the measure μ_1. Moreover, $T_1 = \sigma T_1^-$. By the invariance of the measure μ with respect to the geodesic flow, the flux of the vector field is preserved, i.e., $\int_C d\mu_1 = \int_{T_1^- C} d\mu_1^-$, and since σ is isometric, we have

$$\int_{T_1^- C} d\mu_1^- = \int_{\sigma T_1^- C} d\mu_1^- = \int_{T_1 C} d\mu_1.$$

The lemma is proved. □

This lemma implies that we can find a measurable subset $M_1' \subset M_1$, $\mu_1(M_1') = \mu_1(M_1)$ with respect to T_1 on which the transformation T_1 is measurable, bijective and preserves the measure μ_1. Since $f(x) > 0$ almost everywhere on M_1', it follows from the Poincaré recurrence theorem (also see the corollary to this theorem) that $\sum_{k=0}^{n-1} f(T_1^k x)$ tends to infinity when $n \to \infty$ for almost all $x \in M_1'$.

Now choose a point $x \in \operatorname{Int} M$ and find the nearest point to it $x^- \in \partial M$ on the billiards trajectory so that $x = T^\tau x^-$, $\tau > 0$. We can introduce new coordinates on M by taking the number τ and the coordinates of the point x^- on the boundary to be the coordinates of the point x. Then $d\mu = d\tau\, d\mu_1$. Indeed, $d\mu_1$ is the volume element of an infinitely small surface orthogonal to the vector field X of the geodesic flow, while τ is the coordinate along the trajectory of this flow.

Lemma 2. $\mu(N^{(2)}) = 0$.

Proof. It is clear that if $x = T^\tau x^-$ and $x \in N^{(2)}$, then all the points $x' = T^t x^-$, $0 \le t < f(x^-)$, belong to the set $N^{(2)}$. Therefore

$$\mu(N^{(2)}) = \int_{N^{(2)} \cap M_1} f(x^-)\, d\mu_1(x^-).$$

But the last integral vanishes since $N^{(2)} \cap M_1$ consists of all the points x for which the sums $\sum_{k=0}^{n-1} f(T_1^k x)$ remain bounded when $n \to \infty$. The lemma is proved. □

Lemma 3. $\mu(N^{(1)}) = 0$.

Proof. Suppose $N^{(3)}$ is the set of all $x \in M_1$ for which $\pi(T_1 x)$ is a singular point of the boundary. Then $N^{(3)}$ is the union of a finite number of submanifolds of codimension 1 and therefore $\mu_1(N^{(3)}) = 0$. Hence if $N^{(4)}$ is the set of all $x \in M_1$ such that $T^k x \in N^{(3)}$ for some k, $-\infty < k < \infty$, then $\mu_1(N^{(4)}) = 0$. Further, as before,

$$\mu(N^{(1)}) = \int_{N^{(1)} \cap M_1} f(x^-)\, d\mu_1(x^-) = \int_{N^{(4)}} f(x^-)\, d\mu_1(x^-) = 0.$$

The lemma is proved. □

Since $\mu_1(N^{(4)}) = 0$, we may assume that $M'_1 \cap N^{(4)} = \varnothing$. Now suppose $\tilde{M}$ is the set of all $x \in M$ such that $x^- \in M'_1$. Then $\tilde{M}$ is invariant with respect to the billiards $\{T^t\}$. Moreover $\mu(\tilde{M}) = 1$ and $\tilde{M} \subset M \backslash (N^{(1)} \cup N^{(2)})$.

Lemma 4. *The measure μ is invariant with respect to the group $\{T^t\}$.*

Proof. Choose a point $x \in \operatorname{Int} M \cap \tilde{M}$. As we have already noted, if τ and x^- are the coordinates of the point x, then $d\mu(x) = d\tau\, d\mu_1(x^-)$. Consider the set C of the form $C = (\tau - \varepsilon, \tau + \varepsilon) \times O$, where O is a neighborhood of the point x^- on the boundary and $\varepsilon > 0$. Then for sufficiently small ε and O we have $\mu(C) = 2\varepsilon \cdot \mu_1(0)$. If t is so small that $T^{-t}C$ is not yet on the boundary, the set $T^{-t}C$ will be of the form $T^{-t}C = (\tau - t - \varepsilon, \tau - t + \varepsilon) \times O$. Hence $\mu(T^{-t}C) = 2\varepsilon\mu_1(0) = \mu(C)$. Now suppose t is such that $t > \tau + \varepsilon$ and all the points of C from time $-t$ to time 0 had precisely one reflection from the boundary. Then $T^{-t}C$ is the set whose projection on M_1 is $T_1^{-1}O$; the intersection of $T^{-t}C$ with every geodesic segment whose end points are in M_1 is either empty or is of length 2ε. Hence

$$\mu(T^{-t}C) = 2\varepsilon \cdot \mu_1(T_1^{-1}O) = 2\varepsilon \cdot \mu_1(O).$$

This implies our statement. The lemma is proved. □

Remark. The dynamical systems of the billiards type constructed in this section correspond to the motion of a point with *unit velocity* in the domain Q with piece-wise smooth boundary. Sometimes a slightly more general type of billiards system is considered, for which no conditions are imposed on the absolute value of the velocity. Such systems are constructed similarly, except that instead of the unit tangent bundle one must consider the entire tangent bundle over Q. If $\dim Q = d$, then the phase space in this case will be $2d$-dimensional.

§2. Billiards in Polygons and Polyhedra

Suppose $Q \subset \mathbb{R}^d$ is a convex polyhedron, i.e., a set of the form

$$Q = \{q \in \mathbb{R}^d \colon f_i(q) \geq 0,\ i = 1, \ldots, r\},$$

where the functions $f_1, \ldots, f_r$ are linear. The sets $\Gamma_i = f_i^{-1}(0) \cap Q$ are the faces of the polyhedron Q. The boundary of the polyhedron is the set $\Gamma = \bigcup_{i=1}^r \Gamma_i$. At each interior point $q \in \Gamma_i$ the unit normal vector $n(q)$ to the boundary Γ is the same. It shall be denoted by n_i.

The isometric map σ introduced in the previous section may be defined at each point $x = (q, v)$, $q \in \Gamma_i$ by means of the map $\sigma_i \colon S^{d-1} \to S^{d-1}$, which acts according to the formula $\sigma_i(v) = v - 2(n_i, v)n_i$. The procedure of straightening out polygons, well known from elementary geometry, may be applied in this case, the polygons consisting of the configurational billiard trajectories $\pi(\{T^t x: -\infty < t < \infty\})$, where x is a point of the phase space M of the billiards in the polyhedron Q. Namely, if such a broken line has its vertices on the faces with numbers $i_1, i_2, i_3, \ldots$, then successive reflections of the polyhedron Q with respect to these faces transforms the broken line into a straight line intersecting the polyhedra $Q, Q_{i_1}, Q_{i_1 i_2}, Q_{i_1 i_2 i_3}, \ldots$. Here $Q_{i_1, \ldots, i_k}$ is the polyhedron obtained by reflecting Q with respect to the faces $i_1, \ldots, i_k$.

Let $x_0 = (q_0, v_0) \in M$. The vector $v_0 \in S^{d-1}$ determines the initial velocity of the billiards trajectory originating in q_0. The velocity of motion on the part of the trajectory after the kth reflection is $v_k = (\sigma_{i_k} \sigma_{i_{k-1}}, \ldots, \sigma_{i_1}) v_0$.

Denote by G_Q the subgroup of the isometry group of the sphere S^{d-1} generated by the isometries $\sigma_1, \ldots, \sigma_r$.

Theorem 1. *If G_Q is a finite group, then the billiards in the polyhedron Q are not ergodic. Moreover, to each orbit of the natural action of the group G_Q on S^{d-1} (i.e., to the set $\Omega = \Omega(v_0) = \{gv_0 \in S^{d-1}: g \in G_Q\}, v_0 \in S^{d-1}$), corresponds the set A_Ω, invariant with respect to $\{T^t\}$, consisting of all $x = (q, v) \in M$ such that $v \in \Omega$.*

Proof. Suppose $x = (q, v) \in A_\Omega$, i.e., $v = g_0 v_0, g_0 \in G_Q$. Then for any $t \in \mathbb{R}^1$ we have $(q_t, v_t) = T^t x \in A_\Omega$. Indeed, if, when we went from q to q_k, the reflections with respect to the faces with numbers $i_1, \ldots, i_k$ were carried out, then $v_t = (gg_0)v_0$, where $g = \sigma_{i_k} \sigma_{i_{k-1}}, \ldots, \sigma_{i_1} \in G_Q$.

Since the group G_Q is finite, we can find a measurable set $C \subset S^{d-1}/G_Q$, i.e., a set of orbits (of the group G_Q) whose measure differs from zero or one. The set $A = \bigcup_{\Omega \in C} A_\Omega$ is invariant with respect to $\{T^t\}$ and $\mu(A)$ differs from 0 or 1. The theorem is proved. □

The theorem may be visualized as stating that for a finite group G_Q only a finite number of directions may be obtained when we move along billiards trajectories from the given initial direction. In the two-dimensional case, claiming the fact that the group is finite is equivalent to saying that all the angles of the polyhedron Q are not incommeasurable with π. Then the straightening out procedure for the trajectories described above enables us to study the ergodic property of billiards with finite group G_Q completely.

Example

Billiards in a rectangle. Suppose $Q \subset \mathbb{R}^2$ is the rectangle $ABCD$. Denote by S_1, S_2, S_3, S_4 the symmetries of the plane $\mathbb{R}^2$ with respect to the lines AB, BC, CD, DA respectively and by $\hat{G}_Q$ the discrete subgroup generated by these symmetries. Obviously, Q is a fundamental domain of the group $\hat{G}_Q$.

In other words, the images of the rectangle Q under all possible compositions of the symmetries S_i may be used to fill up the entire plane and, for every point $q \in \mathbb{R}^2$, there exists precisely one point $q' \in Q$ such that $q = g(q')$ for some isometry $g \in \hat{G}_Q$. The procedure of straightening out trajectories may be described in these terms as follows. If $q = q_0 + vt$ is a line on the plane, then its image $q' = \pi(q_0 + vt)$ under the natural projection $\pi\colon \mathbb{R}^2 \to Q$ is a billiards trajectory in Q; conversely, every configurational billiards trajectory in Q can be obtained in the manner just described.

Note that the compositions $S_4 S_2$ and $S_3 S_1$ are parallel translations by the vectors $2\overrightarrow{AB}$ and $2\overrightarrow{AD}$, which we shall denote by τ_1 and τ_2. These translations are the generators of the subgroup of translations $\hat{G}_Q^\tau$ of the group $\hat{G}_Q$. Further, if the lines l and l' are obtained from each other by the translation $\tau \in \hat{G}_Q^\tau$, then the corresponding billiards trajectories $\pi(l)$ and $\pi(l')$ coincide in Q and the converse is also true: if $\pi(l) = \pi(l')$, then $l' = \tau l$, where $\tau \in \hat{G}_Q^\tau$. Therefore the billiards trajectories correspond to the rectilinear trajectories $q = q_0 + vt$ in the quotient space $\mathbb{R}^2/\hat{G}_Q^\tau$, i.e., to the trajectories of a conditionally periodic motion on the torus $\mathrm{Tor}^2 = \mathbb{R}^2/\hat{G}_Q^\tau$. The torus Tor^2 is obtained by identifying the opposite side of the rectangle $\hat{Q}$ spanning the vectors $2\overrightarrow{AB}$ and $2\overrightarrow{AD}$ (these vectors are obtained one from the other by the translations τ_1 and τ_2). $\hat{Q}$ consists of the rectangles Q, $\sigma_2 Q$, $\sigma_3 Q$, $\sigma_2\sigma_3 Q$. If we consider the conditionally periodic motion on the torus corresponding to the velocity $v \in S^1$, then these rectangles should be identified with the rectangles

$$Q \times \{v\}, Q \times \{\sigma_2 v\}, Q \times \{\sigma_3 v\}, Q \times \{\sigma_2\sigma_3 v\},$$

in $Q \times S^1$. Here $\sigma_1 = \sigma_3$, $\sigma_2 = \sigma_4$ are the corresponding reflections in the velocity space S^1. Under the identifications of the boundary of $Q \times S^1$ corresponding to these reflections, we shall precisely obtain a torus from the rectangles listed above, so that the conditionally periodic motion on it indicated above is the flow $\{T_\Omega^t\}$ induced by the flow $\{T^t\}$ on the invariant subset $A_{\Omega(v)}$ constructed in Theorem 1. Note that the group G_Q in this case consists of four elements: $e = Id, \sigma_1, \sigma_2, \sigma_2\sigma_3$.

Thus the phase space of rectangular billiards splits up into subsets A_Ω invariant with respect to the flow $\{T^t\}$, where $\Omega = \Omega(v) = \{v, \sigma_1 v, \sigma_2 v, \sigma_2\sigma_3 v\}$. If $\sigma_1 v \neq v$ and $\sigma_2 v \neq v$, then A_Ω is the torus, and the flow $\{T_\Omega^t\}$ induced on it by the flow $\{T^t\}$ is a one-parameter group of translations of the torus. Introduce cyclic coordinates φ_1, φ_2 on the torus A_Ω in the direction of the generators of the translations $\tau_1 = \tau_{2\overrightarrow{AB}}$, $\tau_2 = \tau_{2\overrightarrow{AD}}$. Then the action of the flow $\{T_\Omega^t\}$ may be written in the form

$$T_\Omega^t(\varphi_1, \varphi_2) = (\varphi_1 + \omega_1 t \ (\mathrm{mod}\ 1), \varphi_2 + \omega_2 t \ (\mathrm{mod}\ 1)),$$

where $\omega_1 = v_1/2a_1$, $\omega_2 = v_2/2a_2$, $a_1 = |AB|$, $a_2 = |AD|$ and v_1, v_2 are the components of the velocity v along the sides AB and AD. If the quotient

$\omega_2/\omega_1 = v_2/v_1 \cdot a_1/a_2$ is irrational, then, as was pointed out in §1, Chap. 3, the flow $\{T^t_\Omega\}$ is ergodic. Therefore the decomposition of the phase space of the rectangular billiards on Q into invariant tori A_Ω is the decomposition into ergodic components: for all $v \in S^1$, except for a countable set, the flow $\{T^t_\Omega\}$ induced on A_Ω is ergodic.

The arguments carried out above may easily be generalized to the multi-dimensional case. For billiards in a rectangular parallelipiped on $Q \subset \mathbb{R}^d$, the ergodic components will be one-parameter groups of translations of the d-dimensional torus.

For arbitrary polygons, and even more so for polyhedra, the study of ergodic properties of the corresponding billiards has not been worked out conclusively. We shall only mention one result, valid in the general case.

Theorem 2. *Suppose $Q \subset \mathbb{R}^2$ is a polygon. For any $q \in Q$ and almost all (with respect to the Lebesgue measure) $v \in S^1$, the closure of the configurational trajectory of the point $x = (q, v)$ with respect to the billiards flow $\{T^t\}$ contains at least one vertex of the polygon.*

Proof. Choose a $q \in Q$ and an arbitrary $\delta > 0$. Suppose C_v, where $v \in S^1$, is the configurational trajectory of the point $x = (q, v)$, Γ_0 is the set of vertices of the polygon Q and ρ is the Lebesgue measure on S^1. To prove the theorem, it obviously suffices to show that $\rho(N_\delta) = 0$, where $N_\delta = \{v \in S^1\colon \operatorname{dist}(C_v, \Gamma_0) \geq \delta\}$.

If, despite our statement, we have $\rho(N_\delta) > 0$, then we can find a density point $v_0 \in N_\delta$. Choose any $\varepsilon > 0$ and consider the points $y = (q, v_0)$, $y' = (q, v_0 + \varepsilon)$ of the phase space of our billiards. Denote their configurational trajectories by C, C' and assume that these trajectories do not intersect on Γ_0. The trajectories C, C' for $t > 0$ intersect the boundary Γ of the polygon Q an infinite number of times. Suppose $q_1, q_2, \ldots, q'_1, q'_2, \ldots$ are the successive intersection points of C, C' with Γ respectively (for $t > 0$) and let q_k belong to the side Γ_{j_k} of the polygon, while $q'_k \in \Gamma_{j'_k}$.

Let us show that there exists a number n such that $j_n \neq j'_n$. Indeed, by applying the straightening out process described above to C, C', we obtain two distinct lines l, l':

$$l = \{q_0 + vt\colon t \in \mathbb{R}^1\},\ l' = \{q + (v_0 + \varepsilon)t\colon t \in \mathbb{R}^1\}.$$

If, for each $k = 1, 2, \ldots$, the points q_k, q'_k were on the same side of the polygon, then this would mean that for all $t > 0$, the inequality $\operatorname{dist}(q + v_0 t, q + (v_0 + \varepsilon)t) \leq \operatorname{diam} Q$ is valid, which is impossible.

Suppose n is the smallest number with the property described above. Consider the polygon Q_n obtained from Q by successive reflections with respect to the sides $\Gamma_{j_1}, \ldots, \Gamma_{j_{n-1}}$. This polygon has at least one vertex z inside the angle formed by the rays $q + v_0 t$, $q + (v_0 + \varepsilon)t$, $t \geq 0$. Construct the circle S_δ of radius δ with centre z. To any line of the form $q + vt$, $t \in \mathbb{R}^1$ intersecting the circle S_δ and lying within this angle corresponds a configurational trajectory of our billiards which passes within a distance of not more than δ from the vertex z. The values $v \in S^1$ corresponding to such lines fill up a certain closed interval $\Delta \subset [v_0, v_0 + \varepsilon]$. It follows from obvious geometric considerations that $\rho(\Delta) \geq \text{const} \cdot \delta \cdot \varepsilon$, const > 0. Hence

$$\frac{\rho(N_\delta \cap [v_0, v_0 + \varepsilon])}{\rho([v_0, v_0 + \varepsilon])} \leq 1 - \text{const} \cdot \delta.$$

Since ε was arbitrary, this contradicts the fact that v_0 was a density point of N_δ. The theorem is proved. □

The billiards in polygons Q with finite group G_Q, i.e., in polygons all of whose angles are commeasurable with π, are intimately related to a class of dynamical systems already considered: interval exchange transformations (see Chap. 5).

Suppose $Q \subset \mathbb{R}^2$ is a convex polygon with r vertices and that its interior angles $\alpha_1, \ldots, \alpha_r$ are commeasurable with π. We can assume that α_i is of the form $\alpha_i = (k_i/n)\pi$, where $n \geq 1$, and the greatest common divisor of the number $n, k_1, \ldots, k_r$ is equal to 1.

Lemma 1. *The group G_Q is isomorphic to the symmetry group of a regular polygon with n vertices.*

Proof. Note that in the velocity space S^1 the composition of the symmetries corresponding to the sides bounding the angle α_i is a rotation $R_{2\alpha_i}$ of the circle S^1 by an angle $2\alpha_i = (2k_i/n) \cdot \pi$. Hence, for any family of integers s_0, $s_1, \ldots, s_r$, the group G_Q contains the rotation by an angle $\alpha = (s_0 n + s_1 k_1 + \cdots + s_r k_r)(2\pi/n)$. But it follows from our assumptions that there exist a family $s_0, s_1, \ldots, s_r$ for which $s_0 n + s_1 k_1 + \cdots + s_r k_r = 1$. Hence the group G_Q contains the rotation $R_{2\pi/n}$ and all the rotations by angles which are multiples of $2\pi/n$. Moreover, the group G_Q contains n symmetries with respect to the axes of the form $R_{2k\pi/n} l$, $k = 0, 1, \ldots, n - 1$, where l is any of the axes of symmetry of $\sigma_1, \ldots, \sigma_r$. The transformations indicated above generate the symmetry group of a regular polygon of n sides. The lemma is proved. □

Now fix a direction l on one of the axes of symmetry generating G_Q and determine the velocities $v \in S^1$ by specifying the angle φ (mod 2π) between the directions l and v. For $\varphi \in [0, 2\pi)$ the orbit of the velocity $v(\varphi)$ consists of $2n$ velocities corresponding to the angles

$$\varphi_k^\pm = \pm\varphi + 2k\pi/n, \qquad k = 0, 1, \ldots, n - 1.$$

If $\varphi \neq s\pi/n$, where s is an integer, then all the angles $\varphi_k^\pm$ are distinct. To these angles correspond $2n$ polygons with r sides.

$$Q_k^\pm = Q \times \{v(\varphi_k^\pm)\} \subset Q \times S^1, \qquad k = 0, 1, \dots, n-1,$$

from which, by pairwise identification of $n \cdot r$ pairs of sides, we can obtain the invariant set $A_\varphi = A_{\Omega(v(\varphi))}$ (of the flow $\{T^t\}$) indicated in Theorem 1. The flow induced on A_φ shall be denoted by $\{T_\varphi^t\}$.

Suppose $\Delta_1, \dots, A_{nr}$ are closed intervals, each of which corresponds to one of the identified pairs of sides of the polygons $Q_k^\pm$. Dispose them on a line one after the other in arbitrary order and put $\Delta = \bigcup_{i=1}^{n \cdot r} \Delta_i$. We may assume that $\Delta = [0, 1]$. The flow $\{T_\varphi^t\}$ induces a transformation $T_\varphi \colon \Delta \to \Delta$. This transformation is constructed from the flow $\{T_\varphi^t\}$ in the same way (see the previous section) as we constructed the transformation T_1 on the boundary of the phase space from the arbitrary billiards flow $\{T^t\}$. The transformation T_φ is obviously piecewise linear. Arguing as in the proof of Lemma 1, §1, we see that T_φ preserves the measure μ_φ on $[0, 1]$ which possesses a piecewise constant density $p(x)$ with respect to the Lebesgue measure: $p(x) = |(v(\varphi_k^\pm), n_i)|$, if x belongs to the ith side of the polygon $Q_k^\pm$; n_i is the normal vector to the side. On the segment $[0, 1]$, introduce a new coordinate z by putting $z(x) = \mu_\varphi([0, x])$. Then it follows from the invariance of the measure μ_φ that, in the metric $\operatorname{dist}(z_1, z_2) = |z_1 - z_2|$, the transformation T_φ is isometric. Finally note that if some orientation is given to the sides of the polygon Q (and therefore to the sides of the polygon $Q_k^\pm$) and this orientation is preserved when we pass to the segments $\Delta_1, \dots, \Delta_{n \cdot r}$, then the transformation T_φ will also be orientation preserving.

Thus T_φ is an isometric piecewise linear transformation of the segment $[0, 1]$ preserving orientation, i.e., it is an interval exchange transformation. Since under the action of the flow $\{T_\varphi^t\}$ each of the r sides of the polygon $Q_k^\pm$ are transferred onto no more than $r - 1$ (other) sides, the number of exchanged intervals is not greater than $n \cdot r(r - 1)$.

The ergodic properties of the flow $\{T_\varphi^t\}$ are determined by the ergodic properties of the interval exchange transformation T_φ. In particular, the number of ergodic components of the flow $\{T_\varphi^t\}$ and the automorphism T_φ are the same. Using the results of Chap. 5, we can now prove the following theorem.

Theorem 3. *If the flow $\{T_\varphi^t\}$ has no periodic trajectories, the number of ergodic normalized invariant measures is no greater than $n \cdot r \cdot (r - 1)$.*

Proof. The interval exchange transformation T_φ corresponding to the flow $\{T_\varphi^t\}$ is aperiodic. Indeed, in the converse case some degree of T_φ would have an entire interval of fixed points, which means that the flow $\{T_\varphi^t\}$ would have a whole family of periodic trajectories. Therefore our statement follows from the Theorem 1 in §2 of Chap. 5. Our theorem is proved. □

Now consider the original polygon Q; Γ_0 is the set of its vertices. By $\hat{\Gamma}_0$ denote the set of vertices of all polygons obtained from Q by successive reflections with respect to the sides under the straightening out procedure of the configurational billiard trajectories, i.e., the set of vertices of the polygons of the form $g(Q)$, $g \in \hat{G}_Q$.

Lemma 2. *If the flow* $\{T_\varphi^t\}$ *possesses periodic trajectories, then the direction* $v(\varphi)$ *coincides with one of the directions of the form* $\overrightarrow{AB}$, *where* $A \in \Gamma_0$, $B \in \hat{\Gamma}_0$.

Proof. Suppose the trajectory of the flow $\{T_\varphi^t\}$ originating at the point $(q_0, v(\varphi))$ is periodic with period τ. Without loss of generality, we may assume that the point q_0 is located on one of the sides AA' of the polygon Q. Applying the straightening out procedure to the corresponding configurational trajectory by reflecting successively with respect to the sides of Q, we see that periodicity implies that the point $q_\tau = q_0 + v(\varphi)\tau$ lies on the side $BB' = g(AA')$ of the polygon $g(Q)$, where $g \in \hat{G}_Q$ is the corresponding composition of reflections, the segments $q_0 A$ and $q_\tau B$ are equal, while the straightened out trajectory makes equal angles with the segments AA' and BB'. Therefore, $Aq_0 q_\tau B$ is a parallelogram and the direction of $v(\varphi)$, i.e., the direction of $q_0 q_\tau$, coincides with the direction of $\overrightarrow{AB}$. The lemma is proved. □

Corollary. *For all directions, except a countable number,* $v(\varphi) \in S^1$, *the flow* $\{T_\varphi^t\}$ *possesses no more than* $n \cdot r \cdot (r - 1)$ *ergodic normalized invariant measures on the invariant set* A_φ.

Proof. Since the group $\hat{G}_Q$ is countable, the set of directions of the form $\overrightarrow{AB}$, where $A \in \Gamma_0$, $B \in \hat{\Gamma}_0$, is also countable. It suffices to apply Theorem 3. The corollary is proved. □

§3. Billiards in Domains with Convex Boundary

In this section we consider billiards on plane convex domains $Q \subset \mathbb{R}^2$ bounded by smooth closed curves $\Gamma = \partial Q$:

$$Q = \{q \in \mathbb{R}^2 \colon f(q) \geq 0\}, \qquad \Gamma = \{q \in \mathbb{R}^2 \colon f(q) = 0\},$$

where $f \in C^\infty(\mathbb{R}^2)$, $\operatorname{grad} f(q) \neq 0$, for $q \in \Gamma$. The phase space M in this case is three-dimensional. Recall that it is obtained from the set $Q \times S^1$ by identifying pairs of points $x' = (q', v')$, $x'' = (q'', v'')$, such that

$$q' = q'' = q \in \Gamma, \qquad x'' = \sigma x' = x' - 2(n(q), x')n(q),$$

where $n(q)$ is the unit normal vector to Γ at the point q.

EXAMPLE

Billiards in an ellipse. Suppose $\Gamma = \Gamma_c$ is an ellipse in the plane $\mathbb{R}^2$ with foci at the points A_1, A_2:

$$\Gamma_c = \{q \in \mathbb{R}^2 : \operatorname{dist}(q, A_1) + \operatorname{dist}(q, A_2) = c\}, \qquad c = \text{const}.$$

For the study of billiards in an ellipse, we shall need the well-known focal property of ellipses and hyperbolas. We shall state it in the form of the following lemma.

Lemma 1. (i) *For any point $P \in \Gamma_c$, the closed intervals PA_1 and PA_2 form equal angles with the tangent l_P to the ellipse at the point P;*

(ii) *the same is true for an arbitrary point P of the hyperbola H_c with foci A_1, A_2:*

$$H_c = \{q \in \mathbb{R}^2 : \operatorname{dist}(q, A_1) \quad \operatorname{dist}(q, A_2) - c\}.$$

Proof. We shall only prove (i), since (ii) is proved in a similar way. For any point $P' \neq P$ on the line l_P, we obviously have the inequality $\operatorname{dist}(P', A_1) + \operatorname{dist}(P', A_2) > c$. Therefore the broken line A_1PA_2', where the point A_2' is symmetric to A_2 with respect to l_P, is part of a straight line—this implies statement (i). The lemma is proved. □

Theorem 1 (The Main Property of Billiards in an Ellipse). *Suppose the broken line $\gamma = P_1P_2, \ldots$ is the configurational trajectory of the billiards in the domain Q bounded by the ellipse Γ_c, where γ does not pass through the foci A_1, A_2. Then either all the line segments P_iP_{i+1} are tangent to one and the same ellipse Γ (with the same foci as Γ_c), or all the P_iP_{i+1} are tangent to a hyperbola H with the same foci (in this case the tangent points are not necessarily located on the segments P_iP_{i+1} themselves, but lie on their extensions (the lines passing through P_i and P_{i+1}).*

Proof. Suppose P_1P, PP_2 are two successive links of the configurational billiards trajectory. It follows from Lemma 1 that either both the segments do not intersect the line segment A_1A_2 joining the foci of the ellipse, or both intersect it. We begin with the first case. Denote by $\Gamma_{c_1}, \Gamma_{c_2}$ the ellipses with foci A_1, A_2, tangent respectively to the segments P_1P, PP_2, and let D_1, D_2 be the tangent points. Let us show that $\Gamma_{c_1} = \Gamma_{c_2}$, i.e.,

$$\operatorname{dist}(D_1, A_1) + \operatorname{dist}(D_1, A_2) = \operatorname{dist}(D_2, A_1) + \operatorname{dist}(D_2, A_2).$$

Suppose B_1, B_2 are the images of the foci A_1, A_2 under symmetries with respect to the lines P_1P, PP_2 respectively. From the focal property of the ellipses $\Gamma_{c_1}, \Gamma_{c_2}$, we may write

$$c_1 = \operatorname{dist}(D_1, A_1) + \operatorname{dist}(D_1, A_2) = \operatorname{dist}(D_1, B_1) + \operatorname{dist}(D_1, B_2) = \operatorname{dist}(A_2, B_1),$$

similarly $c_2 = \operatorname{dist}(A_1, B_2)$. Further it follows from the focal property of the ellipse Γ_c that the angles A_2PB_1 and A_1PB_2 are equal. Since, moreover,

$$\operatorname{dist}(A_1, P) = \operatorname{dist}(B_1, P), \qquad \operatorname{dist}(A_2, P) = \operatorname{dist}(B_2, P),$$

the triangles A_2PB_1 and A_1PB_2 are also congruent. Therefore

$$c_1 = \operatorname{dist}(A_2, B_1) = \operatorname{dist}(A_1, B_2) = c_2,$$

i.e., $\Gamma_{c_1} = \Gamma_{c_2} = \Gamma$.

The second case, when P_1P and PP_2 intersect the closed interval A_1A_2, is considered in a similar way, except that instead of the ellipses $\Gamma_{c_1}, \Gamma_{c_2}$ we must use the hyperbolas H_{c_1}, H_{c_2}. The theorem is proved. □

Corollary. *Billiards in an ellipse are not ergodic.*

Indeed, taking any c_1 and c_2, $0 < c_1 < c_2$, we see that by Theorem 1 the set $A \subset M$ consisting of all the points x of the phase space M for which the configurational trajectories are tangent to the ellipses (or hyperbolas) Γ_c, (H_c), $c_1 < c < c_2$, will be invariant.

For the applications, we shall need the important notion of caustic curve.

Definition 1. A smooth curve γ contained inside the domain $Q \subset \mathbb{R}^2$ is said to be a *caustic* (or a *caustic curve*), when the following condition holds: if at least one of the segments P_kP_{k+1} of an arbitrary configurational trajectory, $\ldots, P_{-1}P_0P_1, \ldots$ of the billiards in Q is tangent to γ, then all the other segments of this trajectory are also tangent to γ.

The theorem just proved above shows that in the case of an ellipse there exists two families of caustics: the ellipses Γ_c and the hyperbolas H_c with the same foci as Γ.

Let us prove one more general statement, relating to billiards in an arbitrary convex domain with smooth boundary.

Theorem 2. *Any billiards of an arbitrary convex domain $Q \subset \mathbb{R}^2$ bounded by a closed curve Γ of class C^1 possesses periodic trajectories with any number of links $n \geq 3$.*

Proof. Consider the set Π_n of all closed broken lines γ inscribed in Γ whose number of links is no greater than n. Obviously Π_n is compact in the natural topology and the perimeter $d(\gamma)$ of the broken line γ, viewed as a function on the space Π_n, is continuous in γ. Therefore $d(\gamma)$ reaches its maximum on Π_n. The broken line $\gamma_0 \in \Pi_n$ of maximal perimeter must have exactly n links: in the converse case its perimeter could be increased by replacing an arbitrary link P_1P_2 by the broken line $P_1P_0P_2$ with vertex P_0 on the arc between P_1 and P_2. Let us show that γ_0 is the configurational trajectory of billiards on Q, i.e., for all k the links $P_{k-1}P_k$ and P_kP_{k+1} form equal angles with Γ at the point P_k.

It follows from the maximality of the perimeter of γ_0 that the sum

$$|PP_{k-1}| + |PP_{k+1}|,$$

for points P on the arc $P_{k-1}P_kP_{k+1}$ of the curve Γ, is maximal when $P = P_k$. Considering the family of ellipses

$$\Gamma_c = \{q \in \mathbb{R}^2 : \text{dist}(q, P_{k-1}) + \text{dist}(q, P_{k+1}) = c\},$$

with foci P_{k-1} and P_{k+1}, we see that P_k is a tangency point of the curve Γ with one of the ellipses from this family. The fact that the angles formed by the line segments $P_{k-1}P_k$ and P_kP_{k+1} with the tangent to Γ at the point P_k are equal follows from the focal property of this ellipse. The theorem is proved. □

§4. Systems of One-dimensional Point-like Particles

In this section we give an example of a mechanical system whose study reduces to a problem in billiards.

Suppose that, on the closed interval $[0, 1]$, we are given $d \geq 2$ mass points $q_1, q_2, \ldots, q_d$, whose masses equal $m_1, m_2, \ldots, m_d$ and whose velocities are $v_1, v_2, \ldots, v_d$. Assume that these points move freely on $[0, 1]$ and reflect elastically when they hit each other or hit the end points of the closed interval. We shall assume that, at the initial moment, $q_1 < q_2 < \cdots < q_d$. Since the order of the points on the interval does not change during the motion, the configurational space of our mechanical system will be the d-dimensional simplex

$$Q = \{q = (q_1, \ldots, q_d) \in \mathbb{R}^d : 0 \leq q_1 \leq \cdots \leq q_d \leq 1\}.$$

Introduce new coordinates in $\mathbb{R}^d$: $q'_i = \sqrt{m_i}\, q_i$, $1 \leq i \leq d$. In these coordinates the simplex Q may be written in the form

$$Q' = \left\{(q'_1, \ldots, q'_d) : 0 \leq q'_i \leq \sqrt{m_i}, \frac{q'_i}{\sqrt{m_i}} \leq \frac{q'_{i+1}}{\sqrt{m_{i+1}}}\right\}.$$

Note that the new velocities $v'_i = \sqrt{m_i}\, v_i$ correspond to the new coordinates. We shall now show that the trajectory of the motion of a particle coincides with the configurational trajectory of the billiards in the simplex Q' determined by the initial point $x = (q', v')$, where $q' = (q'_1, \ldots, q'_d)$, $v' = (v'_1, \ldots, v'_d)$. In the interior points of Q', the trajectories of the billiards, as well as the trajectories of the mechanical system, are rectilinear and, at the initial moment of time, they are given by one and the same velocity vector v'. Therefore it suffices to consider only the moments when the particles hit each other or reflect against the end points of the interval; these moments correspond to the intersection of the configurational trajectory of the billiards with the boundary $\Gamma = \partial Q$. Then we must show that the rule of elastic reflection of the particles when they hit each other results in the reflection of the configurational trajectory from the boundary Γ under which the tangent component of the velocity is preserved, while the normal component changes sign.

We shall only consider the case of two particles hitting each other (the reflection from the end points of the interval is even simpler).

Assume that at the moment t_0 the particles with numbers k and $k + 1$ hit each other. The velocities of all the particles immediately before this event in the coordinate system $(q_1, \ldots, q_d)$, will be denoted by $\bar{v}_1, \ldots, \bar{v}_d$, while the velocities after reflection will be denoted by $\bar{\bar{v}}_1, \ldots, \bar{\bar{v}}_d$. In the coordinate system $(q'_1, \ldots, q'_d)$, the same velocities are denoted respectively by $\bar{v}'_1, \ldots, \bar{v}'_d$ and $\bar{\bar{v}}'_1, \ldots, \bar{\bar{v}}'_d$. It follows from the law of conservation of energy that

$$\sum_{j=1}^{d} m_j \bar{v}_j^2 = \sum_{j=1}^{d} m_j \bar{\bar{v}}_j^2, \qquad \text{i.e.} \quad \sum_{j=1}^{d} (\bar{v}'_j)^2 = \sum_{j=1}^{d} (\bar{\bar{v}}'_j)^2.$$

Therefore the absolute value of the velocity vector $v' = (v'_1, \ldots, v'_d)$ is preserved under reflection. Hence it suffices to show that the tangent component of the velocity v' is also preserved, i.e., the projection of the vector v' on the hyperplane L given by the equation $(1/\sqrt{m_k})q'_k = (1/\sqrt{m_{k+1}})q'_{k+1}$ is preserved (this hyperplane in the old coordinates is given by the equation $q_k = q_{k+1}$).

The law of the conservation of impulse for our system means that

$$\sum_{j=1}^{d} m_j \bar{v}_j = \sum_{j=1}^{d} m_j \bar{\bar{v}}_j, \qquad \text{i.e.} \quad \sum_{j=1}^{d} \sqrt{m_j}\, \bar{v}'_j = \sum_{j=1}^{d} \sqrt{m_j}\, \bar{\bar{v}}'_j.$$

This implies that the scalar product of the vector v' by the vector

$$e = (\underbrace{0, \ldots, 0}_{j}, \sqrt{m_k}, \sqrt{m_{k+1}}, 0, \ldots, 0)$$

is preserved. Moreover, the velocities of all the particles, except the kth and the $(k + 1)$st one, are preserved under reflection. This means that the scalar product of the vector v' by the vectors

$$e_j = (\underbrace{0, \ldots, 0}_{j-1}, 1, 0, \ldots, 0), j = 1, 2, \ldots, k - 1, k + 2, \ldots, d,$$

is preserved. Since the vectors e_j together with the vector e constitutes a basis in the hyperplane L, the preservation of the tangent component of the velocity vector has been proved.

Thus our mechanical system does indeed correspond to billiards in the simplex Q.

Using the results of §2, we can obtain a necessary condition for the ergodicity of the motion considered. Namely, if the group G_Q for the simplex Q (see the definition in §2) is finite, then the motion cannot possibly be ergodic. In the case of two mass points, this condition becomes especially simple: the finiteness of the group G_Q is equivalent to the commeasurability of the numbers $\arctan\sqrt{m_1/m_2}$ and π.

§5. Lorentz Gas and Systems of Hard Spheres

In connection with certain problems of nonequilibrium statistical mechanics, Lorentz introduced a dynamical system, which since then is known as a Lorentz gas. A complete description of this system will be given in §1, Chap. 10, while now we shall consider one of its simplest modifications.

Suppose U is a compact domain with a piecewise smooth boundary in the space $\mathbb{R}^d$, $d \geq 1$; $B_1, \ldots, B_r$ is a system of nonintersecting d-dimensional balls contained in U. Consider the domain $Q = U \setminus \bigcup_{i=1}^r B_i$. Billiards in the domain Q are said to be a Lorentz gas. In the applications, the B_i are viewed as motionless heavy ions, while the moving particle is the classical electron which bounces off the ions and off the boundary of the domain U. Actually, the domain U contains many electrons, but in the approximations corresponding to the Lorentz gas, their interactions may be neglected. Therefore, in the case of compact domains, we can consider the motion of a single electron.

Now consider another dynamical system which appears in physical problems—the system of absolutely elastic spheres.

Suppose once again U is a compact domain with piece-wise boundary in the space $\mathbb{R}^d$, $d \geq 1$. Assume that U contains r hard spheres of radius ρ and mass 1, moving uniformly and rectilinearly within U and bouncing off each other and off the boundary ∂U in accordance to the rule of elastic reflection. We will show that this dynamical system is a system of the billiards type in some domain $Q \subset \mathbb{R}^{d \cdot r}$.

The position of the ith sphere is well determined by the coordinates of its centre $q^{(i)}$, which are denoted by $q_j^{(i)}$, $1 \leq i \leq r$, $1 \leq j \leq d$. Suppose U^- is the subset of the set U consisting of all the points whose distance from the boundary ∂U is no less than ρ. Consider the Cartesian product

$$U^{(r)} = \underbrace{U^- \times U^- \times \cdots \times U^-}_{r \text{ factors}} \subset \mathbb{R}^{d \cdot r}$$

and delete from $U^{(r)}$ the interior of the $r(r-1)/2$ sets

$$C_{i_1,i_2} = \left\{q \in U^{(r)}: \sum_{j=1}^{d} (q_j^{(i_1)} - q_j^{(i_2)})^2 \le (2\rho)^2\right\}, \qquad 1 \le i_1, i_2 \le r, i_1 \ne i_2.$$

Denote the remaining set by Q. It is clear that Q is a domain in $\mathbb{R}^{d\cdot r}$ with piecewise smooth boundary (or the union of several such domains). To every configuration of r spheres of radius ρ in U, we can naturally assign a point $q \in Q$. Therefore the motion of the spheres described above generates a transformation group on the set Q. To prove that this group is billiards in Q, it suffices to show that the reflection of a moving point q from the boundary ∂Q occurs exactly as in the case of billiards. We shall consider only those reflections from the boundary which correspond to two spheres hitting each other (in the case when one sphere reflects from the boundary ∂U, the argument is even simpler).

Assume that r spheres during their motion in the domain U have reached such a position that the i_1st and the i_2nd spheres are tangent to each other, while the other spheres are not tangent among themselves or to the two given spheres. Construct the line segment $l_{i_1 i_2}$ joining the centres of the i_1st and i_2nd spheres, as well as the hyperplane $L_{i_1 i_2}$ orthogonal to this interval and passing through its centre. Decompose the velocities of the i_1st and the i_2nd spheres into two components, one of which (the tangent one) is parallel to L_{i_1, i_2} and the other (the normal one) is parallel to l_{i_1, i_2}. The rule of elastic reflection asserts that after our reflection the tangent components of the velocities of the i_1st and the i_2nd spheres are preserved, while the normal velocities of these spheres are interchanged. Obviously, at this moment the velocities of the other spheres do not change.

To the position of the spheres described above corresponds the point $q = \{q_j^{(i)}\} \in Q$, such that

$$\sum_{j=1}^{d} (q_j^{(i_1)} - q_j^{(i_2)})^2 = (2\rho)^2, \qquad \sum_{j=1}^{d} (q_j^{(k)} - q_j^{l})^2 > (2\rho)^2,$$

when $(k, l) \ne (i_1, i_2)$. The point q is a regular point of the boundary ∂Q. Introduce new coordinates $\bar{q}_j^{(i)}$ by putting $\bar{q}_j^{(i)} = q_j^{(i)}$, when $i \ne i_1, i_2$, and

$$\bar{q}_j^{(i_1)} = \tfrac{1}{2}(q_j^{(i_1)} + q_j^{(i_2)}), \qquad \bar{q}_j^{(i_2)} = \tfrac{1}{2}(q_j^{(i_1)} - q_j^{(i_2)}).$$

To these coordinates correspond the new velocities

$$\bar{v}_j^{(i)} = v_j^{(i)}, \quad \text{when } i \ne i_1, i_2,$$

$$\bar{v}_j^{(i_1)} = \tfrac{1}{2}(v_j^{(i_1)} + v_j^{(i_2)}), \qquad \bar{v}_j^{(i_2)} = \tfrac{1}{2}(v_j^{(i_1)} - v_j^{(i_2)}).$$

For any $d \cdot r$-dimensional vector z with coordinates $z_j^{(i)}$, $1 \le i \le r$, $1 \le j \le d$, let us denote by $z^{(i)}$ the d-dimensional vector whose coordinates are the coordinates of the vector z with superscript i, $i = 1, \ldots, r$. Suppose $n(q)$ is the unit normal vector to ∂Q at the point q and $(\overline{n(q)})_j^{(i)}$ are the new coordinates of this vector. Clearly $(\overline{n(q)})_j^{(i)} = 0$, when $i \neq i_2$, and

$$(\overline{n(q)})_j^{(i)} = \frac{\bar{q}_j^{(i_2)}}{\sqrt{\sum_j (\bar{q}_j^{(i_2)})^2}}, \qquad j = 1, \ldots, d.$$

Note that the d-dimensional vector $(\overline{n(q)})^{(i_2)}$ is parallel to the line segment l_{i_1, i_2} joining the centres of the i_1st and i_2nd balls.

Now let us see how the coordinates of the velocity vector $\bar{v} = (\bar{v})_j^{(i)}$ of the point q change at the moment of reflection. From the description of the change of the d-dimensional velocity vectors of the balls under elastic reflection given above, it immediately follows that:

(1) the vectors $(\bar{v})^{(i)}$ for $i \neq i_2$ are the same before and after the reflection;

(2) if the vector $(\bar{v})^{(i_2)}$ is represented in the form of the sum $(\bar{v})^{(i_2)} = (\bar{v})^{(i_2, 1)} + (\bar{v})^{(i_2, 2)}$, where $(\bar{v})^{(i_2, 1)}$ is orthogonal to the vector $(\overline{n(q)})^{(i_2)}$, while $(\bar{v})^{(i_2, 2)}$ is parallel to $(\overline{n(q)})^{(i_2)}$, then the component $(\bar{v})^{(i_2, 1)}$ does not change under reflection while $(\bar{v})^{(i_2, 1)}$ is replaced by

$$(\bar{v})^{(i_2, 2)} - 2((\overline{n(q)})^{(i_2)}, (\bar{v})^{(i_2)}) \cdot (\overline{n(q)})^{(i_2)}.$$

Since $((\overline{n(q)})^{(i_2)}, (\bar{v})^{(i_2)}) = (\overline{n(q)}, \bar{v})$, it follows from (1) and (2) that the velocity vector $\bar{v}$ changes to $\bar{v} - 2(\overline{n(q)}, \bar{v}) \cdot \overline{n(q)}$, i.e., the reflection rule is the same as in the case of billiards.

Chapter 7

Dynamical Systems in Number Theory

§1. Uniform Distribution

Many problems in number theory may be stated as problems relating to the uniform distribution of certain numerical sequences. Recall that the sequence $x_1, x_2, \ldots, 0 \le x_n \le 1$, is *uniformly distributed* on the closed interval $[0, 1]$, if for any function $f \in C([0, 1])$ we have the relation

$$\lim_{n \to \infty} \frac{1}{n} \sum_{k=1}^{n} f(x_k) = \int_0^1 f(x)\, dx.$$

Similarly, we can define the uniform distribution on an arbitrary closed interval $[a, b]$, $a < b$.

In the sequel we shall need the following:

Lemma 1. *The following statements are equivalent*:

(i) *the sequence* $\{x_n\}$ *is uniformly distributed on* $[0, 1]$;

(ii) *if* $\nu_n([b_1, b_2))$, *for all* b_1, b_2, $0 \le b_1 < b_2 \le 1$, *is the number of those* k, $1 \le k \le n$ *such that* $b_1 \le x_k < b_2$, *then* $\lim_{n \to \infty}(1/n)\nu_n([b_1, b_2)) = b_2 - b_1$;

(iii) $\lim_{n \to \infty}(1/n) \sum_{k=1}^{n} \exp(2\pi i s x_k) = 0$ *for any integer* $s \ne 0$.

Proof. The implication (i) $\Rightarrow$ (iii) is obvious. The implication (iii) $\Rightarrow$ (i) follows from the fact that trigonometric polynomials are dense in $C([0, 1])$. Let us now show that (i) $\Rightarrow$ (ii). Suppose $f_\varepsilon^+, f_\varepsilon^- \in C([0, 1])$ are of the form

$$f_\varepsilon^+(x) = \begin{cases} 1, & \text{if } b_1 \le x < b_2 \\ (1/\varepsilon)[x - (b_1 \quad \varepsilon)], & \text{if } \max(0, b_1 - \varepsilon) \le x < b_1, \\ -(1/\varepsilon)[x - (b_2 + \varepsilon)], & \text{if } b_2 \le x \le \min(b_2 + \varepsilon, 1), \\ 0 & \text{for all other } x, \end{cases}$$

$$f_\varepsilon^-(x) = \begin{cases} 1, & \text{if } b_1 + \varepsilon \le x \le b_2 - \varepsilon, \\ (1/\varepsilon)(x - b_1), & \text{if } b_1 \le x < b_1 + \varepsilon, \\ (1/\varepsilon)(b_2 - x), & \text{if } b_2 - \varepsilon \le x \le b_2, \\ 0 & \text{for all other } x. \end{cases}$$

Suppose further $\chi_{[b_1, b_2)}$ is the indicator of $[b_1, b_2)$. Then $f_\varepsilon^- \leq \chi_{[b_1, b_2)} \leq f_\varepsilon^+$ and for any $n \geq 1$, we have

$$\frac{1}{n}\sum_{k=1}^{n} f_\varepsilon^-(x_k) \leq \frac{\nu_n([b_1, b_2))}{n} = \frac{1}{n}\sum_{k=1}^{n} \chi_{[b_1, b_2)}(x_k) \leq \frac{1}{n}\sum_{k=1}^{n} f_\varepsilon^+(x_k).$$

When $n \to \infty$, we get:

$$b_2 - b_1 - 2\varepsilon \leq \int_0^1 f_\varepsilon^- \, dx \leq \varliminf_{n\to\infty} \frac{\nu_n([b_1, b_2))}{n}$$
$$\leq \varlimsup_{n\to\infty} \frac{\nu_n([b_1, b_2))}{n} \leq \int_0^1 f_\varepsilon^+ \, dx \leq b_2 - b_1 + 2\varepsilon.$$

Since ε was arbitrary,

$$\varliminf \frac{\nu_n([b_1, b_2))}{n} = \varlimsup \frac{\nu_n([b_1, b_2))}{n} = \lim_{n\to\infty} \frac{\nu_n([b_1, b_2))}{n} = b_2 - b_1.$$

The implication (ii) ⇒ (i) is now easily proved by using uniform approximations of arbitrary continuous functions by means of finite sums of indicators of semi-intervals. The lemma is proved. □

The notion of uniform distribution introduced above has a natural generalization. Namely, suppose M is a compact metric space, μ is a normalized Borel measure on M and $\{x_n\}$ is a sequence of elements of M, $n \geq 1$.

Definition 1. The sequence $\{x_n\}$ is *μ-uniformly distributed* on M, if for any $f \in C(M)$ we have

$$\lim_{n\to\infty} \frac{1}{n}\sum_{k=1}^{n} f(x_k) = \int_M f(x)\, d\mu.$$

In several cases it is possible to establish μ-uniform distribution by using ergodic theory. Suppose that T is a uniquely ergodic homeomorphism of the compact metric space M, $F: M \to \mathbb{R}^1$ is a continuous function on M, and the sequence of numbers $\{x_n\}$ may be represented in the form $x_n = F(T^n z_0)$ for some $z_0 \in M$. Denote by μ the unique normalized Borel invariant measure for T. By means of the function F, it induces a certain measure μ_F on $\mathbb{R}^1$: $\mu_F(A) = \mu(\{x \in M: F(x) \in A)\}$ for any Borel set $A \subset \mathbb{R}^1$. It is clear that μ_F is concentrated on the closed interval $[m', m'']$, where $m' = \min F$, $m'' = \max F$.

Lemma 2. *The sequence $\{x_n\}$ is μ_F-uniformly distributed on $[m', m'']$.*

Proof. Suppose g is a continuous function on $[m', m'']$. Then

$$\frac{1}{n}\sum_{k=1}^{n} g(x_k) = \frac{1}{n}\sum_{k=1}^{n} g(F(T^k z_0)).$$

But $f(z) = g(F(z))$ is a continuous function on M. By Theorem 2, §8, Chap. 1, the following limit

$$\lim_{n\to\infty} \frac{1}{n}\sum_{k=1}^{n} g(F(T^n z_0)) = \int_M g(F(z))\, d\mu = \int_{m'}^{m''} g(x)\, d\mu_F(x).$$

exists. Therefore

$$\lim_{n\to\infty} \frac{1}{n}\sum_{k=1}^{n} g(x_k) = \int_{m'}^{m''} g(x)\, d\mu_F.$$

The lemma is proved. □

§2. Uniform Distribution of Fractional Parts of Polynomials

In this section, the following theorem, due to Weyl [1], will be proved using ergodic theory.

Theorem 1. *Suppose $P(x) = a_0 x^r + a_1 x^{r-1} + \cdots + a_r$, $r \geq 1$, is a polynomial with real coefficients such that at least one of the coefficients a_s, $0 \leq s \leq r-1$, is irrational. Then the sequence $x_n = \{P(n)\}$, $n \geq 1$, is uniformly distributed.*[1]

Proof. This proof will be split up into several steps.

1. First suppose that a_0 is irrational. Consider the transformation T of the space $\mathbb{R}^r$ given by

$$T(x_1, \ldots, x_r) = (x_1 + \alpha, x_2 + p_{21}x_1, x_3 + p_{31}x_1 + p_{32}x_2, \ldots, x_r + p_{r1}x_1 + \cdots + p_{r,r-1}x_{r-1}), \quad (1)$$

where α is an irrational number and the p_{ij}, $1 \leq j < i \leq r$, are natural numbers. First let us deduce a useful formula for the iterations of this transformation. Suppose $P = \|p_{ij}\|$ is a square matrix of order r whose elements for $i > j$ appear in (1) and, for $i \leq j$, are equal to zero; denote by $p_{ij}^{(n)}$ the elements of the matrix P^n; $n = 0, 1, 2, \ldots$; $1 \leq i,j \leq r$. Clearly $p_{ij}^{(n)} = 0$ for $i - j \leq n - 1$.

[1] Here the figure brackets denote the fractional part of a number $\{x\} = x - [x]$, where $[x]$ is the integer part of x.

Also put $T^n(x_1, \ldots, x_r) = (x_1^{(n)}, \ldots, x_r^{(n)})$. Let us show that, for $n = 0, 1, 2, \ldots$, we have the equality

$$x_1^{(n)} = x_1 + n\alpha; \tag{2}$$

$$x_l^{(n)} = x_l + \sum_{i=1}^{l-1} x_i \sum_{q=1}^{l-i} \binom{n}{q} p_{li}^{(q)} + \alpha \sum_{q=1}^{l-1} \binom{n}{q+1} p_{l1}^{(q)}, \qquad 2 \le l \le r. \tag{3}$$

Here $\binom{n}{q}$ are binomial coefficients and, as usual, we assume that $\binom{n}{q} = 0$ for $q > n \ge 1$. The equality (2) is obvious, while (3) will be proved by induction.

For $n = 1$, (3) coincides with (1). Clearly for any $n \ge 1$ we can represent $x_l^{(n)}$ in the form

$$x_l^{(n)} = x_l + \sum_{l'=1}^{l-1} \lambda_{l,i}^{(n)} x_i + \lambda_l^{(n)} \alpha,$$

where the coefficients $\lambda_{l,i}^{(n)}, \lambda_l^{(n)}$ do not depend on x.

Now assume that (3) has been proved for some n; let us prove it for $(n + 1)$. From the induction hypothesis

$$\lambda_{l,i}^{(n)} = \sum_{q=1}^{l-i} \binom{n}{q} p_{li}^{(q)}, \qquad 1 \le i \le l-1, \tag{4}$$

$$\lambda_l^{(n)} = \sum_{q=1}^{l-1} \binom{n}{q+1} p_{l1}^{(q)}.$$

Put $\lambda_{l,i}^{(n)} = 0$ for $l > i$, $\lambda_{l,l}^{(n)} = 1$. (1) implies

$$x_l^{(n+1)} = x_l^{(n)} + \sum_{s=1}^{r-1} p_{rs} x_s^{(n)},$$

and therefore

$$\lambda_{l,i}^{(n+1)} = \lambda_{l,i}^{(n)} + \sum_{s=1}^{l-1} p_{r,s} \lambda_{s,i}^{(n)}. \tag{5}$$

Substituting the expressions for $\lambda_{l,i}^{(n)}, \lambda_{s,i}^{(n)}$ into (5), we get, in accordance with (4):

$$\begin{aligned}\lambda_{l,i}^{(n+1)} &= \sum_{q=1}^{l-1} \binom{n}{q} p_{li}^{(q)} + \sum_{s=1}^{l-1} p_{l,s} \sum_{q=1}^{s-i} \binom{n}{q} p_{si}^{(q)} + p_{li} \\ &= \sum_{q=1}^{l-i} \binom{n}{q} p_{li}^{(q)} + \sum_{s=1}^{l-1} p_{ls} \sum_{q=0}^{s-i} \binom{n}{q} p_{si}^{(q)}.\end{aligned}$$

Changing the order of summation in the second sum, we get

$$\lambda_{li}^{(n+1)} = \sum_{q=1}^{l-i} \binom{n}{q} p_{li}^{(q)} + \sum_{q=0}^{l-i-1} \binom{n}{q} \sum_{s=q+i}^{l-1} p_{ls} p_{si}^{(q)}.$$

Using $p_{si}^{(q)} = 0$ for $s < q + i$, we see that

$$\sum_{s=q+i}^{l-1} p_{ls} p_{si}^{(q)} = \sum_{s=1}^{r} p_{ls} p_{si}^{(q)} = p_{li}^{(q+1)},$$

and therefore

$$\begin{aligned}\lambda_{l,i}^{(n+1)} &= \sum_{q=1}^{l-i} \binom{n}{q} p_{li}^{(q)} + \sum_{q=0}^{l-i-1} \binom{n}{q} p_{li}^{(q+1)} \\ &= \sum_{q=1}^{l-i} \binom{n}{q} p_{li}^{(q)} + \sum_{q=1}^{l-i} \binom{n}{q-1} p_{li}^{(q)} = \sum_{q=1}^{l-i} \binom{n+1}{q} p_{li}^{(q)}.\end{aligned}$$

The last relation follows from the fact that

$$\binom{n}{q} + \binom{n}{q-1} = \binom{n+1}{q}.$$

In a similar way, it can be proved that

$$\lambda_{l}^{(n+1)} = \sum_{q=1}^{l-1} \binom{n+1}{q+1} p_{l1}^{(q)}.$$

Thus formula (3) is proved.

2. Now suppose M is the r-dimensional torus with cylindrical coordinates, i.e.,

$$M = \{x = (x_1, \ldots, x_r): 0 \le x_l < 1, 1 \le l \le r\}.$$

The same letter T will denote the transformation of M given by

$$\begin{aligned} Tx = (x_1 + \alpha)(\text{mod } 1), (x_2 + p_{21}x_1)(\text{mod } 1), \\ (x_3 + p_{31}x_1 + p_{32}x_2)(\text{mod } 1), \ldots, \\ (x_r + p_{r1}x_1 + \cdots + p_{r,r-1}x_{r-1})(\text{mod } 1)). \end{aligned} \tag{6}$$

Then the result of the previous step implies that $T^n x = (x_1^{(n)}, \ldots, x_r^{(n)})$, where the $x_l^{(n)}$ may be computed by means of formulas (2) and (3), except that the right-hand sides of these formulas must be taken mod 1.

Let us return to the polynomial $P(x) = a_0x^r + a_1x^{r-1} + \cdots + a_r$ for which a_0 is irrational. Let us prove that there exist natural numbers p_{ij}, $1 \le j < i \le r$, an irrational number α and a point $x = (x_1, \ldots, x_r) \in M$ such that for the transformation T defined by formula (6) we have $x_r^{(n)} = \{P(n)\}$.

Put $p_{ij} = 1$ for $1 \le j < i \le r$. The expression $\varphi_q(n) = \binom{n}{q}$, $0 \le q \le r$, viewed as a function of n, is a polynomial of degree q, and the functions $\varphi_q(n)$, $0 \le q \le r$, form a basis in the space of polynomials of degree no greater than r. For any n we can write

$$P(n) = b_0\binom{n}{r} + b_1\binom{n}{r-1} + \cdots + b_r\binom{n}{0},$$

where $b_0 = r!\, a_0$ and, since a_0 is irrational, so is b_0. On the other hand, by (3):

$$x_r^{(n)} = x_r + \sum_{i=1}^{r-1} x_i \sum_{q=1}^{r-i} \binom{n}{q} p_{ri}^{(q)} + \alpha \sum_{q=1}^{r-1} \binom{n}{q+1} p_{r1}^{(q)} (\mathrm{mod}\ 1).$$

Setting the coefficients of the expressions $\varphi_q(n) = \binom{n}{q}$ in the last two equations (for $P(n)$ and $x_r^{(n)}$) equal to each other, we obtain the following system for determining $\alpha, x_1, \ldots, x_r$:

$$\begin{cases} p_{r,1}^{(r-1)}\alpha = b_0, \\ p_{r,1}^{(r-1)}x_1 + \cdots = b_1, \\ p_{r,2}^{(r-2)}x_2 + \cdots = b_2, \\ \cdots\cdots\cdots\cdots\cdots \\ x_r + \cdots = b_r \end{cases}$$

Here the dots in the sth equation denote terms containing $x_1, x_2, \ldots, x_{s-2}$, α $(s = 2, \ldots, r)$.

The system above is triangular. It follows from the definition of p_{ij} that $p_{r,s}^{(r-s)} = 1$, and therefore the determinant of the system is equal to 1. Hence the unknowns $x_1, x_2, \ldots, x_r, \alpha$ may be found, and then

$$x_r^{(n)} = P(n)(\mathrm{mod}\ 1) = \{P(n)\}.$$

Now note that the transformation T is a compound skew translation on the torus and, for irrational α, by Theorem 1, §2, Chap. 4, T is uniquely ergodic. Therefore the statement of the uniform distribution of the sequence $\{x_n\}$ follows from the lemma below.

Lemma 1. *Suppose T is a uniquely ergodic homeomorphism of the r-dimensional torus* Tor^r *with invariant Lebesgue measure,* $\pi_i\colon \mathrm{Tor}^r \to S^1$ *are the natural projections*

$$\pi_i(x_1, \ldots, x_r) = x_i, \qquad 1 \leq i \leq r.$$

Then for any point $x = (x_1, \ldots, x_r) \in \mathrm{Tor}^r$ *and any* i, $1 \leq i \leq r$, *the sequence* $\{x^{(n)}\} \in [0, 1)$, *where* $x^{(n)} = \pi_i(T^{n-1}x)$, *is uniformly distributed.*

The proof of this lemma is an exact repetition of the proof of Lemma 2, §1, and will be omitted.

3. Consider the general case, when the polynomial $P(x) = a_0x^r + a_1x^{r-1} + \cdots + a_r$ has rational coefficients $a_0, a_1, \ldots, a_{s-1}$, while the coefficient a_s, $s < r$, is irrational. Put $P(x) = P_1(x) + P_2(x)$, where

$$P_1(x) = a_0x^r + a_1x^{r-1} + \cdots + a_{s-1}x^{r-s-1},$$
$$P_2(x) = a_sx^{r-s} + a_{s+1}x^{r-s-1} + \cdots + a_r$$

Present P_1 in the form

$$P_1(x) = \frac{1}{q}(m_0x^r + \cdots + m_{s-1}x^{r-s-1}),$$

where $m_0, \ldots, m_{s-1}, q$ are integers, $q \neq 0$. The polynomial $Q(x) = qP_1(x)$ has integer coefficients. Hence the value of $Q(n)(\mathrm{mod}\ q)$ is entirely determined by the values $n(\mathrm{mod}\ q)$, $n = 1, 2, \ldots$. Thus it follows that the fractional part of P_1 is constant on each residue class mod q. Put $\{P_s(n)\} = d_j$ for $n \in D_j$, where D_j is the jth residue class, $0 \leq j < q$. Further, for any j, $0 \leq j < q$, the polynomial $Q_j(x) = P_2(j + qx)$ has, as can be easily checked, an irrational coefficient in the term of highest degree. Therefore, by the previous step, the sequence $\{Q_j(n)\} = \{P_2(j + qn)\}$ is uniformly distributed for all j, $0 \leq j < q$.

Now it is easy to conclude the proof of the theorem. Suppose $f \in C(S^1)$. Extend this function periodically to the entire real axis. Then

$$\Sigma_N \stackrel{\text{def}}{=} \frac{1}{N}\sum_{k=0}^{N-1} f(\{P(k)\}) = \frac{1}{N}\sum_{k=0}^{N-1} f(P(k)) = \frac{1}{N}\sum_{j=0}^{q-1}\sum_{\substack{0 \leq k \leq N-1\\ k \in D_j}} f(P(k))$$

$$= \frac{1}{q}\sum_{j=0}^{q-1}\frac{q}{N}\sum_{\substack{0 \leq k \leq N-1\\ k \in D_j}} f(P_1(k) + P_2(k)) = \frac{1}{q}\sum_{j=0}^{q-1}\frac{q}{N}\sum_{\substack{0 \leq k \leq N-1\\ k \in D_j}} f(d_j + P_2(k)).$$

Since the sequences $P_2(k)$, $k \in D_j$ are uniformly distributed, we have, for any j, $0 \leq j \leq q - 1$:

$$\lim_{N \to \infty} \frac{q}{N}\sum_{\substack{0 \leq k \leq N-1\\ k \in D_j}} f(d_j + P_2(k)) = \int_0^1 f(x + d_j)\,dx = \int_0^1 f(x)\,dx.$$

Therefore

$$\lim_{N\to\infty} \Sigma_N = \int_0^1 f(x)\,dx.$$

The theorem is proved. □

§3. Uniform Distribution of Fractional Parts of Exponential Functions

In this section, we shall prove a statement similar to Theorem 1, §2, but relating to the exponential function $F(z) = u \cdot \lambda^z$, $u \in \mathbb{R}^1$, instead of polynomials. Fix $\lambda > 1$ and, for any $u \in \mathbb{R}^1$, consider the sequence $x_n = x_n(u) = u\lambda^n$, $n = 1, 2, \ldots$.

Theorem 1. *For almost all u (with respect to the Lebesgue measure) the sequence $\{x_n(u)\}$ is uniformly distributed.*

The proof is based on the following lemma, which is concerned with an arbitrary sequence $\{x_n\}$, $0 \le x_n < 1$.

Lemma 1. *Assume that for any integer $s \neq 0$ there is a sequence of numbers $n_1^{(s)} < \cdots < n_j^{(s)} < \cdots$ such that*

$$\lim_{j\to\infty} \frac{n_{j+1}^{(s)}}{n_j^{(s)}} = 1, \tag{1}$$

$$\lim_{j\to\infty} V_{n_j}^{(s)} = 0, \tag{2}$$

where

$$V_n^{(s)} = V_n^{(s)}(\{x_n\}) = \frac{1}{n}\sum_{k=1}^{n} \exp(2\pi i s x_k).$$

Then the sequence $\{x_n\}$ is uniformly distributed.

Proof. It suffices to check condition (iii) of Lemma 1, §1, i.e., to prove that $V_n^{(s)} \to 0$, when $n \to \infty$, for all $s \neq 0$. Choose an $s \neq 0$ and agree to omit the superscript s in $V_n^{(s)}$, $n_j^{(s)}$. Suppose n satisfies $n_j < n < n_{j+1}$. Then

$$\delta_n \stackrel{\text{def}}{=} \left|\frac{n}{n_j} V_n - V_{n_j}\right| = \left|\frac{1}{n_j}\sum_{k=1}^{n} \exp(2\pi i s x_k) - \frac{1}{n}\sum_{k=1}^{n_j} \exp(2\pi i s x_k)\right|$$

$$= \frac{1}{n_j}\left|\sum_{k=n_j+1}^{n} \exp(2\pi i s x_k)\right| \le \frac{n - n_j}{n_j} \le \frac{n_{j+1}}{n_j} - 1.$$

It follows from (1) that $\delta_n \to 0$, so that by (2), $V_n \to 0$ when $n \to \infty$. The lemma is proved. □

Proof of Theorem 1. Once again choose an $s \neq 0$ and introduce, following Lemma 1, the notation

$$V_n = V_n(u) = \frac{1}{n}\sum_{k=1}^{n} \exp[2\pi i s x_k(u)] = \frac{1}{n}\sum_{k=1}^{n} \exp(2\pi i s \lambda^k u).$$

Then, for all a, b, $-\infty < a < b < \infty$, we have

$$\begin{aligned} I_n = \int_a^b |V_n(u)|^2\, du &= \frac{1}{n^2}\sum_{k,l=1}^{n} \int_a \exp[2\pi i s(\lambda^k - \lambda^l)u]\, du \\ &= \frac{b-a}{n} + \frac{2}{n^2}\sum_{k=2}^{n}\sum_{l=1}^{k-1} \frac{\exp\{[2\pi i s(\lambda^k - \lambda^l)]b\} - \exp\{[2\pi i s(\lambda^k - \lambda^l)]a\}}{2\pi i s(\lambda^k - \lambda^l)} \end{aligned}$$

Hence

$$|I_n| \leq \frac{b-a}{n} + \frac{2}{\pi s n^2}\sum_{k=1}^{n}\sum_{l=1}^{k-1} \frac{1}{\lambda^k - \lambda^l}. \tag{3}$$

But

$$\begin{aligned} \lambda^k - \lambda^l &= (\lambda^k - \lambda^{k-1}) + (\lambda^{k-1} - \lambda^{k-2}) + \cdots + (\lambda^{l+1} - \lambda^l) \\ &= \lambda^{k-1}(\lambda - 1) + \cdots + \lambda^l(\lambda - 1) \geq (k - l)(\lambda - 1). \end{aligned} \tag{4}$$

It follows from (3) and (4) that

$$\sum_{l=1}^{k-1} \frac{1}{\lambda^k - \lambda^l} \leq \frac{1}{\lambda - 1}\sum_{l=1}^{k-1} \frac{1}{k-l} \leq \text{const}\cdot\log k \leq \text{const}\cdot\log n.$$

Thus $I_n \leq (b-a)/n + \text{const}\cdot\log n/n$. Therefore for $n_j = j^2$ the series $\sum_{j=1}^{\infty} I_{n_j}$ converges. Hence $\sum_{j=1}^{\infty} |V_{n_j}(u)|^2 < \infty$ for almost all $u \in [a, b]$, and therefore $V_{n_j}(u) \to 0$ for almost all $u \in [a, b]$. Since $n_{j+1}/n_j \to 1$, Lemma 1 implies that the sequence $\{x_n(u)\}$ is uniformly distributed for almost all $u \in [a, b]$. Since the closed interval $[a, b]$ was arbitrary, the theorem is proved. □

§4. Ergodic Properties of Decompositions into Continuous Fractions and Piecewise-monotonic Maps

1. In various questions of number theory, analysis and probability theory, the decomposition of real numbers into continuous fractions is often used. In this section the methods of ergodic theory are applied to study the metric properties of such fractions.

We begin with an exposition of the necessary information concerning continued fractions (see Khinchin [2]). Any real number $x \in (0, 1)$ can be uniquely presented in the form

$$x = \cfrac{1}{a_1 + \cfrac{1}{a_2 + \cdots}}, \tag{1}$$

where $a_1, a_2, \ldots$ are natural numbers. The expression (1) is said to be a *continued fraction* and is denoted by $[a_1, a_2, \ldots]$, (usually, somewhat more more general fractions, of the form $a_0 + [a_1, a_2, \ldots]$ (where a_0 is an integer), are considered, but we shall not need them).

The decomposition (1) is finite, if x is rational and infinite if x is irrational. In the last case $[a_1, a_2, \ldots] \stackrel{\text{def}}{=} \lim_{n\to\infty}[a_1, a_2, \ldots]$ and the limit always exists. A finite continued fraction can be uniquely written in the form

$$[a_1, a_2, \ldots, a_n] = \frac{p}{q},$$

where p, q are natural numbers relatively prime to each other. The fractions

$$\frac{p_1}{q_1} = [a_1], \frac{p_2}{q_2} = [a_1, a_2], \ldots, \frac{p_k}{q_k} = [a_1, \ldots, a_k],$$

for $k \le n$ are called convergents for the continued fraction $[a_1, a_2, \ldots, a_n]$, $1 \le n \le \infty$. For the numerators and denominators of the convergents, the following recurrent relations hold

$$\begin{aligned} p_k &= a_k p_{k-1} + p_{k-2} \\ q_k &= a_k q_{k-1} + q_{k-2}, \qquad k = 1, 2, \ldots \end{aligned} \tag{2}$$

Here we put $p_0 = 0, p_{-1} = 1, q_0 = 1, q_{-1} = 0$. It follows by induction from (2) that

$$p_k \ge 2^{(k-2)/2}, \qquad q_k \ge 2^{(k-1)/2} \tag{3}$$

The following estimate for the deviation in the approximation of the number $x = [a_1, a_2, \ldots] \in (0, 1)$ by the convergents:

$$\left| x - \frac{p_k}{q_k} \right| \le \frac{1}{q_k q_{k+1}} \tag{4}$$

holds.

2. Just as the decomposition of real numbers into a q-adic fractions (where $q > 1$ is an integer) is related to the transformation T_q which acts according to the formula $T_q x = \{qx\} = qx(\text{mod } 1)$ for any $x \in (0, 1)$, there is a transformation T similarly related to continued fractions. It is said to be the *Gauss transformation* and is given by the formula $Tx = \{1/x\}, x \in (0, 1)$. Let us consider in more detail the relationship between this transformation and the decomposition into a continued fractions.

Let us agree to use the notation $[a_1, a_2, \ldots]$ for expressions of the form (1), where $a_i \neq 0$ are real (not necessarily natural) numbers.

Suppose $x \in (0, 1)$. Then $1/x > 1$,

$$\frac{1}{x} = \left[\frac{1}{x}\right] + \left\{\frac{1}{x}\right\} = \left[\frac{1}{x}\right] + Tx.$$

Therefore

$$x = [a_1 + Tx], \quad \text{where } a_1 = \left[\frac{1}{x}\right] \tag{5}$$

If $Tx = 0$, then our decomposition is finished. If $Tx \neq 0$, then

$$\frac{1}{Tx} = \left[\frac{1}{Tx}\right] + \left\{\frac{1}{Tx}\right\} = \left[\frac{1}{Tx}\right] + T^2x \tag{6}$$

It follows from (5) and (6) that

$$x = [a_1, a_2 + T^2x], \quad \text{where } a_k = \left[\frac{1}{T^k x}\right], \quad k = 1, 2.$$

This process may be continued indefinitely, as long as $Tx, T^2x, \ldots \neq 0$. After the nth step, we get

$$x = [a_1, a_2, \ldots, a_{n-1}, a_n + T^n x],$$

where

$$a_k = a_k(x) = \left[\frac{1}{T^k x}\right], \qquad k = 1, 2, \ldots, n.$$

It follows that

$$a_k(Tx) = a_{k+1}(x), \tag{7}$$

i.e., in the expression of the numbers $x \in (0, 1)$ in the form of continued fractions (with natural elements) the Gauss transformation acts as the shift $T([a_1, a_2, \ldots]) = [a_2, a_3, \ldots]$.

3. The Gauss transformation is a particular case of the so-called piecewise monotonic transformations of the interval. This class includes many other examples, important for ergodic theory; therefore we shall now study the properties of general piecewise monotonic transformations, and the results for Gauss transformations (and, therefore, for continued fractions) will then be obtained as corollaries.

Thus, assume that the space M is the interval (0, 1) with Lebesgue measure ρ; let $\mathfrak{S}$ be the σ-algebra of Borel sets. The function $\varphi(x)$, $x \in (0, 1)$, is said to be *piecewise monotonic*, if the interval (0, 1) may be split up into a finite or countable number of intervals $\Delta_1, \Delta_2, \ldots$ such that on each Δ_i the function φ is strictly monotonic (on certain Δ_i it may be increasing, on others—decreasing). We shall assume that the function φ is not defined at the end points of the Δ_i.

If $0 < \varphi(x) < 1$, then φ defines a measurable transformation T which acts according to the formula $Tx = \varphi(x)$. The transformation T and all of its degrees T^n, $n \geq 0$, are defined everywhere on M, except at a countable number of points. We then have $T^n x = \varphi^{(n)}(x)$, where

$$\varphi^{(n)} = \underbrace{\varphi \circ \varphi \circ \cdots \circ \varphi}_{n}.$$

Such transformations T are called *piecewise monotonic*. Everywhere where the converse is not explicitly stated, we shall assume that T is a transformation of class C^2, i.e., the function which defines φ is twice continuously differentiable on every interval Δ_i. We also introduce the following supplementary conditions:

(i) for any $n = 0, 1, 2, \ldots$, the interval (0, 1) may be split up into a finite or countable number of intervals

$$\Delta^{(n)}_{i_1, i_2, \ldots, i_n} (1 \leq i_1, \ldots, i_n < \infty)$$

such that

(1) $\bigcup_{i_n} \Delta^{(n)}_{i_1, i_2, \ldots, i_{n-1}, i_n} = \Delta^{(n-1)}_{i_1, i_2, \ldots, i_{n-1}} \qquad (n \geq 1)$;

(2) $T(\Delta^{(n)}_{i_1, i_2, \ldots, i_n}) \subset \Delta^{(n-1)}_{i_2, \ldots, i_n} \qquad (n \geq 1)$;

(3) $\Delta^0_1 = (0, 1); \qquad \Delta^{(1)}_i = \Delta_i$

It is easy to check that (i) always holds in the following two cases:

(ii) the number of intervals Δ_i is finite;

(iii) $T(\Delta_i) = (0, 1), \qquad i = 1, 2, \ldots.$

Theorem 1. *Suppose condition* (i) *holds and, moreover:*

(1) *there exists a natural number s such that*

$$\inf_{\Delta_i} \inf_{x \in \Delta_i} \left| \frac{d\varphi^{(s)}}{dx} \right| = \lambda > 1; \tag{8}$$

(2)
$$\sup_{\Delta_i} \sup_{x_1, x_2 \in \Delta_i} \frac{|(d^2\varphi/dx^2)(x_1)|}{[(d\varphi/dx)(x_2)]^2} = C < \infty. \tag{9}$$

Then for the transformation T there exists an invariant normalized Borel measure μ, absolutely continuous with respect to the Lebesque measure ρ, and such that $d\mu/d\rho \le K < \infty$.

The proof is based on the following two lemmas.

Lemma 1. *Suppose T is an arbitrary piecewise monotonic transformation of the interval* (0, 1). *If there exists a constant $K > 0$ such that $\rho(T^{-n}A) \le K \cdot \rho(A)$, $n = 1, 2, \ldots$ for any Borel set $A \subset (0, 1)$, then for the transformation T there is an invariant normalized Borel measure μ, absolutely continuous with respect to ρ, and satisfying $d\mu/d\rho \le K$.*

Lemma 2. *If the transformation T satisfies the conditions of Theorem* 1, *then there exists a constant $K > 0$ such that for all $n = 0, 1, \ldots$ we have the inequality*

$$M(\varphi) \stackrel{\text{def}}{=} \sup_{\Delta^{(n)}_{i_1, \ldots, i_n}} \sup_{x, y \in \Delta^{(n)}_{i_1, \ldots, i_n}} \left| \frac{(d\varphi^{(n)}/dx)(x)}{(d\varphi^{(n)}/dx)(y)} \right| \le K.$$

The proofs of the lemmas will be given below.

Proof of Theorem 1. By Lemma 1, it suffices to show that, for any set $A \in \mathfrak{S}$, we have

$$\rho(T^{-n}A) \le K\rho(A), \qquad n = 1, 2, \ldots.$$

Denote $A_{i_1, \ldots, i_n} = T^{-n}A \cap \Delta^{(n)}_{i_1, \ldots, i_n}$. Since $T^n A_{i_1, \ldots, i_n} \subset A$, we have

$$\begin{aligned} \rho(A) \ge \rho(T^n A_{i_1, \ldots, i_n}) &= \int_{A_{i_1, \ldots, i_n}} \left| \frac{d\varphi^{(n)}}{dx} \right| dx \\ &\ge \rho(A_{i_1, \ldots, i_n}) \inf_{x \in A_{i_1, \ldots, i_n}} \left| \frac{d\varphi^{(n)}}{dx} \right|. \end{aligned} \tag{10}$$

Further

$$\int_{\Delta^{(n)}_{i_1,\ldots,i_n}} \left|\frac{d\varphi^{(n)}}{dx}\right| dx = \rho(T^n \Delta^{(n)}_{i_1,\ldots,i_n}) \leq 1,$$

Therefore we can find a point $\xi \in \Delta^{(n)}_{i_1,\ldots,i_n}$ such that

$$\left|\frac{d\varphi^{(n)}}{dx}(\xi)\right| \geq \frac{1}{\rho(\Delta^{(n)}_{i_1,\ldots,i_n})}.$$

For any $x \in A_{i_1,\ldots,i_n}$, by Lemma 2, we now get

$$\begin{aligned}\left|\frac{d\varphi^{(n)}}{dx}(x)\right| &\geq \left|\frac{d\varphi^{(n)}}{dx}(\xi)\right| \cdot \inf_{x\in\Delta^{(n)}_{i_1,\ldots,i_n}} \left|\frac{(d\varphi^{(n)}/dx)(x)}{(d\varphi^{(n)}/dx)(\xi)}\right| \\ &= \left|\frac{d\varphi^{(n)}}{dx}(\xi)\right| \cdot \left(\sup_{x\in\Delta^{(n)}_{i_1,\ldots,i_n}} \left|\frac{(d\varphi^{(n)}/dx)(\xi)}{(d\varphi^{(n)}/dx)(x)}\right|\right)^{-1} \geq \frac{1}{K\cdot\rho(\Delta^{(n)}_{i_1,\ldots,i_n})}.\end{aligned}$$

From (10) we can now deduce

$$\rho(A_{i_1,\ldots,i_n}) \leq K\cdot\rho(A)\cdot\rho(\Delta^{(n)}_{i_1,\ldots,i_n}),$$

and finally

$$\rho(T^{-n}A) = \sum_{i_1,\ldots,i_n} \rho(A_{i_1,\ldots,i_n}) \leq K\cdot\rho(A)\cdot \sum_{i_1,\ldots,i_n} \rho(\Delta^{(n)}_{i_1,\ldots,i_n}) = K\cdot\rho(A)$$

The theorem is proved. □

Proof of Lemma 1. Define the measures μ_n, $n = 1, 2, \ldots$ by the equality

$$\mu_n(A) = \frac{1}{n}\sum_{k=0}^{n-1} \rho(T^{-k}A),$$

where A is a Borel subset of (0, 1). Then $d\mu_n/d\rho \leq K$ for all n. From the sequence of measures $\{\mu_n\}$, we can choose a subsequence $\{\mu_{n_s}\}$ weakly converging to some normalized measure μ. Clearly μ is absolutely continuous with respect to ρ and $d\mu/d\rho \leq K$. Let us show that the measure μ is invariant with respect to T. It suffices to prove the equality $\mu(A) = \mu(T^{-1}A)$ for any set A which is the union of a finite number of intervals. It follows from the weak convergence of the measures that, for such A, we have

$$\mu(A) = \lim_{s\to\infty} \mu_{n_s}(A), \qquad \mu(T^{-1}A) = \lim_{s\to\infty} \mu_{n_s}(T^{-1}A).$$

But

$$|\mu_{n_s}(A) - \mu_{n_s}(T^{-1}A)| = \frac{1}{n_s}\left|\sum_{k=0}^{n_s-1} \rho(T^k A) - \sum_{k=0}^{n_s-1} \rho(T^{k-1}A)\right| \le \frac{2}{n_s}.$$

When $s \to \infty$, we see that $\mu(A) = \mu(T^{-1}A)$. The lemma is proved. □

Remark. If, in the conditions of Lemma 1, we replace the inequality $\rho(T^{-n}A) \le K \cdot \rho(A)$ by the inequality $(1/K)\rho(A) \le \rho(T^{-n}A) \le K \cdot \rho(A)$, the invariant measure μ constructed will be equivalent to the Lebesgue measure, and its density $d\mu/d\rho$ will satisfy the inequality $1/K \le d\mu/d\rho \le K$.

Proof of Lemma 2. Suppose the points x, y belong to $\Delta^{(n)}_{i_1, \ldots, i_n}$. According to the chain rule,

$$\frac{d\varphi^{(n)}}{dx}(x) = \prod_{k=0}^{n-1} \frac{d\varphi}{dx}(T^k x); \qquad \frac{d\varphi^{(n)}}{dx}(y) = \prod_{k=0}^{n-1} \frac{d\varphi}{dx}(T^k y).$$

Therefore

$$\begin{aligned}\left|\frac{(d\varphi^{(n)}/dx)(x)}{(d\varphi^{(n)}/dx)(y)}\right| &= \prod_{k=0}^{n-1}\left|\frac{(d\varphi/dx)(T^k x)}{(d\varphi/dx)(T^k y)}\right| \\ &= \exp\left[\sum_{k=0}^{n-1} \log\left|1 + \frac{(d\varphi/dx)(T^k x) - (d\varphi/dx)(T^k y)}{(d\varphi/dx)(T^k y)}\right|\right] \\ &\le \exp\left[\sum_{k=0}^{n-1}\left|\frac{(d\varphi/dx)(T^k x) - (d\varphi/dx)(T^k y)}{(d\varphi/dx)(T^k y)}\right|\right]. \qquad (11)\end{aligned}$$

Since the points $T^k x$, $T^k y$, $0 \le k \le n-1$ are located in the same interval Δ_i, Lagrange's theorem implies

$$\left|\frac{d\varphi}{dx}(T^k x) - \frac{d\varphi}{dx}(T^k y)\right| = \left|\frac{d^2\varphi}{dx^2}(\xi_k)\right| \cdot |T^k x - T^k y|, \qquad \xi_k \in \Delta_i. \quad (12)$$

We can now apply the Lagrange theorem once more in the following way

$$\begin{aligned}|T^{k+1}x - T^{k+1}y| &= |\varphi(T^k x) - \varphi(T^k y)| \\ &= \left|\frac{d\varphi}{dx}(\zeta_k)\right| \cdot |T^k x - T^k y|, \qquad \zeta_k \in \Delta_i,\end{aligned}$$

i.e.,

$$|T^k x - T^k y| = \frac{|T^{k+1}x - T^{k+1}y|}{|(d\varphi/dx)(\zeta_k)|}. \qquad (13)$$

Comparing (12) and (13), we get

$$\left|\frac{d\varphi}{dx}(T^k x) - \frac{d\varphi}{dx}(T^k y)\right| = \frac{|(d^2\varphi/dx^2)(\xi_k)|}{|(d\varphi/dx)(\zeta_k)|} \cdot |T^{k+1}x - T^{k+1}y|.$$

Now substitute this expression into (11):

$$\left|\frac{(d\varphi^{(n)}/dx)(x)}{(d\varphi^{(n)}/dx)(y)}\right| \leq \exp\left[\sum_{k=0}^{n-1} \frac{|(d^2\varphi/dx^2)(\xi_k)|}{|(d\varphi/dx)(T^k y)| \cdot |(d\varphi/dx)(\zeta_k)|} \cdot |T^{k+1}x - T^{k+1}y|\right]$$

$$\leq \exp(C) \cdot \exp\left(\sum_{k=1}^{n} |T^k x - T^k y|\right), \qquad C < \infty. \tag{14}$$

The last inequality follows from (9).

The right-hand side of (14) may be estimated by means of (8):

$$|T^k x - T^k y| \leq \frac{1}{\lambda}|T^{k+s}x - T^{k+s}y| \leq \frac{1}{\lambda^2}|T^{k+2s}x - T^{k+2s}y|$$

$$\leq \cdots \leq \frac{1}{\lambda^{[(n-k)/s]}} \cdot |T^{k+[(n-k)/s]s}x - T^{k+[(n-k)/s]s}y| \leq \frac{1}{\lambda^{[(n-k)/s]}}. \tag{15}$$

Therefore

$$\sum_{k=1}^{n} |T^k x - T^k y| = \sum_{j=0}^{s-1} \sum_{\substack{k=j(\text{mod } s)\\ 1\leq k\leq n}} |T^k x - T^k y|$$

$$\leq \sum_{j=0}^{s-1}\left(1 + \frac{1}{\lambda} + \frac{1}{\lambda^2} + \cdots\right) = \frac{s\lambda}{\lambda - 1}. \tag{16}$$

Put $K = \exp[C + s\lambda/(\lambda - 1)]$; then it follows from (14) and (16) that $M(\varphi) \leq K$. The lemma is proved. □

Theorem 2. *Assume that the transformation T is the same as in Theorem* 1, *but the condition* (i) *is replaced by the stronger condition* (iii); *let μ be the measure invariant with respect to T obtained by Theorem* 1. *Then*

(a) *the measure μ is equivalent to the Lebesgue measure ρ; moreover, there exists a constant $K > 0$ such that $1/K \leq d\mu/d\rho \leq K$;*
(b) *the endomorphism T with invariant measure μ is mixing.*

The proof of this theorem will be given in Chap. 10: in §8 of that chapter a stronger statement will be proved (Theorem 4).

4. Now let us return to the study of continued fractions and show, first of all, that the Gauss transformation relating to them satisfies the conditions of Theorems 1 and 2.

Thus, assume that the function $\varphi(x)$, $x \in (0, 1)$ is given by the formula $\varphi(x) = Tx = \{1/x\}$. On each of the intervals of the form $\Delta_{a_1} = (1/(a_1 + 1), 1/a_1)$, $a_1 = 1, 2, \ldots$, the function $\varphi(x)$ is monotonic decreasing and yields a one-to-one map of Δ_{a_1} onto $(0, 1)$. Moreover, $\varphi \in C^2$ for $x \in \Delta_{a_1}$, $a_1 = 1, 2, \ldots$.

The transformation T^2 is given by the function $\varphi^{(2)} = \varphi \circ \varphi$ and is defined everywhere on $(0, 1)$, except at the end points of the intervals Δ_{a_1}. Further, each interval Δ_{a_1} may be subdivided into a countable number of smaller intervals

$$\Delta_{a_1, a_2} = \left(\frac{1}{a_1 + \dfrac{1}{a_2}}, \frac{1}{a_1 + \dfrac{1}{a_2 + 1}} \right)$$
$$= ([a_1, a_2], [a_1, a_2 + 1]), \qquad a_2 = 1, 2, \ldots.$$

Any interval Δ_{a_1, a_2} is mapped bijectively onto Δ_{a_2} by T, and therefore T^2 is a one-to-one map of Δ_{a_1, a_2} onto $(0, 1)$. For the transformation T^3, one similarly defines the intervals

$$\Delta_{a_1, a_2, a_3} = ([a_1, a_2, a_3 + 1], [a_1, a_2, a_3]),$$

etc. Thus condition (iii) of Theorem 2 (and therefore condition (i) of Theorem 1) holds.

Now consider the function $\varphi^{(2)} = \varphi \circ \varphi$ determining the transformation T^2, in order to show that $|d\varphi^{(2)}/dx| \geq \lambda > 1$, everywhere where $d\varphi^{(2)}/dx$ is defined. Indeed, $|d\varphi/dx| = 1/x^2 \geq 1$ for $x \in (0, 1)$ and $|d\varphi/dx| \geq 9/4$ if $x \leq 2/3$. But if $2/3 < x < 1$, then $0 < Tx < 1/2$, so that for any $x \in (0, 1)$ we have

$$\left| \frac{d\varphi^{(2)}}{dx}(x) \right| = \left| \frac{d\varphi}{dx}(x) \right| \cdot \left| \frac{d\varphi}{dx}(Tx) \right| \geq \frac{9}{4}.$$

Now let us compute the number

$$C = \sup_{\Delta_{a_1}} \sup_{x, y \in \Delta_{a_1}} \frac{|\varphi''(x)|}{|\varphi'(y)|^2}.$$

Since $|\varphi''(x)| = 2/x^3$, $|\varphi'(x)| = 1/x^2$, on the interval $(1/(n + 1), 1/n)$, we have

$$\frac{|\varphi''(x)|}{|\varphi'(y)|^2} \leq \frac{|\varphi''(1/(n + 1))|}{|\varphi'(1/n)|^2} = \frac{2(n + 1)^3}{n^4}$$

Therefore, $C = \sup_n[2(n + 1)^3/n^4] = 16$.

Thus the transformation T satisfies all the requirements of Theorem 2, so that we have the following:

Theorem 3. *For the Gauss transformation T there exists an invariant measure μ equivalent to the Lebesgue measure, while the endomorphism T with measure μ is mixing.*

5. It turns out—and this was already known to Gauss—that an explicit form of the normalized invariant measure for the transformation T may be indicated:

$$\mu(A) = \frac{1}{\log 2}\int_A \frac{dx}{1+x}, \qquad A \in \mathfrak{S}.$$

To prove the invariance of μ, it suffices to check that $\mu(T^{-1}\Delta) = \mu(\Delta)$ for any interval $\Delta = (0, \alpha) \subset (0, 1)$, This may be done by means of the following computation

$$\begin{aligned}
\mu(T^{-1}(0, \alpha)) &= \mu\left(\bigcup_{k=1}^{\infty}\left(\frac{1}{k+\alpha}, \frac{1}{k}\right)\right) = \sum_{k=1}^{\infty}\mu\left(\left(\frac{1}{k+\alpha}, \frac{1}{k}\right)\right)\\
&= \frac{1}{\log 2}\sum_{k=1}^{\infty}\int_{1/(k+\alpha)}^{1/k}\frac{dx}{1+x}\\
&= \frac{1}{\log 2}\sum_{k=1}^{\infty}\left[\log\left(1+\frac{1}{k}\right) - \log\left(1+\frac{1}{k+\alpha}\right)\right]\\
&= \frac{1}{\log 2}\sum_{k=1}^{\infty}\left[\log\left(1+\frac{\alpha}{k}\right) - \log\left(1+\frac{\alpha}{k+1}\right)\right]\\
&= \frac{1}{\log 2}\sum_{k=1}^{\infty}\int_{\alpha/(k+1)}^{\alpha/k}\frac{dx}{1+x}\\
&= \frac{1}{\log 2}\int_0^{\alpha}\frac{dx}{1+x} = \mu((0, \alpha)).
\end{aligned}$$

Since, by Theorem 3, the endomorphism T with invariant measure μ is ergodic, we can write out the Birkhoff–Khinchin ergodic theorem for T in the following form—*for any function $f(x) \in L^1(0, 1)$*

$$\lim_{n\to\infty}\frac{1}{n}\sum_{k=0}^{n-1} f(T^k x) = \frac{1}{\log 2}\int_0^1 \frac{f(x)}{1+x}\,dx \tag{17}$$

almost everywhere with respect to the Lebesgue measure (or with respect to μ, which is the same thing, since they are equivalent).

Applying formula (17) to various functions f, it is easy to obtain many important corollaries of the decomposition into continued fractions.

Theorem 4. *For almost all* $x = [a_1, a_2, \ldots] \in (0, 1)$ *we have the following relations*:

(1) $\lim\limits_{n\to\infty} \frac{1}{n}(a_1 + \cdots + a_n) = \infty$;

(2) $\lim\limits_{n\to\infty} \sqrt[n]{a_1 \cdot \ldots \cdot a_n} = \prod\limits_{k=1}^{\infty} \left(1 + \frac{1}{k^2 + 2k}\right)^{\log k/\log 2}$;

(3) $\lim\limits_{n\to\infty} \frac{\log q_n}{n} = \frac{\pi^2}{12 \log 2}$;

where q_n *is the denominator of the nth convergent for* x.

Proof. (1) Take $f(x) = a_1(x)$, i.e. $f(x) = k$ for $x \in (1/(k+1), 1/k)$, $k = 1, 2, \ldots$ Since, by (7), we have $a_k(x) = a_1(T^{k-1}x)$, it follows that

$$\frac{1}{n}\sum_{k=1}^{n} a_k(x) = \frac{1}{n}\sum_{k=0}^{n-1} f(T^k x).$$

However $f \notin L^1(0, 1)$, so that we cannot directly apply the Birkhoff–Khinchin theorem. Introduce the truncated functions

$$f_N(x) = \begin{cases} f(x), & \text{if } f(x) \le N, \\ 0 & \text{if } f(x) > N, \ N = 1, 2, \ldots \end{cases}$$

By (17), for any N, we have

$$\begin{aligned} \varliminf_{n\to\infty} \frac{1}{n}\sum_{k=0}^{n-1} f(T^k x) &\ge \varliminf_{n\to\infty} \frac{1}{n}\sum_{k=0}^{n-1} f_N(T^k x) \\ &= \lim_{n\to\infty} \frac{1}{n}\sum_{k=0}^{n-1} f_N(T^k x) = \frac{1}{\log 2}\int_0^1 \frac{f_N(x)}{1 + x}\,dx, \end{aligned}$$

almost everywhere.

Now statement (1) follows from the fact that $\int_0^1 f_N(x)\,dx \to \infty$ when $N \to \infty$.

(2) Take $f(x) = \log a_1(x)$, i.e., $f(x) = \log k$ for $x \in (1/(k+1), 1/k)$, $k = 1, 2, \ldots$. By (17) we have

$$\begin{aligned} \frac{1}{n}\sum_{k=1}^{n} \log a_k(x) - \frac{1}{n}\sum_{k=0}^{n-1} f(T^k x) &\xrightarrow[n\to\infty]{} \frac{1}{\log 2}\int_0^1 \frac{f(x)}{1 + x}\,dx \\ &= \frac{1}{\log 2}\sum_{k=1}^{\infty}\int_{1/(k+1)}^{1/k} \frac{\log k}{1 + x}\,dx \\ &= \sum_{k=1}^{\infty} \frac{\log k}{\log 2} \cdot \log\left(1 + \frac{1}{k^2 + 2k}\right), \end{aligned}$$

almost everywhere. This implies statement (2).

(3) Now let us derive an auxiliary formula. Suppose $p_n(x)$, $q_n(x)$ are the numerator and denominator of the nth convergent for $x = [a_1, a_2, \ldots] \in (0, 1)$. Then

$$\frac{p_n(x)}{q_n(x)} = [a_1, \ldots, a_n] = \frac{1}{1 + [a_2, \ldots, a_n]} = \frac{1}{1 + [p_{n-1}(Tx)/q_{n-1}(Tx)]}$$
$$= \frac{q_{n-1}(Tx)}{p_{n-1}(Tx) + q_{n-1}(Tx)}. \tag{18}$$

The fractions in the left and right-hand sides of (18) cannot be simplified, therefore, in particular, $p_n(x) = q_{n-1}(Tx)$. This implies

$$\frac{p_n(x)}{q_n(x)} \cdot \frac{p_{n-1}(Tx)}{q_{n-1}(Tx)} \cdot \ldots \cdot \frac{p_1(T^{n-1}x)}{q_1(T^{n-1}x)} = \frac{1}{q_n(x)},$$

i.e.,

$$-\frac{1}{n} \log q_n(x) = \frac{1}{n} \sum_{k=0}^{n-1} \log\left[\frac{p_{n-k}}{q_{n-k}}(T^k x)\right].$$

Take $f(x) = \log x$. Then

$$-\frac{1}{n} \log q_n(x) = \frac{1}{n} \sum_{k=0}^{n-1} f(T^k x) + \frac{1}{n} \sum_{k=0}^{n-1} \left[\log(T^k x) - \log\left(\frac{p_{n-k}}{q_{n-k}}(T^k x)\right)\right]$$
$$\stackrel{\text{def}}{=} \frac{1}{n} \Sigma_n' + \frac{1}{n} \Sigma_n''.$$

By (17) we get

$$\lim_{n \to \infty} \frac{1}{n} \Sigma_n' = \frac{1}{\log 2} \int_0^1 \frac{\log x}{1 + x}\, dx,$$

almost everywhere. Integrating by parts:

$$\frac{1}{\log 2} \int_0^1 \frac{\log x}{1 + x}\, dx = -\frac{1}{\log 2} \int_0^1 \frac{\log(1 + x)}{x}\, dx$$
$$= -\frac{1}{\log 2} \sum_{k=0}^{\infty} (-1)^k \int_0^1 \frac{x^k}{k + 1}\, dx$$
$$= -\frac{1}{\log 2} \sum_{k=0}^{\infty} \frac{(-1)^k}{(k + 1)^2} = -\frac{\pi^2}{12 \log 2}.$$

In order to prove statement (3), it now suffices to show that $|\Sigma_n''| \leq \text{const.}$ It follows from (3) and (4) that for any $x \in (0, 1)$ we have

$$\left|\frac{x}{p_k/q_k} - 1\right| = \frac{q_k}{p_k}\left|x - \frac{p_k}{q_k}\right| \leq \frac{1}{p_k q_{k+1}} \leq \frac{1}{2^{k-1}};$$

therefore

$$|\Sigma_n''| \leq \sum_{k=0}^{n-1}\left|\log\frac{T^k x}{(p_{n-k}/q_{n-k})(T^k x)}\right| \leq \sum_{k=0}^{n-1}\left|\frac{T^k x}{(p_{n-k}/q_{n-k})(T^k x)} - 1\right|$$
$$\leq \sum_{k=0}^{n-1}\frac{1}{2^{n-k-1}} \leq \text{const.}$$

The theorem is proved. □

Chapter 8

Dynamical Systems in Probability Theory

§1. Stationary Random Processes and Dynamical Systems

Suppose M is the set of all sequences, infinite in both directions $x = (\ldots, y_{-1}, y_0, y_1, \ldots)$, whose coordinates y_i are points of a fixed measurable space $(Y, \mathfrak{A})$. M possesses a natural σ-algebra $\tilde{\mathfrak{S}}$ generated by cylindrical sets, i.e., sets of the form

$$A = \{x = (\ldots, y_{-1}, y_0, y_1, \ldots) \in M: y_{i_1} \in C_1, \ldots, y_{i_r} \in C_r\}, \tag{1}$$

where $1 \leq r < \infty$, $i_1, \ldots, i_r$ are integers and $C_1, \ldots, C_r \in \mathfrak{A}$. Suppose μ is a normalized measure on $\tilde{\mathfrak{S}}$ and $\mathfrak{S}$ is the completion of $\tilde{\mathfrak{S}}$ with respect to the measure μ. In probability theory the triple $(M, \mathfrak{S}, \mu)$ is said to be a *discrete time random process* and the space $(Y, \mathfrak{A})$ is the *state space* of this process.

An important class of processes is the class of *stationary* random processes. The stationarity condition requires that, for any set A of the form (1), the measure $\mu(\{x \in M: y_{i_1+n} \in C_1, \ldots, y_{i_r+n} \in C_r\})$ does not depend on n, $-\infty < n < \infty$.

Let us express this condition in another way. Suppose T is the shift on M, i.e., $Tx = x'$, where

$$x = (\ldots, y_{-1}, y_0, y_1, \ldots), \qquad x' = (\ldots, y'_{-1}, y'_0, y'_1, \ldots)$$

and

$$y'_i = y_{i+1}, \qquad -\infty < i < \infty.$$

Then if A is a set of the form (1), then $T^{-n}A = \{x \in M: y_{i_1+n} \in C_1, \ldots, y_{i_r+n} \in C_r\}$, and the stationarity condition may be written in the form $\mu(T^{-n}A) = \mu(A)$, $-\infty < n < \infty$. Since the measure μ is uniquely determined by its values on cylindrical sets, stationarity means that the shift transformation T preserves the measure μ, i.e., T is an automorphism of the space $(M, \mathfrak{S}, \mu)$.

Now let us show that an arbitrary automorphism T' of the measure space $(M', \mathfrak{S}', \mu')$ naturally gives rise to stationary random processes.

Consider some partition ξ of the space M', i.e., a family of nonintersecting measurable subsets whose union covers M'. We shall assume that ξ is finite or countable, i.e., that

$$\xi = (C_1, \dots, C_m), 1 \le m \le \infty, \qquad C_k \in \mathfrak{S} \quad \text{for } 1 \le k \le m.$$

For the state space choose the set $Y = \{1, 2, \dots, m\}$ and put $M = \prod_{n=-\infty}^{\infty} Y^{(n)}$, $Y^{(n)} = Y$. Consider the map $\varphi: M' \to M$ defined as follows: the nth coordinate of the point $\varphi x'$ equals k if and only if $(T')^n x' \in C_k$, $1 \le k \le m$. The map φ is measurable. It transforms the measure μ' on M' into a certain measure μ on M: $\mu(A) \overset{\text{def}}{=} \mu'(\varphi^{-1}A)$, $A \in \mathfrak{S}$, while the transformation T' is the shift on M. From the fact that T' is an automorphism it follows that the random process $(M, \mathfrak{S}, \mu)$ is stationary. Thus every finite or countable partition of a space provided with an automorphism generates a stationary random process. One of the fundamental problems of ergodic theory is the problem of describing the class of stationary processes which correspond to a given automorphism.

Definition 1. The partition $\xi = (C_1, \dots, C_m)$ is said to be *generating* for the automorphism T' if the map φ constructed above determines an isomorphism (mod 0) of the spaces M' and M.

The last statement means that we can eliminate a subset N', $\mu'(N') = 0$ from the space M' so that any point $x' \in M' \setminus N'$ is well defined by the inclusions $T^n x' \in C_{i_n}$, $-\infty < n < \infty$.

If ξ is a generating partition, then the corresponding shift automorphism T is metrically isomorphic to the automorphism T'.

EXAMPLES

1. *Bernoulli automorphisms.* Suppose the phase space $(M, \mathfrak{S}, \mu)$ is a Cartesian product

$$(M, \mathfrak{S}, \mu) = \prod_{n=-\infty}^{\infty} (Y^{(n)}, \mathfrak{A}^{(n)}, \sigma^{(n)}),$$

where $(Y^{(n)}, \mathfrak{A}^{(n)}, \sigma^{(n)}) = (Y, \mathfrak{A}, \sigma)$ is a measure space. The measure μ is the product-measure generated by the measure σ, i.e.,

$$\mu = \bigotimes_{n=-\infty}^{\infty} \sigma. \tag{2}$$

Definition 2. The shift automorphism T with invariant measure μ defined by formula (2) is said to be the *Bernoulli automorphism* with state space $(Y, \mathfrak{A}, \sigma)$. The invariant measures μ corresponding to Bernoulli automorphisms are said to be *Bernoulli measures*.

Theorem 1. *Any Bernoulli automorphism is ergodic and mixing.*

Proof. Suppose $A \in \mathfrak{S}$ is a set invariant with respect to T. For any $\varepsilon > 0$ we can find a positive integer r and a finite number of cylindrical sets $A_i^{(r)}$,

$$A_i^{(r)} = \{x \in M: y_{-r} \in C_{-r}^{(i)}, \ldots, y_r \in C_r^{(i)}\},$$

such that $\mu(A \triangle A^{(r)}) < \varepsilon$ where $A^{(r)} = \bigcup_i A_i^{(r)}$. By the invariance of A for all n we have

$$\mu(A) = \mu(T^{-n}A \cap A) = \mu(T^{-n}A^{(r)} \cap A^{(r)}) + \delta_n,$$

where $|\delta_n| < 2\varepsilon$. Choose an n such that $|n| \geq 2r + 1$. Then

$$\begin{aligned} &T^{-n}A^{(r)} \cap A^{(r)} \\ &\quad = \bigcup_{i,j} \{x \in M: y_{-r} \in C_{-r}^{(i)}, \ldots, y_r \in C_r^{(i)}, y_{-r+n} \in C_{-r}^{(j)}, \ldots, y_{r+n} \in C_r^{(j)}\}, \end{aligned}$$

and all the coordinates $y_{-r}, \ldots, y_r, y_{-r+n}, \ldots, y_{r+n}$ differ. Since μ is the product-measure, we have

$$\mu(T^{-n}A^{(r)} \cap A^{(r)}) = \mu(T^{-n}A^{(r)}) \cdot \mu(A^{(r)}) = [\mu(A^{(r)})]^2 = [\mu(A)]^2 + \delta,$$

where $|\delta| < \varepsilon$. Thus we finally obtain:

$$|\mu(A) - [\mu(A)]^2| < 4\varepsilon.$$

Since ε was arbitrary, $\mu(A)$ equals 0 or 1. Thus the first statement of the theorem is proved.

Let us prove the second one. Choose two sets $A_1, A_2 \in \mathfrak{S}$. For an arbitrary $\varepsilon > 0$ find a positive integer r and sets $A_1^{(r)}$, $A_2^{(r)}$ belonging to the σ-algebra $\bigotimes_{k=-r}^{r} \mathfrak{A}^{(k)}$ such that

$$\mu(A_1 \triangle A_1^{(r)}) < \varepsilon, \qquad \mu(A_2 \triangle A_2^{(r)}) < \varepsilon. \tag{3}$$

Since μ is the product measure, we have for $|n| \geq 2r + 1$

$$\mu(T^{-n}A_1^{(r)} \cap A_2^{(r)}) = \mu(T^{-n}A_1^{(r)}) \cdot \mu(A_2^{(r)}) = \mu(A_1^{(r)}) \cdot \mu(A_2^{(r)}). \tag{4}$$

It follows from (3) and (4) that, for the values of n indicated above, we have

$$\begin{aligned} |\mu(T^{-n}A_1 \cap A_2) - \mu(A_1) \cdot \mu(A_2)| &\leq |\mu(T^{-n}A_1 \cap A_2) - \mu(T^{-n}A_1^{(r)} \cap A_2)| \\ &\quad + |\mu(T^{-n}A_1^{(r)} \cap A_2) - \mu(T^{-n}A_1^{(r)} \cap A_2^{(r)})| \\ &\quad + |\mu(T^{-n}A_1^{(r)} \cap A_2^{(r)}) - \mu(A_1^{(r)}) \cdot \mu(A_2^{(r)})| \\ &\quad + |\mu(A_1^{(r)}) \cdot \mu(A_2^{(r)}) - \mu(A_1) \cdot \mu(A_2)| < 4\varepsilon. \end{aligned}$$

Since ε was arbitrary, the theorem is proved. □

Bernoulli automorphisms are dynamical systems with the strongest mixing properties. These automorphisms play an important role in the entropy theory of dynamical systems (see Part II).

In many questions it turns out to be useful to have a "more abstract" and "more invariant" definition of Bernoulli automorphisms which we shall give now.

Definition 3. Let T be an automorphism of the measure space $(M, \mathfrak{S}, \mu)$. A σ-algebra $\mathfrak{A} \subset \mathfrak{S}$ is called *Bernoulli generating σ-algebra* if σ-algebras $T^n\mathfrak{A}$, $-\infty < n < \infty$, are mutually independent and the least σ-algebra containing all of them coincides with $\mathfrak{S}$. The automorphisms having Bernoulli generating σ-algebras are called *Bernoulli automorphisms*.

It is easy to see that any Bernoulli automorphism defined in this way is metrically isomorphic to some Bernoulli automorphism in the sense of Definition 2.

Meshalkin's Example of Metric Isomorphism of two Bernoulli Automorphisms with Different State Spaces.

Let us consider two Bernoulli automorphisms T_1, T_2 with state spaces (Y_1, σ_1), (Y_2, σ_2), where $Y_1 = \{a_1, a_2, a_3, a_4\}$, $\sigma_1(\{a_i\}) = \frac{1}{4}$, $1 \le i \le 4$ and $Y_2 = \{b_0, b_1, b_2, b_3, b_4, b_5\}$, $\sigma_2(\{b_0\}) = \frac{1}{2}$, $\sigma_2(\{b_j\}) = \frac{1}{8}$, $1 \le j \le 4$. We shall denote the corresponding measure spaces where T_1 and T_2 act by M_1, M_2. The mapping φ: $M_1 \to M_2$ is defined in the following way. Let $x^{(1)} = \{\ldots a_{i_{-1}}, a_{i_0}, a_{i_1}, \ldots\} \in M_1$. For $\varphi(x^{(1)}) = x^{(2)} = \{\ldots b_{j_{-1}}, b_{j_0}, b_{j_1}, \ldots\} \in M_2$ we take the sequence for which $j_n = 0$ iff $i_n = 1$ or 2. If n is such that $i_n = 3$ (4) then we consider the largest $n_1 < n$ for which the cardinality of k, $n_1 \le k \le n$, where $i_k = 1$ or 2 is equal to the cardinality of k, $n_1 \le k \le n$, where $i_k = 3$ or 4. It is easy to see that $i_{n_1} = 1$ or 2. We put $j_n = 1$ (3) if $i_{n_1} = 1$ and $j_n = 2$ (4) if $i_{n_1} = 2$.

It follows from the well-known facts of probability theory that φ is defined almost everywhere. Also it is easy to see from the construction that φ commutes with the shift. We shall omit the proof of the assertion that φ transforms the measure μ_1 in M_1 into the measure μ_2 in M_2.

2. *Markov automorphisms.* Suppose that, as in the previous example, the points x of the space M are of the form $x = (\ldots, y_{-1}, y_0, y_1, \ldots)$, where $y_i \in Y$ and $(Y, \mathfrak{A})$ is a measurable space. A *stochastic operator* on the space $(Y, \mathfrak{A})$ is a function $P(y, C)$ of the variables $y \in Y$, $C \in \mathfrak{A}$ with the following properties:

(1) $P(y, C)$ for any fixed $y \in Y$ is a normalized measure on $(Y, \mathfrak{A})$;
(2) $P(y, C)$ for any fixed $C \in \mathfrak{A}$ is a measurable function on Y.

The normalized measure σ on $(Y, \mathfrak{A})$ is said to be an *invariant measure for the stochastic operator* P if for any $C \in \mathfrak{A}$ we have

$$\sigma(C) = \int_Y P(y, C)\, d\sigma(y).$$

Given a stochastic operator P and an invariant normalized measure σ, we can define a measure μ in the space M in the following way. First for the cylindrical sets

$$A = \{x \in M: y_i \in C_0, y_{i+1} \in C_1, \ldots, y_{i+r} \in C_r\},$$

where $-\infty < i < \infty$, $r \geq 0$, $C_0, \ldots, C_r \in \mathfrak{A}$, put

$$\mu(A) = \int_{C_0} d\sigma(y_i) \int_{C_1} P(y_i, dy_{i+1}) \ldots \int_{C_r} P(y_{i+r-1}, dy_{i+r}). \tag{5}$$

Then, using the Kolmogorov theorem, extend the measure μ to the entire σ-algebra $\mathfrak{S}$. The invariance of σ implies that the measure μ is stationary.

Definition 4. A shift automorphism T with invariant measure μ defined by formula (5) is said to be *Markov automorphism*. The invariant measures μ corresponding to Markov automorphisms are called *Markov measures*.

The ergodic properties of Markov automorphisms may differ for various σ and P. For example, it is easy to construct examples of nonergodic Markov automorphisms. Let us consider in more detail the case when the state space Y is a finite set: $Y = \{e_1, \ldots, e_m\}$. In this case any stochastic operator P is well defined by matrix Π with elements $\pi_{ij} = P(e_i, \{e_j\})$, $1 \leq i, j \leq m$. Here $\{e_j\}$ denotes the one point set consisting of the element e_j. The stochasticity of P is equivalent to the stochasticity of the matrix Π: $\pi_{ij} \geq 0$, $\sum_j \pi_{ij} = 1$. The invariant measure σ may be identified with an m-dimensional vector $\pi = (\pi_1, \ldots, \pi_m)$, $\pi_k = \sigma(\{e_k\})$, such that $\pi_k = \sum_{i=1}^m \pi_i \pi_{ik}$, $1 \leq k \leq m$.

Assume that the matrix Π is such that for some $n_0 \geq 1$ the matrix Π^{n_0} consists of strictly positive elements. The ergodic theorem for Markov chains claims that in this case the invariant measure π is unique and if the $\pi_{ij}^{(n)}$ are the matrix elements of the matrix Π^n, then $\pi_k = \lim_{n\to\infty} \pi_{ik}^{(n)}$, $k = 1, \ldots, m$ (this limit does not depend on i).

Theorem 2. *Under the above conditions, a Markov automorphism is ergodic and mixing.*

Proof. Let us immediately establish that the Markov automorphism T is mixing. Arguing as in the proof of the mixing condition for Bernoulli automorphisms, we see that the argument reduces to proving the mixing property for two cylindrical sets. Each cylindrical set is of the form

$$A = \{x = (\ldots, y_{-1}, y_0, y_1, \ldots) \in M: y_{-r} = e_{i_{-r}}, \ldots, y_0 = e_{i_0}, \ldots, y_r = e_{i_r}\},$$

$e_{i_{-r}}, \ldots, e_{i_r} \in Y$. Therefore it suffices to prove the relation

$$\lim_{|n| \to \infty} \mu(T^{-n}A_1 \cap A_2) = \mu(A_1) \cdot \mu(A_2), \tag{6}$$

for sets A_1, A_2 of the form

$$A_1 = \{x \in M: y_{-r} = e_{i_{-r}}, \ldots, y_r = e_{i_r}\},$$
$$A_2 = \{x \in M: y_{-r} = e_{j_{-r}}, \ldots, y_r = e_{j_r}\}.$$

It follows from the definition of the measure μ that

$$\begin{aligned} \mu(A_1) &= \pi_{i_{-r}} \cdot \pi_{i_{-r}, i_{-r+1}} \cdot \ldots \cdot \pi_{i_{r-1}, i_r}, \\ \mu(A_2) &= \pi_{j_{-r}} \cdot \pi_{j_{-r}, j_{-r+1}} \cdot \ldots \cdot \pi_{j_{r-1}, j_r}. \end{aligned} \tag{7}$$

If n satisfies $|n| \geq 2r + 1$, then

$$T^{-n}A_1 \cap A_2 = \{x \in M: y_{-r+n} = e_{i_{-r}}, y_{r+n} = e_{i_r}, y_{-r} = e_{j_{-r}}, \ldots, y_r = e_{j_r}\}.$$

Therefore

$$\begin{aligned} \mu(T^{-n}A_1 \cap A_2) = \pi_{i_{-r}} \cdot \pi_{i_{-r}, i_{-r+1}} \cdot \ldots \cdot \pi_{i_{r-1}, i_r} \\ \cdot \pi_{i_r, j_{-r}}^{(-n-2r)} \cdot \pi_{j_{-r}, j_{-r+1}} \cdot \ldots \cdot \pi_{j_{r-1}, j_r}. \end{aligned} \tag{8}$$

Since under our assumptions $\lim_{|n| \to \infty} \pi_{i_r, j_{-r}}^{(-n-2r)} = \pi_{j_{-r}}$, (6) follows from (7) and (8). The theorem is proved. □

It is clear that any Bernoulli automorphism is also a Markov automorphism (in this case $P(y, C) = \sigma(C)$, where σ is the measure on the state space). Now let us show that under the assumptions of Theorem 2 a Markov automorphism may be represented as an integral automorphism over a Bernoulli automorphism. In the theory of Markov chains, such a representation is the foundation of the so-called "Doeblin method."

Suppose $Y = \{e_1, \ldots, e_m\}$ is the state space of the Markov automorphism T. Choose an element $e = e_l \in Y$ and put

$$E = \{x = (\ldots, y_{-1}, y_0, y_1, \ldots) \in M: y_0 = e\}.$$

Under our assumptions $\mu(E) > 0$. Denote by T_E the induced automorphism constructed from the automorphism T and the set E. The action of this automorphism may be described as follows. Take a point $x = (\ldots, y_{-1}, y_0, y_1, \ldots) \in E$ and mark off all the places where the coordinate e appears: $\ldots, i_{-2}, i_{-1}, 0, i_1, i_2, \ldots$ (This sequence will be infinite in both directions for almost all $x \in E$). Then $T_E x = T^{i_1}x$ and generally $T_E^n x = T^{i_n}x$,

$-\infty < n < \infty$. Let us show that T_E is metrically isomorphic to some Bernoulli automorphism with a countable state space Y_E. For Y_E take the set consisting of all sequences of the form $z = \{e, e_{i_1}, \ldots, e_{i_r}\}$ such that $1 \le r < \infty$, $e_{i_s} \ne e$ for $1 \le s \le r$, and also of the trivial sequence $\{e\}$. Any point $x = \{y_n\}$ of the phase space of the automorphism T_E can be uniquely written in the form of a sequence of elements $\{z_n\}$, $z_n \in Y_E$, infinite in both directions, while the automorphism T_E acts as a shift on the set of all such sequences. The map φ sending $\{y_n\}$ into $\{z_n\}$ is one-to-one, and we shall show that it provides the desired metric isomorphism.

Define the measure σ_E on the state space Y_E by the relation

$$\sigma_E(\{e, e_{i_1}, \ldots, e_{i_r}\}) = \pi_{l, i_1} \cdot \pi_{i_1, i_2} \cdot \ldots \cdot \pi_{i_{r-1}, i_r} \cdot \pi_{i_r, l}.$$

It suffices to prove that the invariant measure μ_E of the induced automorphism T_E is transformed by the map φ into the product measure generated by the measure σ_E.

For any sequence $z = \{e, e_{i_1}, \ldots, e_{i_r}\}$ put

$$A_z = \{x = (\ldots, y_{-1}, y_0, y_1, \ldots) \in M : y_0 = e, y_1 = e_{i_1}, \ldots, y_r = e_{i_r}, y_{r+1} = e\}.$$

It follows from the definition of Markov measure that $\mu_E(A_z) = \sigma_E(z)$. Choosing a finite number of sequences $z_k = \{e, e_{i_1^{(k)}}, \ldots, e_{i_{r_k}^{(k)}}\}$, $1 \le k \le n < \infty$, we obtain

$$\mu_E(A_{z_1} \cap T_E^{-1} A_{z_2} \cap \cdots \cap T_E^{-(n-1)} A_{z_n})$$

$$= \frac{1}{\pi_l} \cdot \pi_l \cdot \pi_{l, i_1^{(1)}} \cdot \pi_{i_1^{(1)}, i_2^{(1)}} \cdot \ldots \cdot \pi_{i_r^{(1)}-1, i_r^{(1)}} \cdot \pi_{i_r^{(1)}, l} \cdot \ldots \cdot \frac{1}{\pi_l} \cdot \pi_l \cdot \pi_{l, i_1^{(n)}} \cdot \pi_{i_1^{(n)}, i_2^{(n)}}$$

$$\cdot \ldots \cdot \pi_{i_{r_n}^{(n)}-1, i_{r_n}^{(n)}} \cdot \pi_{i_{r_n}^{(1)}, l} = \prod_{k=1}^{n} \mu_E(A_{z_k}).$$

This relation shows that the measure μ_E, viewed as a measure on the set of infinite sequences from Y_E, is Bernoulli. Therefore the Markov automorphism T is metrically isomorphic to an integral automorphism over a Bernoulli automorphism.

Now let us see how Markov automorphisms appear in the study of dynamical systems which have nothing to do with probability theory. Suppose the phase space $(M', \mathfrak{S}', \mu')$ is the two-dimensional torus Tor^2 with normalized Haar measure, T' is its group automorphism

$$T'(x_1, x_2) = (ax_1 + bx_2 \,(\mathrm{mod}\ 1), cx_1 + dx_2 \,(\mathrm{mod}\ 1), \quad \text{where } A = \begin{Vmatrix} a & b \\ c & d \end{Vmatrix}$$

is an integer matrix and $\det A = 1$. Assume that the matrix A has two eigenvalues λ_1, λ_2 satisfying $\lambda_1 > 1$, $\lambda_2 = \lambda_1^{-1} < 1$ and denote the corresponding eigen-vectors by e_1, e_2. At the beginning of this section, we showed how an arbitrary finite or countable partition ξ of the space M' generates a map φ sending T' into a stationary random process or, in other words, into a shift automorphism on the space $(M, \mathfrak{S}, \mu)$. We shall construct a finite partition of the torus such that the corresponding shift automorphism T will turn out to be a Markov automorphism with a finite number of states.

Consider the plane $\mathbb{R}^2$ with coordinates x_1, x_2 and the natural covering map $\pi\colon \mathbb{R}^2 \to \mathrm{Tor}^2$. We shall also consider closed parallelograms (in the plane $\mathbb{R}^2$) mapped onto Tor^2 bijectively, whose sides are parallel to e_1, e_2. The images of these parallelograms under π will also be called parallelograms. The sides of parallelograms parallel to e_1 will be called unstable (dilating) and the sides parallel to e_2—stable (contracting). If C is a parallelogram, then $T'C, (T')^{-1}C$ are also parallelograms. The intersection of two parallelograms is also a parallelogram.

For each parallelogram C, denote by $\Gamma^{(u)}(C)$ (respectively $\Gamma^{(s)}(C)$) the part of the boundary consisting of unstable (respectively stable) edges. By a partition of M' into parallelograms, we mean a finite family of parallelograms $C_1, \ldots, C_r$ such that

$$\bigcup_{i=1}^{r} C_i = M', \qquad \mathrm{Int}\, C_i \cap \mathrm{Int}\, C_j = \varnothing \quad \text{for } i \neq j.$$

For the $\xi = (C_1, \ldots, C_r)$ put

$$\Gamma^{(u)}(\xi) = \bigcup_{i=1}^{r} \Gamma^{(u)}(C_i), \qquad \Gamma^{(s)}(\xi) = \bigcup_{i=1}^{r} \Gamma^{(s)}(C_i).$$

Clearly $\Gamma^{(u)}(\xi)(\Gamma^{(c)}(\xi))$ is a finite set of intervals parallel to $e_1(e_2)$.

Definition 5. ξ is said to be a *Markov partition* if $T^{-1}\Gamma^{(u)}(\xi) \subset \Gamma^{(u)}(\xi)$, $T\Gamma^{(s)}(\xi) \subset \Gamma^{(s)}(\xi)$.

The meaning of this definition will be clarified by Lemma 1 proved below. Note that in our case the map $\varphi\colon M' \to M$ defined at the beginning of this section acts as follows: the nth coordinate of the point $\varphi(x_1, x_2)$ equals k if $T^n(x_1, x_2) \in \mathrm{Int}\, C_k$ (this map is defined almost everywhere on M').

Lemma 1. *If $\xi = (C_1, \ldots, C_r)$ is a Markov partition, then the map φ sends the measure μ' into a Markov measure μ.*

Proof. Choose any finite sequence of integers $(i_1, \ldots, i_n)$, $1 \leq n < \infty$, $1 \leq i_k \leq r$ for $k = 1, 2, \ldots, n$ and consider the corresponding sequence of parallelograms $C_{i_1}, \ldots, C_{i_n}$. From the Markov property of the partition ξ, it follows that the intersection

$$C_{i_1} \cap T'C_{i_2} \cap \cdots \cap (T')^{n-1}C_{i_n}$$

is still a parallelogram or finite number of parallelograms contained within C_{i_1} and each of the stable sides of these parallelograms is contained in a stable side of the parallelogram C_{i_1}. Choose an arbitrary parallelogram $C_{i_0} \in \xi$. Under $(T')^{-1}$ it is dilated in the direction of e_2, therefore the intersection $(T')^{-1}C_{i_0} \cap C_{i_1}$ is one or several parallelograms whose unstable (dilated) sides lie on unstable sides of the parallelogram C_{i_1}.

This means that the conditional measure $\mu((T')^{-1}C_{i_0}|C_{i_1})$ is the quotient of lengths of the unstable sides of the parallelograms $(T')^{-1}C_{i_0} \cap C_{i_1}$ and C_{i_1}. But it follows from the above that the conditional measure

$$\mu((T')^{-1}C_{i_0}|C_{i_1} \cap T'C_{i_2} \cap \cdots \cap (T')^{n-1}C_{i_n})$$
$$= \frac{\mu((T')^{-1}C_{i_0} \cap C_{i_1} \cap T'C_{i_2} \cap \cdots \cap (T')^{n-1}C_{i_n})}{\mu(C_{i_1} \cap T'C_{i_2} \cap \cdots \cap (T')^{n-1}C_{i_n})},$$

equals the quotient of the same sides, i.e., equals $\mu((T')^{-1}C_{i_0}|C_{i_1})$. The lemma is proved. □

Now we shall construct, for a given T', a Markov partition ξ into two parallelograms for which $\Gamma^{(u)}(\xi)$ and $\Gamma^{(s)}(\xi)$ will both be closed intervals directed along e_1, e_2 respectively and containing the point $O = (0, 0)$.

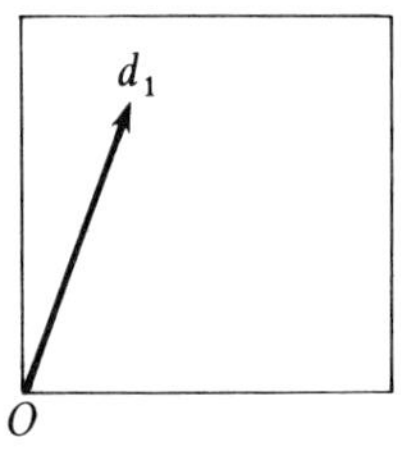

Figure 2

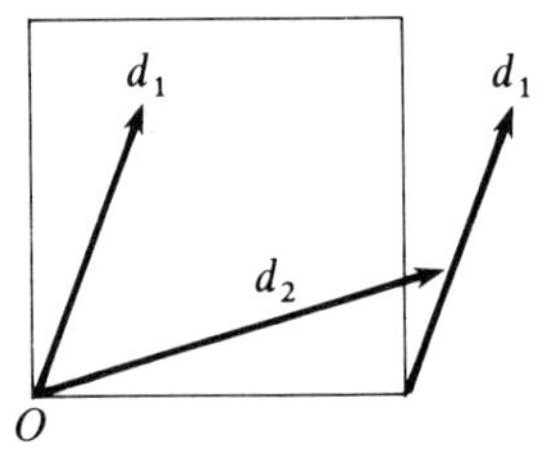

Figure 3

Choose an arbitrary segment d_1 with origin at O in the direction of e_1 (see Fig. 2). Starting from the point O in the direction of e_2, construct the line segment limited by the first intersection with d_1 (Fig. 3). Denote the part of the segment d_1 beginning at the point O and ending at the first intersection with d_2 by d_3 (Fig. 4). Further, construct the segment d_4 beginning from the point O in the direction opposite to that of d_3 until its intersection with d_2, and put $d_5 = d_3 \cup d_4$. Finally, extend d_2 in the first direction until its intersection with d_5, and denote by d_6 the segment thus obtained (Fig. 5). It is easy to see that M' is partitioned into two parallelograms C_1, C_2 and, for the partition $\xi = (C_1, C_2)$, we have $\Gamma^{(u)}(\xi) = d_4$, $\Gamma^{(s)}(\xi) = d_2$. Figure 6

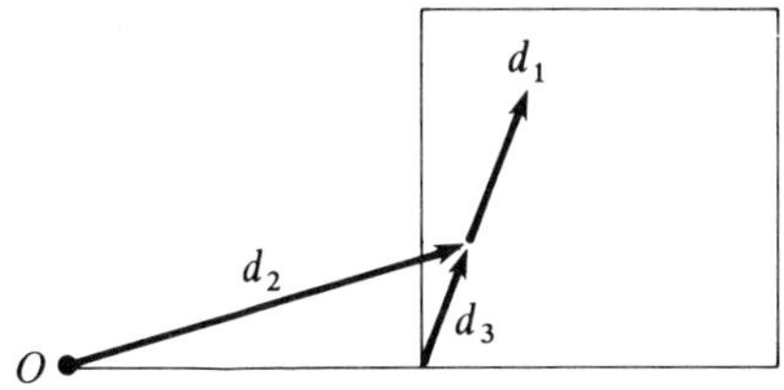

Figure 4

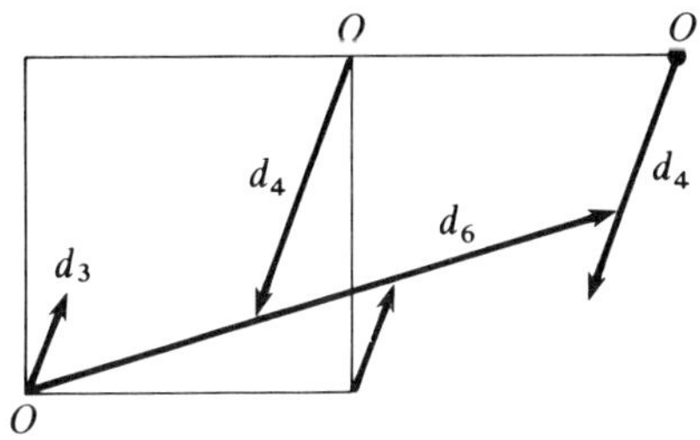

Figure 5

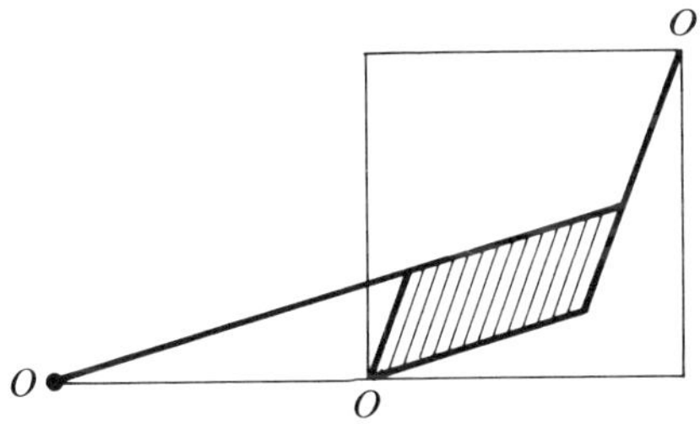

Figure 6

shows one of these parallelograms. It follows from the construction that the partition ξ is Markov. The partition ξ is not necessarily a generating one. To obtain a generating Markov partition, it is necessary to consider parallelograms which are connected components of the intersections $C_{i_0} \cap TC_{i_1}$. We shall not prove this here.

Up to now, in this section, we have been considering the relationship between dynamical systems and discrete time random processes. A similar relationship exists in the case of continuous time. Suppose $(Y, \mathfrak{A})$ is a measurable space. Take, in the role of M, the space of all functions $x(s)$ defined for $s \in \mathbb{R}^1$ and ranging over Y. By $\mathfrak{S}$ denote the minimal σ-algebra containing all the finite dimensional cylinders, i.e., all the sets of the form

$$A = \{x(s) \in M : x(s_1) \in C_1, \ldots, x(s_r) \in C_r\},$$

where $C_1, \ldots, C_r \in \mathfrak{A}$, $s_1, \ldots, s_r \in \mathbb{R}^1$. The triple $(M, \mathfrak{S}, \mu)$, where μ is a normalized measure on $\mathfrak{S}$, is said to be *a continuous time random process* (the completion of the σ-algebra $\mathfrak{S}$ with respect to the measure μ is still denoted by the letter $\mathfrak{S}$).

The one-parameter translation group $\{T^t\}$ acts in the space M in the following way

$$(T^t x)(s) = x(s + t), \qquad -\infty < s, t < \infty.$$

The random process $(M, \mathfrak{S}, \mu)$ is said to be *stationary* if for any cylinder A we have the relation $\mu(T^t A) = \mu(A)$, $-\infty < t < \infty$, i.e., the measure

$$\mu(\{x(s) \in M : x(s_1 + t) \in C_1, \ldots, x(s_r + t) \in C_r\},$$

does not depend on t. In other words, in the case of a stationary random process, the group $\{T^t\}$ is a flow. As in the case of discrete time, it can be shown that any flow generates a class of stationary random processes.

§2. Gauss Dynamical Systems

The Gauss probability distribution often appears in probability theory, statistical physics, and quantum field theory. This is explained, in particular, by the fact that such probability distributions are determined by a small number of parameters and therefore may be easily constructed. An important class of dynamical systems—the so-called Gauss dynamical systems, which shall now be defined—are related to Gauss distributions. Here we assume that the reader is familiar with the properties of multidimensional Gauss distributions.

We begin with the case of discrete time. Consider the space M of sequences of real numbers, infinite in both directions, $x(s)$, where s is an integer and $-\infty < s < \infty$. Suppose $\mathfrak{S}$ is the σ-algebra generated by the cylindrical subsets of the space M. Denote by T the shift transformation in the space M: $(Tx)(s) = x(s + 1)$. The measure μ in $\mathfrak{S}$ is said to be a *Gauss measure* if the joint distribution of any family of variables $x(s_1), x(s_2), \ldots, x(s_r)$ is an r-dimensional Gauss distribution. It is known that such a probability distribution is well defined by the numbers

$$m(s_i) \overset{\text{def}}{=} Ex(s_i) = \int x(s_i)\, d\mu(x), \qquad i = 1, \ldots, r,$$

$$b(s_i, s_j) \overset{\text{def}}{=} E(x(s_i)x(s_j)) = \int x(s_i)x(s_j)\, d\mu(x), \qquad i, j = 1, \ldots, r.$$

If μ is a Gauss measure, then the triple $(M, \mathfrak{S}, \mu)$ is said to be a Gauss discrete time random process. The Gauss measure μ is stationary (invariant with respect to T) if

$$m(s) = m = \text{const}, \qquad b(s_1, s_2) = b(s_1 + t, s_2 + t)$$

for any integer t, i.e.,

$$b(s_1, s_2) = b(0, s_2 - s_1) \overset{\text{def}}{=} b(s_2 - s_1).$$

Further we assume that the mean m vanishes, since the transformation $\{x(s)\} \to \{x(s) - m\}$ maps an arbitrary Gauss measure μ into a Gauss measure with zero mean. The function $b(s) = E((x(t + s)x(t))$ (where s is an integer) is said to be the *correlation function* of the Gauss measure. Clearly, $b(-s) = E(x(t - s)x(t)) = E(x(s + (t - s)x(t + s)) = b(s)$. Moreover, the function $b(s)$ is positive definite, so that by the Bohner–Khinchin theorem it may be presented in the form

$$b(s) = \int_{-\pi}^{\pi} e^{i\lambda s}\, d\sigma(\lambda),$$

where σ is a finite measure on the circle S^1. The measure σ is known as the *spectral measure* of the Gauss measure μ. It is clear that σ completely determines the measure μ (under the assumption $m = 0$). Since $b(s) = b(-s)$, we have $\sigma(\Delta) = \sigma(-\Delta)$ for any Borel set $\Delta \subset S.^1$

Definition 1. The shift transformation in the space M provided with a Gauss stationary measure is said to be a *Gauss automorphism.*

It is useful to give a more abstract and at the same time a more invariant definition of Gauss automorphisms. Suppose T is an automorphism of an arbitrary measure space $(M, \mathfrak{S}, \mu)$. The real element $h_0 \in L^2(M, \mathfrak{S}, \mu)$ is said to be a *Gauss element with zero mean* if for any family of integers $n_1, n_2, \ldots, n_r$ the random variables h_{n_j}, $1 \le j \le r$, where $h_n = U_T^n h_0$ satisfy the joint Gauss probability distribution with zero mean. In nondegenerate case, for any family of Borel sets $C_1, C_2, \ldots, C_r \subset \mathbb{R}^1$, we have

$$\mu(\{x\colon h_{n_1}(x) \in C_1, \ldots, h_{n_r}(x) \in C_r\} = \int_{C_1 \times \cdots \times C_r} p(t^{(1)}, \ldots, t^{(r)})\, dt^{(1)} \ldots dt^{(r)}.$$

Here

$$p(t^{(1)}, \ldots, t^{(r)}) = \text{const} \cdot \exp[-\tfrac{1}{2}(Dt, t)], \qquad t = (t^{(1)}, \ldots, t^{(r)}),$$

D is the matrix inverse to the scalar product matrix $B = \|(h_{n_i}, h_{n_j})\|$ and the constant is determined by the normalization condition.

Definition 1′. The automorphism T is called a *Gauss automorphism* if there is a Gauss element $h_0 \in L^2(M, \mathfrak{S}, \mu)$ with zero mean such that the minimal σ-algebra $\mathfrak{S}_{h_0}$ containing all sets of the form

$$\{x \in M : h_n(x) \in C\},$$

$-\infty < n < \infty$, and $C \subset \mathbb{R}^1$ is a Borel set, coincides with $\mathfrak{S}$.

In the general case, if h_0 is a Gauss element, it is natural to refer to the σ-subalgebra $\mathfrak{S}_{h_0}$ as to the *Gauss subalgebra.*

Let us return to the original Definition 1.

The space $H_1^{(r)}$. Consider finite linear combinations

$$y = \sum_k a_k x(s_k) \in L^2(M, \mathfrak{S}, \mu),$$

where the a_k are real numbers. It follows from the property of Gauss distributions that all such y have the Gauss probability distribution. The closure in $L^2(M, \mathfrak{S}, \mu)$ of the set of random variables y will be denoted by $H_1^{(r)}$. The index r is explained by the fact that at present we are only considering *real* random variables.

Lemma 1. (1) *The space $H_1^{(r)}$ is a closed real subspace of the Hilbert space $L^2(M, \mathfrak{S}, \mu)$ and is invariant with respect to the unitary operator U_T adjoint to the automorphism T.*

(2) *Every random variable $y \in H_1^{(r)}$ has a Gauss probability distribution;*

(3) *There exists an isomorphism $\theta_1^{(r)}$ of the real Hilbert space of functions $\varphi(\lambda) \in L^2(S^1, \sigma)$ satisfying the relation $\varphi(\lambda) = \varphi(-\lambda)$ and space $H_1^{(r)}$ such that*

$$U_T(\theta_1^{(r)}\varphi) = \theta_1^{(r)}(e^{i\lambda}\varphi).$$

Proof. The first statement of the lemma is obvious. To prove the second one, note that if a sequence of random variables $y^{(n)}$ converges to y with respect to the norm in $L^2(M, \mathfrak{S}, \mu)$, then the corresponding distribution functions weakly converge, and therefore the characteristic functions converge uniformly on every finite interval. But characteristic functions of Gauss distributions can only converge to characteristic functions of the same type. Thus the second statement of the lemma is proved.

Now assign the function $\varphi(\lambda) = \sum a_k e^{i\lambda s_k}$ to the random variable $y = \sum a_k x(s_k)$. Clearly $\varphi(\lambda) = \overline{\varphi(-\lambda)}$. Moreover, $Ey^2 = \int |\varphi(\lambda)|^2 \, d\sigma(\lambda)$, which can be checked directly. Thus we have defined a one-to-one linear isometric map $\theta_1^{(r)}$ of the set of trigonometric polynomials $\varphi(\lambda)$ satisfying $\varphi(\lambda) = \overline{\varphi(-\lambda)}$ on the set of random variables y. Obviously $\theta_1^{(r)}(e^{i\lambda}\varphi) = U_T\theta_1^{(r)}(\varphi)$. Extending $\theta_1^{(r)}$ by continuity, we obtain the desired isomorphism. The lemma is proved. □

Denote by $H_1^{(c)}$ the space of complex-valued random variables of the form

$$y = y_1 + iy_2, \qquad y_1 \in H_1^{(r)},\ y_2 \in H_1^{(r)}.$$

To each element $y \in H_1^{(c)}$ assign the function $\varphi_1(\lambda) + i\varphi_2(\lambda)$ where $\varphi_j(\lambda) = \theta_1^{(r)}(y_j)$, $j = 1, 2, \ldots$, while $\theta_1^{(r)}$ was constructed in Lemma 1. Denote the corresponding map by $\theta_1^{(c)}$. The following lemma holds:

Lemma 2. *The map $\theta_1^{(c)}$ establishes an isomorphism of the spaces $L^2(S^1, \sigma)$ and $H_1^{(c)}$. Under this isomorphism $\theta_1^{(c)}(e^{i\lambda}\varphi) = U_T\theta_1^{(c)}(\varphi)$.*

Proof. An arbitrary function $\varphi \in L^2(S^1, \sigma)$ may bc written in the form $\varphi(\lambda) = \varphi_1(\lambda) + i\varphi_2(\lambda)$, where

$$\varphi_1(\lambda) = \frac{1}{2}[\varphi(\lambda) + \overline{\varphi(-\lambda)}], \qquad \varphi_2(\lambda) = \frac{1}{2i}[\varphi(\lambda) - \overline{\varphi(-\lambda)}].$$

Then $\varphi_1(\lambda) = \overline{\varphi_1(-\lambda)}$, $\varphi_2(\lambda) = \overline{\varphi_2(-\lambda)}$.

In view of Lemma 1, the Gauss random variables $y_1, y_2 \in H_1^{(r)}$ correspond to the functions φ_1, φ_2, while the linearity of $\theta_1^{(c)}$ implies that a random variable $y \in H_1^{(c)}$ corresponds to the function φ. The other statement of the lemma follows directly from Lemma 1. The lemma is proved. □

Corollary. *If the spectral measure σ is not continuous, then the automorphism T is nonergodic.*

Proof. Suppose λ_0, $-\pi \le \lambda_0 < \pi$ satisfies $\sigma(\{\lambda_0\}) = \sigma(\{-\lambda_0\}) > 0$. According to Lemma 1, the random variable y_{λ_0} satisfying $[\theta_1^{(c)}]^{-1}y_{\lambda_0} = e_{\lambda_0}$, where $e_{\lambda_0}(\lambda)$ equals 1 for $\lambda = \lambda_0$ and 0 for other λ, is a non-zero complex-valued random variable, whose real and imaginary parts have a non-trivial two-dimensional Gauss distribution. Then

$$[\theta_1^{(c)}]^{-1}U_T y_{\lambda_0} = e^{i\lambda}e_{\lambda_0}(\lambda) = e^{i\lambda_0}e_{\lambda_0}(\lambda) = e^{i\lambda_0}[\theta_1^{(c)}]^{-1}y_{\lambda_0}.$$

This implies $U_T y_{\lambda_0} = e^{i\lambda_0}y_{\lambda_0}$ and therefore $U_T|y_{\lambda_0}| = |y_{\lambda_0}|$, i.e., the function $|y_{\lambda_0}|$ is invariant (mod 0) with respect to T. Since y_{λ_0} satisfies a Gauss distribution, it follows that $|y_{\lambda_0}|$ is not a constant (mod 0). The corollary is proved. □

In Chap. 14 it will be shown that the converse statement is also true: *if the measure σ is continuous, then the Gauss automorphism is ergodic.* It shall also be shown there that *the condition $b(s) \to 0$ for $s \to \infty$ implies that the Gauss automorphism is mixing.*

Now let us consider similar notions in the case of continuous time. Suppose M is the space of all real valued functions $x(s)$ defined for $s \in \mathbb{R}^1$. By $\mathfrak{S}$ denote the minimal σ-algebra containing all the finite dimensional cylinders, i.e., all sets of the form

$$A = \{x(s) \in M: x(s_1) \in C_1, \ldots, x(s_r) \in C_r\},$$

where $C_1, \ldots, C_r$ are Borel subsets of the line and $s_1, \ldots, s_r \in \mathbb{R}^1$. The measure μ on $\mathfrak{S}$ is said to be a *Gauss measure* if the joint distribution of any finite number of random variables $x(s_1), \ldots, x(s_r)$ is an r-dimensional Gauss distribution. If μ is a Gauss measure, then the triple $(M, \mathfrak{S}, \mu)$ is said to be a *Gauss random process with continuous time*. Any Gauss process is determined by its means

$$m(s) \overset{\text{def}}{=} Ex(s), \qquad b(s_1, s_2) \overset{\text{def}}{=} E[x(s_1)x(s_2)].$$

A Gauss process will be called *stationary* if $m(s) = m = \text{const}$, $b(s_1, s_2) = b(s_1 + t, s_2 + t)$ for any t, i.e.,

$$b(s_1, s_2) = b(0, s_2 - s_1) \overset{\text{def}}{=} b(s_2 - s_1).$$

As before, we can assume $m = 0$. In the stationary case under this condition, we have $b(s) = \int_{-\infty}^{\infty} e^{i\lambda s}\, d\sigma(\lambda)$, where σ is an even finite measure on $\mathbb{R}^1$ which is called the *spectral measure* of a Gauss stationary process. The measure σ defines the measure μ entirely.

Definition 2. If μ is a Gauss stationary measure, then the one-parameter group of shifts $\{T^t\}$ on the space M is said to be a *Gauss flow*. As in the case of discrete time it is possible to construct the spaces $H_1^{(r)}$, $H_1^{(c)}$. The ergodicity and mixing criteria for Gauss flows are similar to the corresponding criteria for Gauss automorphisms.

Chapter 9

Examples of Infinite Dimensional Dynamical Systems

§1. Ideal Gas

In this section we consider one of the simplest examples of infinite-dimensional dynamical systems—an ideal gas consisting of an infinite number of noninteracting particles. We begin with the case corresponding to the motion of particles in Euclidian space $\mathbb{R}^d$, $d \geq 1$.

The state of every individual particle is characterized by the sequence of its coordinates $x = (x_1, \ldots, x_d) \in \mathbb{R}^d$ and by the velocity vector $v = (v_1, \ldots, v_d) \in \mathbb{R}^d$, which together constitute a point $(x, v) = (x_1, \ldots, x_d, v_1, \ldots, v_d) \in \mathbb{R}^{2d}$. Further in this section and the following one it shall be convenient to denote the space $\mathbb{R}^d$ of velocities by $\mathbb{R}_v^d$, to avoid confusing it with the coordinate space $\mathbb{R}_x^d$.

The free motion of the particles is described by the system of equations

$$\frac{dx}{dt} = v, \qquad \frac{dv}{dt} = 0.$$

In the space $\mathbb{R}^{2d} = \mathbb{R}_x^d \times \mathbb{R}_v^d$, consider the one-parameter group of translations along the solutions of the system:

$$S^t(x, v) = (x + vt, v) \tag{1}$$

It follows from Liouville's theorem, proved in §2, Chap. 2, that any σ-finite Borel measure ρ on $\mathbb{R}^{2d}$ of the form $d\rho = dx \cdot f(v)\, dv$ is invariant with respect to $\{S^t\}$.

Usually in statistical mechanics one considers the case when $f(v) = \text{const}\, e^{-\beta(v, v)}$, $\beta > 0$. For the sequel we shall only assume that

$$\int_{-\infty}^{\infty} \cdots \int f(v)\, dv < \infty. \tag{2}$$

Now suppose Y is a locally finite subset of the space $\mathbb{R}^{2d}$, i.e., such a countable subset of the space $\mathbb{R}^{2d}$ that for any bounded set $B \subset \mathbb{R}^{2d}$ the set $Y \cap B$ is finite. It is natural to imagine Y as a pair $Y = (X, V)$, where X is a countable set of points $x \in \mathbb{R}_x^d$ while V is a $\mathbb{R}_v^d$-valued function defined on X. The value $v(x) = V(x) \in \mathbb{R}_v^d$ $(x \in X)$ is interpreted as the velocity of the particle at the point x.

The space of all such subsets $Y = (X, V)$ will be denoted by M.

For bounded Borel sets $B \subset \mathbb{R}^{2d}$, we put $\kappa_B(Y) = \text{card}(Y \cap B)$, $Y \in M$. For the σ-algebra $\mathfrak{S}$ of measurable subsets of the space M, we choose the minimal σ-algebra with respect to which all the functions κ_B are measurable. In other words, $\mathfrak{S}$ is the minimal σ-algebra containing all sets of the form

$$C_{B,k} = \{Y \in M: \kappa_B(Y) = k\},$$

where $B \subset \mathbb{R}^{2d}$ is a bounded Borel set and $k \geq 0$ is an integer.

Choose the Poisson measure μ on $\mathfrak{S}$ corresponding to the measure ρ on $\mathbb{R}^{2d}$. It means that

$$\mu(C_{B,k}) = \frac{[\rho(B)]^k}{k!} e^{-\rho(B)},$$

while for disjoint B_1, B_2 the random variables κ_{B_1}, κ_{B_2} are independent, i.e.,

$$\mu(C_{B_1,k_1} \cap C_{B_2,k_2}) = \mu(C_{B_1,k_1}) \cdot \mu(C_{B_2,k_2}).$$

By using this relation, the measure μ may be uniquely extended to the algebra of sets generated by the sets $C_{B,k}$ and further, by using the Kolmogorov theorem, to the entire σ-algebra $\mathfrak{S}$. We now take the completion of the σ-algebra $\mathfrak{S}$ with respect to the measure μ, but retain the notation $\mathfrak{S}$ for the completed σ-algebra.

Now let us define a flow $\{T^t\}$ on the space $(M, \mathfrak{S}, \mu)$ as follows. Let

$$Y = (X, V) = \{(x_1, v_1), (x_2, v_2), \ldots\} \in M.$$

For any $t \in \mathbb{R}^1$, put

$$T^t Y = \{S^t(x_1, v_1), S^t(x_2, v_2), \ldots\}.[1]$$

It is easy to verify that for any t the transformation T^t sends locally finite subsets into locally finite ones: this immediately follows from the fact that bounded sets $B \subset \mathbb{R}^{2d}$ are mapped into bounded ones by S^t.

Therefore an action of the transformation group $\{T^t\}$ on M is defined. It can be shown that the measure μ is invariant with respect to all the transformations T^t. We shall do this a bit later in a more general situation.

[1] The notation $(X, V) = \{(x_1, v_1), (x_2 v_2) \ldots\}$ is somewhat informal, since the points $(x, v) \in (X, V)$ do not constitute an ordered set by any means. We shall use this notation only when no misunderstandings can arise.

The dynamical system $(M, \mathfrak{S}, \mu, \{T^t\})$ constructed in this way is said to be a *d-dimensional ideal gas.*

Now let us prove a lemma which, although it will not be used in the sequel, gives us an idea of the structure of the space M. First consider the natural projection π sending every point $Y \in M$, $Y = \{(x, v_1), (x_2, v_2), \ldots\}$ (i.e., a countable subset of the space $\mathbb{R}^{2d}$) into a countable subset of the space $\mathbb{R}_x^d$:

$$\pi(\{(x_1, v_1), (x_2, v_2), \ldots\}) = \{x_1, x_2, \ldots\}.$$

The space $\mathfrak{Q}$ of all such subsets corresponding to all possible points $Y \in M$ is known as the configuration space of the ideal gas. By means of the projection π, the measure μ induces a certain measure on $\mathfrak{Q}$ which shall be denoted by ν. The points of the space $\mathfrak{Q}$ are not necessarily locally finite subsets of $\mathbb{R}_x^d$, although their inverse images by π were locally finite in $\mathbb{R}^{2d}$. Nevertheless the following statement is valid:

Lemma 1. *If condition* (2) *holds, then almost all* (*with respect to the measure* ν) *points of the space* $\mathfrak{Q}$ *are locally finite subsets of* $\mathbb{R}_x^d$.

Proof. Suppose E is a bounded Borel subset of $\mathbb{R}_x^d$; $Q_{E,k} \subset \mathfrak{Q}$ is the set of all points $X \in \mathfrak{Q}$, $X = (x_1, x_2, \ldots)$ such that

$$\operatorname{card}(X \cap E) = k, \quad k = 0, 1, \ldots, \infty; \qquad C_{E,k} \stackrel{\text{def}}{=} \pi^{-1}(Q_{E,k}).$$

We shall show that $\mu(C_{E,\infty}) = 0$, i.e., $\nu(Q_{E,\infty}) = 0$.

Obviously, for any $k < \infty$, we have

$$C_{E,k} = \bigcup_{n=1}^{\infty} C_{B_n,k}, \text{ where } B_n = E \times \{v \in \mathbb{R}_v^d : |v| \le n\},$$

i.e., B_n is a bounded subset of $\mathbb{R}^{2d}$. Therefore $C_{E,k} \in \mathfrak{S}$ and

$$\mu(C_{E,k}) = \lim_{n \to \infty} \mu(C_{B_n,k}) = \lim_{n \to \infty} \frac{[\rho(B_n)]^k}{k!} e^{-\rho(B_n)} = \frac{[\rho(D)]^k}{k!} e^{-\rho(D)},$$

where $D = E \times \mathbb{R}_v^d$.

By condition (2), $\rho(D) = \int_E dx \int_{-\infty}^{\infty} f(v)\, dv < \infty$; therefore also $\mu(C_{E,k}) < \infty$ and

$$\sum_{k=0}^{\infty} \mu(C_{E,k}) = \sum_{k=0}^{\infty} \frac{[\rho(D)]^k}{k!} e^{-\rho(D)} = 1.$$

But then $\mu(C_{E,\infty}) = \mu(M \setminus \bigcup_{k=0}^{\infty} C_{E,k}) = 0$. To complete the proof, it suffices to consider all possible balls $E = E_m = \{x \in \mathbb{R}_x^d : |x| \leq m\}, m = 1, 2, \ldots$. The lemma is proved. □

We now pass to a more general definition of *ideal gases*. Suppose $(R, \mathfrak{A}, \rho)$ is a measurable space with finite or σ-finite measure ρ. Choose an increasing sequence of subsets $R_i \in \mathfrak{A}$, $1 \leq i < \infty$ such that $\rho(R_i) < \infty$, $\bigcup_{i=1}^{\infty} R_i = R$. The sets $B \in \mathfrak{A}$ for which $B \subset R_i$ for some i will be referred to as *bounded*.

Consider a new space M whose points are locally finite subsets $Y \subset R$, i.e., subsets such that $\text{card}(Y \cap R_i) < \infty$ for all i. For the σ-algebra $\mathfrak{S}$ of measurable sets, take the minimal σ-algebra containing all sets of the form

$$C_{B,k} = \{Y \in M : \text{card}(Y \cap B) = k\},$$

where $B \subset R$ is a bounded measurable set and $k \geq 0$ is an integer. Define a measure μ on $\mathfrak{S}$ by putting

$$\mu(C_{B,k}) = \frac{[\rho(B)]^k}{k!} e^{-\rho(B)},$$

and for nonintersecting $B_1, B_2, \ldots, B_s$ put

$$\mu(C_{B_1,k_1} \cap \cdots \cap C_{B_s,k_s}) = \mu(C_{B_1,k_1}) \cdot \ldots \cdot \mu(C_{B_s,k_s}).$$

Now suppose $\{S^t\}$ is a flow on the space $(R, \mathfrak{A}, \rho)$, i.e., for any $A \in \mathfrak{A}$ such that $\rho(A) < \infty$ and any $t \in \mathbb{R}^1$, the set $S^t A$ belongs to $\mathfrak{A}$, $\rho(S^t A) = \rho(A)$ while for $t_1, t_2 \in \mathbb{R}^1$, we have $S^{t_1 + t_2} = S^{t_1} \cdot S^{t_2}$. Assume that all the transformations S^t send bounded sets into bounded ones and hence locally finite sets into locally finite ones. Then we can define a one-parameter group $\{T^t\}$ of transformations of the space M according to the formula

$$T^t(\{y_1, y_2, \ldots\}) = \{S^t y_1, S^t y_2, \ldots\}, \qquad y_i \in R.$$

Let us show that the measure μ is invariant with respect to the transformations T^t. It suffices to establish that $\mu(T^t C_{B,k}) = \mu(C_{B,k})$ for all $C_{B,k}$ and $t \in \mathbb{R}^1$. But this follows from the relation $T^t C_{B,k} = C_{S^t B,k}$ and the invariance of the measure ρ with respect to S^t.

Definition 1. The dynamical system $\{T^t\}$ on the space $(M, \mathfrak{S}, \mu)$ is said to be the *Poisson suspension* over the dynamical system $\{S^t\}$.

The usual ideal gas can be obtained as the Poisson suspension over the flow $\{S^t\}$ corresponding to the free motion of particles in $\mathbb{R}^d$ described by equation (1).

Now we begin the study of ergodic properties of Poisson suspensions. To do this we shall need the following definition.

Definition 2. Suppose $\{T^t\}$ is a Poisson suspension over $\{S^t\}$. We shall say that the trajectories of the system $\{T^t\}$ *move away to infinity* if there exists a set $A_0 \in \mathfrak{A}$ and a number t_0 with the following properties:

(1) all the unions $\bigcup_{\tau_1 \leq t \leq \tau_2} S^t A_0$ are measurable;
(2) $\bigcup_{-\infty < t < \infty} S^t A_0 = R \pmod 0$;
(3) $A_0 \cap S^t A_0 = \varnothing$ for all t, $|t| > t_0$.

Theorem 1. *If the trajectories of a Poisson suspension move away to infinity, then the flow* $\{T^t\}$ *is mixing.*

Proof. Suppose $B_1, B_2 \subset R$ are bounded sets; let C_{B_1,k_1}, C_{B_2,k_2} be two subsets of M and χ_1, χ_2 be their indicators. We shall show that

$$\lim_{|t| \to \infty} \int_M \chi_1(T^t Y)\chi_2(Y)\, d\mu(Y) = \mu(C_{B_1,k_1}) \cdot \mu(C_{B_2,k_2}),$$

which implies the statement of the theorem, since the linear combinations of such indicators are dense in $L^2(M, \mathfrak{S}, \mu)$. For the same reason we may assume that

$$B_1, B_2 \subset \bigcup_{|t| \leq \tau} S^t A_0, \tag{3}$$

for some $\tau > 0$.

Indeed, if we put

$$B_i^\tau = B_i \cap \bigcup_{|t| \leq \tau} S^t A_0, \qquad i = 1, 2,$$

the sets B_i^τ verify condition (3) and, by (2),

$$\lim_{\tau \to \infty} \rho(B_i \triangle B_i^\tau) = 0$$

Therefore

$$\lim_{\tau \to \infty} \mu(C_{B_i,k_i} \triangle C_{B_i^\tau,k_i}) = \lim_{\tau \to \infty} (1 - e^{-\rho(B_i \triangle B_i^\tau)}) = 0.$$

But if the inclusion (3) holds, then, according to condition (3), we shall have, for $|t| > t_0 + \tau$, the relation $S^t B_1 \cap B_2 = \varnothing$. By the definition of the measure μ, for such t, we have

$$\begin{aligned}\int \chi_1(T^t Y)\chi_2(Y)\,d\mu &= \mu(C_{S^{-t}B_1, k_1} \cap C_{B_2, k_2}) \\ &= \mu(C_{S^{-t}B_1, k_1}) \cdot \mu(C_{B_2, k_2}) = \mu(C_{B_1, k_1}) \cdot \mu(C_{B_2, k_2}) \\ &= \int \chi_1\,d\mu \cdot \int \chi_2\,d\mu.\end{aligned}$$

The theorem is proved. □

Corollary. *For $d \geq 1$ a d-dimensional ideal gas is a mixing dynamical system.*

Proof. Let us show that the trajectories of a d-dimensional ideal gas move away to infinity, i.e., let us check conditions (1) and (2) for a dynamical system $\{S^t\}$ acting in $\mathbb{R}^{2d}$ in accordance to formula (1). Put

$$\begin{aligned}R' &= \{(x_1, \ldots, x_d, v_1, \ldots, v_d) \in \mathbb{R}^{2d} : v_1 \neq 0\} \\ A_0 &= R' \cap \{(x_1, \ldots, x_d, v_1, \ldots, v_d) \in \mathbb{R}^{2d} : |x_1| \leq \min(1, |v_1|)\}.\end{aligned}$$

Clearly $\rho(R') = 1$ and a straightforward verification shows that

$$\bigcup_{-\infty < t < \infty} S^t A_0 \supset R'.$$

Moreover $S^t A_0 \cap A_0 = \varnothing$ if $|t| \geq 2$. The corollary is proved. □

Commentary. At first glance it might seem strange that an ideal gas corresponding to the motion of noninteracting particles is a mixing system, since in statistical mechanics relaxation and irreversibility are usually the result of the interactions of molecules. The consideration of this example enables us to note certain specific traits of dynamical systems of statistical mechanics. As we could see from the previous arguments, it suffices to consider functions $f((X, V))$ depending on the coordinates and velocities of a finite number of molecules situated in a bounded part of space (such functions are dense in $L^2(M, \mathfrak{S}, \mu)$. Under evolution the functions $f((X, V)_t)$ become dependent on the coordinates and velocities of molecules which are far from this part and therefore, in view of the properties of the invariant measure μ, turn out to be statistically independent of the initial functions $f((X, V))$. This explains why Theorem 2 is valid.

Theorem 2. *Suppose $\{T^t\}$ is the Poisson suspension over $\{S^t\}$. If for some $\tau \in \mathbb{R}^1$ there exist a set $C \in \mathfrak{A}$ for which:*

(1) $S^{k\tau}C \cap S^{l\tau}C = \varnothing$ *if $k \neq l$, k, l are integers*;
(2) $\bigcup_{-\infty<k<\infty} S^{k\tau}C = R \pmod 0$;

then T^τ is a Bernoulli automorphism.

Proof. Let us consider the state space Z whose points are of the form $z = Y \cap C$, $Y \in M$. In other words the points of Z are restrictions to C of points Y. The natural σ-algebra of measurable subsets of Z is denoted by $\mathfrak{Q}$. The measure μ induces the measure on $\mathfrak{Q}$ which we shall denote by σ. We shall show that T^τ is metrically isomorphic to the Bernoulli automorphism for which $(Z, \mathfrak{Q}, \sigma)$ is the state space.

For each measurable $D \in R$ we introduce a σ-algebra $\mathfrak{S}_D$ generated by sets $C_{B,k}$ for $B \subset D$, $k \geq 0$. It follows from (1), (2) and the properties of the Poisson measure that the σ-algebras $\mathfrak{S}_{S^{-k\tau}C} = T^{k\tau}\mathfrak{S}_C$, $-\infty < k < \infty$, are mutually independent and the least σ-algebra containing all of them coincides with $\mathfrak{S}$. Therefore, $\mathfrak{S}_C$ is a Bernoulli generating σ-algebra for T^τ, and this fact gives us the desired isomorphism. The theorem is proved. □

Corollary. *If $\{T^t\}$ is the d-dimensional ideal gas, $d \geq 1$, then for any $\tau \neq 0$ the automorphism T^τ is Bernoulli.*

Proof. For $\tau > 0$ put

$$C = C_\tau = \{(x, v) \in \mathbb{R}^{2d} : v_1 > 0, 0 \leq x_1 < \tau v_1\}$$
$$\cup \{(x, v) \in \mathbb{R}^{2d} : v_1 < 0, \tau v_1 < x \leq 0\}.$$

It is easy to check that (1), (2) of Theorem 2 hold.

For $\tau < 0$ the argument is similar. □

Lorentz Gas. One particular case of the construction of an ideal gas described above leads to a model, well known in statistical mechanics under the name of Lorentz gas. Suppose $d > 1$ and assume that the space Q was obtained from $\mathbb{R}^d$ by eliminating an arbitrary set of nonintersecting balls. Consider the dynamical system $\{S^t\}$ corresponding to the motion of a single particle in Q with unit velocity. For the phase space of the system, take the phase space of the particle, i.e., the Cartesian product $Q \times S^{d-1}$, where the second factor corresponds to the velocity space of this particle. For the measure ρ, take the measure $d\rho = dx\, d\omega$, where ω is the Lebesgue measure on the sphere S^{d-1} and dx is the Lebesgue measure on Q. Then ρ is an invariant measure for the dynamical system $\{S^t\}$ and the ideal gas constructed in this case in the manner described above is known as *Lorentz gas.*

§2. Dynamical Systems of Statistical Mechanics

The ideal gas considered in the previous section is the simplest dynamical system of statistical mechanics. This example already demonstrates the distinctive features of such systems—the infinite total number of degrees of freedom and the related infinite-dimensionality of the phase space, as well as the fact that the different degrees of freedom are of equal importance. Mathematically this last fact can be expressed by saying that a group of space translations, commuting with the dynamics, acts in the phase space. In this section we shall construct a wider class of dynamical systems possessing both of the properties indicated above.

We begin by constructing the phase space. Let us consider an infinite (countable) set X of particles $x \in \mathbb{R}_x^d$ which interact with each other by means of pairwise interaction forces defined by the potential $U(r)$, $r \in \mathbb{R}_+^1$. This means that any particle $y \in X$ acts on a given particle $x \in X$ with the force $F = -\operatorname{grad}_x U(\|y - x\|)$, where $\|\cdot\|$ is the norm in the space $\mathbb{R}_x^d$ and the particles belonging to X move in accordance to the Newton equations. (The mass of each particle is assumed equal to 1). As a rule, in statistical mechanics, one considers interactions for which the interacting forces are repulsing at small distances and rapidly decrease at large distances. A sufficiently interesting class of examples arises if we assume that:

$$\text{(a)} \quad U(r) \equiv \infty \qquad \text{for } 0 < r \leq r_0, r_0 > 0 \tag{1}$$

$$\text{(b)} \quad U(r) \equiv 0 \qquad \text{for } r \geq r_1, r_1 > r_0 \tag{2}$$

$$\text{(c)} \quad \frac{\text{const}_1}{(r - r_0)^{\gamma_1}} \leq U(r) \leq \frac{\text{const}_2}{(r - r_0)^{\gamma_2}} \quad \text{for } r_0 < r < r_1, \gamma_1, \gamma_2 > 0 \tag{3}$$

$$\text{(d)} \quad |U'(r)| \leq \text{const}\,|U(r)|^{\gamma_3}, \qquad |U''(r)| < \text{const}\,|U(r)|^{\gamma_3} \tag{4}$$

$$\text{for } r_0 < r < r_1, \gamma_3 > 0.$$

Condition (a) means that the particles must necessarily be situated at a distance greater than r_0 from each other. Therefore it is natural to imagine that we are dealing with solid spheres of radius $r_0/2$, rather than with point-like particles. Condition (b) means that the particles no longer interact if the distance between them is greater than r_1. The number r_1 is called the *radius of action* of the potential $U(r)$. Conditions (c) and (d) simplify subsequent arguments and, in fact, can be considerably weakened. By $\mathfrak{Q}$ we denote the set whose points are countable subsets $X \subset \mathbb{R}_x^d$ such that

$$\|x' - x''\| > r_0, \tag{5}$$

for all $x', x'' \in X''$, $x' \neq x''$. Then X can be considered as a configuration of an infinite number of solid spheres of radius $r_0/2$ in $\mathbb{R}_x^d$. The space $\mathfrak{Q}$ is the configuration space of the future dynamical system.

Let us transform $\mathfrak{Q}$ into a measurable space by assuming that the σ-algebra $\mathfrak{S}_{\mathfrak{Q}}$ is the smallest σ-algebra containing all the subsets of the form

$$C_{E,k} = \{X \in \mathfrak{Q}: \operatorname{card}(X \cap E) = k\},$$

where E is a Borel subset of $\mathbb{R}^d_x$ and $k \geq 0$ is an integer. The phase space M of our dynamical system consists of pairs (X, V), where V is a $\mathbb{R}^d_v$-valued function defined on X. As in the case of an ideal gas, the value

$$v(x) = V(x) \in \mathbb{R}^d_x (x \in X)$$

is interpreted as the velocity of the particle located at the point $x \in \mathbb{R}^d_x$.

By $\mathfrak{S}_M = \mathfrak{S}$ we denote the smallest σ-algebra containing the sets

$$C_{E,P} = \{(X, V) \in M: \operatorname{card}(X \cap E) = 1, v(x) \in P \text{ for } x \in X \cap E\},$$

where $E \subset \mathbb{R}^d_x$, $P \subset \mathbb{R}^d_v$ are Borel sets. The measurable space $(M, \mathfrak{S})$ will be the phase space of our system.

Before beginning the construction of the measure, let us introduce some notations. Suppose π is a natural projection of M onto $\mathfrak{Q}$: $\pi((X, V)) = X$ for $(X, V) \in M$. For any $E \subset \mathbb{R}^d_x$, let X_E, (X_E, V_E) be the restrictions of X, (X, V) respectively to E. By $\mathfrak{S}(E) \subset \mathfrak{S}$ denote the smallest σ-subalgebra of the σ-algebra $\mathfrak{S}$ containing all sets of the form $C_{E',P}$, $E' \subset E$. By $\mathfrak{S}_{\mathfrak{Q}}(E) \subset \mathfrak{S}_{\mathfrak{Q}}$ denote the smallest σ-subalgebra of the σ-algebra $\mathfrak{S}_{\mathfrak{Q}}$ containing all the sets $C_{E',k}$, $E' \subset E$.

Now let us construct a family of normalized measures on $(M, \mathfrak{S})$, called the *Gibbs states*. Each such state is given by three parameters $\lambda > 0$, $\beta > 0$, $\bar{v} \in \mathbb{R}^d_v$. To construct it, take a bounded Borel set $E \subset \mathbb{R}^d_v$ and denote $\bar{E} = \mathbb{R}^d_v \backslash E$.

Fix $(X_{\bar{E}}, V_{\bar{E}})$, i.e., the coordinates and the velocities of particles outside E, and denote by $\nu_{\lambda,E} = \nu_\lambda$ the Poisson measure on $\mathfrak{S}_{\mathfrak{Q}}(E)$ with parameter λ, i.e., for $E' \subset E$, $k = 0, 1, \ldots$, put

$$\nu_\lambda(C_{E',k}) = \frac{\lambda^k[\rho(E')]^k}{k!} e^{-\lambda\rho(E')},$$

where ρ is the Lebesgue measure in $\mathbb{R}^d_x$; by $\bar{\nu}_\lambda = \bar{\nu}$ let us denote the normalized measure (on $\mathfrak{S}(E)$) whose image under the projection π is the measure ν_λ and, for a fixed configuration $X \cap E$, the conditional distribution of velocities $v(x)$, $x \in X \cap E$ is the Cartesian product of normalized measures (corresponding to individual particles) whose density with respect to the Lebesgue measures equals $\sqrt{(\beta/\pi)^d} e^{-\beta\|v(x) - \bar{v}\|^2}$.

Now for a fixed $(X_{\bar{E}}, V_{\bar{E}})$, consider the function

$$g(X_E) = g((X_E, V_E)) = \exp\left\{-\beta\left[\sum_{\substack{x, x' \in X_E \\ x' \neq x''}} U(\|x' - x''\|) + \sum_{\substack{x \in X_E \\ y \in X_{\bar{E}}}} U(\|x - y\|)\right]\right\}. \tag{6}$$

The right-hand side of (6) is meaningful only for $\operatorname{card}(X_E) \geq 2$. If

$$\operatorname{card}(X_E) < 2,$$

then we put $g(X_E) = 1$.

It is easy to check that g is measurable. Moreover, the sums in the right-hand side of (6) contain only a finite number of nonvanishing terms: indeed, by (2) we can limit ourselves to the points $x, y \in X$ in an r_1-neighborhood E, the number of such points being finite by (1). Therefore $g > 0$ and

$$0 < \Xi \stackrel{\text{def}}{=} \int g((X_E, V_E))\, d\bar{\nu} < \infty.$$

Definition 1. A *Gibbs conditional state* (with parameters $\lambda, \beta, \bar{v}$) in the domain E with boundary conditions $(X_{\bar{E}}, V_{\bar{E}})$ is a normalized measure $\mu(\cdot|(X_{\bar{E}}, V_{\bar{E}}))$ on $\mathfrak{S}(E)$, equivalent to the measure $\bar{\nu}$ whose density with respect to the measure $\bar{\nu}$ equals $(1/\Xi)g((X_E, V_E))$.

Definition 1′. A *Gibbs configurational conditional state* in the domain E is the normalized measure on $\mathfrak{S}_{\mathfrak{Q}}(E)$ whose density with respect to the Poisson measure ν_λ is proportional to $g(X_E)$.

Further we shall need an estimate from below of the normalizing coefficient Ξ, obtained in the following way. The value of Ξ can only decrease if the integral $\int g\, d\bar{\nu}$ is taken only over an "empty" configuration, i.e., over the configuration $(X^{(0)}, V^{(0)})$ which has no points in E. Then $g(X_E^{(0)}, V_E^{(0)}) = 1$ and we put

$$\Xi \geq \nu_\lambda(C_{E,0}) = e^{-\lambda\rho(E)}. \tag{7}$$

It follows from the definition above that a Gibbs conditional state in fact depends only on $X_{\bar{E}}$ (and not on $V_{\bar{E}}$), and not even on the entire configuration $X_{\bar{E}}$, but only on its part contained in an r_1-neighborhood of E.

Definition 2. A *Gibbs equilibrium state* (with parameters $\lambda, \beta, \bar{v}$) is a normalized measure μ on $(M, \mathfrak{S})$ such that for an arbitrary bounded Borel set $E \subset \mathbb{R}_x^d$ the conditional (with respect to the measure μ) distribution on $\mathfrak{S}(E)$ for μ-almost every condition $(X_{\bar{E}}, V_{\bar{E}})$ is a Gibbs conditional state for the domain E under the boundary conditions $(X_{\bar{E}}, V_{\bar{E}})$.

Definition 2′. A *Gibbs configurational state* (with parameters λ, β) is a normalized measure $\bar{\mu}$ on the space $(\mathfrak{Q}, \mathfrak{S}_{\mathfrak{Q}})$ such that for any bounded Borel subset $E \subset \mathbb{R}_x^d$ the induced conditional distribution on the configurations $X_E \subset \mathfrak{Q}_E$ for $\bar{\mu}$-almost every condition $X_{\bar{E}}$ coincides with the Gibbs configurational conditional state.

Theorem 1. *Under conditions* (a)–(d) *concerning the potential* $U(r)$, *there exists at least one Gibbs equilibrium state (with the given parameters* λ, β, $\bar{v}$).

Proof. It is easy to see that it suffices to prove the existence of the Gibbs configurational state corresponding to the given potential, since for a fixed configuration X_E the velocities v_x, $x \in X_E$ are independent of each other for any Gibbs conditional state.

Choose an increasing sequence of balls $D^{(m)} \subset \mathbb{R}_x^d$, $\bigcup_m D^m = \mathbb{R}_x^d$. For every m the configuration space $X_{D^{(m)}}$ has a natural compact metric space structure. By the usual diagonal process, we can construct a sequence of conditions $X_{\bar{D}^{(m)}}$ such that if $\bar{\mu}_m$ is a Gibbs conditional state on the configurations $X_{D^{(m)}}$ under the condition $X_{\bar{D}^m}$, then for any k the restriction $\bar{\mu}_m | \mathfrak{S}_{\mathfrak{Q}}(D^{(k)})$ weakly converges when $m \to \infty$. The limits for various k are obviously compatible and, by the Kolmogorov theorem, they define a unique measure $\bar{\mu}$ on the σ-algebra $\mathfrak{S}_{\mathfrak{Q}}$. Let us show that $\bar{\mu}$ will be a Gibbs configurational state.

Consider fixed balls D', D'', $D' \cap D'' = \varnothing$ and continuous functions $f'(X), f''(X)$ which depend only on the configurations $X_{D'}$, $X_{D''}$ respectively. It suffices to show that

$$\int f'(X) f''(X)\, d\bar{\mu} = \int f'(X) E(f''(X) | X_{\bar{D}''})\, d\bar{\mu},$$

where $E(f''(X) | X_{D''})$ is the expectation computed for the Gibbs configurational conditional state under the condition $X_{\bar{D}''}$.

Take an m such that $D' \subset D^{(m)}$, $D'' \subset D^{(m)}$. Then it follows from the form of the Gibbs configurational conditional state that

$$\int f'(X) f''(X)\, d\bar{\mu}_m = \int f'(X) E(f''(X) | X_{\bar{D}''})\, d\bar{\mu}_m.$$

Now note that the function $E(f''(X) | X_{\bar{D}''})$ is continuous and depends only on the configurations contained in the ball $D''' \subset D''$ whose boundary is situated at a distance equalling r_1 from D''. Therefore, by the weak convergence of $\bar{\mu}_m$, we have

$$\begin{aligned} \int f'(X) f''(X)\, d\bar{\mu} &= \lim_{m \to \infty} \int f'(X) f''(X)\, d\bar{\mu}_m \\ &= \lim_{m \to \infty} \int f'(X) E(f''(X) | X_{\bar{D}''})\, d\bar{\mu}_m \\ &= \int f'(X) E(f''(X) | X_{\bar{D}''})\, d\bar{\mu}. \end{aligned}$$

The theorem is proved. □

The parameters λ, β, $\bar{v}$ of the Gibbs equilibrium state are related by the so-called additive integrals of motion: the density, the specific energy, and the mean velocity. The questions of uniqueness and nonuniqueness of Gibbs states for given λ, β, $\bar{v}$ are related to one of fundamental problems of equilibrium statistical mechanics—the phase transition problem. We shall not be concerned here with this question.

Further in this section we shall only consider the case $\bar{v} = 0$. Correspondingly, the Gibbs equilibrium state will be characterized by two parameters: λ, β.

Let us write out the formal expression for the Hamiltonian

$$H(X, V) = \frac{1}{2} \sum_{x \in X} [v(x)]^2 + \sum_{x, y \in X} U(\|y - x\|).$$

The expression H cannot be viewed as a function on the phase space, since the series in the right-hand side generally diverges. However, H may be used to write out the Hamiltonian equations of motion

$$\begin{aligned} \frac{dx}{dt} &= \frac{\partial H}{\partial v(x)} = v(x) \\ \frac{dv(x)}{dt} &= -\frac{\partial H}{\partial x} = -\sum_{y \in X, y \neq x} \operatorname{grad}_x U(\|y - x\|). \end{aligned} \tag{8}$$

The right-hand sides make sense for any point (X, V). Suppose $D \subset \mathbb{R}^d$ is a domain with smooth boundary, $f_D(X, V)$—a complex-valued function of class C^1 in the coordinates and velocities of the particles located in D. For all such functions define the linear differential operator L which acts according to the formula

$$L f_D = -i \left[\sum_{x \in X \cap D} \frac{\partial f_D}{\partial x} \cdot v(x) - \sum_{\substack{x \in D \\ y \in X \\ y \neq x}} \frac{\partial f_D}{\partial v(x)} \cdot \operatorname{grad}_x U(\|y - x\|) \right].$$

(The products which appear here in each of the two sums should be understood as scalar products of vectors).

The reason for introducing such an operator L is the following. Let us assume temporarily that we have constructed the solutions of the system (8) and the one-parameter group of translations $\{T^t\}$ along these solutions which preserves the measure μ. Then the operator L is the generating operator for the adjoint one-parameter group $\{U^t\}$ of unitary operators, since (8) implies that

$$-i \frac{d}{dt} U^t f_D \bigg|_{t=0} = -i \left[\sum_{x \in X \cap D} \left(\frac{\partial f_D}{\partial x} \cdot \frac{dx}{dt} + \frac{\partial f_D}{\partial v(x)} \cdot \frac{dv(x)}{dt} \right) \right] = L f_D.$$

Lemma 1. *Suppose μ is the Gibbs equilibrium state with parameters λ, β. Then the operator L is a symmetric operator in $L^2(M, \mathfrak{S}, \mu)$ defined on the set of functions $f_D(X, V)$ which are dense in $L^2(M, \mathfrak{S}, \mu)$.*

The proof of this lemma will be given later.

The following question naturally arises: does L possess self-adjoint extensions? An equivalent statement of this same question is: does there exist, in Hilbert space $L^2(M, \mathfrak{S}, \mu)$, a one-parameter group of unitary operators $\{U^t\}$ such that $U^t = e^{itL}$? Normally, the solution of this question would follow from the existence and uniqueness theorem for the solution of the corresponding system of differential equations of motion. Then U^t is a transformation of Hilbert space adjoint to the translation along the trajectories of the system by the time t. In our case the phase space is infinite-dimensional, and the usual existence and uniqueness theorems cannot be applied. Moreover it is clear that the existence theorem does not hold on the entire phase space. Indeed, if the initial velocities of the particles are directed to the origin and increase sufficiently quickly at infinity, then the system undergoes an instantaneous collapse in which the particles come together in infinitely small time to the minimal admissible distance; after this their motion is undefined. The way out of this situation consists in using the fact that the phase space is a measure space and therefore existence and uniqueness theorems can be proved by restricting ourselves to subsets of full measure. Such an approach to uniqueness and existence theorems may also be useful in other situations.

We now go on to the systematic exposition, and begin by introducing the following:

Definition 3. A *solution of the system of equations of motion* (8) is a one-parameter family of points $(X_t, V_t) \in M$ such that the coordinates and velocities of each particle are continuously differentiable functions of t whose derivatives with respect to t satisfy the system (8).

Now we can state the following fundamental question. Suppose $(M, \mathfrak{S}, \mu)$ is the phase space of our system and μ is a Gibbs equilibrium state on M. Is there a subset $M' \subset M$, $\mu(M') = 1$, depending in general on μ, and a one-parameter group of transformations $\{T^t\}$ leaving M' invariant, preserving the measure μ and such that the one-parameter family of points

$$\{T^t(X, V): (X, V) \in M'\} = \{(X_t, V_t)\}, \qquad -\infty < t < \infty,$$

is a solution of the system (8)?

We shall prove a somewhat weaker statement.

Theorem 2. *There exists a subset $M' \subset M$ possessing the following properties*:

(1) $\mu(M') = 1$ *for any Gibbs equilibrium state* μ;
(2) *for any initial condition* $(X_0, V_0) \in M'$, *there exists a curve* $(X_t, V_t) \in M$, $-\infty < t < \infty$ *which is the solution of the system* (8);
(3) *if* T^t *denotes the map* $M' \to M$ *sending* (X_0, V_0) *to* (X_t, V_t), *then for any* t *this map is measurable and preserves any Gibbs equilibrium state* μ;
(4) *if* $T^{t_1}(X_0, V_0) \in M'$, *then* $T^{t_1+t_2}(X_0, V_0) = T^{t_2}(X_{t_1}, V_{t_1})$.

Remark. The one-parameter family of maps $\{T^t\}$ is an example of a continuous flow (see the definition in §1, Chap. 1).

Proof. 1. *The construction of approximating dynamical systems.* Suppose $D_r = \{x \in \mathbb{R}^d_x : \|x\| < r\}$ is an open ball of radius r with centre at the origin. We will construct a flow $\{S^t(D_r)$ for which the particles outside of D_r are motionless, the particles within D_r move under the action of pairwise intercation forces as well as of "exterior forces" generated by the interaction with exterior fixed particles and elastically reflect off the boundary of the ball D_r, i.e., when they reach the boundary, the tangent component of the velocity of the particle remains unchanged, while the normal component changes its sign. For a formal construction, we set $C_{D_r,k} = \{(X, V) \in M: \operatorname{card}(X \cap D_r) = k\}$. In other words, $C_{D_r,k}$ consists of points of the phase space for which k particles are located inside D_r. For any fixed k, consider the Hamiltonian for $(X, V) \in C_{D_r,k}$:

$$H(X_{D_r}, V_{D_r} | X_{D_r}, V_{D_r}) = \sum_{x \in D_r \cap X} \frac{\|v(x) - \bar{v}\|^2}{2} + \sum_{\substack{x \in X \cap D_r \\ y \in X \cap D_r \\ y \neq x}} U(\|y - x\|)$$

$$+ \sum_{\substack{x \in X \cap D_r \\ y \in X \cap \bar{D}_r}} U(\|y - x\|) + \sum_{x \in X \cap D_r} W_{D_r}(x).$$

Here $\bar{D}_r = \mathbb{R}^d_x \backslash D_r$, $(X_{\bar{D}_r}, V_{\bar{D}_r})$ is the "frozen" family of fixed particles outside of D_r. The function $W_{D_r}(x)$ vanishes for $x \in D_r$ and becomes infinite in the opposite case. For $(X, V) \in C_{D_r,k}$, the expression written above is the Hamiltonian of the system of k particles moving within D_r. By $\{S^t(D_r)\}$ we shall denote the flow generated by this Hamiltonian.

Lemma 2. *Suppose* $\mu(\cdot | X_{\bar{D}_r}, V_{\bar{D}_r})$ *is a Gibbs conditional state in the region* D_r *for fixed* $X_{\bar{D}_r}, V_{\bar{D}_r}$. *Then it is invariant with respect to the flow* $\{S^t(D_r)\}$.

Corollary. *The flow* $\{S^t(D_r)\}$ *preserves any Gibbs equilibrium state.*

Proof of the Corollary. Suppose $f(X, V)$ is a bounded measurable function on the phase space M. Then if $X = (X_{D_r}, X_{\bar{D}_r})$, $V = (V_{D_r}, V_{\bar{D}_r})$, we have

$$(S^t(D_r)(X, V) = (S^t(D_r)X_{D_r}, X_{\bar{D}_r}; S^t(D_r)V_{D_r}, V_{\bar{D}_r}).$$

Further we shall write $f((X_{D_r}, V_{D_r})|(X_{\bar{D}_r}, V_{\bar{D}_r}))$, viewing the boundary conditions $X_{\bar{D}_r}, V_{\bar{D}_r}$ as a parameter. By $M_{\bar{D}_r}$ we denote the set of all pairs $(X_{\bar{D}_r}, V_{\bar{D}_r})$. Using the formula for conditional expectation, Lemma 1 and Fubini's theorem, we get

$$\begin{aligned}
&\int f(S^t(D_r)(X, V)\, d\mu \\
&\quad = \int_{M_{\bar{D}_r}} d\mu \int f(S^t(D_r)(X_{D_r}, V_{D_r})|(X_{\bar{D}_r}, V_{\bar{D}_r}))\, d\mu(X_{D_r}, V_{D_r}|(X_{\bar{D}_r}, V_{\bar{D}_r})) \\
&\quad = \int_{M_{\bar{D}_r}} d\mu \int f((X_{D_r}, V_{D_r})|(X_{\bar{D}_r}, V_{\bar{D}_r}))\, d\mu(X_{D_r}, V_{D_r}|(X_{\bar{D}_r}, V_{\bar{D}_r})) \\
&\quad = \int f((X, V))\, d\mu.
\end{aligned}$$

The corollary is proved. □

By using the flow $\{S^t(D_r)\}$, let us carry out the proof of Lemma 1, and then of Lemma 2.

Proof of Lemma 1. We must show that for two functions f_D, g_D, we have

$$\int Lf_D \cdot \bar{g}_D\, d\mu = \int f_D \cdot L\bar{g}_D\, d\mu.$$

Consider the ball $D_r \supset D$ whose boundary is located at a distance greater than r_1 from D. On the functions f_D, g_D, the operator L coincides with the generating operator of the group of unitary operators adjoint to the flow $\{S^t(D_r)\}$. Therefore we can apply the formula for complete expectation again, getting

$$\begin{aligned}
\int Lf_D \cdot \bar{g}_D\, d\mu &= \int_{M_{\bar{D}_r}} d\mu \int Lf_D \cdot \bar{g}_D\, d\mu((X_{D_r}, V_{D_r})|(X_{\bar{D}_r}, V_{\bar{D}_r})) \\
&= \int_{M_{\bar{D}_r}} d\mu \int f_D \cdot L\bar{g}_D\, d\mu((X_{D_r}, V_{D_r})|(X_{\bar{D}_r}, V_{\bar{D}_r})) = \int f_D \cdot L\bar{g}_D\, d\mu.
\end{aligned}$$

The lemma is proved. □

Proof of Lemma 2. In the ball D_r, consider a system of k particles numbered from 1 to k with the Hamiltonian $H = H((X_{D_r}, V_{D_r})|(X_{\bar{D}_r}, V_{\bar{D}_r}))$. Since H is a first integral, it follows by Liouville's theorem that the normalized measure $\text{const} \cdot e^{-\beta H} \prod_{i=1}^{k} dx_i\, dv_i$ will be invariant with respect to the flow generated by the Hamiltonian function.[1] Note that the restriction of the flow $\{S^t(D_r)\}$ to the set $C_{D_r,k}$ is obtained by taking the quotient of the phase space of our Hamiltonian system by the permutation group of the particles, which commutes with the dynamics. The measure which thus arises on the quotient space will indeed be Gibbs conditional state. The lemma is proved. □

2. *Description of the set M'. Reduction of the system* (8) *to a system of integral equations.* Suppose c_1, c_2 are natural numbers, D_r, as before, is the ball $\{x \in \mathbb{R}_x^d : \|x\| < r\}$. Denote by M_{c_1, c_2} the set of points $(X, V) \in M$ possessing the following property: for any $\tau > 0$ there is a natural number $n_0 = n_0(c_1, c_2, \tau, (X, V))$ such that for all $n \geq n_0$ the flow $\{S^t(D_{2^n})\}$ satisfies the following inequalities:

(i) the velocities of all the particles $x \in X$ for $|t| \leq \tau$ are less than or equal to (with respect to norm) $c_1 \tau \sqrt{n}$;
(ii) the minimal distance between any two moving particles $x, y \in X$ for $|t| \leq \tau$ is no less than $r_0 + n^{-c_2}\tau$.

Put $M' = \bigcup_{c_1, c_2} M_{c_1, c_2}$.

Lemma 3. *For any Gibbs equilibrium state* μ,

$$\mu(M') = 1.$$

The proof of this lemma is given in Subsection 4.

For any $n = 1, 2, \ldots$, let us replace the system of equations of motion corresponding to the flow $\{S^t(D_{2^n})\}$ by the system of integral equations below, formally equivalent to it:

$$x^{(n)}(t) = x + v(x)t + \sum_{\substack{y \in X \\ y \neq x}} \int_0^t (t - t_1) F(y^{(n)}(t_1) - x^{(n)}(t_1))\, dt_1. \tag{9}$$

Here $x^{(n)}(t)$ is the coordinate vector of the particle $x \in X$ at the moment t when it moves in the flow $S^t(D_{2^n})$. Further, x, $v(x)$ are its initial coordinates

[1] Liouville's theorem can be applied, since in any domain of phase space of the form $H <$ const the vector field corresponding to our system is devoid of singularities.

and velocity, and $F(y^{(n)}(t_1) - x^{(n)}(t_1))$ is the force that the particle with coordinate $y^{(n)}(t_1)$ exerts on the particle with coordinate $x^{(n)}(t_1)$. The sum in (9) is taken over all the moving particles $y \in X$ located inside the ball D_{2^n}, as well as over the "frozen" particles $y \in X$ outside the ball D_{2^n}. By (1) and (2), this sum contains only a finite number of nonzero summands. We shall assume that $t > 0$: the case $t < 0$ is similar.

For arbitrary $b > 0$, $t > 0$ and any positive integer n, put $\alpha_b^{(n)}(t) = \alpha_b^{(n)}(t; (X, V)) = \max\{\|x^{(n)}(t') - x^{(n-1)}(t')\|\}$, where the maximum is taken over all t', $0 \leq t' \leq t$ and all the particles $x \in X \cap D_b$.

Lemma 4. *If $(X, V) \in M'$, then $\alpha_b^{(n)}(t) = \alpha_b^{(n)}(t; (X, V)) \to 0$ for $n \to \infty$ faster than any geometric progression, i.e.,*

$$\alpha_b^{(n)}(t) = o(\varepsilon^n), \tag{10}$$

for any $\varepsilon > 0$.

Proof. Let us fix $c_1, c_2 > 0$, $\tau \geq t$ and choose $(X, V) \in M_{c_1, c_2}$, $n \geq n_0(c_1, c_2, \tau, (X, V))$. For any particle $y \in X$ which adds a nonzero contribution to (9), we shall have

$$\|y^{(n)}(t_1) - x^{(n)}(t_1)\| \leq r_1,$$

for some $t_1 \leq t$; thus by (i),

$$\begin{aligned} \|y\| = \|y(0)\| &\leq \|x(0)\| + \|y(0) - x(0)\| \\ &\leq \|x(0)\| + \|y^{(n)}(t_1) - x^{(n)}(t_1)\| + \|y^{(n)}(t_1) - y^{(n)}(0)\| \\ &\quad + \|x^{(n)}(t_1) - x^n(0)\| \\ &\leq b + r_1 + 2c_1\tau\sqrt{n}t_1 \\ &\leq b + r_1 + 2c_1\tau^2 n, \end{aligned}$$

i.e., $y \in D_{b'}$, $b' = b + r_1 + 2c_1\tau^2 n$.

It follows from (9) that:

$$\begin{aligned} &\|x^{(n)}(t) - x^{(n-1)}(t)\| \\ &\leq \int_0^t (t - t_1) \sum_{\substack{y \in X \\ y \neq x}} \|F(y^{(n)}(t_1) - x^{(n)}(t_1)) - F(y^{(n-1)}(t_1) - x^{(n-1)}(t_1))\| \, dt_1, \end{aligned} \tag{11}$$

and it suffices to take the sum only over the points $y \in D_{b'}$.

Using the estimates (3) and (4) for the potential $U(r)$, as well as property (ii), we can write for any such y:

$$\begin{aligned}
&\|F(y^{(n)}(t_1) - x^{(n)}(t_1)) - F(y^{(n-1)}(t_1) - x^{(n-1)}(t_1))\| \\
&\quad \le \max_{r \ge r_0 + n^{-c_2\tau}} |U''(r)|(\|[y^{(n)}(t_1) - x^{(n)}(t_1)] - [y^{(n-1)}(t_1) - x^{(n-1)}(t_1)]\|) \\
&\quad \le \text{const} \cdot n^{\gamma'}(\|y^{(n)}(t_1) - y^{(n-1)}(t_1)\| + \|x^{(n)}(t_1) - x^{(n-1)}(t_1)\|) \\
&\quad \le \text{const} \cdot n^{\gamma'} \cdot \alpha_{b'}^{(n)}(t_1),
\end{aligned} \tag{12}$$

where $\gamma' = c_2\gamma_2\gamma_3$.

By (2), the maximum number of particles y which can interact with the particle x for $t \le \tau$ is no greater than the magnitude $N_n \le \text{const} \cdot \sqrt{n}$. Therefore, if we substitute the inequality (12) into (11), we obtain

$$\alpha_b^{(n)}(t) \le \text{const}\sqrt{n} \cdot n^{\gamma'} \int_0^t (t - t_1)\alpha_{b+r_1+2c_1\tau^2\sqrt{n}}^{(n)}(t_1)\, dt_1.$$

Integrating this last equation p times, where $p = p(n) > 1$ will be chosen somewhat later, we get

$$\begin{aligned}
\alpha_b^{(n)}(t) &\le (\text{const})^p \cdot n^{p[\gamma' + (1/2)]} \int_0^t dt_1(t - t_1) \int_0^{t_1} dt_2(t_1 - t_2) \\
&\quad \times \cdots \int_0^{t_p} (t_p - t_{p+1})\alpha_{b+p(r_1+2c_1\tau^2\sqrt{n})}^{(n)}(t_{p+1})\, dt_p.
\end{aligned} \tag{13}$$

Suppose $p = [(2^n - b)/(r_1 + 2c_1\tau^2\sqrt{n})]$. Then $b + p(r_1 + 2c_1\tau^2\sqrt{n}) \le 2^n$, and, using property (i), we can write the estimate

$$\alpha_{b+p(r_1+2c_1\tau^2\sqrt{n})}(t_1) \le 2c_1\tau^2\sqrt{n}.$$

Substituting this estimate into (13), we finally obtain

$$\begin{aligned}
\alpha_b^{(n)}(t) &\le \frac{t^{p+1}}{(p+1)!} \cdot \exp[p(\text{const} + (\gamma' + \tfrac{1}{2}) \log n + \log(2c_1\tau^2\sqrt{n})] \\
&\le \frac{1}{(p+1)!} \cdot \exp[p(\text{const} + (\gamma' + \tfrac{1}{2}) \log n \\
&\quad + \log(2c_1\tau^2\sqrt{n}) + (p+1)\log\tau].
\end{aligned}$$

This inequality immediately implies relation (10). The lemma is proved. □

Remark. It is clear from the proof of Lemma 4 that all the arguments remain valid if we assume that the radius b of the ball D_b containing the point $x \in X$

is not necessarily constant but is a function of n which does not increase too rapidly for $n \to \infty$, in particular, if $b = b_n \le \text{const} \cdot \sqrt{n}$. We shall also use Lemma 4 in this somewhat strengthened version.

3. *Construction of the limit dynamics.* Suppose $(X, V) \in M_{c_1, c_2}$ for some c_1, c_2, where the point x satisfies $x = x(0) \in X \cap D_b$, $b < \infty$. It follows from Lemma 4 that the series $\sum_n \alpha_b^{(n)}(\tau)$ converges and therefore there exists a limit $x(t) = \lim_{n \to \infty} x^{(n)}(t)$ (uniform with respect to t, $|t| \le \tau < \infty$).

Let us show that the functions $x(t)$ satisfy the system of integral equations

$$x(t) = x(0) + v(x)t + \sum_{\substack{y \in X \\ y \ne x}} \int_0^t (t - t_1)[F(y(t_1) - x(t_1))]\, dt_1. \tag{14}$$

The left-hand side of (14) is, by definition, the limit for $n \to \infty$ of the left-hand side of (9). Concerning the right-hand side of (14), *a priori* it is not even obvious that it is finite (since the sum in (14) formally contains an infinite number of terms). However, the following lemma is valid.

Lemma 5. *The number of points $y \in X$ interacting with the given point $x \in X$ for $0 \le t \le \tau$, i.e., adding a nonzero contribution to* (14), *is finite.*

Proof. Take any point $y \in X$ which contributes to (14). Then for some t_1, $0 \le t_1 \le \tau$, we have $\|y(t_1) - x(t_1)\| < r_1$, so that

$$\begin{aligned} \|y\| = \|y(0)\| &\le \|y(t_1)\| + \|y(0) - y(t_1)\| \\ &\le \|y(t_1) - x(t_1)\| + \|x(t_1)\| + \|y(0) - y(t_1)\| \\ &\le r_1 + B + \|y(0) - y(t_1)\|, \end{aligned}$$

where

$$B = \max_{\substack{x \in X \cap D_b \\ |t_1| < \tau}} \|x(t_1)\| < \infty.$$

Suppose $n_0 = n_0(c_1, c_2, \tau, (X, V))$. Then

$$\begin{aligned} \|y(0) - y(t_1)\| &\le \|y(t_1)\| + \int_0^{t_1} \|\dot{y}(s)\|\, ds \\ &\le \|y(t_1)\| + \int_0^{t_1} \|\dot{y}^{(n_0)}(s)\|\, ds + \sum_{n = n_0 + 1}^{\infty} \int_0^{t_1} \|\dot{y}^{(n)}(s) - \dot{y}^{(n-1)}(s)\| ds \\ &= \Sigma_1 + \Sigma_2 + \Sigma_3, \end{aligned}$$

where

$$\dot{y}(s) = \left.\frac{dy(t)}{dt}\right|_{t=s}, \qquad \dot{y}^{(n)}(s) = \left.\frac{dy^{(n)}(t)}{dt}\right|_{t=s}.$$

We have $|\Sigma_1| \le r_1 + B$, $|\Sigma_2| \le t_1 \cdot c_1\tau\sqrt{n_0} \le c_1\tau^2\sqrt{n_0}$. To estimate Σ_3, let us write out, for any fixed $n > n_0$, an integral equation for $\dot{y}^{(n)}(s)$;

$$\dot{y}^{(n)}(s) = \dot{y}^{(n)}(0) + \int_0^s \sum_{\substack{z \in X \\ z \neq y}} F(z^{(n)}(s_1) - y^{(n)}(s_1))\, ds_1.$$

Thus we get

$$\begin{aligned} &\|\dot{y}^{(n)}(s) - \dot{y}^{(n-1)}(s)\| \\ &\quad \le \int_0^s \sum_{\substack{z \in X \\ z \neq y}} \|F(z^n(s_1) - y^{(n)}(s_1)) - F(z^{(n-1)}(s_1) - y^{(n-1)}(s_1))\|\, ds_1. \end{aligned} \tag{15}$$

Arguing as in the proof of Lemma 4, we obtain, for any z which contributes to (15), the relation

$$\begin{aligned} &\|F(z^{(n)}(s_1) - y^{(n)}(s_1)) - F(z^{(n-1)}(s_1) - y^{(n-1)}(s_1))\| \\ &\qquad \le \text{const} \cdot n^{\gamma'}(\|z^{(n)}(s_1) - z^{(n-1)}(s_1)\| + \|y^{(n)}(s_1) - y^{(n-1)}(s_1)\|). \end{aligned}$$

But (i) implies $\|y(0)\| \le \text{const} \cdot \sqrt{n}$, $\|z(0)\| \le \text{const} \cdot \sqrt{n}$ for all z which contribute to (15). Hence we can use the strengthened version of Lemma 4, obtaining for any $\varepsilon > 0$

$$\|\dot{y}^{(n)}(s) - \dot{y}^{(n-1)}(s)\| \le \text{const} \cdot n^{\gamma''} \cdot o(\varepsilon^n), \qquad \gamma'' < \infty.$$

Hence

$$|\Sigma_3| = \left| \sum_{n=n_0+1}^{\infty} \int_0^{t_1} \|\dot{y}^{(n)}(s) - \dot{y}^{(n-1)}(s)\|\, ds \le \text{const} < \infty.\right.$$

Thus all the points y which interact with x are situated in a certain ball $\|y(0)\| < \text{const}$ and their number is finite. The lemma is proved. □

This lemma implies that the right-hand side of the integral equation (14) is the limit when $n \to \infty$ of the right-hand side of equation (9), i.e., the functions $x(t)$ satisfy the system (14).

We would now like to show that $x(t)$ also satisfy the system of equations (8) obtained from (14) by taking formal derivatives with respect to t. To make this differentiation meaningful, it suffices to show that the expression under the integral sign in (14) is a bounded function of t.

Since the number of points $y \in X$ over which the sum in (14) is taken may be assumed finite (according to Lemma 5), in order to estimate from above the expression under the integral sign it suffices to estimate from below the expression $\|y(t) - x(t)\|$ for $|t| \le \tau$, $x(0), y(0) \in X \cap D_R$, $R < \infty$. For sufficiently large n_0, by the definition of M_{c_1, c_2}, we have

$$\|y^{(n_0)}(t) - x^{(n_0)}(t)\| \ge r_0 + n_0^{-c_2}\tau = r_0 + \varepsilon_{n_0}.$$

But from the estimate of the difference $\|x^{(n)}(t) - x^{(n-1)}(t)\|$ obtained in Lemma 4, it follows that for sufficiently large n_0 we also have

$$\sum_{n=n_0+1}^{\infty} [\|x^{(n)}(t) - x^{(n-1)}(t)\| + \|y^{(n)}(t) - y^{(n-1)}(t)\|] \le \text{const} \cdot \left(\frac{1}{2}\right)^{n_0} < \frac{\varepsilon_{n_0}}{2}.$$

Hence

$$\|y(t) - x(t)\| \ge \|y^{(n_0)}(t) - x^{(n_0)}(t)\| - \sum_{n=n_0+1}^{\infty} [\|y^{(n)}(t) - y^{(n-1)}(t)\| + \|x^{(n)}(t) - x^{(n-1)}(t)\|] \ge r_0 + \frac{\varepsilon_{n_0}}{2}.$$

Thus we have shown that both sides of (14) may be differentiated with respect to t, so that we see that (X_t, V_t) is a solution of the system (8).

Now take an arbitrary $t \in \mathbb{R}^1$, and consider the map $T^t\colon M' \to M$ defined by the relation $T^t((X_0, V_0)) = (X_t, V_t)$. The fact that it is measurable follows from the definition of T^t as the limit of approximating maps $S^t(D_{2^n})$.

Lemma 6. *Any Gibbs equilibrium state is invariant with respect to T^t.*

Proof. First consider the sets $C \in \mathfrak{S}$ of the form

$$C = C_B = \{(X, V) \in M'\colon \text{card}((X, V) \cap B) = 1,$$

where $B \subset \mathbb{R}_x^d \times \mathbb{R}_v^d$ is a domain (with a piecewise smooth boundary ∂B) which is the Cartesian product $B = B_x \times B_v$, $B_x \subset \mathbb{R}_x^d$, $B_v \subset \mathbb{R}_v^d$. We shall also assume that

$$\text{diam}(B_x) < r_0. \qquad (16)$$

This guarantees that any configuration $X \in \Omega$ contains no more than one point in B_x. For any $\delta > 0$, put

$$B_{x,\delta} = \{x \in \mathbb{R}^d_x \colon x \in B_x, \operatorname{dist}(x, \partial B_x) \geq \delta\}; \qquad E_{x,\delta} = B_x \backslash B_{x,\delta};$$
$$B_{v,\delta} = \{v \in \mathbb{R}^d_v \colon v \in B_v, \operatorname{dist}(v, \partial B_v) \geq \delta\}; \qquad E_{v,\delta} = B_v \backslash B_{v,\delta};$$
$$B_\delta = B_{x,\delta} \times B_{v,\delta}; \qquad C_\delta = \{(X, V) \in M \colon \operatorname{card}((X, V) \cap B_\delta) = 1\}.$$

It follows from (16) that $C_\delta \subset C$.

Suppose $\mu(\cdot | X_{\bar{D}_r}, V_{\bar{D}_r})$ is a Gibbs conditional state in some ball $D_r \supset B_x$. Then

$$\begin{aligned} 0 &\leq \mu(C | X_{\bar{D}_r}, V_{\bar{D}_r}) - \mu(C_\delta | X_{\bar{D}_r}, V_{\bar{D}_r}) = \mu(C \backslash C_\delta | X_{\bar{D}_r}, V_{\bar{D}_r}) \\ &\leq \mu(C_{E_{x,\delta} \times \mathbb{R}^d_v, 1} | X_{\bar{D}_r}, V_{\bar{D}_r}) + \mu(C_{B_{x,\delta} \times E_{v,\delta}, 1} | X_{\bar{D}_r}, V_{\bar{D}_r}) = \mu'_\delta + \mu''_\delta. \end{aligned}$$

Since the boundary is piecewise smooth, $\rho(E_{v,\delta}) \to 0$ for $\delta \to 0$, so that $\mu''_\delta \to 0$ for $\delta \to 0$. To estimate μ'_δ, let us write

$$\begin{aligned} &\mu(C_{E_{x,\delta} \times \mathbb{R}^d_v, 1} | X_{\bar{D}_r}, V_{\bar{D}_r}) \\ &\quad = \sum_{k,l \geq 0} \int_{E_{x,\delta}} \frac{dx^{(1)}}{1!} \underset{D_r \backslash E_{x,\delta}}{\iint \cdots \int} \frac{dx^{(2)} \cdots dx^{(k+1)}}{k!} \cdot \underset{\mathbb{R}^d_x \backslash D_r}{\int \cdots \int} \frac{dy^{(1)} \cdots dy^{(l)}}{l!} \\ &\quad \times \exp\left\{-\beta\left[\sum_{\substack{1 \leq i,j \leq k+1 \\ i \neq j}} U(\|x^{(i)} - x^{(j)}\|) + \sum_{\substack{1 \leq i \leq k+1 \\ 1 \leq j \leq l}} U(\|x^{(i)} - y^{(j)}\|)\right]\right\}. \end{aligned}$$

The number of terms in the sum inside the figure brackets is bounded by an expression which depends only on D_r. Hence the fact that $\rho(E_{x,\delta}) \to 0$ for $\delta \to 0$ implies that $\mu'_\delta \to 0$; thus

$$\mu(C | X_{\bar{D}_r}, V_{\bar{D}_r}) - \mu(C_\delta | X_{\bar{D}_r}, V_{\bar{D}_r}) \to 0,$$

when $\delta \to 0$, uniformly with respect to all the boundary conditions $(X_{\bar{D}_r}, V_{\bar{D}_r})$. Taking the mean over the boundary conditions, we see that the Gibbs equilibrium state μ satisfies the following relation: $\mu(C_\delta) \to \mu(C)$ for $\delta \to 0$.

Now take any $\varepsilon > 0$ and choose $\delta > 0$ so as to have $0 \leq \mu(C) - \mu(C_\delta) < \varepsilon/2$. Using Lemma 3, also choose a positive integer k such that $\mu(M^{(k)}) > 1 - \varepsilon/2$, where

$$M^{(k)} = \bigcup_{c_1, c_2 \leq k} M_{c_1, c_2}.$$

It follows from the convergence of the approximating flows that for all sufficiently large n, $n \geq n_0(k)$, we have the inclusion $S^t(D_{2^n})(C_\delta) \subset T^t(C)$, which implies

$$S^t(D_{2^n})[C_\delta \cap S^{-t}(D_{2^n})(M^{(k)})] \subset T^tC \cap M^{(k)}.$$

Now take any such n and, using the invariance of the measure μ with respect to the flow $S^t(D_{2^n})$, write

$$\begin{aligned}\mu(T^tC) &\geq \mu(T^tC \cap M^{(k)}) \geq \mu(S^t(D_{2^n})[C_\delta \cap S^{-t}(D_{2^n})(M^{(k)})]) \\ &= \mu(C_\delta \cap S^{-t}(D_{2^n})(M^{(k)})) \geq \mu(C_\delta) - \mu(M \setminus S^{-t}(D_{2^n})(M^{(k)})) \\ &= \mu(C_\delta) - \mu(M \setminus M^{(k)}) > \mu(C) - \varepsilon/2 - \varepsilon/2 = \mu(C) - \varepsilon.\end{aligned}$$

Since the number ε was arbitrary, we obtain $\mu(T^tC) \geq \mu(C)$. This inequality may be generalized to sums of disjoint sets $C = C_B$ of the type considered and then, by continuity, to the entire σ-algebra $\mathfrak{S}$. Finally, the validity of the inequality $\mu(T^tC) \geq \mu(C)$ for all $C \in \mathfrak{S}$ implies that we actually have the equality $\mu(T^tC) = \mu(C)$. The lemma is proved. □

4. *Proof of Lemma 3.* The set M' is of the form $M' = M^{(i)} \cap M^{(ii)}$, where $M^{(i)} = \bigcup_{c_1} M_{c_1}^{(i)}$, $M'' = \bigcup_{c_2} M_{c_2}^{(ii)}$, while $M_{c_1}^{(i)}$, $M_{c_2}^{(ii)}$ are determined respectively by the conditions (i) and (ii) stated at the beginning of Subsection 2.

We will give a detailed proof only for the part of the statement of this lemma which concerns condition (i), i.e., we shall prove that $\mu(M^{(i)}) = 1$. The arguments concerning condition (ii) are similar.

Suppose μ is a Gibbs equilibrium state (for certain parameters λ, β). For all positive integers c_1, τ, n denote

$$N_{c_1,\tau,n} = \left\{(X, V) \in M : \max_{x \in X} \max_{|t| \leq \tau} \|v(x^{(n)}(t))\| \geq c_1 \tau \sqrt{n}\right\},$$

where $v(x^{(n)}(t))$ is the velocity of the point x at time t in the motion within the flow $\{S^t(D_{2^n})\}$. We shall show that for sufficiently large c_1 (depending only on μ, i.e., on λ, β), we have

$$\sum_{\tau, n=1}^{\infty} \mu(N_{c_1,\tau,n}) < \infty; \tag{17}$$

hence, by the Borel–Cantelli lemma, we get (i). Now denote by $N_{c_1,\tau,n}^{(1)}$ the set of all $(X, V) \in M$ for which the initial velocity $v(x)$ of at least one particle $x \in X \cap D_{2^n}$ satisfies the inequality $\|v(x)\| \geq c_1 \tau \sqrt{n}$; denote by $N_{c_1,\tau,n}^{(2)}$ the set of all $(X, V) \in M$ such that for at least one particle $x \in X \cap D_{2^n}$ we have the equality $\|v(x)\| = c_1 \tau \sqrt{n}$.

Clearly,

$$N_{c_1,\tau,n} \subset N^{(1)}_{c_1,\tau,n} \cup \bigcup_{|t|\le\tau} S^t(D_{2^n})N^{(2)}_{c_1,\tau,n},$$

hence

$$\mu(N_{c_1,\tau,n}) \le \mu(N^{(1)}_{c_1,\tau,n}) + \mu\left(\bigcup_{|t|\le\tau} S^t(D_{2^n})N^{(2)}_{c_1,\tau,n}\right) = \mu^{(1)}_{c_1,\tau,n} + \mu^{(2)}_{c_1,\tau,n}$$

First let us estimate $\mu^{(1)}_{c_1,\tau,n}$. Since for a fixed $x \in X$ the probability of the event $\|v(x)\| \ge c_1\tau\sqrt{n}$ equals

$$\sqrt{\frac{\beta}{\pi}}\int_{\|v\|\ge c_1\tau\sqrt{n}} \exp(-\beta\|v\|^2)\,dv \le \text{const}\cdot\exp[-\beta(c_1\tau\sqrt{n})^2],$$

while the number of points $x \in X \cap D_{2^n}$ is no greater than const 2nd, we have

$$\mu^{(1)}_{c_1,\tau,n} \le \text{const}\cdot 2^{nd}\cdot\exp[-\beta(c_1\tau\sqrt{n})^2],$$

and for sufficiently large τ, n, we shall have

$$\mu^{(1)}_{c_1,\tau,n} \le \text{const}\cdot\exp\left[n\left(d\log 2 - \frac{\beta}{2}c_1^2\tau^2\right)\right].$$

It follows from this inequality that

$$\sum_{\tau,n=1}^{\infty} \mu(N^{(1)}_{c_1,\tau,n}) < \infty, \tag{18}$$

for sufficiently large c_1.

Now let us estimate $\mu^{(2)}_{c_1,\tau,n}$. This estimate is based on a general consideration, which will be stated in the form of Lemma 7 below.

Suppose $G = G^m$ is a differentiable m-dimensional manifold (m is arbitrary); $\{S^t\}$ is a flow in G induced by a vector field α of class C^2; ν is a smooth finite (not necessarily normalized) invariant measure for $\{S^t\}$, i.e.,

$$d\nu = p(x)\,dx, \qquad p \in C^1(G),\ p \ge 0.$$

Consider the subset $\Gamma \subset G$ which is an (open) differentiable submanifold of codimension 1, i.e., is of the form

$$\Gamma = \{x \in G\colon f(x) = 0\}, \qquad f \in C^1(G).$$

The measure $d\nu$ induces a certain measure (on the surface Γ) which shall be denoted by $d\sigma$. Suppose $\alpha(y) = (\alpha_1, \ldots, \alpha_m)$ is a vector of the vector field at the point $y \in \Gamma$, $n(y) = (n_1, \ldots, n_m)$ is the unit normal vector to Γ at the point $y \in \Gamma$ and $(\alpha(y), n(y)) = \sum_{k=1}^{m} \alpha_k n_k$ is their scalar product. Suppose further

$$\Gamma^\tau = \bigcup_{|t| \le \tau} S^t\Gamma, \qquad \tau > 0.$$

Lemma 7. *For any* $\tau > 0$, *we have*

$$\nu(\Gamma^\tau) \le 2\tau \int_\Gamma |\, \alpha(y), n(y) \,|\, d\sigma(y).$$

Proof. We can assume that the manifold G is embedded in $\mathbb{R}^m$. Choose a positive integer r and put

$$\delta = \delta_r = \tau/r, \qquad \Gamma_\delta = \bigcup_{0 \le t < \delta} S^t\Gamma.$$

Then

$$\nu(\Gamma^\tau) = \bigcup_{j=-r}^{r-1} \nu(S^{j\delta}\Gamma_\delta) \le 2r \cdot \nu(\Gamma_\delta) = \frac{2\tau}{\delta} \nu(\Gamma_\delta). \tag{19}$$

It follows from the smoothness of the vector field that for $\delta \to 0$ we have

$$\nu(\Gamma_\delta) = \delta \int_\Gamma |(\alpha(y), n(y))|\, d\sigma(y) + o(\delta). \tag{20}$$

The statement of the lemma now follows from (19) and (20). □

Now let us apply this lemma to estimate $\mu^{(2)}_{c_1, \tau, n}$. Choose a positive integer k and, for $G = G_k$, take the phase space of the pairs $(X_{D_{2^n}}, V_{D_{2^n}})$ such that card $X_{D_{2^n}} = k$. A natural differentiable manifold structure of dimension $m = 2kd$ may be introduced on this set. The dynamical system $\{S^t\}$ is the restriction of the flow $\{S^t(D_{2^n})\}$ to the invariant set G_k, the measure $\nu = \nu_k$ is the restriction to G_k of the Gibbs conditional state $\mu(\cdot | X_{\bar{D}_{2^n}}, V_{\bar{D}_{2^n}})$ (the boundary conditions $(X_{D_{2^n}}, V_{D_{2^n}})$ are chosen arbitrarily).

To apply Lemma 7, it remains to indicate the manifold which will play the role of Γ.

Subdivide the ball D_{2^n} into finite number of domains $E_n^{(i)}$, $1 \le i \le r_n$, such that

(1) $\qquad \operatorname{diam}(E_n^{(i)}) < r_0;$ (21)

(2) $\qquad 0 < \text{const}_1 \le \rho(E_n^{(i)}) \le \text{const}_2 < \infty;$ (22)

(3) $\qquad r_n \le \text{const} \cdot 2^{nd}.$ (23)

Clearly, this can be done for any n, and then each $E_n^{(i)}$ will contain no more than one point $x \in X \cap D_{2^n}$.

For the set $\Gamma = \Gamma_k^{(i)}$, choose the family of points $(X_{D_{2^n}}, V_{D_{2^n}}) \in G_k$ satisfying $\operatorname{card}(X_{D_{2^n}} \cap E_n^{(i)}) = 1$ and $\|v(x)\| = c_1\tau(\sqrt{n})$ for the point $x \in X_{D_{2^n}} \cap E_n^{(i)}$.

Note for the sequel that

$$N_{c_1,\tau,n}^{(2)} \subset \left\{(X, V) \in M\colon (X_{D_{2^n}}, V_{D_{2^n}}) \in \bigcup_{k=1}^{r_n} \bigcup_{i=1}^{r_n} \Gamma_k^{(i)}\right\}. \tag{24}$$

Now suppose $Y = (X_{D_{2^n}}, V_{D_{2^n}}) \in \Gamma_k^{(i)}$ and assume that the particle satisfying $\|v\| = c_1\tau\sqrt{n}$ is located at the point $x \in E_n^{(i)}$. Then the measure element $d\sigma(Y)$ of the measure σ on $\Gamma_k^{(i)}$, which was considered in Lemma 6, equals, according to the definition of Gibbs conditional state,

$$\begin{aligned} d\sigma(Y) = {} & \frac{\lambda^k}{k!}\Xi_n^{-1} \cdot (c_1\tau\sqrt{n})^{d-1} \cdot d\omega \cdot \exp(-\beta\|v(x)\|^2) \\ & \times \exp\left(-\beta \sum_{\substack{y \in X \\ y \neq x}} U(\|y - x\|)\right) \cdot \exp\left(-\beta \sum_{\substack{y \in X_{D_{2^n}} \\ y \neq x}} \|v(y)\|^2\right) \\ & \times \exp\left(-\beta \sum_{\substack{y' \in X \\ y'' \in X_{D_{2^n}} \\ y' \neq x \\ y'' \neq x}} U(\|y' - y''\|)\right) \cdot \prod_{\substack{y \in X_{D_{2^n}} \\ y \neq x}} dv(y) \cdot dx \cdot \prod_{\substack{y \in X_{D_{2^n}} \\ y \neq x}} dy. \end{aligned}$$

Here λ is the parameter of the Poisson measure, $d\omega$ is the surface element of the unit sphere $S^{d-1} = \{v \in \mathbb{R}_v^d\colon \|v\| = 1\}$ and Ξ_n is the normalizing coefficient for the conditional Gibbs state in the volume D_{2^n}.

Now let us estimate the scalar product $(\alpha(Y), n(Y))$ which appears in Lemma 7. The equations determining the motion in the flow $\{S^t(D_{2^n})\}$ are of the form

$$\begin{cases} \dfrac{dx}{dt} = v, \\[2ex] \dfrac{dv}{dt} = -\displaystyle\sum_{\substack{y \in X \\ y \neq x}} \operatorname{grad}_x U(\|y - x\|), \end{cases}$$

where $(x, v) \in (X_{D_{2^n}}, V_{D_{2^n}})$. The vector $\alpha(Y)$ of the vector field α at the point $Y = (X_{D_{2^n}}, V_{D_{2^n}})$ is a $2kd$-dimensional vector constituted by the right-hand sides of these equations corresponding to all $x \in X_{D_{2^n}}$. Now consider the normal vector $n(Y)$. Since the equation of the submanifold $\Gamma_k^{(i)}$ in a neighborhood of the point $Y = (X_{D_{2^n}}, V_{D_{2^n}})$ is of the form

$$f((X, V)) = \|v(x)\|^2 - c_1^2\tau^2 n = 0,$$

it follows that the unit vector $n(Y)$ (of the same dimension $2kd$) has no more than d nonvanishing coordinates, and they are located in the places corresponding to the velocity $v(x)$. We therefore get

$$|(\alpha(Y), n(Y))| \leq \sum_{\substack{y \in X \\ y \neq x}} \|\operatorname{grad}_x U(\|y - x\|)\|.$$

Applying Lemma 7, we obtain:

$$\begin{aligned}
\nu_k(\Gamma_k^{(i),\tau}) &\stackrel{\text{def}}{=} \nu_k\left(\bigcup_{|t| \leq \tau} S^t(D_{2^n})\Gamma_k^{(i)}\right) \\
&\leq \text{const} \cdot 2\tau \cdot \frac{\lambda^k}{k!} \cdot \Xi_n^{-1} \times (c_1 \tau \sqrt{n})^{d-1} \cdot \exp[-\beta(c_1 \tau \sqrt{n})^2] \\
&\quad \times \int_{E_k^{(i)}} dx \int_{D_{2^n} \setminus E_k^{(i)}} dy^{(1)} \cdots \int_{D_{2^n} \setminus E_k^{(i)}} dy^{(k-1)} \left[\sum_{\substack{y \in X \\ y \neq x}} \|\operatorname{grad}_x(\|y - x\|)\|\right] \\
&\quad \times \exp\left[-\beta\left(\sum_{\substack{y \in X \\ y \neq x}} U(\|y - x\|)\right)\right] \exp\left[-\beta\left(\sum_{\substack{y' \in X \\ y'' \in X_{D_{2^n}} \\ y', y'' \neq x \\ y' \neq y''}} U(\|y' - y''\|)\right)\right] \\
&\leq \text{const} \cdot 2\tau \cdot \frac{\lambda^k}{k!} \Xi_n^{-1} (c_1 \tau \sqrt{n})^{d-1} \\
&\quad \times \exp[-\beta(c_1 \tau \sqrt{n})^2] \cdot \int_{D_{2^n}} dx \int_{D_{2^n}} dy^{(1)} \cdots \int_{D_{2^n}} dy^{(k-1)} \\
&\quad \times \left[\sum_{\substack{y \in X \\ y \neq x}} \|\operatorname{grad}_x U(\|y - x\|)\|\right] \\
&\quad \times \exp\left[-\beta\left(\sum_{\substack{y \in X \\ y \neq x}} U(\|y - x\|)\right)\right] \exp\left[-\beta\left(\sum_{\substack{y' \in X \\ y'' \in X_{D_{2^n}} \\ y', y'' \neq x \\ y' \neq y''}} U(\|y' - y''\|)\right)\right]
\end{aligned}$$

Using the expression for the Gibbs conditional state $\mu(\cdot | X_{\bar{D}_{2^n}}, V_{\bar{D}_{2^n}})$, we can continue this inequality

$$\begin{aligned}
\nu_k(\Gamma_k^{(i),\tau}) &= \mu(\Gamma_k^{(i),\tau} | X_{\bar{D}_{2^n}}, V_{\bar{D}_{2^n}}) \\
&\leq \text{const} \cdot (c_1 \tau \sqrt{n})^{d-1} \cdot \exp[-\beta(c_1 \tau \sqrt{n})^2] \\
&\quad \times \int \sum_{\substack{x, y \in X \\ x \neq y}} \|\operatorname{grad} U(\|y - x\|)\| \, d\mu(X_{D_{2^n}}, V_{D_{2^n}} | X_{\bar{D}_{2^n}}, V_{\bar{D}_{2^n}}).
\end{aligned}$$

Now taking the sum over k, i from 1 to r_n and taking into consideration (23) and (24), we get

$$\mu\left(\bigcup_{|t|\le\tau} S^t(D_{2^n})N^{(2)}_{c_1,\tau,n}\,|\,X_{D_{2^n}}, V_{D_{2^n}}\right) \le \text{const}\cdot 4^n\cdot(c_1\tau\sqrt{n})^{d-1}\cdot\exp[-\beta(c_1\tau\sqrt{n})^2]\cdot I_n,$$

where

$$I_n = \int_M \sum_{\substack{x,y\in X\\ x\ne y}} \|\text{grad } U(\|y-x\|)\|\, d\mu(X_{D_{2^n}}, V_{D_{2^n}}\,|\,X_{\bar{D}_{2^n}}, V_{\bar{D}_{2^n}}).$$

To prove the lemma, it now suffices to show that

$$I_n \le \text{const}\cdot 2^{nd} \tag{25}$$

Indeed, in this case, if we integrate over all possible boundary conditions $(X_{\bar{D}_{2^n}}, V_{\bar{D}_{2^n}})$, i.e., if we pass from the Gibbs conditional state to the unconditional one, we obtain from (24):

$$\mu^{(2)}_{c_1,\tau,n} \le \text{const}\cdot 8^n(c_1\tau\sqrt{n})^{d-1}\cdot\exp[-\beta(c_1\tau\sqrt{n})^2]$$

and therefore, for a sufficiently large c_1, we get

$$\sum_{\tau,n=1}^{\infty} \mu^{(2)}_{c_1,\tau,n} < \infty$$

Together with (18), this implies the validity of (17).

Now let us pass to the proof of the estimate (25). To do this, consider once again the partition of the ball D_{2^n} into the domains $E_n^{(i)}$, $1\le i\le r_n$, satisfying conditions (21), (22), and (23). Introduce the functions $\Phi_n^{(i)}((X, V))$, $1\le i\le r_n$,

$$\Phi_n^{(i)}((X,V)) = \begin{cases} 0 & \text{if } X\cap E_n^{(i)} = \varnothing,\\ \sum\limits_{\substack{y\in X\\ y\ne x}} \|\text{grad } U(\|y-x\|)\| & \text{if } X\cap E_n^{(i)} = \{x\}.\end{cases}$$

Then $I_n = \frac{1}{2}\sum_{i=1}^{r_n} I_n^{(i)}$, where

$$I_n^{(i)} = \int_M \Phi_n^{(i)}((X,V))\, d\mu(X_{D_{2^n}}, V_{D_{2^n}}\,|\,X_{\bar{D}_{2^n}}, V_{\bar{D}_{2^n}}).$$

In view of (23) it suffices to show that

$$I_n^{(i)} \le \text{const} \tag{26}$$

We shall use the formula of complete expectation, in other words, write the integral $I_n^{(i)}$ in the form of a repeated integral. In the inner integral, let us fix the configuration and velocities of the particles in the volume $(D_{2^n}\setminus E_n^{(i)})$ and integrate with respect to the conditional measure

$$d\mu(\cdot \mid X_{\bar{D}_{2^n}}, V_{\bar{D}_{2^n}} \mid X_{D_{2^n}\setminus E_n^{(i)}}, V_{D_{2^n}}\setminus E_n^{(i)}),$$

which, as can be easily checked, coincides with the Gibbs conditional state in the volume $E_n^{(i)}$. The result of the inner integration must now be integrated with respect to the measure $d\mu(\cdot \mid X_{\bar{D}_{2^n}}, V_{\bar{D}_{2^n}})$. Thus

$$I_n^{(i)} = \int d\mu(X_{D_{2^n}}, V_{D_{2^n}} \mid X_{\bar{D}_{2^n}}, V_{\bar{D}_{2^n}})$$

$$\times \int d\mu(X_{E_n^{(i)}}, V_{E_n^{(i)}} \mid X_{\bar{E}_n^{(i)}}, V_{\bar{E}_n^{(i)}}) \sum_{\substack{y \in X \\ y \neq x}} \|\operatorname{grad} U(\|y - x\|)\|.$$

Since the Gibbs conditional state $\mu(\cdot \mid X_{\bar{D}_{2^n}}, V_{\bar{D}_{2^n}})$ is a normalized measure, the inequality (26) will follow from the boundedness of the function now being integrated, i.e., from the boundedness of the inner integral

$$I_n^{(i)}((X, V)) = \int \sum_{\substack{y \in X \\ y \neq x}} \|\operatorname{grad} U(\|y - x\|)\| \, d\mu(X_{E_n^{(i)}}, V_{E_n^{(i)}} \mid X_{\bar{E}_n^{(i)}}, V_{\bar{E}_n^{(i)}}) \quad (27)$$

Note that all the points over which the sum in (27) is taken may be assumed lying in $D_{2^n+r_1}$ (here r_1 is the radius of action of the potential $U(r)$). Similarly to the way we partition D_{2^n}, let us partition the ball $D_{2^n+r_1}$ into domains $E_n^{(j)}$, $1 \le j \le \bar{r}_n$ satisfying conditions (21), (22), and (23), where i is replaced by j and r_n by $\bar{r}_n$.

Denote by y_j the unique point in the intersection $X \cap E_n^{(j)}$ (when it is not void). We may assume that to each domain $E_n^{(j)}$, $1 \le j \le \bar{r}_n$, corresponds precisely one summand in (27): if for some j we have $X \cap E_n^{(j)} = \varnothing$ or $y_j = x$, then we formally put $\operatorname{grad} U(\|y_j - x\|) = 0$.

Now consider all possible families of integers $p = (p_1, \ldots, p_{\bar{r}_n})$ of length $\bar{r}_n$ and represent the integral $I_n^{(i)}((X, V))$ in the form of a sum:

$$I_n^{(i)}((X, V)) = \sum_p \int_{M_{n,p}^{(i)}} \sum_{\substack{y \in X \\ y \neq x}} \|\operatorname{grad} U(\|y - x\|)\| \, d\mu(X_{E_n^{(i)}}, V_{E_n^{(i)}} \mid X_{\bar{E}_n^{(i)}}, V_{\bar{E}_n^{(i)}})$$

$$\stackrel{\text{def}}{=} \sum_p I_n^{(i)}((X, V); p_1, \ldots, p_{\bar{r}_n}),$$

where in any fixed summand the integration is carried out over the set $M_{n,p}^{(i)}$ of those configurations $X_{E_n^{(i)}}$ which satisfy $p_j \le U(\|y_j - x\|) < p_j + 1$

for all j, $1 \leq j \leq \bar{r}_n$. Using the expression for the Gibbs conditional state, we can rewrite this summand in the form

$$I_n^{(i)}((X, V), p_1, \ldots, p_{\bar{r}_n}) = \frac{1}{\Xi_n^{(i)}} \int_{E_{n,p}^{(i)}} \sum_{j=1}^{\bar{r}_n} \|\operatorname{grad} U(\|y_j - x\|)\| \times \exp\left[-\beta\left(\sum_{j=1}^{\bar{r}_n} U(\|y_j - x\|)\right)\right] dx, \tag{28}$$

where $E_{n,p}^{(i)}$ is the set of all $x \in E_n^{(i)}$ which are contained in the configurations in $M_{n,p}^{(i)}$ and $\Xi_n^{(i)}$ is the norming coefficient for the Gibbs conditional state in the volume $E_n^{(i)}$. In view of the estimate (7), we have

$$\Xi_n^{(i)} \geq e^{-\rho(E_n^{(i)})} \geq \text{const}. \tag{29}$$

Note that the potential $U(r)$ is bounded from below: $U(r) \geq$ const. Therefore if we replace U by a non-negative potential $U^+(r) = U(r) - \min_r U(r)$, the right-hand side of (28) will be multiplied by const. Taking this into consideration, we can immediately estimate (28), assuming that $U(r) = U^+(r) \geq 0$.

Using (29), we can write

$$I_n^{(i)}((X, V), p_1, \ldots, p_{\bar{r}_n}) \leq \text{const} \int_{E_{n,p}^{(i)}} \sum \|\operatorname{grad} U(\|y_j - x\|)\| \exp[-\beta \sum U(\|y_j - x\|)]\, dx.$$

In this expression, let us take the sum only over those j, $1 \leq j \leq r_n$, for which y_j belongs to the r_1-neighborhood of the set $E_n^{(i)}$, i.e., the number of summands is no greater than const'. Since we assume $U(r)$ non-negative, we have $I_n^{(i)}((X, V), p_1, \ldots, p_{\bar{r}_n}) = 0$, if not all p_j are non-negative. Therefore we shall assume that $p_j \geq 0$ for all j. Denoting $q = \sum p_j$ and using condition (4) for $U(r)$, we finally get

$$\begin{aligned} I_n^{(i)}((X, V), p_1, \ldots, p_{\bar{r}_n}) &\leq \text{const} \sum_{q=0}^{\infty} \sum_{p:\Sigma p_j = q} \int_{E_{n,p}^{(i)}} \left[\sum_j (p_j + 1)^{\gamma_3}\right] e^{-\Sigma p_j}\, dx \\ &\leq \text{const} \sum_{q=0}^{\infty} \sum_{p:\Sigma p_j = q} \int_{E_{n,p}^{(i)}} [q + \text{const}']^{\gamma_3} e^{-q}\, dx \\ &\leq \text{const} \sum_{q=0}^{\infty} \int_{E_n^{(i)}} [q + \text{const}']^{\gamma_3} e^{-q}\, dx \\ &\leq \text{const} \sum_{q=0}^{\infty} [q + \text{const}']^{\gamma_3} e^{-q} \leq C < \infty. \end{aligned}$$

Lemma 3 is proved, and this concludes the proof of Theorem 2. □

Concerning the Difference Between Ergodic Theory and Nonequilibrium Statistical Mechanics. Suppose $(M, \mathfrak{S}, \mu)$ is a measure space and $\{T^t\}$ is a flow on M. As we have already explained in Chap. 1, the mixing property may be stated as follows. Suppose μ_0 is the initial measure (in M) absolutely continuous with respect to μ and $f_0(x) = d\mu_0/d\mu$. Then the measure μ_t, defined by the relation $\mu_t(A) = \mu_0(T^{-t}A)$, $A \in \mathfrak{S}$, is absolutely continuous with respect to μ and its density $d\mu^t/d\mu$ equals $f_0(T^t x)$. The mixing condition means that the measures μ_t weakly converge to the measure μ in the sense that for any bounded measurable function $g(x)$ the integral

$$\int g(x)\, d\mu_t(x) = \int g(x)\, f_0(T^t x)\, d\mu(x),$$

tends to $\int g(x)\, d\mu(x)$ when $t \to \infty$.

For the dynamical systems of statistical mechanics constructed in this section, the presence of the mixing property is of no particular interest from the point of view of statistical mechanics. Indeed, the main example of the measure μ_0 is constructed by using the function f_0 depending on the coordinates and velocities of a finite number of particles. However, such densities are very unnatural from the point of view of statistical mechanics, since there is no reasonable way of disturbing the probability distribution of a finite number of particles contained in an infinite collection of particles with "equal rights." In infinite dimensional phase spaces, the appropriate probability distribution are, overall, singular with respect to μ, but their restrictions to an arbitrary finite set of families (X_D, V_D) are absolutely continuous with respect to the same kind of restrictions of the measures μ. A simple example is obtained if, instead of the Poisson measure ν_λ with constant intensity, we take a Poisson measure with intensity depending periodically or almost periodically on the points. Such classes of measures are naturally constructed in nonequilibrium statistical mechanics. The main problem of nonequilibrium statistical mechanics is the description of the limit state of the evolution of such measures. The problem is extremely difficult in such general form. In the physical literature, techniques for the description of the evolution of so-called local equilibrium states have been developed. We shall not discuss this in more detail here.

§3. Dynamical Systems and Partial Differential Equations

The ergodic theory of dynamical systems corresponding to partial differential equation has not yet been constructed. There are only some separate examples and related partial results.

In this section we shall consider some aspects of the problem related to classical field theory. Suppose D is an open bounded domain in $\mathbb{R}^d$. The configuration of the classical field in the domain D is a function $\varphi(x)$ defined on D and satisfying boundary conditions of one type or another. A trajectory of a field is a function of two variables $\varphi(x, t)$, $-\infty < t < \infty$; $\varphi_t(x, t)$ is the derivative of the field with respect to time at the point x. The equations of motion usually follow from one of the variational principles. To obtain them in Hamiltonian form, one must start from a Hamiltonian of the field. In the simplest case, denote $\pi(x) = \varphi_t(x, t)$, consider the space of pairs $(\pi(x), \varphi(x)) = (\pi, \varphi)$ and the Hamiltonian H of the form

$$H(\pi, \varphi) = \int_D \left[\frac{1}{2}(\pi^2 + (\text{grad } \varphi)^2 + P(\varphi)\right] dx.$$

Here P is a "nice" function of one variable, e.g., $P(\varphi)$ is a polynomial, or $P(\varphi) = \cos \varphi$, $\sin \varphi$, or the like. The summand $\frac{1}{2}\pi^2$ is the density of kinetic energy, while the summand $\frac{1}{2}(\text{grad } \varphi)^2 + P(\varphi)$ constitutes the density of potential energy. The function under the integral sign is known as the density of the Hamiltonian. The equations of motion are of the form

$$\begin{cases} \varphi_t = \pi, \\ \pi_t = \Delta\varphi - P'(\varphi), \end{cases} \tag{1}$$

or $\varphi_{tt} = \Delta\varphi - P'(\varphi)$. The last equation is an equation of the hyperbolic type. If, for the phase space, we take the space of pairs (φ, φ_t) for which the uniqueness and existence theorems of equations (1) hold, then the one-parameter group $\{T^t\}$ of translations along the solutions of (1) naturally arises in this phase space. The constructions of nontrivial invariant measures for $\{T^t\}$ known at present are, unfortunately, based on the explicit expression for the solutions.

As an example, consider the linear case when $P(\varphi) = m\varphi^2$. Suppose the domain D is compact and the boundary conditions are such that the operator $\Delta\varphi + 2m\varphi$ is self-adjoint in $L^2(D)$. Consider an orthonormalized system of eigen-functions

$$\varphi_k \colon \Delta\varphi_k + 2m\varphi_k = \lambda_k \varphi_k,$$

and assume $\lambda_k < 0$. Writing φ, π in the form $\varphi = \sum c_k \varphi_k$, $\pi = \sum d_k \varphi_k$, we see that the equations (1) are replaced by an infinite system of one-dimensional equations

$$\begin{cases} \dot{c}_k = d_k \\ \dot{d}_k = \lambda_k c_k. \end{cases} \tag{2}$$

Then $-\lambda_k c_k^2 + d_k^2 = I_k$ constitutes a complete family of first integrals of our system. Hence in the entire functional space $L^2(D)$ the flow $\{T^t\}$ is nonergodic. When I_k are fixed, the flow $\{T^t\}$ reduces to a conditionally periodic motion on the infinite dimensional torus. Its ergodic properties depend on the commensurability properties of the frequencies $\sqrt{-\lambda_k}$.

Part II

Basic Constructions of Ergodic Theory

In ergodic theory there are relatively few general results concerning arbitrary dynamical systems, proved by means of constructions in general measure spaces. Such results and constructions are collected in this part of the book.

Chapter 10

Simplest General Constructions and Elements of Entropy Theory of Dynamical Systems

§1. Direct and Skew Products of Dynamical Systems

1. *Direct products.* Consider the measure space $(M, \mathfrak{S}, \mu)$ which is the Cartesian product of two other measure spaces

$$(M, \mathfrak{S}, \mu) = (M_1 \times M_2, \mathfrak{S}_1 \times \mathfrak{S}_2, \mu_1 \times \mu_2)$$

Assume that the dynamical systems $\{T_1^t\}$, $\{T_2^t\}$ act in the factors, either both with discrete time or both with continuous time.

Definition 1. The dynamical system $\{T^t\}$ acting in the space M according to the formula $T^t(x_1, x_2) = (T_1^t x_1, T_2^t x_2)$ is said to be the *direct product* of the dynamical systems $\{T_1^t\}$ and $\{T_2^t\}$.

It is clear that $\{T^t\}$ preserves the measure $\mu = \mu_1 \times \mu_2$. Similarly one defines the direct product of any finite or countable number of dynamical systems.

EXAMPLES

1. Suppose M is the torus Tor^m with cyclic coordinates $x_1, \ldots, x_m$ and T is the group translation, i.e.,

$$T(x_1, \ldots, x_m) = (x_1 + \alpha_1 \pmod 1, \ldots, x_m + \alpha_m \pmod 1).$$

Then T is the direct product of m rotations of the circle.

2. Suppose $M = M_1 \times M_2$ and the automorphisms T_1, T_2 are Bernoulli (Markov) automorphisms on M_1, M_2 respectively. Then $T = T_1 \times T_2$ is also a Bernoulli (Markov) automorphism.

3. Suppose the flow $\{S^t\}$ acts in the space $(R, \mathfrak{A}, \rho)$ and $\{T^t\}$ is the Poisson suspension over $\{S^t\}$. If the space R is representable in the form $R = R_1 \cup R_2 \pmod 0$ where $R_i \in \mathfrak{A}$, $\rho(R_i) > 0$, R_i is invariant with respect to $\{S^t\}$ $(i = 1, 2)$ and $R_1 \cap R_2 = \varnothing$, then $\{T^t\} = \{T_1^t\} \times \{T_2^t\}$ where $\{T_i^t\}$ is the Poisson suspension over $\{S_i^t\}$—the restriction of $\{S^t\}$ to R_i $(i = 1, 2)$.

For any automorphism T, denote by $\Lambda_d(T)$ the set of eigen-values of the unitary operator U_T adjoint to T.

Theorem 1. *For a direct product of two ergodic automorphisms $T = T_1 \times T_2$ to be ergodic, it is necessary and sufficient to have the relation $\Lambda_d(T_1) \cap \Lambda_d(T_2) = \{1\}$.*

Proof. Necessity. Denote by U_1, U_2, U the unitary operators adjoint to T_1, T_2, T respectively. Assume that a certain λ, $\lambda \neq 1$, belongs to the intersection $\Lambda_d(T_1) \cap \Lambda_d(T_2)$. Then there exist normalized eigen-functions f_1, f_2 for which $U_1 f_1 = \lambda f_1$, $U_2 f_2 = \lambda f_2$. It follows from the last equality that

$$(U_2 \bar{f}_2)(x_2) = \bar{f}_2(T_2 x_2) = \bar{\lambda}\bar{f}_2(x_2) = \lambda^{-1}\bar{f}_2(x_2).$$

The function $f(x_1, x_2) = f(x_1) \cdot \bar{f}_2(x_2)$ is obviously normalized and orthogonal to the subspace of constants and

$$Uf = U_1 f \cdot U_2 \bar{f}_2 = \lambda\lambda^{-1} f_1 \bar{f}_2 = f,$$

i.e., $f(x_1, x_2)$ is invariant (mod 0). This contradicts the ergodicity of T.

Sufficiency. Suppose

$$\Lambda_d(T_1) = \{\lambda_k^{(1)}\}, \quad k = 0, 1, 2, \ldots;$$

$$\Lambda_d(T_2) = \{\lambda_l^{(2)}\}, \quad l = 0, 1, 2, \ldots; \lambda_0^{(1)} = \lambda_0^{(2)} = 1.$$

Denote the orthonormed systems of eigen-functions for U_1, U_2 by $\{f_k^{(1)}\}$, $\{f_l^{(2)}\}$ respectively. Put

$$f_{k,l}(x_1, x_2) = f_k^{(1)}(x_1) \cdot f_l^{(2)}(x_2).$$

Then $Uf_{k,l} = U_1 f_k^{(1)} \cdot U_2 f_l^{(2)} = \lambda_k^{(1)} f_k^{(1)} \cdot \lambda_l^{(2)} f_l^{(2)} = \lambda_k^{(1)} \lambda_l^{(2)} \cdot f_{k,l}$, i.e., $f_{k,l}$ is an eigen-function for U with eigen-value $\lambda_k^{(1)} \cdot \lambda_l^{(2)}$. It follows from the assumptions of the theorem that $\lambda_k^{(1)} \cdot \lambda_l^{(2)} \neq 1$ as long as one of the following relations holds $k \neq 0$, $l \neq 0$. Indeed, in the converse case, $\bar{\lambda}_l^{(2)} = (\lambda_l^{(2)})^{-1} = \lambda_k^{(1)}$. Hence $\lambda_l^{(2)} \in \Lambda_d(T_1)$ (the corresponding eigen-function is $\bar{f}_k^{(1)}$), which contradicts the assumption.

Now let us show that the operator U has no other eigen-values. This will imply our theorem, since every invariant (mod 0) function for T is an eigen-function for U with eigen-value 1.

Introduce the subspaces $H_1^{(c)}$, $H_2^{(c)}$ of the spaces $L^2(M_1, \mathfrak{S}_1, \mu_1)$, $L^2(M_2, \mathfrak{S}_2, \mu_2)$ (respectively) consisting of all functions orthogonal to all the $\{f_k^{(1)}\}$ (respectively to all the $\{f_l^{(2)}\}$). Suppose H^c is the closure in $L^2(M, \mathfrak{S}, \mu)$ of the set of functions of the form $\sum c_{j_1, j_2} g_{j_1}^{(1)}(x_1) g_{j_2}^{(2)}(x_2)$, the number of coefficients c_{j_1, j_2} being finite. Also introduce the subspaces $H_k^{(1)}$, $H_l^{(2)}$ (of the space $L^2(M, \mathfrak{S}, \mu)$) consisting of functions of the form $f_k^{(1)}(x_1) g(x_2)$, $g \in H_2^{(c)}$, and $g'(x_1) \cdot f_l^{(2)}(x_2)$, $g' \in H_1^{(c)}$; finally, denote by $H_{k,l}$ the one-dimensional subspaces which span the functions $f_{k,l}$. It is easy to see that

$$L^2(M, \mathfrak{S}, \mu) = (1) \oplus \sum H_{k,l} \oplus \sum H_k^{(1)} \oplus \sum H_l^{(2)} \oplus H^{(c)},$$

$$UH_k^{(1)} = H_k^{(1)}, \; UH_l^{(2)} = H_l^{(2)}, \; UH^{(c)} = H^{(c)},$$

since the subspaces $H_1^{(c)}$, $H_2^{(c)}$ are invariant with respect to U_1, U_2 respectively. (Here (1) denotes the subspace spanned by constants).

Now let us show that the subspaces $H_k^{(1)}$, $H_l^{(2)}$, $H^{(c)}$ contain no eigen-functions of the operator U. Consider the case of the subspace $H^{(c)}$; the other cases are simpler and we shall omit them.

For any function

$$f = \sum c_{j_1, j_2} g_{j_1}^{(1)}(x_1) g_{j_2}^{(2)}(x_2)$$

the scalar product $(U^n f, f)$ tends to zero when n ranges over a certain set of density 1. Indeed the fact that it tends to zero for

$$(U^n(g_{j_1}^{(1)} \cdot g_{j_2}^{(2)}), g_{j_1'}^{(1)} \cdot g_{j_2'}^{(2)}) = (U_1^n g_{j_1}^{(1)}, g_{j_1'}^{(2)}) \cdot (U_2^n g_{j_2}^{(2)}, g_{j_2'}^{(2)})$$

follows from the properties of the subspaces $H_1^{(c)}$, $H_2^{(c)}$ (see Appendix 2). Hence all the functions f of the type indicated are orthogonal to the eigen-functions of the operator U and, therefore, this is true for all $f \in H^{(c)}$.

Thus the eigen-functions of the operator U are of the form $f_k^{(1)} \cdot f_l^{(2)}$ and among them only the function $f_0^{(1)} \cdot f_0^{(2)} = \text{const}$ is invariant with respect to T. The theorem is proved. □

Corollary 1. *If T_1 is weak mixing and T_2 is an ergodic automorphism, then $T = T_1 \times T_2$ is an ergodic automorphism.*

Corollary 2. *If T_1 and T_2 are weak mixing automorphisms, then $T = T_1 \times T_2$ is weak mixing automorphism.*

This does not follow from the statement of Theorem 1, but from its proof. Indeed, we showed that in the subspace $H^{(c)}$ (of the space $L^2(M, \mathfrak{S}, \mu)$), which is the orthogonal complement to the orthogonal sum of the invariant subspaces $H_{k,l}$, there are no eigen-functions of the operator U. In our case this means that U has no eigen-functions at all, except for constants.

Theorem 2. *If T_1 and T_2 are mixing automorphisms then so is $T = T_1 \times T_2$.*

Proof. Suppose $f \in L_0^2(M, \mathfrak{S}, \mu)$ is of the form $f = \sum c_{j_1, j_2} f_{j_1}(x_1) \cdot g_{j_2}(x_2)$, where only a finite number of coefficients c_{j_1, j_2} do not vanish. (Here $L_0^2(M, \mathfrak{S}, \mu)$ is the orthogonal complement to the subspace of constants). Then

$$(U^n f, f) = \sum c_{j'_1, j'_2} \cdot \bar{c}_{j''_1, j''_2} (U_1^n f_{j'_1}, f_{j''_1}) \cdot (U_2^n g_{j'_2}, g_{j''_2}) \to 0,$$

when $n \to \infty$. In the general case, for an arbitrary function $f \in L_0^2(M, \mathfrak{S}, \mu)$, $\|f\| = 1$ and any $\varepsilon > 0$, we can find a function $f' \in L_0^2(M, \mathfrak{S}, \mu)$, $\|f'\| = 1$ which may be represented in the form of a finite sum $\sum c_{j_1, j_2} f_{j_1}(x_1) g_{j_2}(x_2)$ for which $\|f - f'\| < \varepsilon/4$. Then for sufficiently large n we will have

$$|(U^n f, f)| \le |(U^n f', f')| + |U^n(f - f'), f')| + |(U^n f', f - f')| + |(U^n(f - f'), f - f')| < \frac{\varepsilon}{4} + 2\|f - f'\| + \|f - f\|^2 < \varepsilon,$$

i.e., $(U^n f, f) \to 0$.

The theorem is proved. □

2. *Factors of dynamical systems.* Suppose we are given the measurable spaces $(M_1, \mathfrak{S}_1)$, $(M_2, \mathfrak{S}_2)$ and, on each of them, a measurable dynamical system $T_1^t: M_1 \to M_1$, $T_2^t: M_2 \to M_2$, the time in both systems is the same.

Consider a measurable map $\varphi: M_1 \to M_2$ which commutes with the action of the dynamical systems, i.e., $\varphi T_1^t x = T_2^t \varphi x$ for any $x \in M_1$ and all t. Hence, in particular, the subset $\varphi(M_1)$ of the space M_2 is invariant with respect to $\{T_2^t\}$ and, for the sequel, it suffices to limit oursleves to the case when $\varphi(M_1) = M_2$. If μ_1 is an invariant measure for $\{T_1^t\}$, then the measure μ_2 defined on $\mathfrak{S}_2$ by the formula $\mu_2(A) = \mu_1(\varphi^{-1}(A))$, $A \in \mathfrak{S}_2$, will be invariant for $\{T_2^t\}$

Definition 2. The dynamical system $\{T_2^t\}$ on the space $(M_2, \mathfrak{S}_2, \mu_2)$ is said to be a *factor* of the dynamical system $\{T_1^t\}$.

EXAMPLES

1. Every dynamical system $\{T^t\}$ has at least two factors: the given system itself (φ is the identity) and the "trivial" system (when M_2 consists of a single point and the map φ sends M_1 into this point).

2. Suppose Y_1, Y_2 are finite sets and

$$M_1 = \prod_{n=-\infty}^{\infty} Y_1^{(n)}, \quad Y_1^{(n)} \equiv Y_1, \qquad M_2 = \prod_{n=-\infty}^{\infty} Y_2^{(n)}, \quad Y_2^{(n)} \equiv Y_2.$$

Let T_1, T_2 be the shifts in M_1 and M_2 respectively. Any map φ_0 of the set Y_1 onto the set Y_2 may be extended in a natural way to a map φ of the space M_1 onto M_2. If normalized measures σ_1, σ_2 were given on Y_1 and Y_2 respectively, and φ_0 maps σ_1 to σ_2, then φ maps the Bernoulli measure μ_1 corresponding to σ_1 into the Bernoulli measure μ_2 corresponding to σ_2, and the Bernoulli automorphism T_2 (see Chap. 8) may be represented as a factor-automorphism of the Bernoulli automorphism T_1.

3. If $M = M_1 \times M_2$ and the dynamical system $\{T^t\}$ on M is the direct product of the dynamical systems $\{T_1^t\}$ and $\{T_2^t\}$, then each of the latter is a factor-system of the dynamical system $\{T^t\}$. It suffices to consider the natural projections $\varphi_i: M \to M_i$. Further this example will be considerably generalized.

Theorem 3. *If the dynamical system $\{T^t\}$ is*

(1) *ergodic, then any of its factors is ergodic;*
(2) *weak mixing, then any of its factors is weak mixing;*
(3) *mixing, then any of its factors is mixing.*

The statements (1), (2), and (3) are proved in similar fashion, hence we shall only prove the first one. If the set $A \in \mathfrak{S}_2$ is invariant with respect to $\{T_2^t\}$, then $\varphi^{-1}(A)$ is invariant with respect to $\{T_1^t\}$. It follows from the ergodicity of $\{T_1^t\}$ that $\mu_2(A) = \mu_1(\varphi^{-1}(A))$ equals 0 or 1. The theorem is proved.

3. *Skew products of dynamical systems.* The construction described below considerably generalizes the direct product of dynamical systems and often appears in ergodic theory.

Suppose $M = (M_1 \times M_2, \mathfrak{S}_1 \times \mathfrak{S}_2, \mu_1 \times \mu_2)$ is the direct product of measure spaces $(M_1, \mathfrak{S}_1, \mu_1)$ and $(M_2, \mathfrak{S}_2, \mu_2)$. Assume that in $(M_1, \mathfrak{S}_1, \mu_1)$ there acts the dynamical system $\{T_1^t\}$ preserving the measure μ_1 and, for every $x_1 \in M_1$, there is a dynamical system $\{T_2^t(x_1)\}$ on $(M_2, \mathfrak{S}_2, \mu_2)$ which measurably depends on x_1 in the following sense: for every measurable function $f(x_1, x_2)$ on M the function $f(T_1^t x_1, T_2^t(x_1)x_2)$ is measurable on the direct product $T \times M$, where T is the common time of the dynamical systems $\{T_1^t\}$ and $\{T_2^t(x_1)\}$. Then the group $\{T^t\}$ given by the formula $T^t(x_1, x_2) = \{T_1^t x_1, T_2^t(x_1)x_2)$ is a measurable dynamical system acting on M. Let us show that $\{T^t\}$ preserves the measure μ. Indeed, for any $A \in \mathfrak{S}$, $A = A_1 \times A_2, A_i \in \mathfrak{S}_i, i = 1, 2$ we have

$$U^t \chi_A = \chi_{T_1^{-t} A_1}(x_1) \cdot \chi_{T_2^{-t}(x_2) A_2}(x_2),$$

where $\{U^t\}$ is the group of unitary operators adjoint to $\{T^t\}$ and

$$\int U^t \chi_A \, d\mu_1 \, d\mu_2 = \int \chi_{T_1^{-t} A_1}(x_1) \, d\mu_1(x_1) \int \chi_{T_2^{-t}(x_1) A_2}(x_2) \, d\mu_2(x_2)$$
$$= \mu_1(A_1) \cdot \mu_2(A_2) = \mu(A).$$

This implies the invariance of the measure μ.

Definition 3. The dynamical system $\{T^t\}$ defined above is said to be a *skew product* of the dynamical systems $\{T_1^t\}$ and $\{T_2^t(x_1)\}$. The space M_1 is called the *base* and $\{T_1^t\}$ is the *dynamical system acting on the base*, while the dynamical systems $\{T_2^t(x_1)\}$ are referred to as *acting on the fibers*.

Sometimes we shall also say that $\{T^t\}$ is a skew product *over* $\{T_1^t\}$.

If $\pi: M \to M_1$ is the natural projection, then obviously $\pi T^t x = T_1^t \pi x$ for all $x \in M$. This means $\{T_1^t\}$ is a factor of the dynamical system $\{T^t\}$. It can be shown that (under fairly extensive assumptions) any ergodic dynamical system $\{T^t\}$ possessing a factor $\{T_1^t\}$ can be presented in the form of a skew product of $\{T_1^t\}$ and the dynamical systems $\{T_2^t(x_1)\}$, acting on certain measure space.

EXAMPLES

1. Suppose T_1 is an automorphism of the measure space $(M_1, \mathfrak{S}_1, \mu_1)$, and $(M_2, \mathfrak{S}_2, \mu_2)$ is the space consisting of r points of equal measure. Every measure-preserving transformation of M_2 is a permutation. Therefore, if $(M, \mathfrak{S}, \mu) = (M_1, \mathfrak{S}_1, \mu_1) \times (M_2, \mathfrak{S}_2, \mu_2)$, then every skew product in M over T_1 is given by a measurable map of the space M_1 into the permutation group of the set of r elements.

2. Suppose $M = S^1 \times S^1$ is the two-dimensional torus, T_1 is the group translation in S^1, i.e., $T_1 x_1 = x_1 + \alpha (\mathrm{mod}\ 1)$ while $T_2(x_1)$ acts according to the formula $T_2(x_1)x_2 = x_2 + f(x_1) (\mathrm{mod}\ 1)$, where f is a measurable function on S^1. The skew product $T(x_1, x_2) = (x_1 + \alpha (\mathrm{mod}\ 1), x_2 + f(x_1)(\mathrm{mod}\ 1))$ is often referred to as a *skew translation* on the torus.

In relation to the notion of skew product, let us introduce the following definition.

Definition 4. Suppose T is an automorphism of the space $(M, \mathfrak{S}, \mu)$; G is a measurable group, and $\mathbb{Z}$ is the additive group of integers. By a *cocycle* for T with values in the group G we mean a measurable map $S: M \times \mathbb{Z} \to G$, which satisfies the condition

$$S(x, m + n) = S(x, m) \cdot S(T^m x, n).$$

EXAMPLE

Suppose $G = \mathbb{R}^1$ is the additive group of real numbers. Any measurable function $f(x)$ on M generates a cocycle with values in G according to the formula

$$S(x, n) = \begin{cases} \sum_{k=0}^{n-1} f(T^k x) & \text{for } n \geq 1, \\ 0 & \text{for } n = 0, \\ -\sum_{k=1}^{-n} f(T^{-k} x) & \text{for } n < 0. \end{cases}$$

For the group G we can take the group $\mathfrak{M}$ of all automorphisms of some measure space $(M_2, \mathfrak{S}_2, \mu_2)$. A topology may be introduced into this group by specifying a basis of open sets of the form $\{T: \mu_2(T^{-1}A \triangle T_0^{-1}A) < \varepsilon\}$, $T_0 \in \mathfrak{M}$, $A_0 \in \mathfrak{S}_2$, $\varepsilon > 0$. The measurable structure in $\mathfrak{M}$ is the Borel structure induced by this topology.

If T_1 is an automorphism of the space $(M_1, \mathfrak{S}_1, \mu_1)$, then any skew product of the automorphisms T_1 and $T_2(x_1)$, acting in the space $M = (M_1 \times M_2, \mathfrak{S}_1 \times \mathfrak{S}_2, \mu_1 \times \mu_2)$, generates a cocycle for T_1 with values in the group $\mathfrak{M}$. This cocycle is defined by the formula

$$S(x_1, n) = T_2(x_1) \cdot T_2(T_1 x_1) \cdot \ldots \cdot T_2(T_1^{n-1} x_1).$$

§2. Metric Isomorphism of Skew Products. Equivalence of Dynamical Systems in the Sense of Kakutani

1. Suppose $M = M_1 \times M_2$ is the product of measure spaces, T_1 is an automorphism of the space M_1, $\tilde{T}_2(x_1)$, $\tilde{\tilde{T}}_2(x_1)$, $x_1 \in M$ are two measurable families of automorphisms of the space M_2. Construct the skew products $\tilde{T}$, $\tilde{\tilde{T}}$ over T_1 which act according to the formulas

$$\tilde{T}(x_1, x_2) = (T_1 x_1, \tilde{T}_2(x_1) x_2), \qquad \tilde{\tilde{T}}(x_1, x_2) = (T_1 x_1, \tilde{\tilde{T}}_2(x_1) x_2).$$

Introduce the cocycles $\tilde{S}(x_1, n)$, $\tilde{\tilde{S}}(x_1, n)$ with values in the group $\mathfrak{M}$ of automorphisms of the space M_2 corresponding to $\tilde{T}$, $\tilde{\tilde{T}}$ respectively (see §1).

Definition 1. The cocycles $\tilde{S}(x_1, n)$, $\tilde{\tilde{S}}(x_1, n)$ are called *cohomological*, if there exists a measurable family $R(x_1)$, $x_1 \in M_1$, of automorphisms of the space M_2 which satisfy

$$\tilde{S}(x_1, n) = R^{-1}(T_1^n x_1) \cdot \tilde{\tilde{S}}(x_1, n) \cdot R(x_1).$$

Theorem 1. *If the cocycles* $\tilde{S}$, $\tilde{\tilde{S}}$ *are cohomological, then the skew products* $\tilde{T}$, $\tilde{\tilde{T}}$ *are metrically isomorphic.*

Proof. Suppose φ is the automorphism of the space $M = M_1 \times M_2$ defined by the formula $\varphi(x_1, x_2) = (x_1, R(x_1)x_2)$ then

$$\begin{aligned}\tilde{\tilde{T}}^n \varphi(x_1, x_2) &= (T_1^n x_1, \tilde{\tilde{S}}(x_1, n) R(x_1) x_2) \\ &= (T_1^n x_1, R(T_1^n x_1) \cdot \tilde{S}(x_1, n) x_2) = \varphi \tilde{T}^n(x_1, x_2).\end{aligned}$$

The theorem is proved. □

The converse statement is not true in general.

2. *The equivalence of dynamical systems in the sense of Kakutani.* The problem of the metric classification of dynamical systems, i.e., the classification up to a metric isomorphism, is extremely complicated and will hardly

ever be solved completely, because of the large variety of possibilities which arise here. In this connection, various attempts to weaken the isomorphism condition were undertaken in order to obtain a more compact picture. One of the more successful ones was the notion (due to Kakutani) which, in recent years, has been the center of interest of new research work in ergodic theory.

Definition 2. Suppose T_1, T_2 are ergodic automorphisms of the measure space $(M, \mathfrak{S}, \mu)$. We shall say that they are *equivalent in the sense of Kakutani* or simply *equivalent*, if there exists an automorphism R of an arbitrary measure space $(M_1, \mathfrak{S}_1, \mu_1)$ such that each of the automorphisms T_1, T_2 is metrically isomorphic to some integral automorphism constructed over R.

Theorem 2. *The following statements are equivalent:*

(1) *T_1, T_2 are equivalent in the sense of Kakutani;*
(2) *there exists an automorphism T of the measure space $(M_2, \mathfrak{S}_2, \mu_2)$ and subsets $E_1 \in \mathfrak{S}_1$, $E_2 \in \mathfrak{S}_2$ such that $T_1(T_2)$ is metrically isomorphic to the induced automorphism $T_{E_1}(T_{E_2})$.*

Proof. (1) $\Rightarrow$ (2). Suppose T_1 and T_2 are realized as integral automorphisms over R and $r_1(x_1)$, $r_2(x_2)$ are the corresponding functions. Put $r(x_1) = \max\{r_1(x_1), r_2(x_1)\} + 1$ and denote by T the integral automorphism constructed over R by means of the function $r(x_1)$. The points of the space M_2 are of the form (x_1, i), $1 \le i \le r(x_1)$. Therefore choosing

$$E_1 = \{(x_1, i): 1 \le i \le r_1(x_1)\}, \qquad E_2 = \{(x_1, i): 1 \le i \le r_2(x_1)\}$$

we obtain the necessary statement.

(2) $\Rightarrow$ (1). First assume that $\mu(E_1 \cap E_2) > 0$. Putting $E = E_1 \cap E_2$, we see that T_E is the induced automorphism of the automorphism T_{E_1} as well as of the automorphism T_{E_2}. Therefore T_{E_1} and T_{E_2} are representable as integral automorphisms over T_E. In the general case, note that $T_{T^nE_2}$ and T_{E_2} are metrically isomorphic for any n. It follows from the ergodicity of T that we can find an n such that $\mu(E_1 \cap T^nE_2) > 0$, after which the previous arguments are applicable. The theorem is proved. □

Now we can show that Definition 2 does indeed define a partition of the space of all automorphisms into equivalence classes. The reflexivity and symmetry conditions are obvious. To check transitivity, suppose T_1 is equivalent to T_2 and T_2 is equivalent to T_3. Then there is an automorphism R_1 such that T_1 and T_2 are represented as integral automorphisms over R_1, and an automorphism R_2 such that T_2 and T_3 are represented as integral automorphisms over R_2. But then R_1 and R_2 are representable as automorphisms induced from T_2 and, according to the theorem just proved above, there is an automorphism R_3 such that R_1 and R_2 are representable as integral automorphisms over R_3. In this case, T_1, being integral over R_1 is representable as an integral automorphism over R_3 and, similarly, T_2, T_3 are representable as integral ones over R_3. Therefore T_1, T_2, T_3 are equivalent.

§3. Time Change in Flows

For smooth dynamical systems on differentiable manifolds we have defined, in §2, Chap. 2, a "system obtained from the given one by means of a change of time determined by a smooth positive function." There we displayed a formula expressing the invariant measure for such a system in terms of the invariant measure of the given dynamical system.

Now we describe a similar construction for arbitrary flows on measure spaces.

Suppose $\{T^t\}$ is a flow on the space $(M, \mathfrak{S}, \mu)$ and a function $w(x) \in L^1(M, \mathfrak{S}, \mu)$, $w(x) > 0$ is given on M. From the flow $\{T^t\}$ and the function w we shall construct a new flow $\{\overline{T}^t\}$ "obtained from the original one by means of the change of time determined by the function w." This means that the trajectories of the flow $\{\overline{T}^t\}$ are the same as those of the flow $\{T^t\}$, while the "velocity of motion" along the trajectory at the point $x \in M$ of the flow $\{\overline{T}^t\}$ is $w(x)$ times greater than the one of $\{T^t\}$. Now let us pass to a more precise exposition.

For any point $x \in M$ and any $t \geq 0$, consider the equation (in the unknown u):

$$\int_0^u w(T^s x)\, ds = t. \tag{1}$$

Lemma 1. *For almost all $x \in M$ the equation* (1) *for any $t \geq 0$ has a unique solution $u = u_t(x)$, $u \geq 0$.*

Proof. According to the Birkhoff–Khinchin theorem, the limit

$$\overline{w}^+(x) = \lim_{u \to \infty} \frac{1}{u} \int_0^u w(T^s x)\, ds,$$

exists on a certain invariant set $M' \subset M$, $\mu(M') = 1$ and

$$\int_M \overline{w}^+\, d\mu = \int_M w\, d\mu. \tag{2}$$

Let us show that the function $\overline{w}^+$ is strictly positive almost everywhere. Let

$$E = \{x \in M' : \overline{w}^+(x) = 0\}, \qquad M_1 = M' \setminus E.$$

The set M_1 is obviously invariant with respect to $\{T^t\}$. Applying the Birkhoff–Khinchin theorem to the function $w \cdot \chi_{M_1}$, we get

$$\int_{M_1} \overline{w}^+ \, d\mu = \int_M (\overline{w \cdot \chi_{M_1}})^+ \, d\mu = \int_M w \cdot \chi_{M_1} \, d\mu = \int_{M_1} w \, d\mu. \tag{3}$$

From (2) and (3) we get the relation

$$\int_E w \, d\mu = \int_E \overline{w}^+ \, d\mu = 0.$$

Since w was positive, it follows that $\mu(E) = 0$, i.e., $\mu(M_1) = 1$.

For any point $x \in M_1$ we obviously have

$$\lim_{u \to \infty} \int_0^u w(T^s x) \, ds = \infty.$$

Since this last integral is a continuous and strictly monotonic function of the upper limit, there is a unique value of u satisfying (1). The lemma is proved. □

For $t < 0$, consider the equation

$$-\int_u^0 w(T^s x) \, ds = t.$$

It also possesses a unique solution $u = u_t(x)$ for almost all $x \in M$, and we have $u < 0$.

Thus for almost all $x \in M$ a function $u = u_t(x)$ is defined for any t, $-\infty < t < \infty$. By throwing out from M, if necessary, a certain null set invariant with respect to $\{T^t\}$, we may assume that $u_t(x)$ is defined for all x.

It is clear that, for a fixed x, $u_t(x)$ is a strictly increasing function of t and $u_0(x) = 0$,

$$\lim_{t \to +\infty} u_t(x) = +\infty, \qquad \lim_{t \to -\infty} u_t(x) = -\infty.$$

For a fixed t, the function $u_t(x)$ is measurable. This follows (for $t > 0$) from the equality

$$\{x \in M : u_t(x) < a\} = \left\{x \in M : \int_0^a w(T^s x) ds > t\right\}, \qquad a > 0,$$

and from the measurability of $w(T^s x)$ as a function of two variables x, s. A straightforward verification proves the relation

$$u_{t_1}(x) + u_{t_2}(T^{t_1}x) = u_{t_1+t_2}(x), \qquad x \in M, t_1, t_2 \in \mathbb{R}^1. \tag{4}$$

Now define the transformation $\bar{T}^t$ of the space M by putting

$$\bar{T}^t x = T^{u_t(x)}x, \qquad x \in M.$$

Formula (4) shows that the set of transformations T^t for all $t \in \mathbb{R}^1$ is a one-parameter group: $\bar{T}^{t_1} \cdot \bar{T}^{t_2} = \bar{T}^{t_1+t_2}$. Let us show that $\{\bar{T}^t\}$ is a measurable dynamical system on the measurable space $(M, \mathfrak{S})$.

Introduce the map $\psi: M \times \mathbb{R}^1 \to M \times \mathbb{R}^1$ by means of the formula $\psi(x, t) = (x, u_t(x))$. This map is measurable. Indeed, it suffices to check that for any measurable set $A \subset M \times \mathbb{R}^1$ of the form $A = C \times (a, b)$, $C \in \mathfrak{S}$ the inverse image $\psi^{-1}A$ is measurable. This, in its turn, follows (for $a, b > 0$) from the relation

$$\psi^{-1}A = \left\{(x, t) \in M \times \mathbb{R}^1 : x \in C, \int_0^a w(T^s x)\, ds < t < \int_0^b w(T^s x)\, ds\right\}.$$

Now take an arbitrary set $C \in \mathfrak{S}$. In view of the measurability of $\{T^t\}$, the set

$$A = \{(x, t) \in M \times \mathbb{R}^1 : T^t x \in C\},$$

is measurable. Put $B = \{(x, t) \in M \times \mathbb{R}^1 : \bar{T}^t x \in C\}$. Then

$$B = \{(x, t): T^{u_t(x)}(x) \in C\} = \psi^{-1}A,$$

which implies the measurability of B and therefore that of the dynamical system $\{\bar{T}^t\}$.

On $(M, \mathfrak{S})$ introduce the normalized measure $\bar{\mu}$ (absolutely continuous with respect to μ) such that $d\bar{\mu}/d\mu = w(x)/\int_M w\, d\mu$.

Theorem 1. *The measure $\bar{\mu}$ is invariant with respect to $\{\bar{T}^t\}$. In other words, $\{T^t\}$ is a flow on the space $(M, \mathfrak{S}, \bar{\mu})$.*

Proof. Suppose $f(x)$ is a measurable bounded function on M. For any $t > 0$, consider the time mean

$$f_t(x) = \frac{1}{t}\int_0^t f(\bar{T}^\tau x)\, d\tau = \frac{1}{t}\int_0^t f(T^{u_\tau(x)}x)\, d\tau.$$

Carry out the change of variables $u_\tau(x) = v$ in the last integral. Since by (1) we have

$$\frac{du_\tau(x)}{d\tau} = \frac{1}{w(T^\tau x)},$$

it follows that

$$f_t(x) = \frac{1}{t}\int_0^{u_t(x)} f(T^v x) w(T^v x)\, dv.$$

Putting $g(x) = f(x) \cdot w(x)$, we can write

$$f_t(x) = \frac{1}{t} u_t(x) \cdot \frac{1}{u_t(x)} \int_0^{u_t(x)} g(T^v x)\, dv. \tag{5}$$

Since $g \in L^1(M, \mathfrak{S}, \mu)$ and $u_t(x) \to \infty$ when $t \to \infty$, it follows from the Birkhoff–Khinchin ergodic theorem that the limit

$$\lim_{t\to\infty} \frac{1}{u_t(x)} \int_0^{u_t(x)} g(T^v x)\, dv = \bar{g}^+(x), \tag{6}$$

exists almost everywhere and

$$\int_M \bar{g}^+\, d\mu = \int_M g\, d\mu = \int_M f \cdot w\, d\mu = \int_M f\, d\bar{\mu} \cdot \int_M w\, d\mu. \tag{7}$$

According to the same ergodic theorem, the next limit also exists almost everywhere

$$\lim_{t\to\infty} \frac{u_t(x)}{t} = \lim_{u\to\infty} \frac{u}{\displaystyle\int_0^u w(T^s x)\, ds} = \frac{1}{\bar{w}^+(x)}. \tag{8}$$

It follows from (5), (6), and (8) that

$$\lim_{t\to\infty} f_t(x) = \frac{\bar{g}^+(x)}{\bar{w}^+(x)}, \qquad \text{i.e.} \quad \lim_{t\to\infty} \bar{w}^+(x) \cdot f_t(x) = \bar{g}^+(x),$$

almost everywhere. Taking into consideration (7), we get

$$\lim_{t\to\infty} \int_M \bar{w}^+(x) \cdot f_t(x)\, d\mu = \int_M f(x)\, d\bar{\mu} \cdot \int_M w\, d\mu. \tag{9}$$

Replacing in the last relation the function $f(x)$ by $f(\bar{T}^s x)$, where $s \in \mathbb{R}^1$ is arbitrary, we get

$$\lim_{t\to\infty} \int_M \bar{w}^+(x) \cdot f_t(\bar{T}^s x)\, d\mu = \int_M f(\bar{T}^s x)\, d\bar{\mu} \cdot \int_M w\, d\mu. \tag{10}$$

Let us estimate the difference between the integrals in the left-hand sides of (9) and (10)

$$\left| \int_M \bar{w}^+(x) \cdot f_t(\bar{T}^s x)\, d\mu - \int_M \bar{w}^+(x) \cdot f_t(x)\, d\mu \right|$$

$$\leq \frac{1}{t} \int_M \bar{w}^+(x) \left| \int_0^t f(\bar{T}^{\tau+s} x)\, d\tau - \int_0^t f(\bar{T}^\tau x)\, d\tau \right| d\mu$$

$$\leq \frac{2s \cdot \max|f(x)|}{t} \cdot \int_M \bar{w}^+\, d\mu \to 0 \quad \text{for } t \to \infty.$$

Therefore the right-hand sides of (9) and (10) coincide, so that

$$\int f(\bar{T}^s x)\, d\bar{\mu} = \int f(x)\, d\bar{\mu}.$$

Since f was arbitrary, the last relation means that the measure $\bar{\mu}$ is invariant with respect to $\{\bar{T}^t\}$. The theorem is proved. □

Remark. Suppose the function $w(x)$ satisfies $1/w \in L^1(M, \mathfrak{S}, \mu)$. Then the original flow $\{T^t\}$ can be obtained from $\{\bar{T}^t\}$ by the change of time determined by the function $1/w$.

If the flow $\{T^t\}$ is ergodic, then any flow $\{\bar{T}^t\}$ obtained from it by change of time is obviously also ergodic. The mixing property is in general not preserved by change of time.

§4. Endomorphisms and Their Natural Extensions

Consider the measure space $(M_0, \mathfrak{S}_0, \mu_0)$ whose points are of the form $x = (y_0, y_1, y_2, \ldots)$, $y_i \in Y$, where Y is a measurable space with the σ-algebra $\mathfrak{A}$ of measurable sets. Assume that the measure μ_0 has the following property: $\mu_0(\{x \in M_0 : y_{i_1+n} \in C_1, \ldots, y_{i_r+n} \in C_r\})$ does not depend on $n (n \geq 0)$ for arbitrary $i_1, \ldots, i_r \geq 0$ and all $C_1, \ldots, C_r \in \mathfrak{A}$. From the point of view of probability theory, we are considering a one-sided stationary sequence of Y-valued random variables. In the space M_0, we have the shift endomorphism

$$T_0\colon T_0(y_0, y_1, \ldots) = (y_1, y_2, \ldots)$$

preserving the measure μ_0.

In many questions it is natural to extend the space M_0 and consider the space M of sequences infinite in both directions: $x = (\ldots, y_{-1}, y_0, y_1, \ldots)$, where $y_i \in Y$, while the measure μ is defined from the measure μ_0:

$$\mu(\{x \in M\colon y_{i_1} \in C_1, \ldots, y_{i_r} \in C_r\}) = \mu_0(\{x \in M_0\colon y_{i_1+n} \in C_1, \ldots, y_{i_r+n} \in C_r\});$$

here n satisfies $i_k + n \geq 0$, $k = 1, \ldots, r$. The measure μ is obviously invariant with respect to the shift transformation T in the space M acting according to the formula

$$T(\ldots y_{-1}, y_0, y_1, \ldots) = (\ldots y'_{-1}, y'_0, y'_1, \ldots),$$

where $y'_i = y_{i+1}$ for all i. The ergodic properties of T are determined by the ergodic properties of T_0.

We shall now describe a general construction which allows us to find, for an arbitrary endomorphism T_0 of any measure space $(M_0, \mathfrak{S}_0, \mu_0)$, an automorphism of a new space $(M, \mathfrak{S}, \mu)$ naturally related to T. In the case of a shift, this construction reduces to the one described above.

The point x of the space M may be constructed as an infinite sequence $x = (x^{(0)}, x^{(1)}, x^{(2)}, \ldots)$, where $x^{(i)} \in M_0$, $T_0 x^{(i)} = x^{(i-1)}$ for all $i > 0$. Let us transform M into a measure space. For the σ-algebra $\mathfrak{S}$ choose the minimal σ-algebra containing all the sets of the form

$$A = A_{i,C} = \{x = (x^{(0)}, x^{(1)}, \ldots) \in M\colon x^{(i)} \in C\},$$

where $i \geq 0$, $C \in \mathfrak{S}_0$. For any such A put $\mu(A) = \mu_0(C)$. It follows from the definitions that

$$\begin{aligned}\mu(\{x = (x^{(0)}, x^{(1)}, \ldots) \in M\colon x^{(0)} \in C_0, \ldots, x^{(r)} \in C_r\} \\ = \mu_0(T_0^{-r}C_0 \cap T_0^{-r+1}C_1 \cap \cdots \cap C_r).\end{aligned}$$

This relation gives a compatible family of finite-dimensional probability distributions which, according to Kolmogorov's theorem, may be extended to a normalized measure μ defined on the σ-algebra $\mathfrak{S}$.

Consider the transformation T of the space M given by the formula $Tx = (T_0 x^{(0)}, T_0 x^{(1)}, \ldots)$ for $x = (x^{(0)}, x^{(1)}, \ldots)$. This transformation is invertible $T^{-1}x = (x^{(1)}, x^{(2)}, \ldots)$ for $x = (x^{(0)}, x^{(1)}, \ldots)$. It can be checked directly that the measure μ is invariant with respect to T, i.e., T is an automorphism.

Definition 1. The automorphism T of the space M is called the *natural extension* of the endomorphism T_0 of the space M_0.

The relationship between the ergodic properties of T_0 and T can be seen from the following theorem.

Theorem 1. (i) *The automorphism T is ergodic if and only if T_0 is ergodic;*
(ii) *The automorphism T is mixing if and only if T_0 is mixing.*

Proof. If $C \subset M_0$ is an invariant (mod 0) set for T_0, then $A = \{x = (x^{(0)}, x^{(1)}, \ldots) \in M: x^{(0)} \in C\}$ will obviously be an invariant set for T. It follows from the definition of the measure μ that $\mu(A) = \mu_0(C)$. Therefore, if T_0 is nonergodic, then so is T.

Now suppose T_0 is ergodic. Let us show that so is T. By von Neumann's ergodic theorem, for any function $f \in L^1(M_0, \mathfrak{S}_0, \mu_0)$ we have

$$\lim_{n\to\infty} \frac{1}{n} \sum_{k=0}^{n-1} f(T_0^k x^{(0)}) = \int_{M_0} f \, d\mu_0,$$

where the convergence takes place with respect to the norm in the space $L^1(M_0, \mathfrak{S}_0, \mu_0)$.

This immediately implies that for any function $F \in L^1(M, \mathfrak{S}, \mu)$ of the form $F(x) = f(x^{(i)})$, $x = (x^{(0)}, x^{(1)}, \ldots)$ we shall also have the relation

$$\lim_{n\to\infty} \frac{1}{n} \sum_{k=0}^{n-1} F(T^k x) = \int_M F \, d\mu, \tag{1}$$

in the sense of convergence in $L^1(M, \mathfrak{S}, \mu)$. But by the condition $T_0 x^{(i)} = x^{(i-1)}$, $i > 0$, any function $G(x^{(0)}, x^{(1)}, \ldots, x^{(i)})$ may be represented in the form $G(x^{(0)}, x^{(1)}, \ldots, x^{(i)}) = F(x^{(i)})$. Therefore functions of this type constitute a dense set in $L^1(M, \mathfrak{S}, \mu)$ and therefore the relation (1) is valid for any function $F \in L^1(M, \mathfrak{S}, \mu)$. Thus T is ergodic and the statement (i) is proved.

The statement (ii) may be proved in a similar way. Namely, if

$$f \in L^2(M_0, \mathfrak{S}_0, \mu_0), \qquad \int f \, d\mu_0 = 0 \quad \text{and} \quad F(x) = f(x^{(i)}),$$

then

$$(U_T^n F, F) = (U_{T_0}^n f, f)$$

and, when $n \to \infty$, the left-hand side and the right-hand side of this relation tend to zero simultaneously. The general case once again reduces to the one considered, since the set of functions F is dense in $L^2(M, \mathfrak{S}, \mu)$. The theorem is proved. $\square$

§5. The Rohlin–Halmos Lemma

Suppose T is an automorphism of the measure space $(M, \mathfrak{S}, \mu)$.

Definition 1. The point $x \in M$ is said to be *a periodic point* of the automorphism T, if $T^n x = x$ for some integer $n \neq 0$. The automorphism T is called *aperiodic* if the set of its periodic points is of zero measure.

The following statement is often used in ergodic theory. It relates to the automorphisms of *Lebesgue spaces* (see Appendix 1).

Theorem 1 (the Rohlin–Halmos Lemma). *If T is an aperiodic automorphism of the Lebesgue space $(M, \mathfrak{S}, \mu)$, then for all $\varepsilon > 0$ and every positive integer n there is a set $E \in \mathfrak{S}$ such that*

(1) *the sets $E, TE, \ldots, T^{n-1}E$ are disjoint;*

(2) $\mu\left(\bigcup_{i=0}^{n-1} T^i E\right) > 1 - \varepsilon$.

Proof. Our argument shall be split up into separate steps.

1. For any positive integer n there is a $F_n \in \mathfrak{S}$, $\mu(F_n) > 0$, such that the sets $F_n, TF^n, \ldots, T^{n-1}F_n$ are two-by-two disjoint.

Let us prove this by induction on n. For $n = 1$ the statement is trivial. In passing from n to $n + 1$, consider the set F_n already constructed. Let us show that there exist $F_n' \in \mathfrak{S}$, $F_n' \subset F_n$ satisfying $\mu(F_n' \triangle T^n F_n') > 0$. If this were not so, the transformation T^n would map any measurable $A \subset F_n$ into a set A' satisfying $\mu(A \triangle A') = 0$. Since M is a Lebesgue space, this will imply that T^n is the identity (mod 0) on F_n (see Appendix 1) which contradicts the aperiodicity of T. Put $F_{n+1} = F_n' \setminus T^n F_n'$. Obviously $F_{n+1}, TF_{n+1}, \ldots, T^n F_{n+1}$ are pairwise disjoint and $\mu(F_{n+1}) > 0$.

Note that if some set $A \in \mathfrak{S}$, $\mu(A) > 0$ is invariant with respect to T, we can construct an $F_n \subset A$.

2. Identify any subsets $F', F'' \in \mathfrak{S}$ such that $\mu(F' \triangle F'') = 0$, and consider the corresponding equivalence classes of subsets. For the sets F_n (constructed in step 1 of the proof) such that $F_n, TF_n, \ldots, T^{n-1}F_n$ are pairwise disjoint, the corresponding equivalence classes shall also be denoted by F_n. Introduce a partial order on the set of these classes by inclusion (mod 0). Then every increasing chain $\{F_n^{(\alpha)}\}$ has an upper bound, namely $F_n = \bigcup_\alpha F_n^{(\alpha)}$. Since any strictly increasing chain $\{F_n^{(\alpha)}\}$, indexed by transfinite numbers, ends on a countable transfinite, we can assume that α runs over a merely countable set of values, so that $F_n \in \mathfrak{S}$. By Zorn's lemma, there is a maximal element; in other words, we can find an $F_n^{(\max)} \subset \mathfrak{S}$ such that any $F_n \in \mathfrak{S}$, satisfying $\mu(T^i F_n \cap T^j F_n) = 0$ for $0 \leq i < j \leq n - 1$ and $F_n \supseteq F_n^{(\max)}$ coincides (mod 0) with $F_n^{(\max)}$. We also have $\mu(T^i F_n^{(\max)} \cap T^j F_n^{(\max)}) = 0$, $0 \leq i < j \leq n - 1$.

3. Suppose $F_m^{(\max)}$ is the set constructed according to step 2 for some m. For any $x \in F_m^{(\max)}$ put

$$r(x) = \min\{r \geq 1 \colon T^r x \in F_m^{(\max)}\};$$

let us show that $m \leq r(x) \leq 2m - 1$ for almost all such x.

The inequality $r(x) \geq m$ follows from the construction of $F_m^{(\max)}$. Suppose $G \subset F_m^{(\max)}$ consists of all those x which satisfy $r(x) \geq 2m$. Put $\tilde{F}_m = F_m^{(\max)} \cup T^m G$. Then $\mu(T^i \tilde{F}_m \cap T^j \tilde{F}_m) = 0$ for all $0 \leq i < j \leq m - 1$ and, if $\mu(G) > 0$, then $\mu(\tilde{F}_m) > \mu(F_m^{(\max)})$ although $F_m^{(\max)}$ is maximal. Hence $\mu(G) = 0$.

Now put

$$F_{m,k}^{(\max)} = \{x \in F_m^{(\max)} \colon r(x) = k\}.$$

By the above

$$\bigcup_{k=m}^{2m-1} F_{m,k}^{(\max)} = F_m^{(\max)} (\text{mod } 0).$$

4. Suppose $M_1 = \bigcup_{k=m}^{2m-1} \bigcup_{i=0}^{k-1} T^i F_{m,k}^{(\max)}$. Let us prove $M_1 = M$ (mod 0). Note first of all that the set M_1 is invariant. Indeed, if $x \in T^i F_{m,k}^{(\max)}$ for $i < k - 1$, then $Tx \in T^{i+1} F_{m,k}^{(\max)}$. But if $x \in T^{k-1} F_{m,k}^{(\max)}$, then $Tx \in F_m^{(\max)} = \bigcup_{k=m}^{2m-1} F_{m,k}^{(\max)} \subset M_1$.

If $\mu(M \setminus M_1) > 0$, then $M \setminus M_1$ is an invariant set of positive measure. By the remark at the end of step 1, in this case we can find an $F_m \subset M \setminus M_1$, $\mu(F_m) > 0$ such that the sets $F_m, TF_m, \ldots, T^{m-1} F_m$ are pairwise disjoint. Putting $\tilde{\tilde{F}}_m = F_m^{(\max)} \cup F_m$, we obtain a contradiction to the maximality of $F_m^{(\max)}$.

5. Choose m so as to have $n/m < \varepsilon$ and put

$$E = \bigcup_{k=m}^{2m-1} \bigcup_{\substack{i \equiv 0 (\text{mod } n) \\ 0 \leq i \leq k-1}} T^i F_{m,k}^{(\max)}.$$

The set E is the one we need. Indeed, it follows from the construction that $E, TE, \ldots, T^{n-1}E$ are disjoint. Further

$$\bigcup_{i=0}^{n-1} T^i E = \bigcup_{k=m}^{2m-1} \bigcup_{i=0}^{p_k} T^i F_{m,k}^{(\max)},$$

where $p_k = \max\{i \colon 0 \leq i \leq k - 1, i \equiv (n - 1)(\text{mod } n)\}$. On the other hand, by step 4 we have

$$\bigcup_{k=m}^{2m-1} \bigcup_{i=0}^{k-1} T^i F_{m,k}^{(\max)} = M (\text{mod } 0).$$

For each k, $m \le k \le 2m - 1$ the summands in the inner sum have the same measure and are disjoint. Therefore

$$\frac{\mu\left(\bigcup_{i=0}^{p_k} T^i F_{m,k}^{(\max)}\right)}{\mu\left(\bigcup_{i=0}^{k-1} T^i F_{m,k}^{(\max)}\right)} = \frac{p_k + 1}{k} \ge \frac{k - n}{k} = 1 - \frac{n}{k} \ge 1 - \frac{n}{m} > 1 - \varepsilon,$$

hence

$$\mu\left(\bigcup_{i=0}^{n-1} T^i E\right) = \frac{\mu\left(\bigcup_{k=m}^{2m-1} \bigcup_{i=0}^{p_k} T^i F_{m,k}^{(\max)}\right)}{\mu\left(\bigcup_{k=m}^{2m-1} \bigcup_{i=0}^{k=1} T^i F_{m,k}^{(\max)}\right)} = \frac{\sum_{k=m}^{2m-1} \mu\left(\bigcup_{i=0}^{p_k} T^i F_{m,k}^{(\max)}\right)}{\sum_{k=m}^{2m-1} \mu\left(\bigcup_{i=0}^{k-1} T^i F_{m,k}^{(\max)}\right)} > 1 - \varepsilon.$$

The theorem is proved. □

Remark. It follows from the constructions of the set E that $\bigcup_{i=0}^{2n-1} T^i E = M \pmod 0$.

The Strong Form of the Rohlin–Halmos Lemma. Suppose $\xi = (C_1, \ldots, C_r)$ is a finite partition of the Lebesgue space $(M, \mathfrak{S}, \mu)$. Then for any $\varepsilon > 0$ and any positive integer n, we can find an $E \in \mathfrak{S}$ such that $E, TE, \ldots, T^{n-1}E$ are pairwise disjoint, $\mu(\bigcup_{i=0}^{n-1} T^i E) > 1 - \varepsilon$ and $\mu(E \cap C_j) = \mu(E) \cdot \mu(C_j)$ for $j = 1, 2, \ldots, r$.

We shall not give the proof of this statement.

An Application of the Rohlin–Halmos Lemma. Suppose T is an arbitrary automorphism of the Lebesgue space $(M, \mathfrak{S}, \mu)$. Consider the space M' of all sequences infinite in both directions $x' = (\ldots, x'_{-1}, x'_0, x'_1, \ldots)$ where every coordinate x'_n may assume r values $1, 2, \ldots, r$ $(r \ge 2)$ and also the shift transformation T' on the space M'. Take some normalized measure ν on M' invariant with respect to T' and put

$$\nu(k_0, k_1, \ldots, k_n) \overset{\text{def}}{=} \nu(\{x' \in M' \colon x'_0 = k_0, x'_1 = k_1, \ldots, x'_n = k_n\})$$

for all $n \ge 0$, $1 \le k_1, \ldots, k_n \le r$.

Theorem 2. *For any $\varepsilon > 0$ and all positive integers n there exists a partition ξ of the space M into r sets $C_1, \ldots, C_r$ such that*

$$|\mu(C_{k_0} \cap T^{-1}C_{k_1} \cap \cdots \cap T^{-n}C_{k_n}) - \nu(k_0, k_1, \ldots, k_n)| < \varepsilon.$$

Proof. For any $\delta > 0$ and any natural N let us find a set $E \in \mathfrak{S}$ such that $T^iE \cap T^jE = \varnothing$ when $0 \le i < j \le N$, and $\mu(\bigcup_{i=0}^N T^iE) > 1 - \delta$. Put $B = M \setminus \bigcup_{i=0}^N T^iE$. Now construct the partitions $\xi^{(s)}$, $s = 0, 1, \ldots, N$ of the set E into subsets $C_1^{(s)}, \ldots, C_r^{(s)}$ so that

$$\frac{1}{\mu(E)} \cdot \mu(C_{k_0}^{(0)} \cap C_{k_1}^{(1)} \cap \cdots \cap C_{k_N}^{(N)}) = \nu(k_0, k_1, \ldots, k_N),$$

for all $k_0, k_1, \ldots, k_N$, $1 \le k_i \le r$. Define the necessary sets $C_1, \ldots, C_r$ by putting

$$C_i = C_i^{(0)} \cup TC_i^{(1)} \cup \cdots \cup T^N C_i^{(N)} \cup B_i, \qquad i = 1, \ldots, r,$$

where $B_i \subset B$ and the sets $B_1, \ldots, B_r$ constitute an arbitrary partition of the set B. We now have

$$\begin{aligned}
\Sigma &\stackrel{\text{def}}{=} |\mu(C_{k_0} \cap T^{-1}C_{k_1} \cap \cdots \cap T^{-n}C_{k_n}) - \nu(k_0, k_1, \ldots, k_n)| \\
&= \Bigg| \sum_{s=0}^{N} \mu(C_{k_0} \cap T^{-1}C_{k_1} \cap \cdots \cap T^{-n}C_{k_n} | T^sE) \cdot \mu(T^sE) \\
&\quad + \mu(C_{k_0} \cap T^{-1}C_{k_1} \cap \cdots \cap T^{-n}C_{k_n} | B) \cdot \mu(B) - \nu(k_0, k_1, \ldots, k_n) \Bigg| \\
&= \Bigg| \sum_{s=0}^{N} [(\mu(C_{k_0} \cap T^{-1}C_{k_1} \cap \cdots \cap T^{-n}C_{k_n} | T^sE) - \nu(k_0, k_1, \ldots, k_n)) \\
&\quad \cdot \mu(T^sE)] + [\mu(C_{k_0} \cap T^{-1}C_{k_1} \cap \cdots \cap T^{-n}C_{k_n} | B) - \nu(k_0, k_1, \ldots, k_n)] \\
&\quad \cdot \mu(B) \Bigg| \\
&\le 2n \cdot \mu(E) + 2\mu(B) + \sum_{s=0}^{N-n} |\mu(C_{k_0} \cap T^{-1}C_{k_1} \cap \cdots \cap T^{-n}C_{k_n} | T^sE) \\
&\quad - \nu(k_0, k_1, \ldots, k_n)| \cdot \mu(T^sE).
\end{aligned}$$

The last sum vanishes since for $s \le N - n$ we have the equality

$$\begin{aligned}
&\mu(C_{k_0} \cap T^{-1}C_{k_1} \cap \cdots \cap T^{-n}C_{k_n} | T^sE) \\
&\quad = \frac{1}{\mu(T^sE)} \cdot \mu(T^sC_{k_0}^{(s)} \cap T^sC_{k_1}^{(s+1)} \cap \cdots \cap T^sC_{k_n}^{(s+n)} \cap T^sE) \\
&\quad = \frac{1}{\mu(E)} \cdot \mu(C_{k_0}^{(s)} \cap C_{k_1}^{(s+1)} \cap \cdots \cap C_{k_n}^{(s+n)} \cap E) = \nu(k_0, k_1, \ldots, k_n).
\end{aligned}$$

Therefore $\Sigma \le 2n/N + 2\delta$ and we can choose N and δ so that the last expression is less than ε. The theorem is proved. $\square$

§6. Entropy

To each automorphism T of the Lebesgue space $(M, \mathfrak{S}, \mu)$ it is possible to assign a number $h(T)$ known as the entropy of the automorphism T. The expression $h(T)$ for various T may assume any value from 0 to $+\infty$ (including 0 and $+\infty$) and we have $h(T_1) = h(T_2)$ if the automorphisms T_1 and T_2 are metrically isomorphic. This means that the entropy $h(T)$ is a metric invariant.

The entropy theory of dynamical systems is a large and important branch of ergodic theory. In this book, however, it will play a relatively small role. We now give the necessary facts concerning the entropy of partitions and will give the definition of the entropy of an automorphism.

Suppose ξ is a partition of the space M (mod 0) into measurable subsets $C_1, \ldots, C_r$, i.e., $\bigcup_{i=1}^r C_i = M$ (mod 0), $C_i \cap C_j = \varnothing$, for $i \neq j$, $1 \leq i, j \leq r$.

Definition 1. The *entropy of the partition* ξ is the number

$$H(\xi) = -\sum_{i=1}^{r} \mu(C_i) \log \mu(C_i).$$

If $\mu(C_i) = 0$, we assume that $\mu(C_i)\log \mu(C_i) = 0$. Further, as usual, all logarithms are to the base e.

Denote by $T^k\xi$, for the partition $\xi = (C_1, \ldots, C_r)$, the partition into the sets $T^kC_1, \ldots, T^kC_r$ $(k = 0, \pm 1, \ldots)$. For any sequence of partitions $\xi^{(j)}$, $j = 1, \ldots, N$, $\xi^{(j)} = (C_1^{(j)}, \ldots, C_r^{(j)})$ denote by $\xi^{(1)} \vee \xi^{(2)} \vee \cdots \vee \xi^{(N)}$ the partition whose elements are all possible sets of the form $C_{i_1}^{(1)} \cap C_{i_2}^{(2)} \cap \cdots \cap C_{i_N}^{(N)}$, $1 \leq i_1, \ldots, i_N \leq r$.

Simplest Properties of the Entropy of Partitions

1. $0 \leq H(\xi) \leq \log r$ for any partition $\xi = (C_1, \ldots, C_r)$. The first inequality is obvious, the one on the right follows from the fact that the expression $H(\xi)$ assumes its maximum if $\mu(C_i) = 1/r$ for all i, $1 \leq i \leq r$.

2. Suppose $\xi = (C_1, \ldots, C_r)$, $\eta = (D_1, \ldots, D_r)$ are two partitions of the space M. Then

$$H(\xi \vee \eta) \leq H(\xi) + H(\eta).^1 \tag{1}$$

[1] So as to avoid complicating the notation here (and in certain other cases), we shall assume that the number of elements y in the partitions ξ, η is the same. The general case is readily reduced to this one.

Indeed,

$$\begin{aligned}
H(\xi \vee \eta) &= -\sum_{i,j=1}^{r} \mu(C_i \cap D_j)\log \mu(C_i \cap D_j) \\
&= -\sum_{i,j=1}^{r} \mu(C_i \cap D_j)\log[\mu(C_i)\mu(D_j \mid C_i)] \\
&= -\sum_{i,j=1}^{r} \mu(C_i \cap D_j)\log \mu(C_i) - \sum_{i,j=1}^{r} \mu(C_i)\mu(D_j \mid C_i)\log \mu(D_j \mid C_i) \\
&= H(\xi) - \sum_{j=1}^{r} \left[\sum_{i=1}^{r} \mu(C_i)\mu(D_j \mid C_i)\log \mu(D_j \mid C_i)\right]. \qquad (2)
\end{aligned}$$

Applying the Jensen inequality for convex functions to the function $f(x) = -x \ln x$, $0 < x < 1$, we get

$$-\sum_{i=1}^{r} \mu(C_i)\mu(D_j \mid C_i)\log \mu(D_j \mid C_i) \le -\mu(D_j)\log \mu(D_j). \qquad (3)$$

Comparing (2) and (3), we get the inequality (1).

It follows from the proof that the equality in (1) takes place iff the partitions ξ and η are independent, i.e., iff

$$\mu(C_i \cap D_j) = \mu(C_i) \cdot \mu(D_j), \qquad 1 \le i, j \le r.$$

Assume again that $\xi = (C_1, \ldots, C_r)$, $\eta = (D_1, \ldots, D_r)$ are two partitions of the space M. We will suppose that $\mu(D_j) > 0$, $j = 1, \ldots, r$.

Definition 2. The *conditional entropy of the partition ξ with respect to η* is the number

$$H(\xi \mid \eta) = -\sum_{j=1}^{r} \mu(D_j) \sum_{i=1}^{r} \mu(C_i \mid D_j)\log \mu(C_i \mid D_j).$$

Consider each D_j as the space with measure $\mu(\cdot \mid D_j)$. Introduce the partition ξ_{D_j} induced by the partition ξ on the set D_j. We then get

$$H(\xi \mid \eta) = \sum_{j=1}^{r} \mu(D_j) H(\xi_{D_j}). \qquad (4)$$

Simplest Properties of Conditional Entropy

1. For all partitions ξ_1, ξ_2, η

$$H(\xi_1 \vee \xi_2) = H(\xi_1) + H(\xi_2 \mid \xi_1),$$

$$H(\xi_1 \vee \xi_2 \mid \eta) \le H(\xi_1 \mid \eta) + H(\xi_2 \mid \eta).$$

These properties are proved exactly in the same way as property 2 of the entropy of a partition. It is clear that we have an equality in the last relation only if the partitions ξ_1 and ξ_2 are *independent with respect to* η, i.e., if for every element $D \in \eta$ the partitions $(\xi_1)_D$, $(\xi_2)_D$ are independent.

2. For all partitions ξ, η_1, η_2, we have

$$H(\xi | \eta_1 \vee \eta_2) \leq H(\xi | \eta_1).$$

Indeed

$$\begin{aligned} H(\xi | \eta_1 \vee \eta_2) &= H(\xi \vee \eta_1 \vee \eta_2) - H(\eta_1 \vee \eta_2) \\ &= H(\xi \vee \eta_2 | \eta_1) + H(\eta_1) - H(\eta_2 | \eta_1) - H(\eta_1) \\ &\leq H(\xi | \eta_1) + H(\eta_2 | \eta_1) - H(\eta_2 | \eta_1) = H(\xi | \eta_1). \end{aligned}$$

We have the equality only if ξ and η_2 are independent with respect to η_1. This immediately implies the following property of conditional entropy:

3. If the partitions ξ, η_1, η_2 satisfy the equality

$$H(\xi | \eta_1 \vee \eta_2) = H(\xi | \eta_1)$$

and ζ is a partition such that $\zeta \preccurlyeq \xi$, then

$$H(\zeta | \eta_1 \vee \eta_2) = H(\zeta | \eta_1).$$[1]

Now let us generalize the notion of conditional entropy in the following way. Suppose ξ is a finite partition of the space M; $\mathfrak{A}$ is a complete σ-subalgebra of the σ-algebra $\mathfrak{S}$. Since M is a Lebesgue space, to the σ-algebra $\mathfrak{A}$ corresponds a measurable (but not necessarily finite) partition $\eta = \eta(\mathfrak{A})$ (see Appendix 1). On each element $D \in \eta$ the partition ξ induces a finite partition ξ_D. Moreover, on almost all elements $D \in \eta$ we have the conditional measure μ_D. Following (4) let us introduce a definition.

Definition 3. *The conditional entropy of the finite partition* ξ *with respect to the measurable partition* η is the number

$$H(\xi | \eta) = \int_{M/\eta} H(\xi_D) \, d\mu_D.$$

It can be shown that properties 1–3 of conditional entropy remain valid in this case as well.

[1] The notation $\zeta \preccurlyeq \xi$ means that each element of the partition ζ is the union of some elements of the partition ξ (see Appendix 1).

Suppose k, l are integers $k \leq l$ and ξ is a finite partition of the space M. Introduce the partition

$$\xi_k^l = T^k\xi \vee T^{k+1}\xi \vee \cdots \vee T^l\xi.$$

The numbers $H_n = H(\xi_0^{n-1})$, $n \geq 0$ satisfy the inequality $H_{n+m} \leq H_n + H_m$. Indeed,

$$\begin{aligned} H_{n+m} &= H(\xi_0^{n+m-1}) \leq H(\xi_0^{n-1}) + H(\xi_n^{n+m-1}) \\ &= H_n + H(\xi_n^{n+m-1}) = H_n + H_m. \end{aligned}$$

Therefore, the following limit exists

$$h(T, \xi) \stackrel{\text{def}}{=} \lim_{n\to\infty} \frac{1}{n} H(\xi_0^{n-1}).$$

There is another expression for $h(T, \xi)$ containing conditional entropy:

$$h(T, \xi) = H\left(\xi \,\middle|\, \bigvee_{k=1}^{\infty} T^{-k}\xi\right). \tag{5}$$

(The conditional entropy in the right-hand side of this relation should be understood in the sense of Definition 3).

Indeed, for any n, we have

$$\begin{aligned} \frac{1}{n} H(\xi \vee T\xi \vee \cdots \vee T^{n-1}\xi) &= \frac{1}{n}[H(T^{n-1}\xi \mid T^{n-2}\xi \vee \cdots \vee \xi) \\ &\quad + H(T^{n-2}\xi \mid T^{n-3}\xi \vee \cdots \vee \xi) + \cdots \\ &\quad + H(T\xi \mid \xi) + H(\xi)] \\ &= \frac{1}{n}[H(\xi) + H(\xi \mid T^{-1}\xi) \\ &\quad + H(\xi \mid T^{-1}\xi \vee T^{-2}\xi) + \cdots \\ &\quad + H(\xi \mid T^{-1}\xi \vee \cdots \vee T^{-n+1}\xi)]. \end{aligned}$$

This implies

$$h(T, \xi) = \lim_{n\to\infty} \frac{1}{n} \sum_{k=0}^{n-1} H\left(\xi \,\middle|\, \bigvee_{l=1}^{k-1} T^{-l}\xi\right). \tag{6}$$

The sequence of numbers $\alpha_n = H(\xi \mid \bigvee_{k=1}^{n} T^{-k}\xi)$ is monotonic decreasing and therefore has a limit for $n \to \infty$. This limit obviously coincides with the limit in the right-hand side of (6). On the other hand, by Doob's theorem on the convergence of conditional expectations, we have

$$\lim_{n\to\infty} H\left(\xi \,\middle|\, \bigvee_{k=1}^{n} T^{-k}\xi\right) = H\left(\xi \,\middle|\, \bigvee_{k=1}^{\infty} T^{-k}\xi\right).$$

Thus (5) is proved.

Definition 4. The *entropy of an automorphism* T is the number $h(T) = \sup h(T, \xi)$, where the supremum is taken over all finite partitions ξ.

It immediately follows from the definition that $h(T)$ is a metric invariant. We shall give a theorem which will allow us to compute the entropy $h(T)$ in many cases. For this we shall need the notion of generating partition, which we met with in Chap. 8. We shall now give a somewhat modified definition, which coincides with the one in Chap. 8 in the case of Lebesgue spaces.

Definition 5. The partition $\xi = (C_1, \ldots, C_r)$ is said to be *generating* for the automorphism T if the minimal complete σ-algebra containing all the sets

$$T^n C_i, \qquad i = 1, \ldots, r, \ -\infty < n < \infty$$

coincides with $\mathfrak{S}$.

Theorem 1. *If the automorphism T possesses a finite generating partition η, then $h(T) = h(T, \eta)$.*

Proof. Suppose $\xi = (C_1, \ldots, C_r)$ is an arbitrary finite partition and $\varepsilon > 0$ is given. We will show that

$$h(T, \xi) \le h(T, \eta) + \varepsilon;$$

hence, ε being arbitrary, it will follow that $h(T, \xi) \le h(T, \eta)$.

For all $m, n > 0$ we have

$$\begin{aligned} H(\xi \vee T\xi \vee \cdots \vee T^{n-1}\xi) &\le H(T^{-m}\eta \vee \cdots \vee T^{n+m}\eta \vee \xi \vee \cdots \vee T^{n-1}\xi) \\ &\le H(T^{-m}\eta \vee \cdots \vee T^{n+m}\eta) \\ &\quad + H(\xi \vee \cdots \vee T^{n-1}\xi \mid T^{-m}\eta \vee \cdots \vee T^{n+m}\eta) \\ &\le H(T^{-m}\eta \vee \cdots \vee T^{n+m}\eta) \\ &\quad + \sum_{k=0}^{n-1} H(T^k\xi \mid T^{-m}\eta \vee \cdots \vee T^{n+m}\eta) \\ &= H(T^{-m}\eta \vee \cdots \vee T^{n+m}\eta) \\ &\quad + \sum_{k=0}^{n-1} H(\xi \mid T^{-m-k}\eta \vee \cdots \vee T^{n+m-k}\eta). \end{aligned} \tag{7}$$

We will show that for sufficiently large m and all $k, n, 0 \leq k \leq n - 1$ we have the inequality

$$H(\xi \,|\, T^{-m-k}\eta \vee \cdots \vee T^{n+m-k}\eta) < \varepsilon. \tag{8}$$

Denote by $\eta^{(m)}$ the partition $T^{-m}\eta \vee \cdots \vee T^{m}\eta$. Then

$$H(\xi \,|\, T^{-m-k}\eta \vee \cdots \vee T^{n+m-k}\eta) \leq H(\xi \,|\, \eta^{(m)}).$$

Hence it suffices to show that $H(\xi \,|\, \eta^{(m)}) < \varepsilon$ for sufficiently large m.

Since η is a generating partition, for any element C_i of the partition ξ and every $\delta > 0$, for a sufficiently large m, we can find a set C_i', consisting of elements of the partition $\eta^{(m)}$, satisfying $\mu(C_i \triangle C_i') < \delta$. The sets $C_i \cap C_i'$ are disjoint for distinct i and we have

$$\mu\left(\bigcup_{i=1}^{r} (C_i \cap C_i')\right) > 1 - r\delta.$$

Suppose C_i'' is the set consisting of all elements D of the partition $\eta^{(m)}$ such that $\mu(C_i \,|\, D) > 1 - \sqrt{\delta}$. Then $\mu(C_i'') \geq \mu(C_i') - \sqrt{\delta}$. Indeed, if $D \subset C_i' \setminus C_i''$, then $\mu(M \setminus C_i \,|\, D) \geq \sqrt{\delta}$. Therefore

$$\begin{aligned}\mu(C_i' \setminus C_i'') &= \sum_{D \subset C_i' \setminus C_i''} \mu(D) \\ &\leq \frac{1}{\sqrt{\delta}} \sum_{D \subset C_i' \setminus C_i''} \mu(M \setminus C_i \,|\, D) \cdot \mu(D) \leq \frac{1}{\sqrt{\delta}} \mu(C_i' \setminus C_i) < \sqrt{\delta}.\end{aligned}$$

We now have

$$H(\xi \,|\, \eta^{(m)}) = -\sum_{D \in \eta^{(m)}} \mu(D) \sum_{i=1}^{r} \mu(C_i \,|\, D) \log \mu(C_i \,|\, D) = -\Sigma^{(1)} - \Sigma^{(2)},$$

where the sum in $\Sigma^{(1)}$ is taken over those elements D (of the partition $\eta^{(m)}$) which belong to the set $\bigcup_{i=1}^{r} C_i''$, and, in $\Sigma^{(2)}$, over the remaining elements of the partition $\eta^{(m)}$.

Since

$$\mu\left(M \setminus \bigcup_{i=1}^{r} C_i''\right) \leq 2r\sqrt{\delta},$$

we have $|\Sigma^{(2)}| \leq 2r\sqrt{\delta} \ln r$. As to $\Sigma^{(1)}$, for a sufficiently small δ and $D \subset C_i''$, we will have

$$\left| \sum_{i=1}^{r} \mu(C_i \,|\, D) \log \mu(C_i \,|\, D) \right| < \frac{\varepsilon}{2},$$

i.e., $|\Sigma^{(1)}| < \varepsilon/2$. Hence for sufficiently small δ the entire sum will be less than ε.

Now take an m such that (8) holds. Then, dividing both parts of inequality (7) by n and assuming that n tends to infinity, we get $h(T, \xi) \leq h(T, \eta) + \varepsilon$. The theorem is proved. □

EXAMPLES OF ENTROPY COMPUTATION

1. *The entropy of periodic automorphisms.* Suppose T is a periodic automorphism, i.e., there exists a positive integer m such that $T^m x = x$ for almost all $x \in M$. Then for any finite partition $\xi = (C_1, \ldots, C_r)$ of the space M and for any $n > 0$, the number of elements in the partition ξ_0^{n-1} is no greater than r^m, so that

$$H(\xi_0^{n-1}) \leq m \log r, \qquad n = 1, 2, \ldots$$

Hence

$$0 \leq h(T, \xi) = \lim_{n \to \infty} \frac{H(\xi_0^{n-1})}{n} \leq \lim_{n \to \infty} \frac{m \log r}{n} = 0,$$

i.e., $h(T) = \sup_\xi h(T, \xi) = 0$.

2. *The entropy of circle rotations.* Suppose

$$T_\alpha x = x + \alpha \qquad (\text{mod } 1),\ x \in S^1.$$

By 1, we have $h(T_\alpha) = 0$ for rational α. If α is irrational, then T_α is a minimal homeomorphism of the circle. This immediately implies that the partition $\xi = (C_1, C_2)$, where $C_1 = [0, 1/2)$, $C_2 = [1/2, 1)$, is generating for T_α. Moreover, for any $n > 0$ the partition ξ_0^{n-1} is a partition of the circle S^1 into $2n$ segments. This statement can easily be proved by induction.

Indeed, for $n = 1$ it follows from the definition of ξ and, when we pass from $n - 1$ to n, we must add to the end points of the segments belonging to ξ_0^{n-1} precisely 2 new points: $T_\alpha^n(0)$, $T_\alpha^n(1/2)$. Therefore

$$0 \leq h(T_\alpha, \xi) = \lim_{n \to \infty} \frac{H(\xi_0^{n-1})}{n} \leq \lim_{n \to \infty} \frac{\log(2n)}{n} = 0,$$

hence $h(T_\alpha) = h(T_\alpha, \xi) = 0$.

The same arguments are applicable to translations on the torus Tor^m, $m > 1$.

3. *Entropy of Bernoulli automorphisms.* Suppose T is a Bernoulli automorphism with the state space

$$(Y, \sigma),\ y = \{a_1, \ldots, a_r\},\ \sigma(\{a_k\}) = p_k, \qquad 1 \le k \le r.$$

The phase space of this automorphism is $M = \prod_{n=-\infty}^{\infty} Y^{(n)}$, $Y^{(n)} \equiv Y$. The partition $\xi = (C_1, \ldots, C_r)$ of the space M where

$$C_k = \{x = (\ldots y_{-1}, y_0, y_1, \ldots) \in M : y_0 = a_k\},$$

is obviously a generating one. For $n > 0$ let us compute the entropy $H(\xi_0^{n-1})$:

$$\begin{aligned} H(\xi_0^{n-1}) &= - \sum_{k_1, \ldots, k_n = 1}^{r} p_{k_1} \cdot p_{k_2} \cdot \ldots \cdot p_{k_n} \cdot \log(p_{k_1} \cdot p_{k_2} \cdot \ldots \cdot p_{k_n}) \\ &= - \sum_{k_1, \ldots, k_n = 1}^{r} p_{k_1} \cdot \ldots \cdot p_{k_n} \log p_{k_1} - \sum_{k_1, \ldots, k_n = 1}^{r} p_{k_1} \cdot \ldots \cdot p_{k_n} \log p_{k_2} - \cdots \\ &\quad - \sum_{k_1, \ldots, k_n = 1}^{r} p_{k_1} \cdot \ldots \cdot p_{k_n} \log p_{k_n} = -n \sum_{k_1, \ldots, k_n = 1}^{r} p_{k_1} \cdot \ldots \cdot p_{k_n} \log p_{k_1} \\ &= -n \sum_{k_2 = 1}^{r} p_{k_2} \sum_{k_3 = 1}^{r} p_{k_3} \cdot \ldots \cdot \sum_{k_n = 1}^{r} p_{k_n} \sum_{k_1 = 1}^{r} p_{k_1} \log p_{k_1} = -n \sum_{k=1}^{r} p_k \log p_k. \end{aligned}$$

Hence

$$h(T) = h(T, \xi) = \lim_{n \to \infty} \frac{H(\xi_0^{n-1})}{n} = - \sum_{k=1}^{r} p_k \log p_k.$$

4. *Entropy of Markov automorphisms.* Suppose the phase space M is the same as in Example 3, T is a Markov automorphism with transition matrix $\Pi = \|p_{ij}\|$, $1 \le i, j \le r$ and stationary probabilities $\sigma = (p_1, \ldots, p_r)$. The partition ξ given in Example 3 is still a generating one in this case. Similarly to the above, we can show that, for $n > 1$,

$$H(\xi_0^{n-1}) = -(n-1) \sum_{i,j=1}^{r} p_i p_{ij} \log p_{ij} - \sum_{k=1}^{r} p_k \log p_k.$$

Hence

$$h(T) = h(T, \xi) = \lim_{n \to \infty} \frac{H(\xi_0^{n-1})}{n} = - \sum_{i,j=1}^{r} p_i p_{ij} \log p_{ij}.$$

Now let us list the basic properties of the entropy of automorphisms.

Theorem 2. (1) *$h(T^m) = |m| \cdot h(T)$, where m is an integer.*

(2) *If $T = T_1 \times T_2$ is the direct product of two automorphisms, then $h(T) = h(T_1) + h(T_2)$.*

(3) *If T_E is an induced automorphism constructed from the ergodic automorphism T and the set E, $\mu(E) > 0$, then $h(T_E) = [1/\mu(E)] \cdot h(T)$.*

(4) *If T^f is the integral automorphism constructed from the ergodic automorphism T and the integer-valued function $f \in L^1(M, \mathfrak{S}, \mu)$, and $f > 0$, then $h(T^f) = [1/(\int_M f\,d\mu)] \cdot h(T)$.*

Proof. We shall only prove the first statement of the theorem. The other ones will not be used in this book. Their proofs may be found, for example, in Brown [1].

First let us show that $h(T^{-1}) = h(T)$. For any finite partition ξ and any positive integer n, we have the identity

$$\begin{aligned} H(\xi \vee T^{-1}\xi \vee \cdots \vee T^{-n+1}\xi) &= H(T^n(\xi \vee T^{-1}\xi \vee \cdots \vee T^{-n+1}\xi)) \\ &= H(\xi \vee T\xi \vee \cdots \vee T^{n-1}\xi). \end{aligned}$$

This implies $h(T^{-1}, \xi) = h(T, \xi)$ and

$$h(T^{-1}) = \sup_\xi h(T^{-1}, \xi) = \sup_\xi h(T, \xi) = h(T).$$

We can now assume that $m > 0$. For a finite partition ξ put

$$\eta = \eta(m, \xi) = \xi \vee T\xi \vee \cdots \vee T^{m-1}\xi.$$

Then, for any $n > 0$, we have

$$H(\eta \vee T^m\eta \vee \cdots \vee T^{m(n-1)}\eta) = H(\xi \vee T\xi \vee \cdots \vee T^{mn-1}\xi).$$

Therefore, when $n \to \infty$, we get $h(T^m, \eta) = m \cdot h(T, \xi)$. Hence

$$\begin{aligned} h(T^m) &= \sup_\xi h(T^m, \xi) = \sup_\xi h(T^m, \eta(m, \xi)) \\ &= m \cdot \sup_\xi h(T, \xi) = m \cdot h(T). \end{aligned}$$

The theorem is proved. □

We now introduce the notion of the entropy of a flow.

Definition 6. Suppose $\{T^t\}$ is a flow on the Lebesgue space $(M, \mathfrak{S}, \mu)$. The *entropy of the flow* $\{T^t\}$ is the number $h(\{T^t\}) \overset{\text{def}}{=} h(T^1)$.

The following theorem shows that this definition is quite natural.

Theorem 3. *Suppose $\{T^t\}$ is a flow in the Lebesgue space $(M, \mathfrak{S}, \mu)$. Then $h(T^t) = |t| \cdot h(T^1)$ for all $t \in \mathbb{R}^1$.*

Proof. Since, by the first statement of Theorem 2, $h(T^{-t}) = h(T^t)$ we may assume that $t > 0$. We shall first prove that $0 < u < t$ implies $h(T^t) \le (t/u) \cdot h(T^u)$.

Suppose m is a positive integer, $\delta = 1/m$ and ξ is a finite partition of the space M. Put

$$\eta = \xi \vee T^{\delta u}\xi \vee T^{2\delta u}\xi \vee \cdots \vee T^{(m-1)\delta u}\xi.$$

Further, fix a positive integer n and denote by $k = k(n)$ some natural number such that $nt \le ku < (n+1)t$. For $p = 1, 2, \ldots, n$ denote by $r(p)$ the natural number satisfying $r(p)\delta u \le pt < [r(p)+1]\delta u$. Using the properties of the entropy of a partition, we can write

$$\begin{aligned} H(T^t\xi \vee \cdots \vee T^{nt}\xi) &\le H(T^t\xi \vee \cdots \vee T^{nt}\xi \vee \eta \vee T^u\eta \vee \cdots \vee T^{ku}\eta) \\ &= H(\eta \vee T^u\eta \vee \cdots \vee T^{ku}\eta) \\ &\quad + H(T^t\xi \vee \cdots \vee T^{nt}\xi \,|\, \eta \vee T^u\eta \vee \cdots \vee T^{ku}\eta) \\ &= H(\eta \vee T^u\eta \vee \cdots \vee T^{ku}\eta) \\ &\quad + H(T^t\xi \vee \cdots \vee T^{nt}\xi \,|\, \xi \vee T^{\delta u}\xi \vee T^{2\delta u}\xi \vee \cdots \\ &\quad \vee T^{[k(m+1)-1]\delta u}\xi) \\ &\le H(\eta \vee T^u\eta \vee \cdots \vee T^{ku}\eta) + \sum_{p=1}^{n} H(T^{pt}\xi \,|\, T^{r(p)\delta u}\xi). \end{aligned} \tag{9}$$

But $H(T^{pt}\xi \,|\, T^{r(p)\delta u}\xi) = H(T^\tau\xi \,|\, \xi)$, where $\tau = pt - r(p)\delta u < \delta u$. Choose an arbitrary $\varepsilon > 0$. Since M is a Lebesgue space, $L^2(M, \mathfrak{S}, \mu)$ is separable and therefore the group $\{U^t\}$ of unitary operators adjoint to the flow $\{T^t\}$ is continuous. This implies $\lim_{\delta \to 0} \mu(T^\delta A \triangle A) = 0$ for any $A \in \mathfrak{S}$. Therefore for any sufficiently small $\delta > 0$ we have the inequality

$$H(T^\tau\xi \,|\, \xi) < \varepsilon. \tag{10}$$

Comparing (9) and (10), we get

$$H(T^t\xi \vee \cdots \vee T^{nt}\xi) \le H(\eta \vee \cdots \vee T^{k(n)u}\eta) + n\varepsilon.$$

Since $\lim_{n\to\infty}[k(n)/n] = t/u$, the last inequality implies $h(T^t) \le (t/u)\cdot h(T^u) + \varepsilon$. Since ε was arbitrary, we get $h(T^t) \le (t/u)\cdot h(T^u)$. Now suppose the positive integer r satisfies $t/r < u$. By Theorem 2, $h(T^t) = r\cdot h(T^{t/r})$. It follows from what we have proved above that $h(T^u) \le (ur/t)\cdot h(T^{t/r})$. Therefore $h(T^t) = (t/u)\cdot h(T^u)$. The theorem is proved. □

Now suppose T is a Bernoulli automorphism, $\xi = (C_1, \ldots, C_r)$ is the generating partition for T considered in Example 3. The following important theorem specifies the character of the convergence (when $n \to \infty$) of

$$\frac{H(\xi^{n-1})}{n} \to h(T, \xi) = h(T). \tag{11}$$

For any point $x = (\ldots, y_{-1}, y_0, y_1, \ldots) \in M$ suppose

$$C^{(n)}(x) = \{x' = (\ldots, y'_{-1}, y'_0, y'_1, \ldots) \in M : y'_k = y_k, 0 \le k \le n-1\}.$$

In other words, $C^{(n)}(x)$ is the element of the partition ξ_0^{n-1} which contains x.

The expression $H(\xi_0^{n-1})$ can now be written in the form

$$H(\xi_0^{n-1}) = \int_M -\log \mu(C^{(n)}(x))\, d\mu.$$

We shall show that not only the integral of the function

$$f^{(n)}(x) = -\frac{1}{n}\log \mu(C^{(n)}(x))$$

tends to $h(T, \xi) = h(T)$ when $n \to \infty$ (this is precisely what relation (11) means), but that the function $f^{(n)}(x)$ itself tends to $h(T, \xi)$ almost everywhere.

Theorem 4. *For almost all $x \in M$ we have*

$$\lim_{n\to\infty}\left[-\frac{1}{n}\log \mu(C^{(n)}(x))\right] = h(T, \xi).$$

Proof. On M consider the function

$$f(x) = f(\ldots, y_{-1}, y_0, y_1, \ldots) = -\log \sigma(y_0).$$

Then

$$\int_M f\, d\mu = \sum_{k=1}^{r}\int_{C_k} f\, d\mu = -\sum_{k=1}^{r}\mu(C_k)\log \sigma(a_k) = -\sum_{k=1}^{r} p_k \log p_k.$$

Applying the Birkhoff–Khinchin ergodic theorem to the function $f(x)$, we get, using the fact that the automorphism T is ergodic,

$$\lim_{n\to\infty} \frac{1}{n} \sum_{k=0}^{n-1} f(T^k x) = -\sum_{k=1}^{r} p_k \log p_k = h(T, \xi) = h(T).$$

But

$$\begin{aligned}\frac{1}{n} \sum_{k=0}^{n-1} f(T^k x) &= -\frac{1}{n} \sum_{k=0}^{n-1} \log \sigma(y_k) \\ &= -\frac{1}{n} \log \prod_{k=0}^{n-1} \sigma(y_k) = -\frac{1}{n} \log \mu(C^{(n)}(x)).\end{aligned}$$

The theorem is proved. □

Since convergence almost everywhere implies convergence in measure, we come to the following statement.

Corollary. *For any* $\varepsilon > 0$ *denote by* $E_\varepsilon^{(n)}$ *the union of all elements* $C^{(n)}(x) = C^{(n)} \in \xi_0^n$ *such that*

$$\exp\{-n[h(T, \xi) + \varepsilon]\} < \mu(C^{(n)}) < \exp\{-n[h(T, \xi) - \varepsilon]\}. \tag{12}$$

Then $\lim_{n\to\infty} \mu(E_\varepsilon^{(n)}) = 1$.

It is sometimes useful to state this corollary in a somewhat different form. For a fixed $\varepsilon > 0$, partition the set of all cylinders $C^{(n)} \subset M$ of length n into two classes. The first class contains all the $C^{(n)}$ satisfying condition (12), the second one—all the others. Then the total measure of all cylinders of length n contained in the second class tends to zero when $n \to \infty$.

Theorem 4 is a very particular case of an important general theorem which we shall now state.

Theorem 5 (the Shannon–McMillan–Breiman Theorem). *Suppose* T *is an ergodic automorphism of the Lebesgue space* $(M, \mathfrak{S}, \mu)$, ξ *is an arbitrary finite partition of the space* M. *Then for almost all* $x \in M$, *we have*

$$\lim_{n\to\infty} \left[-\frac{1}{n} \mu(C^{(n)}(x)) \right] = h(T, \xi),$$

where $C^{(n)}(x)$ *is the element of the partition* $\xi \vee T\xi \vee \cdots \vee T^{n-1}\xi$ *which contains* x.

We shall not give the proof of this theorem (see, for example, Billingsley [1]).

§7. Metric Isomorphism of Bernoulli Automorphisms

The question of finding conditions under which two Bernoulli automorphisms are metrically isomorphic is one of the oldest questions of ergodic theory. Only when the notion of entropy was introduced, was this question extricated from the dead end which it had reached. Clearly, for any $h > 0$, there are Bernoulli automorphisms with entropy h, and for distinct h such automorphisms are not metrically isomorphic. In Chap. 8, §1 we gave an example of metric isomorphism of two Bernoulli automorphisms T_1, T_2. It is easy to check that for this example $h(T_1) = h(T_2) = \log 4$.

Suppose (Y_1, σ_1) and (Y_2, σ_2) are two finite measure spaces. Without loss of generality we may assume

$$Y_1 = \{0, 1, \ldots, m_1 - 1\};\ p_k^{(1)} = \sigma_1(\{k\}) > 0, \qquad 0 \le k \le m_1 - 1$$

and

$$Y_2 = \{0, 1, \ldots, m_2 - 1\};\ p_k^{(2)} = \sigma_2(\{k\}) > 0, \qquad 0 \le k \le m_2 - 1.$$

Consider the Bernoulli automorphisms T_1, T_2 generated by (Y_1, σ_1), (Y_2, σ_2). Their phase spaces will be (respectively)

$$M_1 = \prod_{n=-\infty}^{\infty} Y_1^{(n)}, \quad Y_1^{(n)} \equiv Y_1, \qquad M_2 = \prod_{n=-\infty}^{\infty} Y_2^{(n)}, \quad Y_2^{(n)} \equiv Y_2.$$

The measures μ_1, μ_2 in M_1, M_2 will be the product-measures of the measures σ_1, σ_2 respectively. The relation $h(T_1) = h(T_2)$ in the given case may be written in the form

$$-\sum_{k=0}^{m_1-1} p_k^{(1)} \log p_k^{(1)} = -\sum_{k=0}^{m_2-1} p_k^{(2)} \log p_k^{(2)}.$$

The isomorphism $\psi: M_1 \to M_2$ may be viewed as a one-to-one coding procedure which sends the infinite sequence $x^{(1)} \in M_1$, written in the alphabet Y_1 into an infinite sequence $x^{(2)} \in M_2$, written in the alphabet Y_2. In this case we require that the measure μ_1 be transformed into the measure μ_2, while the shifted sequence $T_1 x^{(1)}$ is mapped to the shifted sequence $T_2 x^{(2)}$. In information theory such a coding procedure is known as *stationary*. The stationary code ψ which we shall construct in this section will possess an additional remarkable property: in order to determine the values of the coordinates $y_0^{(2)}$ of the point $x^{(2)} = (\ldots, y_{-1}^{(2)}, y_0^{(2)}, y_1^{(2)}, \ldots) = \psi(x^{(1)})$, it suffices to know only a finite number of coordinates of the point $x^{(1)}$. Such a code is called *finitary*.

The main goal of this section is the proof of the following theorem.

Theorem 1. *Suppose T_1, T_2 are two Bernoulli automorphisms with finite state spaces. If $h(T_1) = h(T_2) = h$, then T_1, T_2 are metrically isomorphic. In other words, the equality $h(T_1) = h(T_2)$ implies the existence of a stationary code.*

The proof of this theorem is fairly long. We shall consider the case $m_1 \geq 3$, $m_2 \geq 3$ in detail and, at the end, we shall describe how this restriction may be lifted.

The proof will be split up into several steps.

1. We begin with several lemmas and auxiliary constructions.

Lemma 1. *If $m_1, m_2 \geq 3$, then there exists a probability vector $(q_0, q_1, \ldots, q_{m-1})$, $m \geq 3$ such that*

$$-\sum_{k=0}^{m-1} q_k \log q_k = h$$

and $q_0 = p_k^{(1)}$, $q_1 = p_{k'}^{(2)}$ for some k, k',

Proof. Rearranging, if necessary, the $p_k^{(1)}$ and $p_k^{(2)}$, we may assume

$$p_0^{(1)} \geq p_1^{(1)} \geq \cdots \geq p_{m_1-1}^{(1)},$$

$$p_0^{(2)} \geq p_1^{(2)} \geq \cdots \geq p_{m_2-1}^{(2)} \quad \text{and} \quad p_{m_1-1}^{(1)} \geq p_{m_2-1}^{(2)}.$$

Put $q_0 = p_0^{(1)}$, $q_1 = p_{m_2-1}^{(2)}$, $q_2' = 1 - q_0 - q_1$; it easily follows from the properties of the function $H = -\sum x_k \log x_k$ that $H(q_0, q_1, q_2') \leq h$. Since

$$\lim_{n\to\infty} H\Bigg(q_0, q_1, \underbrace{\frac{q_2'}{n}, \frac{q_2'}{n}, \ldots, \frac{q_2'}{n}}_{n}\Bigg) = \infty,$$

we can find (by continuity) a vector

$$(q_0, q_1, q_2, \ldots, q_{m-1}), \qquad \sum_{k=2}^{m-1} q_k = q_2',$$

such that $H(q_0, q_1, q_2, \ldots, q_{m-1}) = h$. The lemma is proved. □

On the basis of this lemma, we can assume in the sequel that at least one of the letters in the alphabets Y_1, Y_2 has the same probability. Indeed, if Theorem 1 is proved in this particular case, then it suffices to consider the auxiliary Bernoulli automorphism T with alphabet (Y, σ),

$$Y = \{0, 1, \ldots, m-1\}, \qquad \sigma(\{k\}) = q_k, \qquad 0 \leq k \leq m-1.$$

Then, from what we have proved, it follows that any of the automorphisms T_1, T_2 is metrically isomorphic to T so that T_1 is metrically isomorphic to T_2. To be definite, let us assume that the letter with the same probability is 0, i.e., $p_0^{(1)} = p_0^{(2)}$.

The subsequent constructions will be more motivated if we describe at this point the main property of the code being constructed. For any $x \in M_i$, $i = 1, 2$; $x = (\ldots, y_{-1}, y_0, y_1, \ldots)$ put $N_x = \{n: y_n = 0\}$; by the *fiber containing* x, we mean the set $M_x^{(i)} = \{x' \in M_i: N_{x'} = N_x\}$. The above-mentioned property of the isomorphism ψ is the following: ψ is defined on a set $\overline{M}_1 \subset M_1$ of measure 1 consisting of entire fibers and each separate fiber in $\overline{M}_1$ will be mapped onto a separate fiber in $\overline{M}_2$, i.e., $N_{\psi(x)} = N_x$, $\psi(M_x^{(1)}) = M_{\psi(x)}^{(2)}$.

2. *Skeletons.* We now introduce a notion which will play the role of "finite-dimensional model" of the fibers M_x. Fix an increasing sequence of natural numbers $N_1 < N_2 < \cdots < N_r < \cdots$—the explicit form of this sequence will be indicated later. Consider the alphabet consisting of two letters 0 and $\times$ ("zeros" and "crosses") and finite sequences Σ of letters of this alphabet beginning and ending with zero i.e. of the form

$$\Sigma:\ \underbrace{0\ 0 \cdots 0}_{n_0}\ \underbrace{\times\ \times \cdots \times}_{l_1}\ \underbrace{0\ 0 \cdots 0}_{n_1}\ \underbrace{\times\ \times \cdots \times}_{l_2} \cdots \underbrace{\times\ \times \cdots \times}_{l_k}\ \underbrace{0\ 0 \cdots 0}_{n_k}. \tag{1}$$

The pair $s = (r, \Sigma)$ consisting of a natural number r and a sequence Σ of the form (1) will be called a *skeleton of rank* r, if

(1) $l_t \geq 1, \quad 1 \leq t \leq k$;
(2) $n_t \geq 1, \quad 0 \leq t \leq k$;
(3) $n_t < N_r \leq \min(n_0, n_k), \quad 1 \leq t \leq k - 1$.

The sequence Σ will be known as the *configuration* of the skeleton s. Note that the same configuration Σ may, in general, form skeletons with different r (skeletons of different ranks) and then these skeletons are viewed as different. The rank of a skeleton will be denoted by $\rho(s)$ and its length in the configuration (1) is the number $l(s) = l_1 + \cdots + l_k$. We shall say that the skeleton $s' = (r', \Sigma')$ is a *subskeleton* of the skeleton $s = (r, \Sigma)$, if its configuration is of the form

$$\Sigma:\underbrace{0\ 0 \cdots 0}_{n_t}\ \underbrace{\times\ \times \cdots \times}_{l_{t+1}}\ \underbrace{0\ 0 \cdots 0}_{n_{t+1}} \cdots \underbrace{\times\ \times \cdots \times}_{l_{t'}}\ \underbrace{0\ 0 \cdots 0}_{n_{t'}}$$

for some $t, t', 0 \leq t, t' \leq k$.

Two subskeletons of skeleton s will be called *disjoint* if the corresponding sets $\{t + 1, \ldots, t'\}$ are disjoint.

If $s_1, \ldots, s_q$ are subskeletons of the skeleton s, we shall say that they constitute a partition of s and write $s = s_1 \oplus s_2 \oplus \cdots \oplus s_q$, if the s_i are pairwise disjoint and the union of the corresponding sets $\{t + 1, \ldots, t'\}$ is $\{1, 2, \ldots, k\}$.

Lemma 2. *Any skeleton $s = (r, \Sigma)$ of rank $r > 1$ possesses a unique partition*

$$s = s_1 \oplus s_2 \oplus \cdots \oplus s_q$$

into subskeletons of rank $(r - 1)$.

Proof. If $(r - 1)$ is a rank possible for the configuration Σ, i.e. $(r - 1, \Sigma)$ is a skeleton of rank $(r - 1)$, then we can take $q = 1$, $s_1 = (r - 1, \Sigma)$. If $(r - 1)$ is not a possible rank for Σ, then the set $\{t: 1 \le t \le k - 1, n_t \ge N_{r-1}\}$ is nonempty and gives us a partition of the configuration Σ into parts $\Sigma_1, \ldots, \Sigma_q$, which are configurations of skeletons of rank $(r - 1)$. The partition is unique since s has no other subskeleton of rank $(r - 1)$. The lemma is proved. □

The partition constructed in this lemma will be referred to as the *rank partition* of s.

Now let us relate the notion of skeleton introduced above with points of the phase spaces of the automorphisms T_1, T_2. Suppose the measure space M is M_1 or M_2 and $x = (\ldots, y_{-1}, y_0, y_1, \ldots) \in M$. Consider a skeleton s. We shall say that the skeleton s *appears* in x at the points $n', n' + 1, \ldots, n''$ (or on the segment $\Delta = [n', n'']$) if:

(1) when we replace all the nonzero coordinates y_n, $n' \le n \le n''$ by the letter $\times$, the sequence $y_{n'}, \ldots, y_{n''}$ is transformed into the configuration Σ of the skeleton s;

(2) $y_{n'-1} \ne 0$, $y_{n''+1} \ne 0$.

Denote $M^* = \{x \in M: y_0 \ne 0\}$.

Lemma 3. (1) *For almost all $x = (\ldots, y_{-1}, y_0, y_1, \ldots) \in M^*$ there exists a unique sequence of skeletons $s_1(x), \ldots, s_r(x), \ldots$ such that $\rho(s_r(x)) = r$ and $s_r(x)$ appears in x on the segment $\Delta_r = [n'_r, n''_r]$, $n'_r < 0$, $n''_r > 0$.*

(2) *For any sequence $L_1, L_2, \ldots, L_r, \ldots$ of natural numbers tending to infinity we can choose the numbers $N_1, N_2, \ldots, N_r, \ldots$ (which appear in the definition of a skeleton) so that the inequality*

$$l(s_r(x)) \ge L_r$$

holds for almost all $x \in M^*$ whenever r is sufficiently large (this number, of course, depends on x).

Proof. (1) In the role of Δ_r, choose the segment between the first appearance of a sequence containing $\geq N_r$ zeros to the left of the 0th coordinate of the point x (including the sequence itself) and the first appearance of such a sequence to the right of the 0th coordinate (also inclusive). This can obviously be done for almost all x. Replacing, in the sequence $\{y_n\}$, $n \in \Delta_r$, all nonzero letters by the letter $\times$, we obtain the configuration of the given skeleton $s_r(x)$.

(2) Suppose the numbers $N_1, \ldots, N_r$ and the skeletons $s_1(x), \ldots, s_r(x)$ $(x \in M^*)$ (depending only on these numbers) have been defined so as to satisfy

$$\mu(E_k) \leq 1/2^k, \qquad 1 \leq k \leq r,$$

where $E_k = \{x \in M^*: l(s_k(x)) < L_k\}$. If we now change the number $N = N_{r+1}$, then the skeleton $s_{r+1}(x)$ $(x \in M^*)$ will, of course, depend on $N: s_{r+1}(x) = s_{r+1}^{(N)}(x)$ and

$$\lim_{N \to \infty} l(s_{r+1}^{(N)}(x)) = \infty \quad \text{for all } x \in M^*.$$

Therefore we can take an N_{r+1} for which $\mu(E_{r+1}) \leq 1/2^{r+1}$. By induction $\mu(E_r) \leq 1/2^r$, $r = 1, 2, \ldots$ Since $\sum_{r=1}^{\infty} \mu(E_r) < \infty$, it follows from the Borel–Cantelli lemma that almost all points $x \in M^*$ belong to a finite number of the E_r. The lemma is proved. $\square$

3. *Fillers.* Suppose s is a skeleton, Y is one of the alphabets Y_1, Y_2. The set of all finite sequences (words) in the alphabet Y, obtained from s by replacing every occurrence of the letter $\times$ by an arbitrary (but individual in each case) nonzero letter from Y, will be called a *filler* of the skeleton s with respect to the alphabet Y.

If the length of the skeleton $l(s)$ equals l (recall that the length of a skeleton is the number of letters $\times$ which occur in s), then the set of all fillers of s may be naturally identified with the set

$$F(s) = \underbrace{Y^{(0)} \times \cdots \times Y^{(0)}}_{l} = (Y^{(0)})^l = \{f = \{y_1, \ldots, y_l\}, y_i \in Y^{(0)}\},$$

where $Y^{(0)} = Y \backslash \{0\}$. $Y^{(0)}$ shall be referred to as *the filler alphabet*. On the alphabet $Y^{(0)}$, consider the measure μ_0

$$\mu_0(\{k\}) = \frac{p_k}{1 - p_0}, \qquad k \in Y^{(0)},$$

and, using it, construct the product measure $\underbrace{\mu_0 \times \cdots \times \mu_0}_{l}$ on $F(s)$. This measure shall also be denoted by μ_0. Note that if $s = s_1 \oplus \cdots \oplus s_q$, then the measure space $F(s)$ may be naturally identified with the space $\prod_{i=1}^{q} F(s_i)$. For the sequel we shall need the notation

$$\eta = \max \mu_0(\{k\}), \qquad k \in Y^{(0)},$$

$$\theta = \max \mu_0(\{k\}), \qquad k \in Y^{(0)},$$

$$h_0 = -\sum_{k \in Y^{(0)}} \mu_0(\{k\}) \log \mu_0(\{k\}).$$

4. *Classes of equivalent fillers.* For any skeleton s of arbitrary rank r we would like to construct an equivalence relation $\sim$ on the set $F(s)$. This shall be done by induction on the rank $r = \rho(s)$. Fix a sequence $\varepsilon_1 > \varepsilon_2 > \cdots > 0$, $\lim_{r \to \infty} \varepsilon_r = 0$. Suppose $\rho(s) = 1$, $l(s) = l$. For $f = \{y_1, \ldots, y_l\} \in F(s)$ put

$$I(f) = \left\{k : 1 \le k \le l, \mu_0(y_1) \cdot \mu_0(y_2) \cdot \ldots \cdot \mu_0(y_{k-1}) \ge \frac{1}{\eta} \exp[-h_0(1 - \varepsilon_1)l]\right\}.$$

Clearly $I(f)$ is of the form $\{1, 2, \ldots, k_f\}$.[1] Roughly speaking, $I(f)$ shows for what k the measure $\prod_{i=1}^{k} \mu_0(y_i)$ assumes its typical value. We shall write $f_1 \sim f_2$, if $I(f_1) = I(f_2) = I$ and the coordinates of the fillers f_1, f_2 located at the places contained in I coincide. Obviously, the relation just introduced is an equivalence relation. Assume that for all skeletons s of rank $<r$ and any filler $f \in F(s)$ we have already defined the set $I(f) \subset \{1, 2, \ldots, l(s)\}$ so that the relation $f_1 \sim f_2$ defined by $I(f_1) = I(f_2) = I$ and by the fact that the coordinates at the places in I of the fillers f_1, f_2 coincide is an equivalence relation on the set $F(s)$. Suppose s is a skeleton of rank r and

$$s = s_1 \oplus s_2 \oplus \cdots \oplus s_q$$

is the rank partition of s,

$$f = (f_1, f_2, \ldots, f_q) = \{y_1, \ldots, y_l\} \in F(s), \qquad \text{where} \quad f_i \in F(s_i),\ 1 \le i \le q.$$

First put

$$\bar{I}(f) = \bigcup_{i=1}^{q} I(f_i) \subset \{1, 2, \ldots, l\}.$$

[1] If $\mu_0(y_1) < (1/\eta)\exp[-h_0(1 - \varepsilon_1)l]$, then we put by definition $I(f) = \{1\}$, i.e., $k_f = 1$.

Arrange the elements of the complement $\{1, 2, \ldots, l\} \setminus \bar{I}(f)$ in increasing order $k_1 < k_2 < \cdots < k_d$. Put

$$I(f) = \bar{I}(f) \cup \left\{k_u: 1 \leq u \leq d, \left(\prod_{k \in \bar{I}(f)} \mu_0(y_k) \right) \right.$$
$$\left. \times\ \mu_0(y_{k_1}) \cdot \mu_0(y_{k_2}) \cdots \mu_0(y_{k_{u-1}}) \geq \frac{1}{\eta} \exp[-h_0(1 - \varepsilon_r)l] \right\}.$$

In this case it is obvious that the relation $f_1 \sim f_2$, defined by $I(f_1) = I(f_2) = I$ and by the fact that the coordinates of the fillers f_1, f_2 located at the places which are contained in I coincide, is an equivalence relation.

The equivalence class containing the filler $f = f(s)$ will be denoted by $\tilde{f}$, while the set of equivalence classes will be written $\tilde{F}(s)$. For the filler $f = f(s)$ the coordinates which are included in the set $I(f)$ will be referred to as *essential*, the others—as *inessential*. Note for the sequel that for any skeleton s of rank $r > 1$ together with the equivalence relation $\sim$ we can consider on the set $F(s)$ another equivalence relation: write $f_1 \leftrightarrow f_2$ if $\bar{I}(f_1) = \bar{I}(f_2)$ and the coordinates located at the places included in $\bar{I}$ of the fillers f_1, f_2 coincide.

The relation $\sim$ is finer than the relation $\leftrightarrow$, i.e., the $\leftrightarrow$-equivalence classes are unions of $\sim$-equivalence classes; in the case $r = 1$ both relations coincide by definition. The set of $\leftrightarrow$-equivalence classes will be denoted by $\bar{F}$ and the classes themselves by $\bar{f}$.

Lemma 4. (1) *Suppose r is an arbitrary natural number and l satisfies*

$$\frac{1}{\eta} \exp[-h_0(1 - \varepsilon_1)l] < 1;$$

if $\rho(s) = r$, $l(s) = l$ and $f \in E(s)$, then $\mu_0(\tilde{f}) \geq \exp[-h_0(1 - \varepsilon_r)l]$.

(2) *For any $\delta > 0$ and any r we can find an $l_1 = l_1(\delta, r)$ such that if the skeleton s is of rank $\rho(s) = r$ and its length satisfies $l(s) = l \geq l_1$, then for all $f \in F(s)$, except for a certain set of fillers, the total μ_0-measure of which is less than δ we shall have*

$$\mu_0(\tilde{f}) \leq \frac{1}{\eta} \exp[-h_0(1 - \varepsilon_r)l]. \tag{2}$$

(3) *For any $\delta > 0$ and any r we can find a $l_2 = l_2(\delta, r)$ such that if the skeleton s is of rank $\rho(s) = r$ and the length satisfies $l(s) = l \geq l_2$, then for all $f \in F(s)$ except for a set of fillers whose total μ_0-measure is less than δ, we shall have*

$$\frac{\operatorname{card} I(f)}{l} \geq 1 - \frac{2h_0}{|\log \theta|} \varepsilon_r.$$

Proof. (1) Let us carry out an induction over r. If $r = \rho(s) = 1$, $I(f(s)) = \{1, 2, \ldots, d\}$, then

$$\mu_0(\tilde{f}) = \prod_{k=1}^{d} \mu_0(y_k) = \prod_{k=1}^{d-1} \mu_0(y_k) \cdot \frac{\mu_0(y_d)}{\eta} \cdot \eta$$

$$\geq \frac{1}{\eta} \exp[-h_0(1 - \varepsilon_1)l] \cdot \eta = \exp[-h_0(1 - \varepsilon_1)l].^1$$

Now suppose $r > 1$ and assume that our statement is proved for ranks less than r. Consider the rank partition $s = s_1 \oplus \cdots \oplus s_q$; $l(s_i) = l_i$, $1 \leq i \leq q$, and write $f = (f_1, \ldots, f_q)$. If $I(f) = \bigcup_{i=1}^{q} I(f_i)$, then

$$\mu_0(\tilde{f}) = \prod_{i=1}^{q} \mu_0(\tilde{f}_i) \geq \prod_{i=1}^{q} \exp[-h_0(1 - \varepsilon_{r-1})l_i]$$

$$= \exp[-h_0(1 - \varepsilon_{r-1})l] \geq \exp[-h_0(1 - \varepsilon_r)l],$$

if $I(f) \neq \bigcup_{i=1}^{q} I(f_i)$ and k_d is the maximal element in $I(f) \setminus \bigcup_{i=1}^{q} I(f_i)$, then

$$\mu_0(\tilde{f}) = \left[\prod_{k \in \bar{I}(f)} \mu_0(y_k)\right] \mu_0(y_{k_1}) \mu(y_{k_2}), \ldots, \mu_0(y_{kd})$$

$$\geq \frac{\mu_0(y_{k_d})}{\eta} \exp[-h_0(1 - \varepsilon_r)l] \geq \exp[-h_0(1 - \varepsilon_r)l].$$

(2) Denote $F'(s) = \{f \in F(s)\colon \text{ card } I(f) = l(s)\}$. If $f \notin F'(s)$, i.e., card $I(f) < l$, then, by the definition of $I(f)$, we will have

$$\mu_0(\tilde{f}) \leq \frac{1}{\eta} \exp[-h_0(1 - \varepsilon_r)l].$$

If $f \in F'(s)$ then $\tilde{f} = f$. Therefore it suffices to estimate from above, for large $l(s)$, the measure of the set

$$\left\{f \in F(s)\colon \mu_0(f) > \frac{1}{\eta} \exp[-h_0(1 - \varepsilon_r)l]\right\}.$$

In view of the corollary to Theorem 4, §6, we can find an $l_1 = l_1(\delta, r)$ such that for $l \geq l_1$ this set will be of μ_0-measure less than δ.

[1] If $I(f) = \{1\}$, then in the calculations above we put formally

$$\prod_{k=1}^{d-1} \mu_0(y_k) = \prod_{k=1}^{0} \mu_0(y_k) = 1.$$

(3) Suppose $f \in F(s)$, $l(s) \geq l_1(\delta/2, r)$ and the inequality (2) is satisfied for f. By the Subsection 2, the μ_0-measure of the set of all such f is no less than $1 - \delta/2$. If we also have

$$\frac{\text{card } I(f)}{l} < 1 - \frac{2h_0}{|\log \theta|} \varepsilon_r,$$

then

$$\begin{aligned}\mu_0(f) &\leq \mu_0(\tilde{f}) \cdot \theta^{(2h_0/|\log \theta|)\varepsilon_r l} \\ &\leq \frac{1}{\eta} \exp[-h_0(1 - \varepsilon_r)l] \cdot \exp(-2h_0 \varepsilon_r l) = \frac{1}{\eta} \exp[-h_0(1 + \varepsilon_r)l]. \quad (3)\end{aligned}$$

By the corollary to Theorem 4, §6, we can find an $l_2 \geq l_1$ $(\delta/2, r)$ such that $l \geq l_2$ implies that the total measure of all f satisfying inequality (3) is less than $\delta/2$. The lemma is proved. □

5. *The choice of the sequence* $\{N_r\}$. We can now specify the choice of the sequence $\{N_r\}$ which appears in the definition of a skeleton (see Subsection 2). In Subsection 3 we introduced the notion of a filler $f(s)$ of the skeleton s with respect to the alphabet Y_1 or Y_2. For each of these two alphabets Lemma 4 gives the corresponding numbers $l_1(\delta, r)$, $l_2(\delta, r)$ which shall be denoted respectively by $l_1^{(1)}, l_2^{(1)}, l_1^{(2)}, l_2^{(2)}$. These numbers depend on the choice of the sequence $\{\varepsilon_r\}$, but do not depend on the sequence $\{N_r\}$. We shall assume that for the number $l_i^{(j)}$, $1 \leq i, j \leq 2$ we have the inequality

$$\frac{1}{\eta} \exp[-h_0(1 - \varepsilon_1)l_i^{(j)}] < 1$$

which appears in statement (1) of Lemma 4.

Now choose an arbitrary sequence $\delta_r > 0$, $\lim_{r\to\infty} \delta_r = 0$ and for each r choose $L_r \geq \max\{l_i^{(j)}(\delta_r, r)\}$, $1 \leq i, j \leq 2$, so as to have

$$\lim_{r\to\infty} L_r(\varepsilon_{r-1} - \varepsilon_r) = \infty. \quad (4)$$

Using the sequence $\{L_r\}$, choose the number N_r in accordance to Lemma 3. The meaning of condition (4) will become clear in Subsection 8.

The set of fillers of the skeleton s with respect to the alphabets Y_1, Y_2 will now be denoted by $F_1(s)$, $F_2(s)$; the set of $\sim$-equivalence classes of these fillers by $\tilde{F}_1(s)$, $\tilde{F}_2(s)$; the set of $\leftrightarrow$-equivalence classes by $\bar{F}_1(s)$, $\bar{F}_2(s)$.

6. *Correct maps.* In the sequel we shall need a combinatorial fact which resembles the well-known "transport problem" of linear programming.

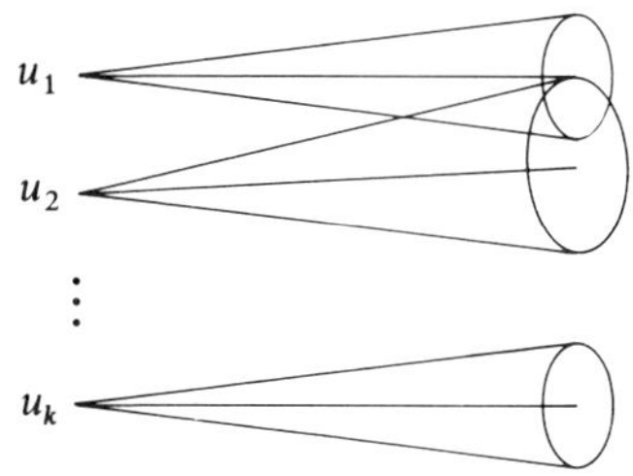

Figure 7

Suppose we are given a finite set of points $U = \{u_1, \ldots, u_k\}$ (factories) and a finite set $V = \{v_1, \ldots, v_l\}$ of points (warehouses); assume that at each factory $u \in U$ there is a known mass $\rho(u) > 0$ of produce, while each warehouse $v \in V$ has a known capacity $\sigma(v) > 0$. Moreover, assume that to each factory $u \in U$ a certain subset of warehouses (those which have roads connecting them with the factory u) has been assigned, i.e., we are given a map $R: U \to 2^V$.[1] As in information theory, it is convenient to imagine the map R in the form of a family of fans (Fig. 7).

It is asked under what conditions it is possible to transport the entire mass of produce from the factories to the warehouses using the existing roads. The proposed transportation method is given by the matrix $\|\alpha_{ij}\|$, $1 \le i \le k$, $1 \le j \le l$, where α_{ij} is the amount of produce transported from the ith factory to the jth warehouse. Clearly we must have the inequalities:

(1) $\alpha_{ij} \ge 0$;
(2) $\sum_{j=1}^{l} \alpha_{ij} = \rho(u_i)$, $1 \le i \le k$;
(3) $\sum_{i=1}^{k} \alpha_{ij} \le \sigma(v_j)$, $1 \le j \le l$;
(4) $\alpha_{ij} = 0$, if $v_j \notin R(u_i)$.

It follows from these inequalities, in particular, that if $R(u_i)$ is a one-point set: $R(u_i) = \{v_j\}$ then $\alpha_{ij} = \rho(u_i) \le \sigma(v_j)$.

We can view (U, ρ) and (V, ρ) as two finite spaces with measures (which are not necessarily normalized). Then an obvious necessary condition for solving the transport problem is the following: for any subset $A \subseteq U$ we must have the inequality

$$\rho(A) \ge \sigma(R(A)). \tag{5}$$

A map $R: U \to 2^V$ satisfying condition (5) will be called *correct*.

Lemma 5. *The correctness of the map R is also a sufficient condition for solving the transport problem.*

[1] 2^V denotes the set of all subsets of V.

The proof of Lemma 5 will be given later.

The set of correct maps $R: U \to 2^V$ possesses a natural partial order: $R_1 \preccurlyeq R_2$, if $R_1(u) \subseteq R_2(u)$ for all $u \in U$. By the *defect* of a correct map R we mean the number $\pi(R) = \text{card}\, \Gamma(R)$, where $\Gamma(R)$ is the set of all $v \in V$ for which we can find $u_1, u_2 \in U$, $u_1 \neq u_2$ satisfying $v \in R(u_1) \cap R(u_2)$. The notion of the defect of a correct map will be very important in the sequel. It shows to what extent the map is not invertible.

Lemma 6. *For any correct map $R: U \to 2^V$ there exists a correct map $\Phi \preccurlyeq R$ satisfying $\pi(\Phi) < \text{card}\, U - 1$.*

The proof of the lemma will be given later. Further, for correct maps, we shall often construct minimal correct maps in the sense of the partial order introduced above. Lemma 6 in fact states that if Φ is a minimal correct map, then $\pi(\Phi) < \text{card}\, U - 1$.

For any two maps (not necessarily correct)

$$R_1: U_1 \to 2^{V_1}, \qquad R_2: U_2 \to 2^{V_2}$$

define the map $R_1 \times R_2: U_1 \times U_2 \to 2^{V_1 \times V_2}$ by the formula

$$(R_1 \times R_2)(u^{(1)}, u^{(2)}) = R_1(u^{(1)}) \times R_2(u^{(2)}), \qquad u^{(1)} \in U_1, u^{(2)} \in U_2.$$

Further, for the map $R: U \to 2^V$, introduce the conjugate map $R^*: V \to 2^U$ in the following way: for $v \in V$, $R^*(v)$ is the set of all $u \in U$ satisfying $v \in R(u)$.

Lemma 7. (1) *If R_1, R_2 are correct maps, then $R_1 \times R_2$ is also correct;*

(2) *if $R: U \to 2^V$ is a correct map and the measures in the spaces U, V are normalized, then R^* is also a correct map.*

Proof. (1) Suppose that for $i = 1, 2$ the sequence of numbers $\alpha_i(u^{(i)}, v^{(i)})$, $u \in U_i$, $v \in V_i$ determines a method of transportation corresponding to R_i. Then if we put

$$\alpha((u^{(1)}, u^{(2)}), (v^{(1)}, v^{(2)})) = \alpha_1(u^{(1)}, v^{(1)}) \cdot \alpha_2(u^{(2)}, v^{(2)}),$$

we obtain, as is easily checked, a transportation method corresponding to $R_1 \times R_2$. This means that $R_1 \times R_2$ is a correct map.

(2) Assume, in contradiction to our statement, that $\sigma(B) > \rho(R^*(B))$ for some $B \subset V$. For any element $u \in U \setminus R^*(B)$, by definition, $B \cap R(u) = \varnothing$ so that $R(U \setminus R^*(B)) \subseteq V \setminus B$. Hence

$$\sigma(R(U \setminus R^*(B))) \leq \sigma(V \setminus \{B\}) = 1 - \sigma(B) < 1 - \rho(R^*(B)) = \rho(U \setminus R^*(B)).$$

This contradicts the correctness of the map R. The lemma is proved. $\square$

7. *Sets of correct maps.* In this subsection we shall define a sequence of correct maps which at the limit will give us the desired isomorphism.

For any $r \geq 1$ and any skeleton s of rank $\rho(s) = r$, let us define the correct map Φ_s which, for even r, will be the map $\Phi_s: \tilde{F}_2(s) \to 2^{\tilde{F}_1(s)}$ and, for odd r, the map $\Phi_s: \tilde{F}_1(s) \to 2^{\tilde{F}_2(s)}$. Here the role of the mass of produce and the capacity of warehouses (see Subsection 6) will be played by the measures of the corresponding equivalence classes $\tilde{f}_1(s), \tilde{f}_2(s)$.

The construction of Φ_s will be carried out by induction over r. Suppose at first that s is a skeleton of rank $r = 1$. Define the map $R_s: \tilde{F}_1(s) \to 2^{\tilde{F}_1(s)}$ by putting $R_s(\tilde{f}) = \tilde{F}_2(s)$ for any $\tilde{f} \in \tilde{F}_1(s)$. Clearly R_s is a correct map. For Φ_s take any of the minimal maps $\Phi_s \preccurlyeq R_s$. By Lemma 6, for such a Φ_s we have the inequality $\pi(\Phi_s) < \operatorname{card}(\tilde{F}_1(s)) - 1$. Note that statement (1) of Lemma 4 implies card $\tilde{F}_1(s) \leq \exp[h_0(1 - \varepsilon_r)l]$.

Now suppose s is a skeleton of even rank and assume that for any skeleton s_i in its rank partition $s = s_1 \oplus s_2 \oplus \cdots \oplus s_q$ we have already defined a correct map

$$\Phi_{s_i}: \tilde{F}_1(s_i) \to 2^{\tilde{F}_2(s_i)}, \qquad 1 \leq i \leq q.$$

Put

$$\bar{F}_2(s) = \tilde{F}_2(s_1) \times \cdots \times \tilde{F}_2(s_q), \qquad \bar{F}_1(s) = \tilde{F}_1(s_1) \times \cdots \times \tilde{F}_1(s_q),$$

and define the map $R_s: \bar{F}_2(s) \to 2^{\bar{F}_1(s)}$ by the formula

$$R_s = (\Phi_{s_1} \times \Phi_{s_2} \times \cdots \times \Phi_{s_q})^* = \Phi_{s_1}^* \times \Phi_{s_2}^* \times \cdots \times \Phi_{s_q}^*.$$

Since each element of the set $\bar{F}_1(s)$ is the union of certain $\sim$-equivalence classes, i.e., the union of certain elements of the set $\tilde{F}_1(s)$, it follows that the map R_s may be viewed as a map $R_s: \bar{F}_2(s) \to 2^{\tilde{F}_1(s)}$ (so as to simplify our exposition, we continue to denote this map by R_s). By Lemma 7, R_s is also a correct map. For Φ_s take the minimal map

$$\Phi_s \preccurlyeq R_s, \qquad \Phi_s: \bar{F}_2(s) \to 2^{\tilde{F}_1(s)}.$$

Since each element of the set $\bar{F}_2(s)$ is the union of certain elements of the set $\tilde{F}_2(s)$, it follows that the map Φ_s may be considered as defined on $\tilde{F}_2(s)$, i.e., $\Phi_s: \tilde{F}_s(s) \to 2^{\tilde{F}_1(s)}$. Note, however, that it is not necessarily minimal.

Suppose further that s is a skeleton of odd rank $r > 1$, and $s = s_1 \oplus \cdots \oplus s_q$ is its rank partition; assume that we have already defined the correct map

$$\Phi_{s_i}: \tilde{F}_2(s_i) \to 2^{\tilde{F}_1(s_i)}, \qquad 1 \leq i \leq q.$$

Introduce, as before, the map

$$R_s = (\Phi_{s_1} \times \cdots \times \Phi_{s_q})^* = \Phi_{s_1}^* \times \cdots \times \Phi_{s_q}^*,$$

which again shall be viewed as a map

$$R_s\colon \bar{F}_1(s) \to 2^{\bar{F}_2(s)}.$$

Now choose a minimal (in the sense of the partial order) map $\Phi_s \preccurlyeq R_s$ and consider it as a map $\Phi_s\colon \tilde{F}_1(s) \to 2^{\tilde{F}_2(s)}$. Thus by induction the map Φ_s has been constructed for any s. Note that during our construction we viewed the map Φ_s in two ways (in one the domain of the map was a set of $\leftrightarrow$-equivalence classes, in the other a set of $\sim$-equivalence classes). From now on we shall remain on the first point of view, i.e., will assume that

$$\Phi_s\colon \begin{cases} \bar{F}_1(s) \to 2^{\tilde{F}_2(s)} & \text{for odd } r(s), \\ \bar{F}_2(s) \to 2^{\tilde{F}_1(s)} & \text{for even } r(s). \end{cases}$$

8. We shall need two statements concerning the properties of the maps Φ_s. These statements will be proved in Lemmas 8 and 10 (Lemma 9 will be used only for the proof of Lemma 10).

First let us introduce some notations. Suppose $s = s_r$ is a skeleton of rank r and assume we are given a point $x^{(i)} \in M_i^*$ $(i = 1, 2)$. Denote by $f_r(x^{(i)})$ the filler of the skeleton s_r corresponding to the point $x^{(i)}$; denote by $\tilde{f}_r(x^{(i)})$ the $\sim$-equivalence class containing $f_r(x^{(i)})$. Assume further that $M_i(s_r)$ is the set of all points $x \in M_i$ satisfying $s_r(x) = s_r$. And, finally, for even r, denote

$$E_1(s_r) = \{x \in M_1(s_r)\colon \tilde{f}_r(x) \notin \Gamma(\Phi_{s_r})\}.$$

Lemma 8. *If r is even and $l_r = l(s_r) \geq L_r$, then*

$$\mu_1(E_1(s_r)) \geq (1 - \beta_r)\mu_1(M_1(s_r)).$$

where $\beta_r > 0$ is a certain sequence of numbers, $\beta_r \to 0$ for $r \to \infty$.

Proof. The map Φ_{s_r} is defined on the set $\bar{F}_2(s_r)$. First let us estimate card $\bar{F}_2(s_r)$. Suppose the rank partition of the skeleton s_r is of the form

$$s_r = s_{r-1}^{(1)} \oplus s_{r-1}^{(2)} \oplus \cdots \oplus s_{r-1}^{(q)}, \qquad l(s_{r-1}^{(i)}) = l_{r-1}^{(i)}, \qquad 1 \leq i \leq q.$$

It follows from the first statement of Lemma 4 that

$$\begin{aligned} \operatorname{card} \bar{F}_2(s_r) &= \prod_{i=1}^{q} \operatorname{card} \bar{F}_2(s_{r-1}^{(i)}) \\ &\leq \prod_{i=1}^{q} \exp[h_0(1 - \varepsilon_{r-1})l_{r-1}^{(i)}] = \exp[h_0(1 - \varepsilon_{r-1})l_r]. \end{aligned}$$

Since the map Φ_{s_r} is minimal, Lemma 6 implies that the defiect of this map satisfies

$$\pi(\Phi_{s_r}) = \text{card}\, \Gamma(\Phi_{s_r}) \leq \exp[h_0(1 - \varepsilon_{r-1})l_r].$$

It now follows from the second statement of Lemma 4 that

$$\mu_0^{(1)}(\Gamma(\Phi_{s_r})) \leq \exp[h_0(1 - \varepsilon_{r-1})l_r] \cdot \frac{1}{\eta} \exp[-h_0(1 - \varepsilon_r)l_r] + \delta_r$$

$$\leq \frac{1}{\eta} \exp[-h_0(\varepsilon_{r-1} - \varepsilon_r)L_r] + \delta_r \stackrel{\text{def}}{=} \beta_r. \qquad (6)$$

By (4), $\beta_r \to 0$ when $r \to \infty$. Since $\mu_0^{(1)}$ is obviously a conditional measure under the condition $s_r(x) = s_r$, i.e., under the condition $x \in M_1(s_r)$, it follows that (6) is equivalent to the statement of the lemma. The lemma is proved. □

In the next lemma M, M^*, $M(s_r)$, μ will denote one of M_i, M_i^*, $M_i(s_r)$, μ_i, $i = 1, 2$.

A point $x = (\ldots, y_{-1}, y_0, y_1, \ldots) \in M^*$ will be referred to as r-nice (nice with respect to the rank r) if its 0th coordinate y_0 is essential in the filler $f_r(x) \stackrel{\text{def}}{=} f_r(s_r(x))$. For any skeleton s_r, denote by $\hat{G}(s_r)$ the set of all r-nice points x belonging to $M(s_r)$.

Lemma 9. *If $l_r = l(s_r) \geq L_r$, then*

$$\mu(\hat{G}(s_r)) \geq \left(1 - \delta_r - \frac{2h_0}{|\log \theta|} \varepsilon_r\right)\mu(M(s_r)).$$

Proof. Partition $M(s_r)$ into two sets:

$$M(s_r) = M'(s_r) \cup M''(s_r),$$

where $M'(s_r)$ is the set of all $x \in M(s_r)$ satisfying

$$\text{card}\, I(f_r(x)) \geq \left(1 - \frac{2h_0}{|\log \theta|} \varepsilon_r\right) l_r, \qquad (7)$$

$$M''(s_r) = M(s_r) \backslash M'(s_r).$$

By the third statement of Lemma 4, $\mu(M''(s_r)) < \delta_r$.

Now assume f is a filler, $f \in F(s_r)$. Put

$$M(f) = \{x \in M(s_r) : f_r(x) = f\}.$$

The set $M(s_r)$ may be represented in the form $M(s_r) = \bigcup_{f_r} M(f_r)$ where the sum is taken over all f_r satisfying inequality (7). To prove the lemma, it suffices to show that for any such f_r we have

$$\mu(\hat{G}(s_r) \cap M(f_r)) \geq \left(1 - \frac{2h_0}{\log \theta} \varepsilon_r\right) \mu(M(f_r)). \tag{8}$$

But, in its turn,

$$M(f_r) = \bigcup_{i=1}^{l_r} M^{(i)}(f_r),$$

where $M^{(i)}(f_r)$ is the set of all $x \in M(f_r)$ whose 0th coordinate has the number i in the filler $f_r(x)$. Since the sets $M^{(i)}(f_r)$, for distinct i, are obtained from each other by a shift, we have

$$\mu(M^{(1)}(f_r)) = \mu(M^{(2)}(f_r)) = \cdots = \mu(M^{(l_r)}(f_r)).$$

It follows from the definition of r-nice points that

$$\hat{G}(s_r) \cap M(f_r) = \bigcup_{\substack{1 \leq i \leq l_r \\ i \in I(f_r(x))}} M^{(i)}(f_r).$$

so that inequality (8) immediately follows from (7). The lemma is proved. □

Now assume $r > 1$ is even, and let s_{r-1} be a skeleton of rank $(r - 1)$. Choose an arbitrary point $x^{(2)} \in \hat{G}^{(2)}(s_{r-1})$ and consider its $\leftrightarrow$-equivalence class of rank r; denote it by $\bar{f}_2(s_r)$, $s_r = s_r(x^{(2)})$. Under the map $\Phi_{s_r}: \bar{F}_2(s_r) \to 2^{\tilde{F}_1(s_r)}$ our element $\bar{f}_2(s_r)$ corresponds to a certain set $Q \subset \tilde{F}_1(s_r)$ of $\sim$-equivalence classes.

Now consider the points $x^{(1)} \in M_1(s_r)$ for which the skeleton s_r is located at the same places as for the point $x^{(2)}$ and $\tilde{f}_1(s_r(x^1)) \in Q$. The set of all such points $x^{(1)}$ corresponding to all points $x^{(2)} \in G_2(s_{r-1})$ will be denoted by $G_1(s_{r-1})$.

Lemma 10. *If $l_{r-1} = l(s_{r-1}) \geq L_{r-1}$, then*

$$\mu_1(G_1(s_{r-1})) \geq \left(1 - \delta_{r-1} - \frac{2h_0}{|\log \theta|} \varepsilon_{r-1}\right) \mu_1(M_1(s_{r-1})).$$

Proof. Our statement immediately follows from Lemma 9 if we apply the latter to the rank $(r - 1)$ and the space M_2, and make use of the correctness of the map Φ_{s_r}.

9. *Proof of Theorem 1 for* $m_1, m_2 \geq 3$. To construct the desired isomorphism ψ, we will show, for almost every point

$$x^{(1)} = (\ldots, y_{-1}^{(1)}, y_0^{(1)}, y_1^{(1)}, \ldots) \in M_1,$$

how to find the 0th coordinate $y_0^{(2)}$ of the point $x^{(2)} = \psi(x^{(1)}) \in M_2$. In the case of stationary coding, this is sufficient for the construction of the entire map ψ.

If $y_0^{(1)} = 0$, we put $y_0^{(2)} = 0$.

If $y_0^{(1)} \neq 0$, i.e., $x^{(1)} \in M_1^*$, then, by Lemma 3, there exists with probability 1 an infinite sequence of skeletons $s_1(x^{(1)}), s_2(x^{(1)}), \ldots, s_r(x^{(1)}), \ldots$ of increasing ranks: $\rho(s_r(x^{(1)})) = r$; $s_r = s_r(x^{(1)})$ appears in $x^{(1)}$ in the segment

$$\Delta_r = [n_r', n_r''], \qquad n_r' < 0, n_r'' > 0.$$

Consider the maps

$$\Phi_{s_r}: \bar{F}_2(s_r) \to 2^{\bar{F}_1(s_r)}, \qquad r = 2, 4, 6, \ldots,$$

and the equivalence classes $\tilde{f}_1 = \tilde{f}_1^{(r)} \in \tilde{F}_1(s_r)$. Assume that for some even r there exists a unique equivalence class $\bar{f}_2 \in \bar{F}_2(s_r)$ satisfying $\tilde{f}_1 \in \Phi_{s_r}(\bar{f}_2)$; later, by using Lemma 8, we shall show that this assumption is valid for almost all $x^{(1)} \in M_1^*$.

Now consider all possible points $x^{(2)} \in M_2^*$ for which the skeleton s_r appears at the same places as in $x^{(1)}$, and whose coordinates located at the places which are contained in $\bar{f}_2$, take on the same fixed values as in $\bar{f}_2$. Assume finally that the 0th coordinate is fixed, (i.e., is contained in $\bar{f}_2$) and equals k, $0 \leq k \leq m_2 - 1$; (later, by using Lemma 10, we shall show that this assumption is also valid for almost all $x^{(1)} \in M_1^*$). In this case we put $y_0^{(2)} = k$ and say that the point $x^{(1)}$ is a point of rank no greater than r. If the coordinate $y_0^{(2)}$ has not been determined when the ranks $2, 4, \ldots, r$ were considered, i.e., $x^{(1)}$ is not a point of rank $\leq r$, then we pass to the following even rank $(r + 2)$, etc.

It follows from the construction of the set of correct maps Φ_{s_r} that the values of $y_0^{(2)}$ defined for two distinct ranks must coincide.

Now we pass to the detailed proof; we first show that the procedure described above does indeed give a value of $y_0^{(2)}$ for almost all $x^{(1)} \in M$. Suppose M_1' is the set of all points $x^{(1)} \in M_1^*$ satisfying $l(s_r(x)) \geq L_r$ for all sufficiently large r, i.e., for $r \geq r_0(x^{(1)})$. By Lemma 3, $\mu_1(M_1') = \mu(M_1^*)$. For even values $r = 2, 4, \ldots$ also define the sets $R_1^{(r)}, E_1^{(r)}, G_1^{(r)} \subseteq M_1'$ in the following way:

$$R_1^{(r)} = \{x^{(1)} \in M_1' : r_0(x^{(1)}) < r\};$$

$$E_1^{(r)} = \bigcup_{s_r} E_1(s_r),$$

where the sum is taken over all skeletons of rank r and $E_1(s_r)$ was defined in Subsection 8;

$$G_1^{(r)} = \bigcup_{s_{r-1}} G_1(s_{r-1}),$$

where the sum is taken over all skeletons of rank $(r - 1)$ and $G_1(s_{r-1})$ was defined in Subsection 8.

Clearly,

$$\mu_1(R_1^{(r)}) = (1 - \alpha_r)\mu_1(M_1^*),$$

where $\alpha_r \to 0$ when $r \to \infty$.

Further, it follows from Lemma 8 that

$$\mu_1(E_1^{(r)}) \geq (1 - \beta_r)\mu_1(M_1^*),$$

where $\beta_r \to 0$ when $r \to \infty$. It follows from Lemma 10, that

$$\mu_1(G_1^{(r)}) \geq (1 - \gamma_{r-1})\mu_1(M_1^*),$$

where $\gamma_r \to 0$ when $r \to \infty$.

Thus for the set

$$A_1^{(r)} = R_1^{(r)} \cup E_1^{(r)} \cup G_1^{(r)},$$

we have $\mu(A_1^{(r)}) \to \mu_1(M_1^*)$ when $r \to \infty$, so that almost all points $x^{(1)} \in M_1^*$ belong to at least one of the $A_1^{(r)}$.

On the other hand, if $x^{(1)} \in A_1^{(r)}$ then, by construction, $x^{(1)}$ is a point of rank $\leq r$, i.e., the coordinate $y_0^{(2)}$ is defined at the rank r.

Carry out a similar construction for every other coordinate $y_n^{(2)}$, $-\infty < n < \infty$. Then, to almost every point $x^{(1)} \in M_1$, we shall have assigned the point

$$\psi(x^{(1)}) = x^{(2)} = (\ldots, y_{-1}^{(2)}, y_0^{(2)}, y_1^{(2)}, \ldots) \in M_2.$$

The set $\overline{M}_1 \subseteq M_1$ where ψ is defined is obviously invariant with respect to the shift, and, as can be easily checked,

$$\psi(T_1 x^{(1)}) = T_2 \psi(x^{(1)}), \qquad x^{(1)} \in \overline{M}.$$

Suppose $\overline{M}_2 = \psi(\overline{M}_1)$. We want to prove that $\mu_2(\overline{M}_2) = 1$. This will be done in Lemma 13, which in its turn is based on Lemmas 11 and 12.

Lemma 11 concerns a property of the map ψ which is of interest in itself. Before we state it, let us notice that the spaces M_1, M_2 possess a natural compact metric space structure with the distance

$$\text{dist}\,(x', x'') = 1 : \sup\{k : x_i' = x_i'' \text{ for } |i| \leq k\}.$$

Lemma 11. *The map ψ is continuous on the set $\overline{M}_1$, i.e., for any sequence of points $x^{(1),k} \in \overline{M}_1$ possessing a limit $\lim_{k\to\infty} x^{(1),k} = x^{(1)} \in \overline{M}_1$, we have the relation $\lim_{k\to\infty} x^{(2),k} = x^{(2)}$, where $x^{(2),k} = \psi(x^{(1),k})$, $x^{(2)} = \psi(x^{(1)}) \in \overline{M}_2$.*

Proof. Since $x^{(1)} \in \overline{M}_1$, for any n, $-\infty < n < \infty$, we can find an even r such that the nth coordinate of the point $\psi(x^{(1)})$ is defined at rank r, i.e., from the map $\Phi_{s_r(x^{(1)})}$. For all points $x^{(1),k}$ with a sufficiently large number k, $k > k(n)$ the coordinates included in the skeleton $s_r(x^{(1)})$ are the same. Since n was arbitrary, it follows that the sequence $\psi(x^{(1),k})$ converges and $\lim_{k\to\infty} \psi(x^{(1),k}) = \psi(x^{(1)})$. The lemma is proved. □

For each $k = 0, 1, 2, \ldots$ and every even $r = 2, 4, \ldots$ now denote by $M_1^{r,k} \subset M_1$ the set of all points $x^{(1)}$ whose coordinates with numbers i, $|i| \le k$, are uniquely defined by the coordinates of the point $x^{(2)} = \psi(x^{(1)})$ contained in the skeleton $s_r(x^{(2)})$. In other words, $x^{(1)} \in M_1^{r,k}$, if for each point $x^{(1)'} \in \overline{M}_1$, $\psi(x^{(1)'}) = x^{(2)'}$ satisfying $f_r^{(2)}(x^{(2)'}) = f_r^{(2)}(x^{(2)})$ we have the relation $y_i^{(1)} = y_i^{(1)'}$ for $|i| \le k$.

Lemma 12. *For any k*

$$\lim_{r\to\infty} \mu_1(M_1^{r,k}) = \mu_1(M^*).$$

Proof. We shall only consider the case $k = 0$. For other k the argument is similar. Suppose the point $x^{(1)} \in \overline{M}_1$ is of rank $\le r$ (r is even) and suppose the 0th coordinate is essential in the filler $f_{r-1}^{(1)}(x^{(1)})$. It is easy to check that the measure of the set of all such points tends to $\mu_1(M_1^*)$ and $r \to \infty$.

Suppose $x^{(1)} \notin M^{r,0}$. Then choose a point $x^{(1)'} \in \overline{M}_1$ such that $y_0^{(1)} \ne y_0^{(1)'}$, $f_r^{(2)}(x^{(2)}) = f_r^{(2)}(x^{(2)'})$, where $x^{(2)} = \psi(x^{(1)})$, $x^{(2)'} = \psi(x^{(1)'})$. Suppose s_r is the skeleton of rank r which is common to all these points. Under the map $\varphi_{s_r}\colon \overline{F}_2(s_r) \to 2^{\overline{F}_1(s_r)}$ both equivalence classes $\tilde{f}_r^{(1)}, \tilde{f}_r^{(1)'}$ corresponding to the points $x^{(1)}$, $x^{(1)'}$ are sent into the equivalence class $\bar{f}_r^{(2)} = \bar{f}_r^{(2)}(x^{(2)}) = \bar{f}_r^{(2)}(x^{(2)'})$. Now consider the map $\Phi_{s_{r-1}}\colon \overline{F}_1(s_{r-1}) \to 2^{\overline{F}_2(s_{r-1})}$, where s_{r-1} is a skeleton of rank $(r-1)$ also common to all the points considered. By construction of the system of maps $\{\Phi\}$, the element $\tilde{f}_{r-1}^{(2)}$ containing $f_{r-1}^{(2)}(x^{(2)}) = f_{r-1}^{(2)}(x^{(1)})$ must be mapped into the element $\bar{f}_{r-1}^{(1)} \ni f_{r-1}^{(1)}(x^{(1)})$ as well as into the element $\bar{f}_{r-1}^{(1)'} \ni f_{r-1}^{(1)}(x^{(1)'})$; these are distinct, since $y_0^{(1)} \ne y_0^{(1)'}$ and the 0th coordinate is essential in $f_{r-1}^{(1)}$. Hence $\tilde{f}_{r-1}^{(2)} \in \Gamma(\Phi_{s_{r-1}})$. It follows from Lemma 8 and the correctness of the map $\Phi_{s_{r-1}}$ that

$$\mu_1(M_1^* \setminus M_1^{r,0}) \le \beta_r \to 0 \quad \text{when } r \to \infty.$$

The lemma is proved. □

Lemma 13. $\mu_2(\overline{M}_2) = 1$.

Proof. For any even r consider the set of points $x^{(1)} \in M_1$ of rank $\leq r$. To each such point $x^{(1)}$ assign the unique element $\bar{f}_r^{(2)} \in \bar{F}_2(s_r)$ for which $f_r^{(1)}(x) \in \Phi_{s_r}(\bar{f}_r^{(2)})$ and then consider the set of points $x^{(2)} \in M_2$ such that $s_r(x^{(2)}) = s_r(x^{(1)})$, $\bar{f}_r^{(2)}(x^{(2)}) = \bar{f}_r^{(2)}$ and the skeleton s_r appears at the same places as it does for the point $x^{(1)}$.

The set of such $x^{(2)}$ corresponding to all possible points $x^{(1)}$ of rank $\leq r$ will be denoted by $M_2^{(r)}$.

Arguing as in the proof of Lemma 12, we see that $\mu_2(M_2^{(r)}) \to \mu_2(M_2^*)$ for $r \to \infty$. Therefore almost all the points $x^{(2)} \in M_2^*$ belong to an infinite set of sets $M_2^{(r)}$. To prove the lemma, it suffices to show that any such point $x^{(2)}$ belong to $\bar{M}_2$. For the point $x^{(2)}$, first choose an r_1 such that $x^{(2)} \in M_2^{(r_1)}$. By the construction of $M_2^{(r_1)}$, we can find a set of positive measure $E_1 \subset \bar{M}_1$ of points $x^{(1)}$ whose images $\psi(x^{(1)})$ have the same coordinates as $x^{(2)}$ in the places corresponding to $\bar{f}_{r-1}^{(2)}(x^{(2)})$ (we may assume that the 0th coordinate is among them). By Lemma 12 we can find an $r_1' > r_1$ satisfying $E^{(1)} \cap M_1^{r_1', 1} \neq \varnothing$. Choose a point $x^{(1), 1} \in E^{(1)} \cap E^{r_1', 1}$. Now take an $r_2 > r_1'$ such that $x^{(2)} \in M_2^{(r_2)}$ and consider a set of positive measure $E^{(2)} \subset \bar{M}_1$ consisting of points $x^{(1)}$ whose images at the places corresponding to $f_{r_2}^{(2)}(x^{(2)})$ has coordinates equal to those of $x^{(2)}$. By Lemma 12, we can find an $r_2' > r_2$ such that $E^{(2)} \cap M_1^{r_2', 2} \neq \varnothing$. Choose a point $x^{(1), 2} \in E^{(2)} \cap M_1^{r_2', 2}$, etc.

Thus we will construct a sequence of points $x^{(1), 1}, x^{(1), 2}, \ldots, x^{(1), k}, \ldots$. Since for every i all the points of this sequence with numbers $k \geq |i|$ have the same ith coordinate, there is a limit $x^{(1)} = \lim_{k \to \infty} x^{(1), k} \in \bar{M}_1$. By Lemma 11, $\psi(x^{(1)}) = x^{(2)}$, i.e., $x^{(2)} \in \bar{M}_2$. The lemma is proved. □

Interchanging M_1 and M_2, we can construct by the same method, but by using the sequence $\{\Phi_{s_r}\}$ for odd r, a map $\varphi: M_2 \to M_1$ (defined almost everywhere on M_2) whose image is almost all M_1. It follows immediately from the construction that $\varphi = \psi^{-1}$, i.e., ψ is one-to-one.

To conclude the proof of Theorem 1, it suffices to show that the map ψ sends the measure μ_1 on the space M_1 into the measure μ_2 on M_2, in other words, that for any measurable set $A_2 \subset M_2$ we have the relation

$$\mu_1(A_1) = \mu_2(A_2), \quad \text{where } A_1 = \psi^{-1}(A_2).$$

We may assume that A_2 is a cylindrical set, i.e., is of the form

$$A_2 = \{x^{(2)} \in M_2 \colon y_{i_1}^{(2)} = a_{i_1}, \ldots, y_{i_k}^{(2)} = a_{i_k}\},$$

since such sets generate the entire σ-algebra of the space M_2. Using the fact that ψ commutes with the shift transformation, we may also assume that the 0th coordinate was one of the coordinates $i_1, \ldots, i_k$ defining the cylinder A_2 and $y_0^{(2)} = a_0 \neq 0$.

The set $A_1 = \psi^{-1}(A_2)$ will be partitioned into nonintersecting subsets $A_1 = \bigcup_{r=1}^{\infty} A_1^{(r)}$ in the following way: $A_1^{(r)}$ consists of all points $x^{(1)} \in A_1$ for which the coordinates with numbers $i_1, \ldots, i_k$ of the point $\psi(x^{(1)})$ first appear at rank r (r is even).

For each point $x^{(1)} \in A_1^{(r)}, r = 2, 4, 6, \ldots$, consider the skeleton $s_r = s_r(x^{(1)})$. Note that there exists a unique element $\bar{f} = \bar{f}^{(2)}(s_r) \in \bar{F}_2(s_r)$ satisfying $\bar{f}_1(s_r) \in \Phi_{s_r}(\bar{f})$. Partition $A_1^{(r)}$ into nonintersecting subsets

$$A_1^{(r)} = \bigcup_{\bar{f}} \bigcup_{s_r} A_1^{\bar{f}(s_r)},$$

where the summand $A^{\bar{f}(s_r)}$ corresponds to a fixed skeleton s_r and a fixed element $\bar{f} \in \bar{F}_2(s_r)$. Thus we have, finally, the partition

$$A_1 = \bigcup_{\bar{f}} \bigcup_{s_r} \bigcup_{r} A_1^{\bar{f}(s_r)}.$$

Since ψ is one-to-one, the above partition induces a partition of A_2:

$$A_2 = \bigcup_{\bar{f}} \bigcup_{s_r} \bigcup_{r} A_2^{\bar{f}(s_r)}.$$

It follows from the correctness of the map Φ_{s_r} that for any summand of our partition we have

$$\mu_2^{(s_r)}(A_2^{\bar{f}(s_r)}) \leq \mu_1^{(s_r)}(A_1^{\bar{f}(s_r)}),$$

where $\mu_i^{(s_r)}$ is the conditional measure in the space M_i under the condition $s_r(x^{(i)}) = s_r$. The relation $p_0^{(1)} = p_0^{(2)}$ implies that the two conditions have the same measure

$$\mu_1(M_1(s_r)) = \mu_2(M_2(s_r)).$$

This implies that $\mu_2(A_2) \leq \mu_1(A_1)$. Since A_2 was arbitrary, we have the equality $\mu_2(A_2) = \mu_1(A_1)$. The theorem is proved.

Remark 1. Theorem 1 was proved under the restrictions $m_1, m_2 \geq 3$. If, say, $m_1 = 2$, then Lemma 1, on which the subsequent constructions were based, is obviously false. However, instead of this lemma we have the following one.

Lemma 1'. *There exists a probability vector $q = (q_0, q_1, \ldots, q_{m-1})$ of length $m \geq 3$, an integer $k \geq 1$, such that $[p_0^{(1)}]^k p_1^{(1)} = q_0^k q_1$.*

Proof. Choose a $q_0 > \max(p_0^{(1)}, p_1^{(1)})$ and take k so large that $q_1 = (p_0^{(1)}/q_0)^k p_1^{(1)}$ implies $q_0 + q_1 < 1$ and $H(q_0, q_1, 1 - q_0 - q_1) < h$. This is possible since $q_1 \to 0$ where $k \to \infty$. Further the argument is the same as in the proof of Lemma 1. □

By using Lemma 1′, we may assume that for the automorphism T_1, T_2 there is a cylindrical set C of the form $\underbrace{0\ 0\ \ldots\ 0}_{k}\ 1$ such that $\mu_1(C) = \mu_2(C)$.

Using this, we can again define skeletons, fillers, equivalence classes. In this construction the role of the letter 0 will be played by the cylinder C.

All the arguments remain the same except that in Subsection 4, instead of Theorem 4, §6, we shall need a more general statement—the Shannon–McMillan–Breiman theorem (Theorem 5, §6).

10. *Proof of Lemma 5.* Carry out an induction over $k = \text{card } U$. For $k = 1$ the statement is obvious. Assume that it has already been proved for $\text{card } U < k$. For $\text{card } U = k$ consider two cases.

(1) For some set $B \subset U$, $B \neq U$, we have

$$\rho(B) = \sigma(R(B)). \tag{9}$$

Then for any $B' \subset U \backslash B$ we have

$$\begin{aligned}\rho(B) + \rho(B') &= \rho(B \cup B') \leq \sigma(R(B \cup B')) \\ &= \sigma(R(B)) + \sigma(R(B \cup B') \backslash R(B)) = \sigma(R(B)) + \sigma(R(B') \backslash R(B)).\end{aligned}$$

Taking into consideration (9), we get $\rho(B') \leq (R(B') \backslash R(B))$. This implies that the map

$$R'(u) = \begin{cases} R(u), & \text{if } u \in B, \\ R(u) \backslash R(B), & \text{if } u \notin B \end{cases}$$

is also correct. Since $\text{card } B < k$, it follows from the induction hypothesis that the transportation may be carried out separately on B and separately on $U \backslash B$.

(2) For any $B \subset U$, $B \neq U$, we have

$$\rho(B) < \sigma(R(B)).$$

Suppose $\varepsilon = \min_{B \neq U}[\sigma(R(B)) - \rho(B)]$. For any $u \in U$, take a rational number $\rho'(u)$:

$$\rho(u) \leq \rho'(u) \leq \rho(u) + \frac{\varepsilon}{2 \text{ card } U},$$

and for any $v \in V$ choose a rational number $\sigma'(v)$:

$$\sigma(v) \geq \sigma'(v) \geq \sigma(v) - \frac{\varepsilon}{2 \text{ card } V}.$$

The map R will be correct for the pair of spaces (U, ρ'), (V, σ') as well. It is also clear that if we can carry out the transportion corresponding to this pair of spaces, then we can do it for the original spaces (U, ρ), (V, σ) as well.

Since the simultaneous multiplication of all the numbers $\rho'(u)$ $(u \in U)$, $\sigma'(v)$ $(v \in V)$ by the same number does not influence the solubility of the transportation problem and since, moreover, these numbers are rational, we may assume that they are integers.

Now let us transport a unit of produce from any $u_0 \in U$ to any $v_0 \in R(u_0)$. In other words put

$$\tilde{\rho}(u) = \begin{cases} \rho'(u), & \text{if } u \neq u_0, \\ \rho'(u) - 1, & \text{if } u = u_0; \end{cases}$$

$$\tilde{\sigma}(v) = \begin{cases} \sigma'(v), & \text{if } v \neq v_0, \\ \sigma'(v) - 1, & \text{if } v = v_0. \end{cases}$$

The map R is obviously correct also for the spaces $(U, \tilde{\rho})$, $(V, \tilde{\sigma})$.

The operation described above may be repeated as long as we meet with case (2). Then either we will have transported the entire mass of produce in a finite number of steps, or we will meet with case (1). But in this case, as was shown above, the transportation problem may be successfully solved. The lemma is proved. □

Proof of Lemma 6. Among the maps Φ satisfying $\Phi \preccurlyeq R$ there is a minimal one in the sense of the partial order. Let us show that Φ is the map we want. A sequence of elements $u_1, \ldots, u_t \in U$ will be called a *cycle* with respect to Φ, if we can find such $v_1, \ldots, v_t \in V$ that

$$v_i \in \Phi(u_i) \cap \Phi(u_{i+1}), \qquad 1 \leq i < t,$$

$$v_t \in \Phi(u_t) \cap \Phi(u_1).$$

Let us show that Φ has no cycles. To do this, assume the converse and, using Lemma 5, consider the transportation corresponding to the map Φ. Suppose $\alpha(u, v)$ is the mass of produce which is then transported from u to v. Choose a small $\varepsilon > 0$ and put

$$\alpha_\varepsilon(u_i, v_i) = \alpha(u_i, v_i) - \varepsilon, \qquad 1 \leq i \leq t,$$

$$\alpha_\varepsilon(u_{i+1}, v_i) = \alpha(u_{i+1}, v_i) + \varepsilon,$$

$$\alpha_\varepsilon(u, v) = \alpha(u, v),$$

for the other u, v. If all the $\alpha(u_i, v_i)$ are positive, then for $\varepsilon \leq \varepsilon_0 = \min_i \alpha(u_i, v_i)$ the numbers $\alpha_\varepsilon(u_i, v_i)$ also define a transportation. But for $\varepsilon = \varepsilon_0$ at least one of the i will satisfy $\alpha_{\varepsilon_0}(u_i, v_i) = 0$. Therefore, if we exclude v_i from the set $\Phi(u_i)$, we obtain a correct map $\Phi' \prec \Phi$ this contradicts the minimality of Φ. The lemma is proved. □

§8. K-systems and Exact Endomorphisms

The notion of entropy is intimately related to an important class of dynamical systems having strong mixing properties. The phase space $(M, \mathfrak{S}, \mu)$ in this section is assumed to be a Lebesgue space and we shall use certain facts and notations from the theory of measurable partitions (see Appendix 1). In particular, all the equalities and inequalities between partitions should be understood mod 0.

Definition 1. The automorphism T is said to be a *K-automorphism* if there exists a σ-subalgebra of measurable sets $\mathfrak{S}^{(0)} \subset \mathfrak{S}$ such that:

(i) $T\mathfrak{S}^{(0)} \supset \mathfrak{S}^{(0)}$;

(ii) $\bigvee_{n=-\infty}^{\infty} T^n\mathfrak{S}^{(0)} = \mathfrak{S}$;

(iii) $\bigwedge_{n=-\infty}^{\infty} T^n\mathfrak{S}^{(0)} = \mathfrak{N}$.

Here $T^n\mathfrak{S}^{(0)}$ is the σ-algebra of sets of the form T^nC, $C \in \mathfrak{S}^{(0)}$ and $\mathfrak{N}$ is the trivial σ-algebra consisting of the sets of measure 0 and 1.

Definition 2. The flow $\{T^t\}$ is said to be a *K-flow* if there exists a σ-subalgebra of measurable sets $\mathfrak{S}^{(0)} \subset \mathfrak{S}$ such that

(i) $T^t\mathfrak{S}^{(0)} \supset \mathfrak{S}^{(0)}$ for any $t > 0$;

(ii) $\bigvee_{-\infty < t < \infty} T^t\mathfrak{S}^{(0)} = \mathfrak{S}$;

(iii) $\bigwedge_{-\infty < t < \infty} T^t\mathfrak{S}^{(0)} = \mathfrak{N}$.

If $\{T^t\}$ is a K-flow, then any of its automorphisms is a K-automorphism. Since M is a Lebesgue space, the σ-algebra $\mathfrak{S}^{(0)}$ corresponds to a certain measurable partition $\xi^{(0)} = \xi(\mathfrak{S}^{(0)})$. It will be called a *$K$-partition*. K-automorphisms and K-flows are called *K-systems*.

EXAMPLES

1. T is a Bernoulli automorphism with state space $(Y, \mathfrak{A}, \nu)$. Introduce the partition ξ_0 (of the phase space M) whose elements are the sets

$$C = C(y) = \{x = (\ldots, y_{-1}, y_0, y_1, \ldots) \in M : y_0 = y\}, \qquad y \in Y.$$

Then the partition $\xi^{(0)} = \bigvee_{n=0}^{\infty} T^n\xi_0$ is a K-partition. This statement is equivalent to the well-known "zero-one" law due to Kolmogorov. Therefore T is a K-automorphism.

2. T is a mixing Markov automorphism with finite state space Y. It follows from ergodic theorem for Markov chains that the partition $\xi^{(0)}$ defined just as in the previous example will be a K-partition in this case as well.

3. Consider an ideal gas in $\mathbb{R}^d$, $d \geq 1$ for which the probability distribution of the velocity of particles is continuous at the origin. Let us show that the corresponding flow is a K-flow by constructing an appropriate K-partition. Suppose $y = (x, v)$ consists of the sequence of coordinates $x = (x_1, \ldots, x_d)$ and of the velocity vector $v = (v_1, \ldots, v_d)$ of some particle. Let us say that the particle x is *distinguished*, if $x_1 < 0$, $v_1 > 0$ or $x_1 > 0$, $v_1 < 0$. Two infinite families

$$(X^{(1)}, V^{(1)}) = \{(x^{(1)}, v^{(1)})\}, \qquad (X^{(2)}, V^{(2)}) = \{(x^{(2)}, v^{(2)})\},$$

from the phase space M will be called *equivalent* if the coordinates and velocities of all the distinguished particles in $(X^{(1)}, V^{(1)})$ and $(X^{(2)}, V^{(2)})$ are the same. This equivalence relation determines a partition of the phase space M which we shall denote by $\xi^{(0)}$. It is easy to see that $\xi^{(0)}$ is measurable. Let us show that it is a K-partition.

Put $\xi^{(t)} = T^t\xi^{(0)}$. The partition ξ^t may also be described by means of an equivalence relation. Namely, let us say that the particle $y = (x, v)$ is *t-distinguished*, if $x_1 < v_1 t$ for $v_1 > 0$ or $x_1 > v_1 t$ for $v_1 < 0$. The families

$$(X^{(1)}, V^{(1)}) = \{(x^{(1)}, v^{(1)})\}, \qquad (X^{(2)}, V^{(2)}) = \{(x^{(2)}, v^{(2)})\},$$

shall be called *t-equivalent* if all their t-distinguished particles have the same coordinates and velocities. It is easy to see that the elements of the partition $\xi^{(t)}$ consist of t-equivalent families (X, V). This immediately implies that $\xi^{(t)} \succcurlyeq \xi^{(0)}$ when $t \geq 0$ and $\bigvee_t \xi^{(t)} = \varepsilon$, where ε is the partition of M into separate points.

Let us show that $\bigwedge_t \xi^{(t)} = \nu$. Indeed, suppose $E_1, \ldots, E_r$ are bounded connected open sets located at a positive distance from the hyperplane $v_1 = 0$ of the space $\mathbb{R}^{2d} = \mathbb{R}^d_x \times \mathbb{R}^d_v$, and C is a subset of the form

$$C = \{(X, V)\colon \operatorname{card}(X \cap E_1) = k_1, \ldots, \operatorname{card}(X \cap E_r) = k_r\}.$$

Then C, for sufficiently large $|t|$, $t < 0$ does not depend on $\xi^{(t)}$, since for such t the sets $E_1, \ldots, E_r$ do not intersect the set

$$\{x_1 < tv_1, v_1 > 0\} \cup \{x_1 > tv_1, v_1 < 0\}.$$

It therefore follows that C does not depend on the partition $\bigwedge_t \xi^{(t)}$. Since the finite union of sets C are dense in the entire σ-algebra $\mathfrak{S}$, we see that $\mathfrak{S}(\bigwedge_t \xi^{(t)})$ does not depend on $\mathfrak{S}$. But this is possible only if $\bigwedge_t \xi^{(t)} = \nu$.

Let us deduce the simplest consequences from the definition of K-automorphisms.

1. Any K-automorphism is Mixing.

Suppose T is a K-automorphism and $\xi^{(0)}$ is its K-partition. It suffices to show mixing for all $f, g \in L_0^2(M, T^r\mathfrak{S}^{(0)}, \mu)$ for some r, since all such functions, by (ii) generate the entire space $L_0^2(M, \mathfrak{S}, \mu)$. Note that

$$f(T^k x), g(T^k x) \in L_0^2(M, T^{r+k}\mathfrak{S}^{(0)}, \mu).$$

Therefore, denoting by P_k the orthogonal projection operator on $L_0^2(M, T^k\mathfrak{S}^{(0)}, \mu)$, we can write

$$\begin{aligned}\int_M f(T^n x)g(x)\,d\mu &= \int_M f(x)g(T^{-n}x)\,d\mu = (f, \overline{g(T^{-n}x)})_{L^2} \\ &= (P_{r-n}f, \overline{g(T^{-n}x)}) \leq \|P_{r-n}f\| \cdot \|g\| \to 0 \quad \text{for } n \to \infty,\end{aligned}$$

since $\|P_{r-n}f\| \to 0$ when $n \to \infty$ by condition (iii).

2. For Any K-automorphism T the Entropy $h(T) > 0$ Is Positive.

Construct a finite nontrivial partition η of the space M such that $T\eta^- \succ \eta^-$ where $\eta^- = \bigvee_{k=0}^{\infty} T^{-k}\eta$.

Suppose $\xi^{(0)}$ is a K-partition for T and $\eta_1 \prec \eta_2 \prec \cdots$ is a sequence of finite partitions, $\bigvee_i \eta_i = \xi^{(0)}$. For every i we obviously have $T\eta_i^- \succcurlyeq \eta_i^-$, where $\eta_i^- = \bigvee_{k=0}^{\infty} T^{-k}\eta_i$. If for all i we would have the relation $T\eta_i^- = \eta_i^-$, this would mean $T\xi^{(0)} = \xi^{(0)}$, which contradicts property (i). Therefore there exists a j for which we have the strict inequality $T\eta_j^- \succ \eta_j^-$. Putting $\eta = \eta_j$, we obtain the required partition. For such a η we have

$$h(T, \eta) = H(\eta \,|\, T^{-1}\eta^-) = H(\eta \,|\, T^{-1}\eta) > 0,$$

therefore $h(T) = \sup h(T, \eta) > 0$.

For K-flows we have the following similar statements:

(1) any K-flow is mixing;
(2) the entropy of a K-flow is positive.

Actually the statements on the mixing property and the fact that the entropy of K-automorphisms is positive may be considerably sharpened. This follows from Theorem 1, which characterizes the basic properties of K-automorphism. Further by Tail(η) we denote the partition

$$\mathrm{Tail}(\eta) = \bigwedge_{n=0}^{\infty} \bigvee_{k=n}^{\infty} T^{-k}\eta.$$

Theorem 1. *The following conditions are equivalent:*

(i) *T is a K-automorphism;*
(ii) $\operatorname{Tail}(\eta) = \nu$ *for any finite partition η of the space M, where ν is the trivial partition, i.e., $\nu = (M, \varnothing)$;*
(iii) *$h(T, \xi) > 0$ for any finite partition $\xi \neq \nu$; in other words any factor automorphism of the automorphism T has positive entropy;*
(iv) *T is K-mixing (see the definition in §6, Chap. 1).*

Proof. We shall carry out the proof under the assumption that T possesses a finite generating partition (see §6). The general case necessitates additional arguments which we shall omit.

(1) The implication (i) $\Rightarrow$ (ii) is based on the following lemma.

Lemma 1. *Suppose ξ_1 is a measurable partition, ξ_2 a finite partition of the space M and $\xi_2 \preccurlyeq \bigvee_{n=-\infty}^{\infty} T^n \xi_1$. Then* $\operatorname{Tail}(\xi_2) \preccurlyeq \operatorname{Tail}(\xi_1)$.

The proof of the lemma will be given later. In the role of ξ_1 take the K-partition ξ for T and put $\xi_2 = \eta$. Then $\eta \preccurlyeq \bigvee_{n=-\infty}^{\infty} T^n \xi$ and therefore by Lemma 1 $\operatorname{Tail}(\eta) \preccurlyeq \operatorname{Tail}(\xi) = \nu$. Thus (i) $\Rightarrow$ (ii) is proved.

(2) Let us prove (ii) $\Rightarrow$ (i). Suppose ζ is a finite generating partition for T; $\xi = \bigvee_{n=0}^{\infty} T^{-n}\zeta$. Then ξ is a K-partition.

(3) Let us prove (ii) $\Rightarrow$ (iii). Suppose $h(T, \xi) = 0$ for some finite $\xi \neq \nu$. Then

$$H\left(\xi \,\middle|\, \bigvee_{k=1}^{\infty} T^{-k}\xi\right) = 0, \qquad \text{i.e.,} \quad \xi \preccurlyeq \bigvee_{k=1}^{\infty} T^{-k}\xi.$$

This implies

$$\bigvee_{k=0}^{\infty} T^{-k}\xi = \bigvee_{k=1}^{\infty} T^{-k}\xi, \quad \bigvee_{k=m}^{\infty} T^{-k}\xi = \bigvee_{k=n}^{\infty} T^{-k}\xi,$$

for all $m, n \geq 0$. But then

$$\bigwedge_{n=1}^{\infty} \bigvee_{k=n}^{\infty} T^{-k}\xi = \bigvee_{k=1}^{\infty} T^{-k}\xi \succcurlyeq T^{-1}\xi \neq \nu,$$

which contradicts (ii).

(4) Let us prove (iii) $\Rightarrow$ (ii). Assume that $\operatorname{Tail}(\eta) \neq \nu$ for some finite partition η. Consider the finite partition ζ, $\zeta \neq \nu$ such that $\zeta \preccurlyeq \operatorname{Tail}(\eta)$.

Lemma 2. *For any pair of finite partitions ξ_1, ξ_2, we have the relation*

$$H\left(\xi_1 \,\middle|\, \bigvee_{n=1}^{\infty} T^{-n}\xi_1 \vee \mathrm{Tail}(\xi_2)\right) = h(T, \xi_1).$$

The proof of the lemma shall be given later. Using Lemma 2 for $\xi_1 = \zeta$, $\xi_2 = \eta$, we can write

$$H\left(\zeta \,\middle|\, \bigvee_{n=1}^{\infty} T^{-n}\zeta \vee \mathrm{Tail}(\eta)\right) = h(T, \zeta).$$

Since $\zeta \preccurlyeq \mathrm{Tail}(\eta)$, the left-hand side of the last relation vanishes, i.e., $h(T, \zeta) = 0$ which contradicts (iii).

(5) Let us prove that (ii) $\Rightarrow$ (iv). Suppose $A_0, A_1, \ldots, A_r$ are the sets which appear in the definition of K-mixing. Without loss of generality, we may assume that $A_0, A_1, \ldots, A_r$ constitute a partition η of the space M. Fix n and choose a set $B = B_n$, $B \in \mathfrak{S}(\eta_n)$, where $\eta_n = \bigvee_{k=n}^{\infty} T^{-k}\eta$.

Then, using the definition of conditional measure

$$\mu(\cdot \mid \mathfrak{S}(\eta_n)) = \mu(\cdot \mid C_{\eta_n}(x)),$$

we may write

$$\begin{aligned}|\mu(A_0 \cap B) - \mu(A_0)\mu(B)| &= \left| \int_B \mu(A_0 \mid C_{\eta_n}(x))\, d\mu(x) - \int_B \mu(A_0)\, d\mu(x) \right| \\ &\leq \int_B |\mu(A_0 \mid C_{\eta_n}(x) - \mu(A_0)| \, d\mu(x).\end{aligned}$$

Since $\bigcap_{n=1}^{\infty} \mathfrak{S}(\eta_n)$ is a trivial σ-algebra containing only sets of measure 0 and 1, it follows from Doob's theorem on the convergence of conditional expectations, that the function under the integral sign in the right-hand side of the last expression tends to zero when $n \to \infty$ for μ-almost all x. Hence, by Lebesgue's theorem,

$$\lim_{n \to \infty} \sup_{B \in \mathfrak{S}(\eta_n)} |\mu(A_0 \cap B) - \mu(A_0)\mu(B)| = 0, \tag{1}$$

so that T is K-mixing.

(6) Let us prove that (iv) $\Rightarrow$ (ii). Suppose $A_0 \in \mathrm{Tail}(\eta)$ for some finite partition η. Then $A_0 \in \mathfrak{S}(\eta_n)$ for any n. Hence, by putting $B = A_0$ in (1), we get

$$\lim_{n \to \infty} |\mu(A \cap A) - \mu(A) \cdot \mu(A)| = 0,$$

i.e., $\mu(A)$ equals 0 or 1. The theorem is proved. $\square$

Now let us prove Lemma 2 and then use it to prove Lemma 1.

Proof of Lemma 2. It follows from the properties of entropy that

$$H\left(\xi_1 \,\middle|\, \bigvee_{n=1}^{\infty} T^{-n}\xi_1 \vee \operatorname{Tail}(\xi_2)\right) \le H\left(\xi_1 \,\middle|\, \bigvee_{n=1}^{\infty} T^{-n}\xi_1\right).$$

To prove the opposite inequality, note that for any finite partition η and any $n \ge 0$, we have

$$H\left(\bigvee_{k=0}^{n-1} T^k\eta \,\middle|\, \bigvee_{l=1}^{\infty} T^{-l}\eta\right) = \sum_{k=0}^{n-1} H\left(T^k\eta \,\middle|\, \bigvee_{l=-k+1}^{\infty} T^{-l}\eta\right)$$

$$= nH\left(\eta \,\middle|\, \bigvee_{l=1}^{\infty} T^{-l}\eta\right) = nh(T, \eta).$$

Applying this last equality for $\eta = \xi_1 \vee \xi_2$, we get

$$h(T, \xi_1 \vee \xi_2) = \frac{1}{n} H\left(\bigvee_{k=0}^{n-1} T^k(\xi_1 \vee \xi_2) \,\middle|\, \bigvee_{l=1}^{\infty} T^{-l}(\xi_1 \vee \xi_2)\right)$$

$$\le \frac{1}{n} H\left(\bigvee_{k=0}^{n-1} T^k(\xi_1 \vee \xi_2) \,\middle|\, \bigvee_{l=1}^{\infty} T^{-l}\xi_1\right)$$

$$\le \frac{1}{n} H\left(\bigvee_{k=0}^{n-1} T^k(\xi_1 \vee \xi_2)\right) = h(T, \xi_1 \vee \xi_2) + \varepsilon_n,$$

where $\varepsilon_n \to 0$ when $n \to \infty$. Further

$$\frac{1}{n} H\left(\bigvee_{k=0}^{n-1} T^k(\xi_1 \vee \xi_2) \,\middle|\, \bigvee_{l=1}^{\infty} T^{-l}(\xi_1 \vee \xi_2)\right)$$

$$= \frac{1}{n} H\left(\bigvee_{k=0}^{n-1} T^k\xi_1 \,\middle|\, \bigvee_{l=1}^{\infty} T^{-l}(\xi_1 \vee \xi_2)\right)$$

$$+ \frac{1}{n} H\left(\bigvee_{k=0}^{n-1} T^k\xi_2 \,\middle|\, \bigvee_{l=1}^{\infty} T^{-l}\xi_2 \vee \bigvee_{l=-n+1}^{\infty} T^{-l}\xi_1\right);$$

$$\frac{1}{n} H\left(\bigvee_{k=0}^{n-1} T^k(\xi_1 \vee \xi_2) \,\middle|\, \bigvee_{l=1}^{\infty} T^{-l}\xi_1\right) = \frac{1}{n} H\left(\bigvee_{k=0}^{n-1} T^k\xi_1 \,\middle|\, \bigvee_{l=1}^{\infty} T^{-l}\xi_1\right)$$

$$+ \frac{1}{n} H\left(\bigvee_{k=0}^{n-1} T^k\xi_2 \,\middle|\, \bigvee_{l=-n+1}^{\infty} T^{-l}\xi_1\right).$$

The limits of the left-hand side are equal by the above, while each summand in the right-hand side in the second equality is not less than the corresponding summand in the first equality. It therefore follows that the limits of the corresponding summands in the right-hand side are also equal, i.e.,

$$\lim_{n\to\infty}\frac{1}{n}H\left(\bigvee_{k=0}^{n-1}T^k\xi_1\,\middle|\,\bigvee_{l=1}^{\infty}T^{-l}(\xi_1\vee\xi_2)\right)=\lim_{n\to\infty}\frac{1}{n}H\left(\bigvee_{k=0}^{n-1}T^k\xi_1\,\middle|\,\bigvee_{l=1}^{\infty}T^{-l}\xi_1\right)$$
$$=h(T,\xi_1).$$

For the left-hand side of the last equality we have:

$$\lim_{n\to\infty}\frac{1}{n}H\left(\bigvee_{k=0}^{n-1}T^k\xi_1\,\middle|\,\bigvee_{l=1}^{\infty}T^{-l}(\xi_1\vee\xi_2)\right)$$
$$=\lim_{n\to\infty}\frac{1}{n}\sum_{k=0}^{n-1}H\left(T^k\xi_1\,\middle|\,\bigvee_{l=-k+1}^{\infty}T^{-l}\xi_1\vee\bigvee_{l=1}^{\infty}T^{-l}\xi_2\right)$$
$$=\lim_{n\to\infty}\frac{1}{n}\sum_{k=0}^{n-1}H\left(\xi_1\,\middle|\,\bigvee_{l=1}^{\infty}T^{-l}\xi_1\vee\bigvee_{l=k+1}^{\infty}T^{-l}\xi_2\right).$$

It follows from the properties of entropy

$$h(T,\xi_1)=\lim_{k\to\infty}H\left(\xi_1\,\middle|\,\bigvee_{l=1}^{\infty}T^{-l}\xi_1\vee\bigvee_{l=k+1}^{\infty}T^{-l}\xi_2\right)$$
$$=H\left(\xi_1\,\middle|\,\bigwedge_{k=1}^{\infty}\left(\bigvee_{l=1}^{\infty}T^{-l}\xi_1\vee\bigvee_{l=k+1}^{\infty}T^{-l}\xi_2\right)\right)$$
$$\leq H\left(\xi_1\,\middle|\,\bigvee_{l=1}^{\infty}T^{-l}\xi_1\vee\mathrm{Tail}(\xi_2)\right)\leq H\left(\xi_1\,\middle|\,\bigvee_{l=1}^{\infty}T^{-l}\xi_1\right)=h(T,\xi_1).$$

Here we have used the fact that

$$\bigwedge_{k=1}^{\infty}\left(\bigvee_{l=1}^{\infty}T^{-l}\xi_1\vee\bigvee_{l=k+1}^{\infty}T^{-l}\xi_2\right)\succcurlyeq\bigvee_{l=1}^{\infty}T^{-l}\xi_1\vee\mathrm{Tail}(\xi_2)\succcurlyeq\bigvee_{l=1}^{\infty}T^{-l}\xi_1.$$

The lemma is proved. □

Proof of Lemma 1. First consider the case of finite ξ_1, ξ_2. It suffices to show that for any finite partition $\eta \preccurlyeq \mathrm{Tail}(\xi_2)$ we have the inequality $\eta \preccurlyeq \mathrm{Tail}(\xi_1)$.

Consider all the partitions ζ such that

$$\zeta\preccurlyeq\bigvee_{k=-n}^{n}T^k\xi_1 \tag{2}$$

for some n; let us show that for such ζ we have

$$H(\zeta \mid \mathrm{Tail}(\xi_1) \vee \eta) = H(\zeta \mid \mathrm{Tail}(\xi_1)). \tag{3}$$

To do this, note the equality

$$H\left(\bigvee_{k=-n}^{n} T^k \xi_1 \,\middle|\, \bigvee_{l=n+1}^{\infty} T^{-l}\xi_1 \vee \eta\right) = H\left(\bigvee_{k=-n}^{n} T^k \xi_1 \,\middle|\, \bigvee_{l=n+1}^{\infty} T^{-l}\xi_1\right). \tag{4}$$

Let us prove this. The left-hand side of (4) can easily be written in the form $nH(\xi_1 | \bigvee_{l=1}^{\infty} T^{-l}\xi_1 \vee \eta)$ and the right-hand side in the form

$$nH\left(\xi_1 \,\middle|\, \bigvee_{l=1}^{\infty} T^{-l}\xi_1\right).$$

By Lemma 2, the last expressions are equal to each other. Using the properties of entropy we can write

$$\begin{aligned} H\left(\zeta \vee \bigvee_{k=-n}^{n} T^k\xi_1 \,\middle|\, \bigvee_{l=n+1}^{\infty} T^{-l}\xi_1 \vee \eta\right) &= H\left(\zeta \,\middle|\, \bigvee_{l=n+1}^{\infty} T^{-l}\xi_1 \vee \eta\right) + H\left(\bigvee_{k=-n}^{n} T^k\xi_1 \,\middle|\, \zeta \vee \bigvee_{l=n+1}^{\infty} T^{-l}\xi_1 \vee \eta\right), \end{aligned}$$

$$\begin{aligned} H\left(\zeta \vee \bigvee_{k=-n}^{n} T^k\xi_1 \,\middle|\, \bigvee_{l=n+1}^{\infty} T^{-l}\xi_1\right) &= H\left(\zeta \,\middle|\, \bigvee_{l=n+1}^{\infty} T^{-l}\xi_1\right) \\ &\quad + H\left(\bigvee_{k=-n}^{n} T^k\xi_1 \,\middle|\, \zeta \vee \bigvee_{l=n+1}^{\infty} T^{l}\xi_1\right). \end{aligned}$$

Each summand in the right-hand side of the second equality is no less than the corresponding summand in the right-hand side of the first equality. Since

$$\zeta \vee \bigvee_{k=-n}^{n} T^k\xi_1 = \bigvee_{k=-n}^{n} T^k\xi_1,$$

it follows from (4) that the left-hand sides of these equalities are the same. Hence the corresponding summands in the right-hand side are equal. In particular,

$$H\left(\zeta \,\middle|\, \bigvee_{l=n+1}^{\infty} T^{-l}\xi_1 \vee \eta\right) = H\left(\zeta \,\middle|\, \bigvee_{l=n+1}^{\infty} T^{-l}\xi_1\right).$$

When $n \to \infty$, we get

$$\begin{aligned}\lim_{n\to\infty} H\left(\zeta \,\Bigg|\, \bigvee_{l=n+1}^{\infty} T^{-l}\xi_1 \vee \eta\right) &= H\left(\zeta \,\Bigg|\, \bigwedge_{n=1}^{\infty} \left(\bigvee_{l=n+1}^{\infty} T^{-l}\xi_1 \vee \eta\right)\right) \\ &\le H\left(\zeta \,\Bigg|\, \bigwedge_{n=1}^{\infty} \bigvee_{l=n+1}^{\infty} T^{-l}\xi_1\right) \vee \eta\Bigg) \\ &\le H(\zeta \,|\, \mathrm{Tail}(\xi_1)) = \lim_{n\to\infty} H\left(\zeta \,\Bigg|\, \bigvee_{l=n+1}^{\infty} T^{-l}\xi_1\right),\end{aligned}$$

which gives us (3).

Since $\eta \preccurlyeq \bigvee_{k=-\infty}^{\infty} T^k\xi_1$, we can approximate η as closely as we wish by partitions ζ of the form (2).[1]

Using the continuity of entropy, we may replace ζ by η in (3), getting

$$0 = H(\eta \,|\, \mathrm{Tail}(\xi_1) \vee \eta) = H(\eta \,|\, \mathrm{Tail}(\xi_1)).$$

But this means $\eta \preccurlyeq \mathrm{Tail}(\xi_1)$.

Let us now pass to the general case (ξ_1 not necessarily finite) and consider the increasing sequence of partitions $\xi_1^{(m)} \to \xi_1$. For a fixed m it follows from $\zeta \preccurlyeq \bigvee_{n=-\infty}^{\infty} T^n\xi_1^{(m)}$ that (using what we have proved above):

$$H(\zeta \,|\, \mathrm{Tail}(\xi_1^{(m)}) \vee \eta) = H(\zeta \,|\, \mathrm{Tail}(\xi_1^{(m)})).$$

Since $\zeta \preccurlyeq \bigvee_{n=-\infty}^{\infty} T^n\xi_1^{(l)}$ for all $l \ge m$ when $m \to \infty$ we get

$$H(\zeta \,|\, \tilde{\xi} \vee \eta) = H(\zeta \,|\, \tilde{\xi}),$$

where $\tilde{\xi} = \bigvee_m \mathrm{Tail}(\xi_1^{(m)})$. As before, using continuity, we replace ζ by η:

$$0 = H(\eta \,|\, \tilde{\xi} \vee \eta) = H(\eta \,|\, \tilde{\xi}),$$

i.e., $\eta \preccurlyeq \tilde{\xi} \preccurlyeq \mathrm{Tail}(\xi_1)$. The lemma is proved. □

A theorem analogous to Theorem 1 for K-flows follows from the next theorem.

Theorem 2. *If at least one automorphism contained in the flow $\{T^t\}$ is a K-automorphism, then $\{T^t\}$ is a K-flow.*

[1] The distance between partitions can be measured as follows: if $\xi = (C_1, C_2, \ldots)$, $\eta = (D_1, D_2, \ldots)$ are two partitions, then $\mathrm{dist}(\xi, \eta) = \inf \sum \mu(C_i \,\Delta\, D_{k_i})$ where the g.l.b. is taken over all possible $(k_1, k_2, \ldots)$. The entropy of a finite partition ξ continuously depends on ξ with respect to this distance.

We shall not give the proof of Theorem 2.

We now pass to the study of endomorphisms and will introduce a notion analogous to K-automorphisms for them. Suppose T_0 is an endomorphism of the space $(M_0, \mathfrak{S}_0, \mu_0)$. Construct the sequence $\mathfrak{S}_n$ of σ-subalgebras of the σ-algebra $\mathfrak{S}_0$, where $C \in \mathfrak{S}_n$ if and only if C is of the form $C = T_0^{-n} C_1$, $C_1 \in \mathfrak{S}^0$. The subalgebras $\mathfrak{S}_n$ satisfy the relation $\mathfrak{S}_{n+1} \subset \mathfrak{S}_n$.

Definition 3. The endomorphism T_0 is called *exact* if the intersection $\bigcap_n \mathfrak{S}_n = \mathfrak{N}_0$ where $\mathfrak{N}_0$ is a trivial σ-algebra, i.e., the σ-algebra of subsets of the space M_0 of measure 0 or 1.

Theorem 3. *If the endomorphism T_0 is exact, then its natural extension T is a K-automorphism.*

Proof. Suppose T is the natural extension of the endomorphism T_0 which acts in the measure space $(M, \mathfrak{S}, \mu)$. Recall that the points of the space M are sequences $x = (x^{(0)}, x^{(1)}, x^{(2)}, \ldots)$, where $T_0 x^{(n+1)} = x^{(n)}$ and $x^{(n)} \in M_0$ $(n = 0, 1, 2, \ldots)$. Take the σ-subalgebra $\mathfrak{S}^{(0)} \subset \mathfrak{S}$ consisting of sets of the form

$$C = \{x \in M : x^{(0)} \in C_0 \in \mathfrak{S}_0\}.$$

Let us show that

(i) $T\mathfrak{S}^{(0)} \supset \mathfrak{S}^{(0)}$;

(ii) $\bigvee_n T^n \mathfrak{S}^{(0)} = \mathfrak{S}$;

(iii) $\bigwedge_n T^n \mathfrak{S}^{(0)} = \mathfrak{N}$.

Proof of (*i*). For the set

$$C \in \mathfrak{S}^{(0)}, \qquad C = \{x \in M : x^{(0)} \in C_0\},$$

we have

$$TC = \{x = (x^{(0)}, x^{(1)}, \ldots) \in M : T^{-1} x \in C\} = \{x \in M : x^{(1)} \in C_0\}.$$

If we can find $C_1 \in \mathfrak{S}_0$ such that $C_0 = T_0^{-1} C_1$ then

$$TC = \{x \in M : x^{(0)} \in C_1\} \in \mathfrak{S}^{(0)}.$$

But this means $T\mathfrak{S}^{(0)} \supset \mathfrak{S}^{(0)}$

Proof of (*ii*). Notice that the σ-subalgebra $T^n \mathfrak{S}^{(0)}$ consists of sets of the form

$$\{x = (x^{(0)}, x^{(1)}, \ldots) \in M : x^{(n)} \in C_0 \in \mathfrak{S}_0\}.$$

By definition, sets of this form generate the entire σ-algebra $\mathfrak{S}$. Thus (ii) is proved.

Proof of (*iii*). For $n \geq 0$ the subsets contained in the σ-algebra $T^{-n}\mathfrak{S}^{(0)}$ are in isometric one-to-one correspondence with the subsets of the σ-algebra $\mathfrak{S}_n$. This implies that the equality $\bigwedge_n \mathfrak{S}_n = \mathfrak{N}_0$ is equivalent to $\bigwedge_n T^n\mathfrak{S}^{(0)} = \mathfrak{N}$. The theorem is proved. □

Important examples of exact endomorphisms are contained in the class of piecewise monotonic maps of the interval, considered in §4 of Chap. 7.

Recall that piecewise monotonic maps $T: (0, 1) \to (0, 1)$ are maps which act according to the formula $Tx = \varphi(x)$, where the interval $(0, 1)$ may be partitioned into a finite or countable number of intervals $\Delta_1, \Delta_2, \ldots$ so that the function $\varphi \in C^2$ is strictly monotonic on every Δ_i. In Theorem 1, §4, Chap. 7, it was proved that under sufficiently general assumptions such transformations have an invariant measure which is absolutely continuous with respect to the Lebesgue measure.

Theorem 2 of the section quoted above was only stated but not proved. Now we will prove a stronger statement.

Theorem 4. *Suppose the transformation T is the same as in Theorem* 1, §4, *Chap.* 7, *but instead of condition* (i) (*see page* 168) *we have the stronger condition* (iii) (*loc. cit.*); *μ is the invariant measure with respect to T obtained in the above-mentioned theorem. Then*:

(a) *the measure μ is equivalent to the Lebesgue measure ρ; moreover, there exists a constant $K > 0$ such that $1/K \leq d\mu/d\rho \leq K$;*

(b) *the endomorphism T with invariant measure μ is exact.*

Proof. (1) We shall use the notation introduced in Theorem 1, §4, Chap. 7. In particular, for a set $A \in \mathfrak{S}$ we put as before $A_{i_1, \ldots, i_n} = T^{-n}A \cap \Delta^{(n)}_{i_1, \ldots, i_n}$. Condition (iii) enables us to estimate the measure $\rho(A_{i_1, \ldots, i_n})$ from below. Indeed, now $T^n A_{i_1, \ldots, i_n} = A$, hence

$$\rho(A) = \int_{A_{i_1, \ldots, i_n}} \left| \frac{d\varphi^{(n)}}{dx} \right| dx \leq \rho(A_{i_1, \ldots, i_n}) \sup_{x \in A_{i_1, \ldots, i_n}} \left| \frac{d\varphi^{(n)}}{dx} \right|. \tag{5}$$

Moreover $T^n(\Delta^{(n)}_{i_1, \ldots, i_n}) = (0, 1)$, hence

$$\int_{\Delta^{(n)}_{i_1, \ldots, i_n}} \left| \frac{d\varphi^{(n)}}{dx} \right| dx = 1,$$

and we can find a point $\xi \in \Delta^{(n)}_{i_1, \ldots, i_n}$ such that

$$\left| \frac{d\varphi^{(n)}}{dx}(\xi) \right| \leq \frac{1}{\rho(\Delta^{(n)}_{i_1, \ldots, i_n})}.$$

By Lemma 1, §4, Chap. 7, for any $x \in A_{i_1,\ldots,i_n}$ we have

$$\left|\frac{d\varphi^{(n)}}{dx}(x)\right| \leq \left|\frac{d\varphi^{(n)}}{dx}(\xi)\right| \cdot \sup_{x \in \Delta^{(n)}_{i_1,\ldots,i_n}} \left|\frac{(d\varphi^{(n)}/dx)(x)}{(d\varphi^{(n)}/dx)(\xi)}\right| \leq \frac{K}{\rho(\Delta^{(n)}_{i_1,\ldots,i_n})},$$

and (5) yields

$$\rho(A_{i_1,\ldots,i_n}) \geq \frac{1}{K}\rho(A)\cdot\rho(\Delta^{(n)}_{i_1,\ldots,i_n}), \tag{6}$$

$$\rho(T^{-n}A) = \sum_{i_1,\ldots,i_n} \rho(A_{i_1,\ldots,i_n}) \geq \frac{1}{K}\rho(A).$$

Statement (a) now follows from the remark after the proof of Lemma 1, §4, Chap. 7.

(2) Assume, in contradiction to statement (b), that the endomorphism T is not exact, i.e., there exists a set B, such that

$$0 < \mu(B) < 1, \qquad B \in \bigwedge_n T^{-n}\mathfrak{S}.$$

Then for any $n \geq 0$ there is a set $A_n \in \mathfrak{S}$, such that $\mu(A_n) = \mu(B)$ and $B = T^{-n}A_n$. Applying inequality (6) to the set A_n, we see that for any interval $\Delta^{(n)}_{i_1,\ldots,i_n}$, $n = 1, 2, \ldots$ we have

$$\rho(B \cap \Delta^{(n)}_{i_1,\ldots,i_n}) \geq \frac{1}{K}\rho(B)\rho(\Delta^{(n)}_{i_1,\ldots,i_n}). \tag{7}$$

Further, from inequality (15) of §4, Chap. 7 for $k = 0$, we get the estimate

$$\rho(\Delta^{(n)}_{i_1,\ldots,i_n}) = \lambda^{-[n/s]}. \tag{8}$$

Suppose $C = (0, 1)\backslash B$. Since $\mu(C) > 0$, we have $\rho(C) > 0$, and there is a density point x_0 (for C) which is not the end point of any of the intervals $\Delta^{(n)}_{i_1,\ldots,i_n}$. Take $\varepsilon = (1/K)\rho(B) > 0$ and find a $\delta > 0$ such that for any interval Δ, $\rho(\Delta) < \delta$, $x_0 \in \Delta$we have

$$\rho(C \cap \Delta) > (1 - \varepsilon)\rho(C)\cdot\rho(\Delta). \tag{9}$$

By (8), in the role of Δ we can take the interval $\Delta^{(n)}_{i_1,\ldots,i_n}$, containing x for n sufficiently large. Then the inequality (9) contradicts (7). The theorem is proved. □

Chapter 11

Special Representations of Flows

§1. Definition of Special Flows

There is a general method in ergodic theory which reduces many problems concerning dynamical systems with continuous time to the corresponding problem for dynamical systems with discrete time. This method goes back to Poincaré; for the study of trajectories of a smooth dynamical system in the neighborhood of a closed trajectory he proposed to consider the "return" map which arises on a transversal surface of codimension 1 to the closed trajectory: the transformation consists in following the trajectory starting at a given point of the surface until its next intersection with the surface.

We begin with the definition of special flows. Suppose that on the measure space $(M_1, \mathfrak{S}_1, \mu_1)$ with a given automorphism T, we have a measurable function $f(x_1) > 0$ satisfying $\int f \, d\mu_1 = 1$. The measure μ_1 is not necessarily normalized but must be finite: $\mu_1(M_1) < \infty$. For almost all $x_1 \in M_1$, in accordance to the Birkhoff–Khinchin theorem, we have

$$\lim_{n\to\infty} \frac{1}{n} \sum_{k=0}^{n-1} f(T^k x_1) > 0$$

and, in particular, $\sum_{k=0}^{\infty} f(T^k x_1) = \infty$. In the sequel it will be convenient to eliminate the invariant set of zero measure on which the last relation does not hold, thus assuming

$$\sum_{k=0}^{\infty} f(T^k x_1) = \infty \quad \text{for all } x_1 \in M_1.$$

Consider the set

$$M_1^f = \{(x_1, s) : x_1 \in M_1, 0 \le s < f(x_1)\}.$$

Sometimes this set will be referred to as the space under the function f. Let us transform $M = M_1^f$ into a measure space by taking for the σ-algebra of measurable sets the σ-algebra formed by the measurable subsets of the Cartesian product $M \times \mathbb{R}^1$ which belong to M_1^f and by putting, for each set A,

$$\mu(A) = \iint_A d\mu_1(x_1)\, ds.$$

In other words, μ is the restriction to M_1^f of the Cartesian product of the measure μ_1 and the Lebesgue measure on $\mathbb{R}^1$. We shall have $\mu(M_1^f) = 1$.

Definition 1. The *special flow* constructed from the automorphism T and the function f is the flow $\{V^t\}$ which acts on M_1^f, for $t \geq 0$, according to the formula

$$V^t(x_1, s) = \left(T^n x_1, s + t - \sum_{k=0}^{n-1} f(T^k x_1)\right),$$

where n is uniquely determined from the inequalities

$$\sum_{k=0}^{n-1} f(T^k x_1) \leq s + t < \sum_{k=0}^{n} f(T^k x_1);$$

and, for $t < 0$, acts according to the formulas:

$$V^t(x_1, s) = (x_1, s + t) \quad \text{if } s + t \geq 0,$$

$$V^t(x_1, s) = \left(T^{-n} x_1, s + t + \sum_{k=-n}^{-1} f(T^k x_1)\right), \quad \text{if } s + t < 0;$$

here n is uniquely determined from the inequality

$$-\sum_{k=-n}^{-1} f(T^k x_1) \leq s + t < -\sum_{k=-n+1}^{-1} f(T^k x_1).$$

Sometimes we shall say that the flow $\{V^t\}$ is a special flow over the automorphism T.

Let us identify the points $(x_1, f(x_1))$ and $(Tx_1, 0)$. Visually, the motion under the action of $\{V^t\}$ for $t \geq 0$ should be pictured as follows: the point (x_1, s) moves vertically upward until it reaches the point $(x_1, f(x_1))$. As the result of identification it turns out to be at the point $(Tx_1, 0)$ from which it continues its motion vertically upward, etc.

The set of points $(x_1, 0)$, $x_1 \in M_1$ is sometimes referred to as the base of the special flow, and T as the base automorphism.

Lemma 1. *The special flow $\{V^t\}$ preserves the measure μ.*

Proof. Suppose $t > 0$ and $\overline{M} = M_1 \times \mathbb{R}^1_+$ is the direct product of the measure space M_1 and the positive half-line $\mathbb{R}^1_+$. Denote by $\bar{\mu}$ and $\{\overline{V}^t\}$ respectively the measure in $\overline{M}$ which is the direct product of the measure μ_1 and the Lebesgue measure on $\mathbb{R}^1_+$, and the semigroup of transformations on $\overline{M}$ acting according to the formula $\overline{V}^t(x_1, s) = (x_1, s + t)$, $t \geq 0$. It is clear that the semigroup $\{\overline{V}^t\}$ preserves the measure $\bar{\mu}$.

Put

$$\overline{M}_0 = M_1^f,\ \overline{M}_k = \left\{(x_1, s): \sum_{j=0}^{k-1} f(T^j x_1) \leq s < \sum_{j=0}^{k} f(T^j x_1)\right\} \qquad k = 1, 2, \ldots$$

The sets $\overline{M}_k$ are pairwise disjoint and $\bigcup_{k=0}^{\infty} \overline{M}_k = \overline{M}$. For $k \geq 1$ introduce the map $\varphi_k: \overline{M}_k \to \overline{M}_0$ by putting

$$\varphi_k(x_1, s) = \left(T^k x_1, s - \sum_{j=0}^{k-1} f(T^j x_1)\right)$$

Clearly, φ_k is measurable and maps $\overline{M}_k$ onto $\overline{M}_0 = M_1^f$ bijectively.

Let us show that φ_k sends the measure $\bar{\mu}$ on $\overline{M}_k$ into the measure μ on M_1^f. It suffices to consider sets A of the form

$$A = \{(x_1, s): x_1 \in E, a < s < b\},$$

where $E \in \mathfrak{S}_1$, a and b are constants. Then

$$\varphi_k(A) = \left\{(x_1, s): x_1 \in T^k E, a - \sum_{j=0}^{k-1} f(T^j x_1) < s < b - \sum_{j=0}^{k-1} f(T^j x_1)\right\},$$

and

$$\mu(\varphi_k(A)) = \mu_1(T^k E)(b - a) = \mu_1(E)(b - a) = \bar{\mu}(A),$$

which was to be proved.

Now for an arbitrary $A \subset M_1^f$ it follows from the definition of special flows that

$$V^t A = \bigcup_{k=0}^{\infty} \varphi_k((\bar{V}^t A) \cap \bar{M}_k)$$

and, as can be easily seen, the sets $\varphi_k((\bar{V}^t A) \cap \bar{M}_k)$ are disjoint. Therefore

$$\begin{aligned} \mu(V^t A) &= \sum_{k=0}^{\infty} \mu(\varphi_k((\bar{V}^t A) \cap \bar{M}_k)) \\ &= \sum_{k=0}^{\infty} \bar{\mu}((\bar{V}^t A) \cap \bar{M}_k) = \bar{\mu}(\bar{V}^t A) = \bar{\mu}(A) = \mu(A). \end{aligned}$$

Similar arguments are valid for $t < 0$. The lemma is proved. □

§2. Statement of the Main Theorem on Special Representation of Flows and Examples of Special Flows

Theorem 1. *Any flow $\{T^t\}$ without fixed points on the Lebesgue space $(M, \mathfrak{S}, \mu)$ is metrically isomorphic to a special flow.*

An isomorphism to a special flow is sometimes called a *special representation.*

The proof of this theorem will be carried out in the next section; now we shall show that special representations naturally arise in a series of examples. But first we shall prove a lemma which explains the meaning of the quantity $\mu_1(M_1)$.

Lemma 1. *Suppose $\{V^t\}$ is an ergodic special flow constructed from the automorphism T and the function f. For any point $x \in M = M_1^f$ denote by $v_t(x)$ the cardinality of the set of all τ, $0 \le \tau < t$, such that $V^\tau x \in M_1$. Then $\lim_{t\to\infty}(1/t)v_t(x) = \mu_1(M_1)$ for almost all $x \in M$.*

Proof. Suppose $x = (x_1, s)$. Put $g(x) = 1/f(x_1)$. Clearly $g(x) > 0$ almost everywhere and $\int g\, d\mu = \mu_1(M_1)$. Further $v_t(x)$ differs from the value $\int_0^t g(V^u x)\, du$ by no more than 2. Hence, by the ergodicity of $\{V^t\}$, we have

$$\lim_{t\to\infty} \frac{1}{t} v_t(x) = \lim_{t\to\infty} \frac{1}{t} g(V^u x)\, du = \int g\, d\mu = \mu_1(M_1).$$

The lemma is proved. □

EXAMPLES OF SPECIAL REPRESENTATIONS

1. Suppose M is the two-dimensional torus with cyclic coordinates x_1, x_2 and with a given system of differential equations

$$\frac{dx_1}{dt} = f_1(x_1, x_2), \qquad \frac{dx_2}{dt} = f_2(x_1, x_2).$$

Assume that $f_1, f_2 \in C^2(M)$ and

$$\frac{\partial}{\partial x}(\rho f_1) + \frac{\partial}{\partial x_2}(\rho f_2) = 0,$$

for some positive function $\rho \in C^1(M)$. Then, according to Liouville's theorem, the one-parameter group $\{S^t\}$ of translations along the solutions of our system preserves the measure $d\mu = \rho\, dx_1\, dx_2$, which, without loss of generality, we may assume normalized.

Suppose $f_1 > 0$. Put $M_1 = \{(x_1, x_2) \in M : x_1 = 0\}$ and introduce the transformation $T: M_1 \to M_1$ as follows. For each point $x \in M_1$ consider the trajectory until its first return to M_1; denote by Tx the point where this trajectory reaches M_1. It follows from the fact that f_1 is positive that the transformation T is continuous.

Introduce the measure μ_1 in M_1 by putting

$$d\mu_1(x_2) = \rho(0, x_2) \cdot f_1(0, x_2)\, dx_2.$$

Let us show that the measure μ_1 is invariant with respect to T. To do this introduce the new flow $\{\bar{S}^t\}$ obtained from $\{S^t\}$ by the time change

$$\bar{S}^t x = S^{\bar{t}} x, \quad \text{where } \bar{t} = \int_0^t [f_1(S^u x)]^{-1}\, du.$$

The flow $\{\bar{S}^t\}$ has the invariant measure $\bar{\mu}$ with density $\rho(x) \cdot f(x)$ (see §3, Chap. 10). The system of differential equations corresponding to $\{\bar{S}^t\}$ is of the form

$$\frac{dx_1}{dt} = 1, \qquad \frac{dx_2}{dt} = \frac{f_2(x_1, x_2)}{f_1(x_1, x_2)}.$$

Clearly $\{\bar{S}^t\}$ sends the meridian $M_1 = \{(0, x_2)\}$ into the meridian $\{(t, x_2)\}$

and $\bar{S}^1(0, x_2) = T(0, x_2)$. Therefore, for any $E \subset M_1$, when $0 < t < 1$, we have

$$\bar{\mu}\left(\bigcup_{0 \le \tau \le t} \bar{S}^\tau E\right) = t \cdot \mu_1(E) = \bar{\mu}\left(\bigcup_{1 \le \tau < 1+t} \bar{S}^\tau E\right)$$

$$= \bar{\mu}\left(\bigcup_{0 \le \tau < t} \bar{S}^\tau(TE)\right) = t\mu_1(TE),$$

i.e., $\mu_1(E) = \mu_1(TE)$.

If $f(x)$, for $x \in M_1$, is the time that it takes to reach M_1, then the flow $\{S^t\}$ may be represented as the special flow over the automorphism T with the function f.

2. Suppose $Q \subset \mathbb{R}^d$ is a compact domain with piecewise smooth boundary, M is the unit tangent bundle over Q and $\{S^t\}$ is the billiards on Q (see Chap. 6). Suppose M_1 denotes the set of unit tangent vectors x whose carriers belong to the boundary ∂Q, the vectors themselves being directed towards the interior of Q. By T denote the transformation of M_1 that consists in moving along the trajectory of the billiards determined by the vector x until the first intersection with the boundary followed by the reflection from it. If $f(x)$ is the time which passes until the next reflection, then the billiards $\{S^t\}$ may be represented as a special flow over the automorphism T with the function f. In Chap. 6 it was shown that the invariant measure for T is of the form

$$d\mu_1 = d\rho(q)\, d\omega_q |(n(q), x)|,$$

where $d\rho$ is the surface element on ∂Q, $d\omega$ is the volume element on the $(d - 1)$ dimensional sphere and $n(q)$ is the unit normal vector at the point $q \in \partial Q$.

3. Let us show how the theory of special flows may be used to derive the well-known formula (in the theory of Gauss stationary processes) giving the mean number of intersections of a random realization of given level. We begin with the statement of the problem.

Suppose M is the space of continuously differentiable real valued functions $x(s)$ defined for $s \in \mathbb{R}^1$. Introduce a Gauss stationary measure μ (see §2, Chap. 8) and denote by σ the corresponding spectral measure on $\mathbb{R}^1$. We shall assume that $\int \lambda^4\, d\sigma(\lambda) < \infty$: this condition guarantees that the measure μ is concentrated on M. Since M consists of real-valued functions, the measure σ is even.

Consider the case when $m(s) \stackrel{\text{def}}{=} \int x(s)\, d\mu(x) = 0$. Then, as can be easily checked, $\int x'(s)x(s)\, d\mu(x) = 0$ for any s. The derivative $\{x'(s): -\infty < s < \infty\}$

is also a stationary Gauss process with spectral measure $\lambda^2\, d\sigma(\lambda)$. Assume that the measure σ is continuous. In Chap. 14 it will be shown that, in this case, the flow of the translations $\{S^t\}$ corresponding to the Gauss stationary process is ergodic.

Fix a number $a > 0$. Then for $x = \{x(s): -\infty < s < \infty\}$ denote by $v_t(x)$ the cardinality of the set of all those s, $0 \leq s < t$ for which $x(s) = a$.

Theorem 2. *With probability* 1, *we have*

$$\lim_{t\to\infty} \frac{1}{t}\, v_t(x) = \frac{1}{\pi}\left(\frac{d_1}{d_0}\right)^{1/2} \exp\left(-\frac{a^2}{2d_0}\right),$$

where

$$d_0 = \int d\sigma(\lambda) = \int x^2(s)\, d\mu(x), \qquad d_1 = \int \lambda^2\, d\sigma(\lambda) = \int (x'(s))^2\, d\mu(x).$$

We shall not give a complete proof of this theorem; we will only show how everything reduces to special flows. Suppose M_1 is the set of all $x \in M$ for which $x(0) = a$. For $x \in M_1$, denote by $f = f(x)$ the least $s > 0$ for which $x(s) = a$ and introduce the transformation T of the set M_1 which acts according to the formula $Tx = S^{f(x)}x$. This allows us to present the flow $\{S^t\}$ as the special flow over the automorphism T with the function f. According to Lemma 1, we have $\lim_{t\to\infty} (1/t)v_t(x) = \mu_1(M_1)$ and everything reduces to calculating $\mu_1(M_1)$.

Take a small $t > 0$ and consider the set $M_1^t = \bigcup_{0\leq u<t} S^u M_1$, consisting of all $x \in M$ which intersect the level a in the semi-interval $[0, t)$. As can easily be shown, $\mu(M_1^t) = t \cdot \mu_1(M_1) + o(t)$.

Introduce the set

$$\tilde{M}_1^t = \left\{x \in M: x(t) \geq a,\ x'(t) > \frac{x(t) - a}{t}\right\} \cup \left\{x \in M: x(t) < a,\ x'(t) < \frac{x(t) - a}{t}\right\}$$

It can be shown that $\lim_{t\to 0} [\mu(\tilde{M}_1^t)/\mu(M_1^t)] = 1$. Further,

$$\mu(\tilde{M}_1^t) = \int_a^\infty \frac{1}{\sqrt{2\pi\, d_0}} \exp\left(-\frac{u^2}{2d_0}\right) du \int_{(u-a)/t}^\infty \frac{1}{\sqrt{2\pi\, d_1}} \exp\left(-\frac{v^2}{2d_1}\right) dv$$
$$+ \int_{-\infty}^a \frac{1}{\sqrt{2\pi\, d_0}} \exp\left(-\frac{u^2}{2\, d_0}\right) du \int_{-\infty}^{(u-a)/t} \frac{1}{\sqrt{2\pi\, d_1}} \exp\left(-\frac{v^2}{2\, d_1}\right) dv.$$

Carry out the substitution $u = a + tw$ in the integrals. Integrating by parts, we get

$$\mu(\tilde{M}_1^t) = t\Bigg[\int_0^\infty \frac{1}{\sqrt{2\pi\, d_0}} \exp\left(-\frac{(a+tw)^2}{2\,d_0}\right) dw \int_w^\infty \frac{1}{\sqrt{2\pi\, d_1}} \exp\left(-\frac{v^2}{2\,d_1}\right) dv$$

$$+ \int_{-\infty}^0 \frac{1}{\sqrt{2\pi\, d_0}} \exp\left(-\frac{(a+tw)^2}{2\,d_0}\right) dw$$

$$\times \int_{-\infty}^w \frac{1}{\sqrt{2\pi\, d_1}} \exp\left(-\frac{v^2}{2\,d_1}\right) dv\Bigg]$$

$$= t\Bigg[\int_0^w \frac{1}{\sqrt{2\pi\, d_0}} \exp\left(-\frac{(a+ty)^2}{2\,d_0}\right) dy \int_w^\infty \frac{1}{\sqrt{2\pi\, d_1}} \exp\left(-\frac{v^2}{2\,d_1}\right) dv \Bigg|_0^\infty$$

$$+ \int_0^\infty \frac{1}{\sqrt{2\pi\, d_1}} \exp\left(-\frac{w^2}{2\,d_1}\right) dw \int_0^w \frac{1}{\sqrt{2\pi\, d_0}} \exp\left(-\frac{(a+ty)^2}{2\,d_0}\right) dy$$

$$- \int_w^0 \frac{1}{\sqrt{2\pi\, d_0}} \exp\left(-\frac{(a+ty)^2}{2\,d_0}\right) dy$$

$$\times \int_{-\infty}^w \frac{1}{\sqrt{2\pi\, d_1}} \exp\left(-\frac{v^2}{2\,d_1}\right) dv \Bigg|_{-\infty}^0 + \int_0^\infty \frac{1}{\sqrt{2\pi\, d_1}} \exp\left(-\frac{w^2}{2\,d_1}\right) dw$$

$$\times \int_w^0 \frac{1}{\sqrt{2\pi\, d_0}} \exp\left(-\frac{(a+ty)}{2\,d_0}\right) dy\Bigg]$$

$$= \frac{t}{2\pi\sqrt{d_0\, d_1}}\Bigg[\int_0^\infty \exp\left(-\frac{w^2}{2\,d_1}\right) dw \int_0^w \exp\left(-\frac{(a+ty)^2}{2\,d_0}\right) dy$$

$$+ \int_{-\infty}^0 \exp\left(-\frac{w^2}{2\,d_1}\right) dw \int_w^0 \exp\left(-\frac{(a+ty)^2}{2\,d_0}\right) dy\Bigg].$$

By Lebesgue's theorem on the passage to the limit under the integral sign the expression in the square brackets has the following limit when $t \to 0$:

$$\exp\left(-\frac{a^2}{2\,d_0}\right)\left[\int_0^\infty \exp\left(-\frac{w^2}{2\,d_1}\right) w\, dw - \int_{-\infty}^0 \exp\left(-\frac{w^2}{2\,d_1}\right) w\, dw\right]$$

$$= 2\exp\left(-\frac{a^2}{2\,d_0}\right) d_1.$$

Thus we get

$$\lim_{t\to 0}\frac{1}{t}\mu(M_1^t)=\lim_{t\to 0}\frac{1}{t}\mu(\tilde{M}_1^t)=\frac{1}{2\pi\sqrt{d_0\,d_1}}\cdot 2\exp\left(-\frac{a^2}{2\,d_0}\right)d_1$$

$$=\frac{1}{\pi}\left(\frac{d_1}{d_0}\right)^{1/2}\exp\left(-\frac{a^2}{2\,d_0}\right),$$

which was to be proved.

§3. Proof of the Theorem on Special Representation

We begin with the following general lemma.

Lemma 1. *For any function $f\in L^2(M,\mathfrak{S},\mu)$, we have*

$$\lim_{\alpha\to 0}\frac{1}{\alpha}\int_0^\alpha f(T^tx)\,dt=f(x),$$

in the sense of convergence with respect to the norm in the space $L^2(M,\mathfrak{S},\mu)$.

Proof. The function $b(t)=(U^tf,f)$ is positive definite and, by the Bochner–Khinchin theorem it may be represented in the form

$$b(t)=\int_{-\infty}^{\infty}e^{i\lambda t}\,d\sigma(t),$$

where σ is the measure on the line. According to spectral theory (see Appendix 2) we have

$$\left\|\frac{1}{\alpha}\int_0^\alpha f(T^tx)\,dt-f(x)\right\|^2=\int_{-\infty}^{\infty}\left|\frac{e^{i\alpha\lambda}-1}{i\alpha\lambda}-1\right|^2 d\sigma(\lambda).$$

By Lebesgue's theorem on the passage to the limit under the integral sign, the last expression vanishes when $\alpha\to 0$. The lemma is proved. □

Corollary. $(1/\alpha)\int_0^\alpha f(T^tx)\,dt$ *for $\alpha\to 0$ converges to $f(x)$ in measure.*

We now pass to the actual proof of the theorem on special representation. In fact the entire proof splits up into two independent statements, each of which is interesting in itself. Hence we begin by stating them, together with the appropriate definitions.

Definition 1. The partition ξ of the space M is said to be a *regular partition* for the flow $\{T^t\}$ if it possesses the following properties:

(1) each element C_ξ of the partition ξ is a semi-interval of the form $\{T^s x: 0 \leq s < f\}$ belonging to a single trajectory of the flow; x is the left end point of this semi-interval and f is its length; we assume that the representation of the point $y \in C$ in the form $y = T^s x$ for $0 \leq s < f$ is unique;
(2) the functions F, G defined for an arbitrary point $y = T^\tau x \in C_\xi$, $0 \leq \tau < f$ by the relations $F(y) = f$, $G(y) = \tau$ are measurable.

The two following statements will be proved.

(A) For any flow $\{T^t\}$ without fixed points, there exists an invariant set of positive measure possessing a regular partition; we then have $F(y) \geq C$ for some constant $C > 0$.
(B) If the flow $\{T^t\}$ has an invariant set of measure 1 possessing a regular partition and $F(y) \geq C > 0$, then this flow is metrically isomorphic to a special flow.

Let us deduce the statement of our theorem from (A) and (B). According to (A), we can find an invariant set E of positive measure possessing a regular partition. Then (B) implies that the restriction of the flow $\{T^t\}$ to E is metrically isomorphic to a special flow. Applying the same argument to the complement $M \backslash E$ and continuing in the same way, we get our theorem by transfinite induction.

The proof of statement (A) will be split up into several steps.

1. Consider the set $C \subset M$ entirely contained in some trajectory of the flow $\{T^t\}$, i.e., the set of points of the form $T^t x_0$, where $x_0 \in M$, while t ranges over some set $E \subset \mathbb{R}^1$. Denote by Cl C (respectively ∂C) the set of points of the form $T^t x_0$, where t ranges over the set Cl E (respectively ∂E).[1]

Definition 2. The *closure* (respectively the *boundary*) of the set $A \subset M$ with respect to the flow $\{T^t\}$ is the set Cl A (respectively ∂A) such that for any $x \in M$ we have the relation

$$(\mathrm{Cl}\, A) \cap \{T^t x: -\infty < t < \infty\} = \mathrm{Cl}[A \cap \{T^t x: -\infty < t < \infty\}],$$

(respectively

$$(\partial A) \cap \{T^t x: -\infty < t < \infty\} = \partial[A \cap \{T^t x: -\infty < t < \infty\}]).$$

Clearly these relations determine the sets Cl(A) and ∂A uniquely.

[1] Here Cl E is the closure of E and ∂E the boundary of E.

Lemma 2. *For any set $A \in \mathfrak{S}$, $\mu(A) > 0$ and any $\varepsilon > 0$, we can find an $A^{(\varepsilon)} \in \mathfrak{S}$ such that*

(1) *for all $x \in M$ the set $\{t: T^t x \in A^{(\varepsilon)}\}$ is open;*
(2) $\mu(\partial A^{(\varepsilon)}) = 0$;
(3) $\mu(A \triangle A^{(\varepsilon)}) < \varepsilon$.

Proof. Put

$$A_{n,\beta} = \left\{x \in M: n \int_0^{1/n} \chi_A(T^t x)\, dt > \beta\right\}.$$

For any β, $0 < \beta < 1$ and n sufficiently large, by Lemma 1, we have

$$\mu(A \triangle A_{n,\beta}) < \varepsilon. \tag{1}$$

Note that for all $x \in M$ and $\alpha > 0$ the function

$$h_\alpha(\tau) = h_\alpha(\tau; x) = \frac{1}{\alpha} \int_0^\alpha \chi_A(T^{t+\tau}x)\, dx,$$

is continuous and even satisfies the Lipshitz condition with constant $2/\alpha$. Indeed

$$\begin{aligned} |h_\alpha(\tau_1) - h_\alpha(\tau_2)| &= \frac{1}{\alpha}\left| \int_0^\alpha \chi_A(T^{t+\tau_1}x)\, dt - \int_0^\alpha \chi_A(T^{t+\tau_2}x)\, dt \right| \\ &= \frac{1}{\alpha}\left| \int_{\tau_1}^{\tau_1+\alpha} \chi_A(T^t x)\, dt - \int_{\tau_2}^{\tau_2+\alpha} \chi_A(T^t x)\, dt \right| \le \frac{2}{\alpha}|\tau_1 - \tau_2|. \end{aligned} \tag{2}$$

This implies that for all n, β, x the set $\{t \in R^1: T^t x \in A_{n,\beta}\}$ is open and we have

$$\partial A_{n,\beta} \subset \left\{x \in M: n \int_0^{1/n} \chi_A(T^t x)\, dt = \beta\right\}.$$

Choose β, $0 < \beta < 1$, so that the set in the right-hand side has μ-measure zero independently of n and then choose n so as to have (1); put $A^{(\varepsilon)} = A_{n,\beta}$. The lemma is proved. □

2. Suppose the set $A \in \mathfrak{S}$ is such that there exists t_0 for which

$$\delta_0 \overset{\text{def}}{=} \mu((M \setminus A) \cap T^{t_0} A) > 0.$$

The existence of such a set follows from the fact that $\{T^t\}$ has no fixed points.

For $\alpha > 0$ consider the function

$$\varphi_\alpha(x) = h_\alpha(0; x) = \frac{1}{\alpha} \int_0^\alpha \chi_A(T^t x)\, dt,$$

and put

$$E_1 = E_1^{(\alpha)} = \{x \in M : \varphi_\alpha(x) < \tfrac{1}{4}\},$$
$$E_2 = E_2^{(\alpha)} = \{x \in M : \varphi_\alpha(x) > \tfrac{3}{4}\},$$
$$E = E^{(\alpha)} = E_1^{(\alpha)} \cap T^{t_0} E_2^{(\alpha)}.$$

Since, by the corollary to Lemma 1, $\varphi_\alpha(x)$ converges in measure to $\chi_A(x)$ when $\alpha \to 0$, α may be chosen so small that

$$\mu((M \setminus A) \triangle E_1) < \tfrac{1}{2}\delta_0, \qquad \mu(A \triangle E_2) < \tfrac{1}{2}\delta_0.$$

This implies

$$\mu(E) \geq \mu((M \setminus A) \cap T^{t_0} A) - \mu((M \setminus A) \triangle E_1) - \mu(A \triangle E_2) > 0.$$

Therefore we can throw out an invariant subset N from the space M so that $\mu(M \setminus N) > 0$ and, for $x \in M \setminus N$, we can find positive and negative t, as large as we wish in absolute value, such that $T^t x \in E$. Now put

$$E_1' = \bigcup_{\substack{0 \leq t \leq \alpha/8 \\ t \text{ rational}}} T^t E_1; \qquad E_2' = \bigcup_{\substack{0 \leq t \leq \alpha/8 \\ t \text{ rational}}} T^t E_2.$$

Let us prove that $E_1' \cap E_2' = \varnothing$. Indeed, suppose the converse: there exists $x \in E_1' \cap E_2'$. Then we can find $x_1 \in E_1$, $x_2 \in E_2$ and numbers τ_1, τ_2, $0 \leq \tau_1$, $\tau_2 \leq \alpha/8$, such that $x = T^{\tau_1} x_1$, $x = T^{\tau_2} x_2$. Hence $x_1 = T^{-\tau_1} x$, $x_2 = T^{-\tau_2} x$.

By (2) we have

$$\frac{1}{2} < |\varphi_\alpha(x_1) - \varphi_\alpha(x_2)| = |h_\alpha(-\tau_1; x) - h_\alpha(-\tau_2; x)|$$
$$\leq \frac{2}{\alpha} |\tau_1 - \tau_2| \leq \frac{2}{\alpha} \cdot \frac{\alpha}{8} = \frac{1}{4}.$$

The contradiction thus obtained proves our statement.

3. Let us show that

$$E_1' = \bigcup_{0 \leq t \leq \alpha/8} T^t E_1, \qquad E_2' = \bigcup_{0 \leq t \leq \alpha/8} T^t E_2.$$

Here the sum is taken over all (not only rational) values $t \in [0, \alpha/8]$.

Let us carry out the argument for E_1'; for E_2' it is the same. Suppose $z = T^{t_0}x$, $x \in E_1$, $0 \le t_0 \le \alpha/8$. By continuity of the function $h(t) = \varphi_\alpha(T^t x)$, there is an $\varepsilon_0 > 0$ such that $|\varepsilon| < \varepsilon_0$ implies $T^\varepsilon x \in E_1$. Choosing this ε to that $r = t - \varepsilon$ is rational, we obtain $z \in E_1'$.

4. For any point $x \in M \setminus N$ consider its trajectory $\{T^t x: -\infty < t < \infty\}$. The intersection of this trajectory with E_1' as well as with E_2' is an unbounded open set on it, whose connected components (intervals) are, by step 3, of length no less than $\alpha/8$.[1] Therefore, for each point of a trajectory, we can indicate the nearest interval to its left, as well as to its right, belonging to the union $E_1' \cup E_2'$.

Now we can construct the set M_1 which will be simultaneously the base of desirable special representation and the set of left end points of the required regular partition. Namely, let M_1 be the set of points $x \in M$ possessing the following properties:

(a) x is the right end point of one of the intervals belonging to E_1';
(b) the nearest interval to the right of the union $E_1' \cup E_2'$ belongs to E_2'.

On each trajectory belonging to $M \setminus N$ there is a countable set of points, unbounded in both directions, belonging to M_1; it follows from step 3 that the distance (along the trajectory) between any such two points is no less than $\alpha/8$.

5. For $x \in M_1 \setminus N$ put $f(x) = \min\{t > 0: T^t x \in M_1\}$. It follows from the previous statements that $\alpha/8 \le f(x) < \infty$. Put

$$M_1^f = \{(x, s): x \in M_1, 0 \le s < f(x)\}.$$

The map $\varphi: M_1^f \to M \setminus N$ given by the formula $\varphi(x, s) = T^s x$, establishes a one-to-one correspondence between M_1^f and $M \setminus N$. For the partition ξ, take the image under φ of the partition of M_1^f into vertical segments of the form

$$C_\xi = \{(x_0, s): 0 \le s < f(x_0)\},$$

and for $y \in M \setminus N$, $y = \varphi(x, s)$, put $F(y) = f(x)$, $G(y) = s$. To conclude the proof of statement (A), it remains to check that the functions F and G are measurable. For all $c > 0$, we have

$$\{y: F(y) > c\} = \left(\bigcup_{\substack{0 \le r \le c \\ r \text{ rational}}} T^{-r}\{y: G(y) > c\} \right) \cup T^{-c}\{y: G(y) \ge c\},$$

[1] Here we identify the point $T^t x$ belonging to the trajectory with the number t, thus carrying over the topology from $\mathbb{R}^1$ onto the trajectory.

and therefore it suffices to prove the measurability of the function G. For this, in its turn, it suffices to prove the measurability of the sets

$$B_k^{(n)} = \left\{x \in M: \frac{k}{n} \leq G(x) < \frac{k+1}{n}\right\}, \qquad k = 0, 1, \ldots; n = 1, 2, \ldots$$

Note that $B_{k+1}^{(n)} = (M \setminus B_0^{(n)}) \cap T^{1/n} B_k^{(n)}$. Therefore it suffices to prove the measurability of the sets $B_0^{(n)}$. Since we need only consider large n, we may assume $n > 8/\alpha$. For such n the condition $x \in B_0^{(n)}$ is equivalent to the conjunction of the following three conditions:

(1) $x \notin E_1'$;
(2) $T^{-1/n} x \in E_1'$;
(3) for some integer $p > 0$ $T^{1/n}x, T^{2/n}x, \ldots, T^{p/n}x \notin E_1' \cup E_2'$, but $T^{(p+1)/n}x \in E_2'$.

Thus the measurability of $B_0^{(n)}$ is established and statement (A) is proved.

Proof of statement (*B*)

1. Suppose $\{T^t\}$ is a flow in the space M and ξ is its regular partion. It is convenient to introduce a change of time and consider the new flow $\{\overline{T}^t\}$ by putting (for $0 \leq t < 1$ and any point $x \in M$):

$$\overline{T}^t x = \begin{cases} T^{t/F(x)} & \text{for } t < \bar{t}(x), \\ T^{[t - \bar{t}(x)]/F(y)}\, y & \text{for } t \geq \bar{t}(x), \end{cases}$$

where $\bar{t}(x) = [F(x) - G(x)]/F(x)$, $y = T^{F(x) - G(x)}x$.

Further $\{\overline{T}^t\}$ can be extended by means of the group relation $\overline{T}^{t_1 + t_2}$ $\overline{T}^{t_1}\overline{T}^{t_2}$. The meaning of this change of time is that now under the action of the flow $\{\overline{T}^t\}$ each point x passes through each element of the partition ξ in unit time.

It is not difficult to check that $\{\overline{T}^t\}$ is a measurable flow. By Theorem 1, §3, Chap. 10, it preserves the measure $\bar{\mu}$ for which $d\bar{\mu}/d\mu = [F(x) \int d\mu/F(x)]^{-1}$. We then have $\int d\mu/F(x) < \infty$, since $F \geq C > 0$.

The partition ξ remains regular with respect to the flow $\{\overline{T}^t\}$. Indeed, the corresponding functions $\overline{F}$ and $\overline{G}$ are of the form $\overline{F}(x) = 1$, $\overline{G}(x) = G(x)/F(x)$.

First we shall show that the flow $\{\overline{T}^t\}$ possesses a special representation.

2. Suppose M_1 is a set whose points are all possible elements C_ξ of the partition ξ. Each point $x \in M_1$ may be given by the left end point of the corresponding C_ξ. Thus M_1 is the space of left end points of the elements of the partition ξ. Denote the natural map of M onto M_1 by π.

Transform M_1 into a measure space by taking the σ-algebra $\mathfrak{S}_1$ of subsets $C \subset M_1$ for which $\pi^{-1}(C) \in \mathfrak{S}$ and by putting $\mu_1(C) = \bar{\mu}(\pi^{-1}(C))$ for any $C \in \mathfrak{S}_1$. Introduce the transformation T of the space M_1 by putting $T(C_\xi) = \overline{T}^1(C_\xi)$. Since the flow $\{\overline{T}^t\}$ preserves the measure $\bar{\mu}$, it follows that

T preserves the measure μ_1 and is an automorphism of the measure space $(M_1, \mathfrak{S}_1, \mu_1)$. The automorphism T will be the base automorphism of the required special representation.

Put $M^{(1)} = M_1 \times [0, 1)$ and define a one-to-one map $\psi: M \to M^{(1)}$ by the formula $\psi(y) = (x, t)$, where $y = \bar{T}^t x = T^{t \cdot F(y)} x$ and x is the left end point of the element C_ξ containing y. Like any one-to-one map, ψ generates the σ-algebra $\mathfrak{S}^{(1)} = \psi(\mathfrak{S})$ of subsets of the space $M^{(1)}$ and the measure $\mu^{(1)} = \psi(\mu)$ on $\mathfrak{S}^{(1)}$. Then a flow $\{\bar{V}^t\}$, $\bar{V}^t = \psi \bar{T}^t \psi^{-1}$, metrically isomorphic to the flow $\{\bar{T}^t\}$, arises on $M^{(1)}$.

We will prove that:

(1) the σ-algebra $\mathfrak{S}^{(1)}$ coincides with the σ-algebra $\mathfrak{S}_0^{(1)}$ of measurable subsets of $M^{(1)}$, viewed as the Cartesian product of the measurable spaces M_1 and $[0, 1)$;

(2) the measure $\mu^{(1)}$ coincides with the Cartesian product of the measure μ_1 and the Lebesgue measure on $[0, 1)$.

3. First let us show that $\mathfrak{S}_0^{(1)} \subset \mathfrak{S}^{(1)}$. Indeed, suppose the set $C \in \mathfrak{S}_0^{(1)}$ is of the form $C = C_1 \times \Delta$, where $C_1 \in \mathfrak{S}_1$, $\Delta = [t_1, t_2) \subset [0, 1)$ is a semi-interval. Then

$$\psi^{-1}(C) = \{y: \pi(y) \in C_1, t_1 \le \bar{G}(y) < t_2\},$$

and therefore $\psi^{-1}(C) \in \mathfrak{S}$. This implies the required inclusion.

In order to prove the converse inclusion, first consider the sets $B^* \in \mathfrak{S}$ such that $B^* \cap C_\xi$ is open for every C_ξ. Put

$$B = \{x \in M_1: B^* \cap C_\xi(x) \ne \varnothing\}.$$

Let us show that $B \in \mathfrak{S}_1$, i.e., that the set $\bar{B} = \pi^{-1}(B)$ belongs to $\mathfrak{S}$. Suppose

$$B^*_{n,k} = \left\{x \in B^*: \frac{k}{2^{n+1}} < \bar{G}(x) < \frac{k+1}{2^{n+1}}\right\},$$

$$B_{n,k} = \pi(B^*_{n,k}),$$

$$\bar{B}_{n,k} = \pi^{-1}(B_{n,k}); \qquad k = 1, 2, \ldots, 2^{n+1} - 2, n = 1, 2, \ldots,$$

$$B^*_{k/2^n} = \{x \in B^*: \bar{G}(x) = k/2^n\},$$

$$B_{k/2^n} = \pi(B^*_{k/2^n}),$$

$$\bar{B}_{k/2^n} = \pi^{-1}(B_{k/2^n}), \qquad k = 0, 1, \ldots, 2^n - 1, n = 1, 2, \ldots.$$

Then

$$\bar{B}_{n,k} = \left[\left(\bigcup_{\substack{0 \le t < 1 - k/2^{n+1} \\ t \text{ rational}}} \bar{T}^t B^*_{n,k}\right) \cap \left\{x \in M: \bar{G}(x) > \frac{1}{2^{n+1}}\right\}\right]$$

$$\cup \left[\left(\bigcup_{\substack{0 \le t < (k+1)/2^{n+1} \\ t \text{ rational}}} T^{-t} B^*_{n,k}\right) \cap \left\{x \in M: \bar{G}(x) < 1 - \frac{1}{2^{n+1}}\right\}\right]$$

$$\bar{B}_{k/2^{n+1}} = \bigcup_{-(k/2^{n+1}) \le t \le 1 - k/2^{n+1}} \bar{T}^t B^*_{k/2^{n+1}},$$

$$\bar{B} = \left(\bigcup_{\substack{k = 1, \dots, 2^{n+1} - 2 \\ n = 1, 2, \dots}} \bar{B}_{n,k}\right) \cup \left(\bigcup_{\substack{k = 0, \dots, 2^n - 1 \\ n = 1, 2, \dots}} \bar{B}_{k/2^n}\right)$$

i.e., $\bar{B} \in \mathfrak{S}$. This implies $B \in \mathfrak{S}_1$ and $\mu_1(B) = \bar{\mu}(\bar{B})$.

4. Continuing to use the notations from step 3, put

$$D_{n,k} = B_{n,k} \times \left(\frac{k}{2^{n+1}}, \frac{k+1}{2^{n+1}}\right),$$

$$D = \left[\bigcap_{l=1}^{\infty} \bigcup_{n=l}^{\infty} \bigcup_{k=1}^{2^{n+1}-2} D_{n,k}\right] \cup \left[\bigcup_{\substack{k = 0, 1, \dots, 2^n - 1 \\ n = 1, 2, \dots}} B^*_{k/2^n}\right]. \tag{3}$$

Then $B^* \subset D \subset \mathrm{Cl}\, B^*$.

Consider the sets $B^*_{n,k} \in \mathfrak{S}$. For each element C_ξ, the intersection $B^*_{n,k} \cap C_\xi$ is open. Therefore we can apply the argument in step 3 to the sets $B^*_{n,k}$ and obtain $B_{n,k} \in \mathfrak{S}_1$. Formula (3) now shows that the set $\psi(D)$ belongs to the σ-algebra $\mathfrak{S}_0^{(1)}$.

5. Consider in M the σ-algebra $\mathfrak{M}$ of sets which is the inverse image under ψ of the σ-algebra $\mathfrak{S}_0^{(1)}$. Suppose the measure $\bar{\nu}$ on $M^{(1)}$ is given by the formula $\bar{\nu} = \mu_1 \times \rho$, where ρ is the Lebesgue measure on $[0, 1)$. Then $\bar{\mu}(E) = \bar{\nu}(\psi(E))$ for any $E \in \mathfrak{M}$. Indeed, this relation can be checked immediately if $\psi(E) = E_1 \times \Delta$, where $E_1 \in \mathfrak{S}_1$, and Δ is an interval from $[0, 1)$, while such sets generate the σ-algebra $\mathfrak{M}$.

Hence, in particular, the σ-algebra $\mathfrak{M}$ is complete with respect to the measure μ (see Appendix 1). It follows from step 4 and Lemma 2 that for any $A \in \mathfrak{S}$ and any $\varepsilon > 0$ we can find an $A^{(\varepsilon)} \in \mathfrak{M}$ such that $\mu(A \triangle A^{(\varepsilon)}) < \varepsilon$. Since M is a Lebesgue space and the σ-algebra $\mathfrak{M}$ is complete, it follows that $\mathfrak{M} = \mathfrak{S}$. Moreover, by the previous remarks, $\bar{\nu} = \psi(\bar{\mu})$.

6. It is now easy to conclude the proof of the theorem on special representation. The map $\psi: M \to M^{(1)}$ introduced in step 2 is not only a one-to-one correspondence between M and $M^{(1)}$, but also establishes a metric isomorphism of the flows $\{\bar{T}^t\}$ and $\{\bar{V}^t\}$.

Now carry out the inverse time change, i.e., consider the flow $\{V^t\}$ on $M^{(1)}$ given by the formulas

$$V^t(x, s) = \bar{V}^{t/f(x)}(x, s) \quad \text{for} \ -\frac{f(x)}{s} \leq t < \frac{f(x)}{1 - s},$$

$$V^{f(x)/1-s}(x, s) = (Tx, 0).$$

Further the flow $\{V^t\}$ may be extended by means of the group relation $V^{t_1+t_2} = V^{t_1} \cdot V^{t_2}$.

The flow $\{V^t\}$ preserves the measure ν given by the formula $d\nu = d\mu_1 \times f(x)\, d\rho$. Note that

$$\nu(M^{(1)}) = \iint d\mu_1 f(x)\, d\rho = \int_0^1 d\rho \int_{M_1} f(x)\, d\mu_1 = 1.$$

Since a measurable change of time transforms metrically isomorphic flows into metrically isomorphic ones, the flow $\{V^t\}$ is metrically isomorphic to the flow $\{T^t\}$. On the other hand, it is obvious that $\{V^t\}$ is metrically isomorphic to the special flow constructed from the automorphism T and the function f. The theorem is proved. □

Remark 1. Note that the phase space $(M_1, \mathfrak{S}_1, \mu_1)$ of the base automorphism of a special representation is a Lebesgue space. Indeed, since $M_1 = M/\xi$ is the quotient space of the Lebesgue space M with respect to the partition ξ, our statement follows from the measurability of the partition ξ. The latter, in its turn, follows from the following fact: if we consider, on each element C_ξ, its own Lebesgue measure μ_C, we obtain a canonical system of conditional measures for ξ (see Appendix 1).

Remark 2. Sometimes it is convenient to consider that the function $f(x)$, $x \in M_1$ which, together with the base automorphism T, determines the special representation of the flow $\{T^t\}$, is bounded not only from below but also from above. Let us show how this may be achieved.

Take $x \in M_1$ and consider the semi-interval $\{(x, s): 0 \leq s < f(x)\}$ which shall be denoted by $[0, f(x))$ for brevity. Subdivide it into a finite number of semi-intervals

$$[0, c), [c, 2c), \ldots, [(k-1)c, kc), [kc, f(x)),$$

where $c = \inf f(x) > 0$ and k is chosen so that the length l of the last semi-interval satisfies $c < l < 2c$. Doing this for all $x \in M$, we get a new regular partition of the phase space which, by statement (B), induces a new special representation. The corresponding function $\tilde{f}(x)$ satisfies the inequalities $0 < x \leq \tilde{f}(x) \leq 2c$.

Remark 3. The following generalization of the theorem on the special representation of flows is valid.

Suppose $\{T^t\}$ is a flow on the Lebesgue space $(M, \mathfrak{S}, \mu)$. Assume that for $\{T^t\}$ there exists an increasing σ-subalgebra $\mathfrak{S}^{(0)}$ of the σ-algebra $\mathfrak{S}$ i.e., a σ-algebra such that $T^t\mathfrak{S}^{(0)} \supset \mathfrak{S}^{(0)}$ for all $t > 0$ and $T^t\mathfrak{S}^{(0)} \neq \mathfrak{S}^{(0)}$.

Theorem 1. *For the flow $\{T^t\}$ we can find a special representation constructed from the automorphism T_1 of the Lebesgue space $(M', \mathfrak{S}', \mu')$ and function f, $f \geq \text{const} > 0$ such that*

(1) *the automorphism T_1 has an increasing σ-subalgebra $\mathfrak{S}_0$, i.e., a subalgebra such that $T_1\mathfrak{S}_0 \supset \mathfrak{S}_0$ and $T_1\mathfrak{S}_0 \neq \mathfrak{S}_0$;*
(2) *f is measurable with respect to $\mathfrak{S}_0$;*
(3) *the minimal σ-subalgebra of the σ-algebra $\mathfrak{S}$ containing all the sets of the form $\bigcup_{0 \leq t \leq t_1} T^tC$, where $C \in \mathfrak{S}_0$ and $f(x) \geq t_1$ for almost all $x \in C$, coincides with $\mathfrak{S}^{(0)}$.*

The proof of this theorem may be obtained by modifying Rohlin's proof of the theorem on the special representation of flows (see Rohlin [5]), based on the theory of measurable partitions. The statement contained in the theorem given above will be needed only once (in §5 of Chap. 12) for the analysis of the spectrum of K-flows.

§4. Rudolph's Theorem

The theorem on the special representation of flows proved in §3 may be sharpened in the following useful way

Theorem 1. *Suppose $\{T^t\}$ is an ergodic flow on the Lebesgue space $(\overline{M}, \overline{\mathfrak{S}}, \bar{\mu})$ and suppose we are given positive numbers p, q, $\tilde{p}$, where p/q is irrational. There exists a special representation of the flow $\{T^t\}$ defined by the objects $(M, \tilde{T}, \mu, f)$, where $\tilde{T}\colon M \to M$ is a base automorphism of the Lebesgue space M with invariant measure μ, while the function $f\colon M \to \mathbb{R}^1$ assumes* (mod 0) *only two values, p, q and $\mu(\{x \in M\colon f(x) = p\}) = \tilde{p} \cdot \mu(\{x \in M\colon f(x) = q\})$.*

The proof of this theorem will be carried out in two steps. First we describe the overall construction, skipping certain estimates and some technical details. Then we shall prove the lemmas and carry out the arguments filling in the missing details.

1. In view of the main theorem on the special representation of flows and of Remark 2 at the end of §3, we may assume that the flow $\{T^t\}$ is already represented as a special flow corresponding to the objects (M_0, T_0, μ_0, f_0) where the function $f_0: M_0 \to \mathbb{R}^1$ is bounded from above and from below: $0 < c_0 < f_0(x) < C_0 < \infty$. We want to construct a special representation of the form shown on Fig. 8. To do this we must "reconstruct" the original regular partition of the space $\overline{M} = \{(x, s): x \in M_0,\ 0 \le s < f_0(x)\}$ into vertical segments, obtaining a partition into the vertical segments shown on Fig. 8, for which the length of each segment either equals p or q and the measures of the p- and q-sets are given. This reconstruction will be carried out by the "method of successive approximations."

At the first step we must choose a sufficiently large number n_1 and, using the Rohlin–Halmos lemma (also see the remark after its proof) take a measurable set $M_1 \subset M_0$ such that the sets $M_1, T_0 M_1, T_0^2 M_1, \ldots, T_0^{n_1-1} M_1$ are disjoint and

$$\bigcup_{k=0}^{2n_1-1} T^k M_1 = M_0 \pmod 0.$$

Suppose

$$\tau_1(x) = \inf\{k \ge 1: T^k x \in M_1\}, \qquad x \in M_1.$$

Put

$$f_1(x) = \sum_{k=0}^{\tau_1(x)-1} f_0(T_0^k x), \qquad x \in M_1.$$

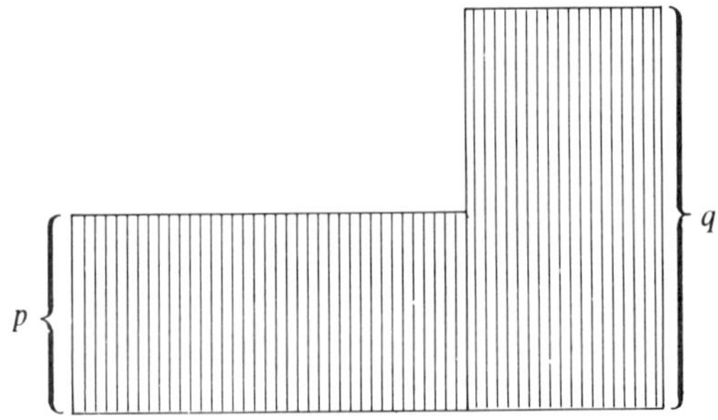

Figure 8

The flow $\{T^t\}$ may now be represented as the special flow constructed from the base automorphism $T_1 = (T_0)_{M_1}$,—the induced automorphism of T_0 on the set M_1—and the function $f_1(x)$.

For $x \in M_1$ put $l = f_1(x)$ and subdivide the semi-interval $\{(x, s): 0 \leq s < l\}$ into intervals $I_n = I_n(x) = [s_n, s_{n+1})$, where $0 = s_0 < s_1 < \cdots < s_{N+1} \leq s_{N+2} = l$. The intervals I_n $(n = 0, 1, \ldots, N)$ are of length p or q, while the remainder I_{N+1} is of length no greater than q (we assume that $p < q$). The intervals of length p (respectively q) will be called p-intervals (q-intervals).

For the function f_1 we have the estimate $c_1 < f_1(x) < C_1$, where $c_1 = n_1 c_0$, $C_1 = 2n_1 C_0$. Since n_1 is large, we may assume c_1 as large as we wish, thus making the partition into intervals I_n such that the number N_p of p-intervals and the number N_q of q-intervals will approximately occur in the required proportion: $N_p \approx \tilde{\rho} \cdot N_q$ (the exact estimates will be provided in the second part of the proof). For the sequel it will be useful to specify an initial part J_n of the interval I_n. Take $\alpha = \frac{1}{2}\min(p, q - p)$ and put

$$J_n = J_n(x) = [s_n, s_n + \alpha) \subset I_n, \qquad 0 \leq n \leq N.$$

Carrying this out for all $x \in M_1$, denote

$$B_1 = \left\{(x, s): x \in M_1, s \in \bigcup_{n=0}^{N} J_n(x)\right\}.$$

The partition into the intervals $I_n(x)$ for distinct $x \in M_1$ must not be carried out independently but, in a certain sense, measurably. This will be described more precisely below. The set B_1 is the "first approximation" to the set $\{(x, s): x \in M, 0 \leq s \leq \alpha\}$. Now let us describe the inductive procedure.

At the ith step of the procedure the following objects shall be defined: the measure space M_i, the automorphism T_i, the function $f_i: M_i \to \mathbb{R}^1$ satisfying $0 < c_i < f_i(x) < C_i$, the set $B_i \in \mathfrak{S}$ and, for each point $x \in M_i$, the partition of the semi-interval $L_x = [0, l)$, where $l = f_i(x)$ into N_p p-intervals, N_q q-intervals and a remainder of length no greater than q. (Of course the numbers l, N_p, N_q depend on x, i but we do not reflect this in the notation so as to simplify the exposition). Also, after the ith step, the relation $N_p \approx \tilde{\rho} \cdot N_q$ must hold with sufficiently small tolerance. We shall write this in the following form: suppose $\rho = \tilde{\rho}(1 + \tilde{\rho})^{-1}$; we require that for sufficiently small $\eta_i > 0$ (they will be defined later) we have the inequality

$$|N_p(N_p + N_q)^{-1} - \rho| < \eta_i.$$

At the $(i + 1)$th step choose a sufficiently large n_{i+1} and a set $M_{i+1} \subset M_i$ so that $M_{i+1}, T_i M_{i+1}, \ldots, T^{n_{i+1}-1} M_{i+1}$ are pairwise disjoint and

$$\bigcup_{k=0}^{2n_{i+1}-1} T_i^k M_{i+1} = M_i \ (\text{mod } 0).$$

Suppose

$$\tau_{i+1}(x) = \inf\{k \geq 1 \colon T_i^k x \in M_{i+1}\}, \qquad x \in M_{i+1}.$$

Put

$$f_{i+1}(x) = \sum_{k=0}^{\tau_{i+1}(x)-1} f_i(T_i^k x), \qquad x \in M_{i+1}.$$

The flow $\{\bar{T}^t\}$ may be represented as a special flow constructed from the base automorphism $T_{i+1} = (T_i)_{M_{i+1}}$—induced from T_i on the set M_{i+1}—and the function $f_{i+1}(x)$. For a fixed $x \in M_{i+1}$ now put $l = f_{i+1}(x)$ and consider the semi-interval $L_x = \{(x, s) \colon 0 \leq s < l\}$. It has already some p-intervals and some q-intervals constructed on the previous ith step of our argument. These "old" intervals are located on the semi-intervals $[0, f_i(x))$, $[f_i(x), f_i(x) + f_i(T_i x))$, ... contained in the "new" semi-interval $[0, l)$. Between the right end point of one of the intervals and the left end point of the next one there may be a remainder not filled up by p- and q-intervals. Let us carry out a certain reconstruction to eliminate these remainders.

Suppose that during the previous steps we had defined the sequence $1 = \varepsilon_1 > \varepsilon_2 > \cdots > \varepsilon_i > 0$, $\varepsilon_i \leq 2^{-i}$. Consider the function $g(\varepsilon)$, $\varepsilon > 0$ which is equal to the greatest lower bound of the $t > 0$ such that for any $s \geq t$ there are positive integers k_1, k_2 satisfying $0 < s - k_1 p - k_2 q < \varepsilon$. Since p/q is irrational, it follows that $g(\varepsilon)$ is finite for all $\varepsilon > 0$.

Suppose $[z_1, z_1')$ is the first of the remainders. Clearly $z_1' = f_i(x)$. Choose ε_{i+1} so that

$$0 < \varepsilon_{i+1} < \min(2^{-i-1}, \varepsilon_i, p) \quad \text{and} \quad g(\varepsilon_{i+1}) \geq q.$$

If the number n_i was chosen sufficiently large during the ith step, then c_i (the lower bound of $f_i(x)$) will be much larger than $g(\varepsilon_{i+1})$. This allows us to carry out the following construction.

Suppose r_1 is the last of the left end points of the p- or q-intervals satisfying $r_1 \leq z_1' - g(\varepsilon_{i+1})$. (Clearly, $r_1 \geq c_i - q - g(\varepsilon_{i+1})$). Using the definition of the function $g(\varepsilon)$ we can now replace the "old" subdivision into p- and q-intervals of the semi-interval $[r_1, z_1')$ by a new one so as to have only one remainder between the last of the intervals and z_1', the length of the remainder being no greater than ε_{i+1}. This reconstruction will be called essential (to differentiate it from a small translation which shall be described below).

Now consider the next "block" of p- and q-intervals, on the semi-interval $[f_i(x), f_i(x) + f_i(T_i x))$. Suppose $[z_2, z_2')$ is the second remainder obtained during the ith step (this means that the semi-interval $[z_1', z_2)$ is entirely constituted by p- and q-intervals). Notice that $|z_2' - z_1'| = |f_i(T_i x)| \geq c_i$.

Translate the entire block of p- and q-intervals contained in $[z_1', z_2)$ to the left by the distance g_1, thus eliminating the empty space between it and the previous block; then we get a longer connected interval of p- and q-intervals. Suppose r_2 is the last of the left end points of the p- and q-intervals of this longer line satisfying $r_2 \leq z_2' - g(\varepsilon_{i+1})$. We may once again redefine the partition into intervals of the semi-interval $[r_2, z_1')$ obtaining a connected block of p- and q-intervals and a single remainder between the last of them and z_2', the length g_2 of the remainder being no greater than ε_{i+1}. Again translate the obtained block to the left by g_2 and continue this process until the connected line constructed will differ from l by a magnitude not exceeding $2^{-i}l$ but larger than $2^{-i-1}l$.

Most of the intervals constructed during the previous step were translated by a distance not exceeding ε_{i+1} and only a small part of them were essentially reconstructed. Therefore although the tolerance η_i of the approximation of the ratio ρ of appearance of p- and q-intervals has been increased, this increase was not too large. So as to "improve" this frequency again, we have saved the last empty space whose size is approximately $2^{-i-1}l$. At the same time, the number of p- and q-intervals on the interval of length 2^{-i-l} is large enough, which is guaranteed by the sufficiently large value of n_{i+1}. The number n_{i+1} must be large also in order that the replacement of one p- or q- interval by another does not change the relative frequency of p- and q-intervals by more than η_{i+1}.

Assume that we have already constructed a new partition of the semi-interval $[0, l)$ into p- and q-intervals and the remainder of length no greater than q for all $x \in M_{i+1}$ and $l = f_{i+1}(x)$. For simplicity, as before, denote the end points of the intervals by $0 = s_0 < s_1 < \cdots < s_{N+1} \leq s_{N+2} = l$. Suppose

$$B_{i+1}\{(x, s): x \in M_{i+1}, s_n \leq s \leq s_n + \alpha \text{ for some } n, 0 \leq n \leq N\}.$$

Then B_{i+1} is a subset of the space

$$\overline{M}_{i+1} = \{(x, s): x \in M_{i+1}, 0 \leq s < f_{i+1}(x)\}.$$

Taking into consideration the fact that the measure spaces $\overline{M}_i$ (for a distinct i) are naturally isomorphic, we may consider all the B_i as subsets of the space $\overline{M}_0 = \overline{M}$.

Since, during each step of the construction, the passage from the old partition into p- and q-intervals to the new one reduces to a small translation everywhere except a relatively small part of $[0, l)$, the sequence of sets B_i, viewed as subsets of $\overline{M}_0$, converges when $i \to \infty$ and the limit set $B = \bigcup_{n=1}^{\infty} \bigcap_{i=n}^{\infty} B_i$ (mod 0) has the following property: almost every trajectory remains in the set B during time α, then during time $p - \alpha$ or $q - \alpha$ remains in $\overline{M}_0 \backslash B$ then again remains in B for time α, etc. A set of zero measure consisting of the trajectories not possessing this property will be excluded from $\overline{M}_0$; this enables us to assume that this property holds everywhere. Put

$$M = \{(x, s) \in \overline{M}_0 \colon T^t(x, s) \in B \text{ for all rational } t,\ 0 < t < \alpha\};$$

$$f(x) = \inf\{t > 0 \colon T^t x \in M\}, \qquad x \in M.$$

By using Lemma 1 on measurability proved below, it is easy to check that the automorphism $\tilde{T}$ which acts on M according to the fomula $\tilde{T}x = T^{f(x)}x$ and the function $f(x)$ do indeed give the required special representation of the flow $\{T^t\}$.

Now let us point out how the reconstruction described above for the ith step may be carried out in a measurable way.

Suppose $\mathfrak{F} \subset \overline{\mathfrak{S}}$ is the σ-subalgebra of all measurable sets A of the space $\overline{M}_0 = \{(x, s) \colon 0 \le s < f_0(x)\}$, such that for any measurable function $h \colon M_0 \to \mathbb{R}^1$ the set $\{x \in M_0 \colon (x, h(x)) \in A\}$ is μ_0-measurable. Assume that we have already shown $B_i \in \mathfrak{F}$. Let us show how the inclusion $B_{i+1} \in \mathfrak{F}$ may be achieved.

The set B_{i+1} may be partitioned into a finite number of measurable subsets E such that the lengths of the semi-intervals $L_x = [0, l)$ and $L_{x'} = [0, l')$, where $l = f_{i+1}(x)$, $l' = f_{i+1}(x')$ are near to each other for all $x, x' \in E$[1] and moreover, the partition of the semi-intervals $[0, l)$ and $[0, l')$ on the ith step is characterized by the same sequence of the form $(p, p, q, q, q, r, p, q, \ldots)$ showing the order of occurrence of the p-intervals, q-intervals and remainders (the remainders are coded by means of the letter r). The lengths of the remainders corresponding to each other on L_x and $L_{x'}$ do not necessarily coincide, but are also near to one another. If the reconstruction of the partition of the intervals L_x at the $(i + 1)$th step is carried out in the same way for all $x \in E$ then, using Lemma 1 proved below, we obtain $B_{i+1} \in \mathfrak{F}$. Similar arguments yield $B_1 \in \mathfrak{F}$, therefore the relation $B_i \in \mathfrak{F}$ has been proved by induction.

Lemma 1. *The σ-algebra $\mathfrak{F} \subset \overline{\mathfrak{S}}$ has the following properties*:

(1) *$\mathfrak{F}$ is invariant with respect to $\{T^t\}$*;
(2) *if $\mathfrak{S}_0$ is the σ-algebra of μ_0-measurable sets of the space M_0, $\mathbb{B}$ is the σ-algebra of Borel sets in $\mathbb{R}^1_+$ then $\mathfrak{F} \supseteq \mathfrak{S}_0 \times \mathbb{B}$*;
(3) *if $A \in \mathfrak{F}$ and for any $\overline{x} \in \overline{M}$ the sets*

$$\{t \in \mathbb{R}^1 \colon T^t\overline{x} \in A\} \quad \textit{and} \quad \{t \in \mathbb{R}^1 \colon T^t\overline{x} \notin A\}$$

[1] The exact meaning of this statement will become clear after we define all the constants required for the construction; this will be done below.

are the union of segments, intervals, and semi-intervals whose lengths are bounded from below, then

$$\bigcup_{t \in (a, b)} T^t A \in \mathfrak{F}, \qquad \bigcap_{t \in (a, b)} T^t A \in \mathfrak{F}$$

(here (a, b) *denotes an arbitrary segment, interval or semi-interval);*

(4) *if* A *is the same as in* (3), *then* $g(x) = \inf\{t > 0: T^t x \in A\}$ *is a measurable function on* M_0.

Proof. (1) Let $A \in \mathfrak{F}$. Represent $T^t A$ in the form $T^t A = \bigcup_{n=0}^{\infty} A_n$ where

$$A_n = T^t\left(\left\{(x, s) \in A: \sum_{k=0}^{n-1} f_0(T_0^k x) \le s + t < \sum_{k=0}^{n} f_0(T_0^k x)\right\}\right).$$

Denote $\tilde{A}_n = T^{-t} A_n$; let us prove that $\tilde{A}_n \in \mathfrak{F}$. Choose a measurable function $g: M_0 \to \mathbb{R}^1$. Then

$$\{x: (x, g(x)) \in \tilde{A}_n\} = \{x: (x, g(x)) \in A\}$$
$$\cap \left\{x: \sum_{k=0}^{n-1} f_0(T_0^k x) \le g(x) + t < \sum_{k=0}^{n} f_0(T_0^k x)\right\}.$$

Since both sets in the right-hand side are μ_0-measurable, we have $\tilde{A}_n \in \mathfrak{F}$. Recalling the definition of special flows, we get

$$A_n = \left\{\left(T_0^{n-1} x, s + t - \sum_{k=0}^{n-1} f_0(T_0^k x)\right): (x, s) \in \tilde{A}_n\right\}$$
$$= \left\{(x, s): \left(T_0^{1-n} x, s + \sum_{k=0}^{n-1} f_0(T_0^{k+n-1} x) - t\right) \in \tilde{A}_n\right\}.$$

Hence the set

$$\{x: (x, g(x)) \in A_n\}$$
$$= T_0^{1-n}\left(\left\{x: \left(x, \left[g(T_0^{n-1} x) + \sum_{k=0}^{n-1} f_0(T_0^{k+2n-2} x) - t\right]\right) \in \tilde{A}_n\right\}\right)$$

is μ_0-measurable so that

$$T^t A = \bigcup_{n=0}^{\infty} A_n \in \mathfrak{F}$$

(2) Suppose $E \in \mathfrak{S}_0$ and (a, b) is an interval. Then

$$\{x\colon (x, g(x)) \in E \times (a, b)\} = E \cap \{x\colon a < g(x) < b, 0 < g(x) < f_0(x)\}.$$

It follows that $\mathfrak{F}$ contains the σ-algebra generated by sets of the form $E \times (a, b)$, i.e., $\mathfrak{F} \supset \mathfrak{S}_0 \times \mathbb{B}$.

(3) Suppose (a, b) is an interval. Then

$$\bigcup_{t \in (a, b)} T^t A = \bigcup_{\substack{t \in (a, b)\\ t \text{ rational}}} T^t A.$$

A similar relation is valid for intersections. In the case of a segment or a semi-interval, we must add summands corresponding to end points.

(4) Suppose A is as in (3), and α is the lower bound of lengths of segments of the form

$$\{t \in \mathbb{R}^1\colon T^t\bar{x} \in A\}, \qquad \{t \in \mathbb{R}^1\colon T^t\bar{x} \notin A\},$$

then

$$\{x\colon g(x) < c\} = \bigcup_{n=0}^{\infty} [T^{-t_n}A \cap M],$$

where $t_n \to c$, $t_{n+1} - t_n < \alpha/2$, $0 = t_0 < t_1 < t_2 < \cdots$. The lemma is proved. □

Lemma 2. *Suppose the interval of length $l \geq 2(p + q)$ is subdivided in two ways into N_p (respectively N'_p) p-intervals and N_q (respectively N'_q) q-intervals and also into a remainder which in both cases is no greater than q. Then*

$$\Delta \overset{\text{def}}{=} |N_p(N_p + N_q)^{-1} - N'_p(N'_p + N'_q)^{-1}| \leq 2ql^{-1}(|N_p - N'_p| + |N_q - N'_q|). \tag{1}$$

Proof. We have

$$\begin{aligned}\Delta &= |(N_pN'_q - N'_pN_q)(N_p + N_q)^{-1}(N'_p + N'_q)^{-1}|\\ &\leq (N_p|N_q - N'_q| + N_q|N_p - N'_p|)(N_p + N_q)^{-1}(N'_p + N'_q)^{-1}\\ &\leq |N_q - N'_q|(N'_p + N'_q)^{-1} + |N_p - N'_p|(N'_p + N'_q)^{-1}.\end{aligned}$$

Taking into consideration $q(N'_p + N'_q) \geq pN'_p + qN'_q \geq \frac{1}{2}l$ we get (1). The lemma is proved. □

Let us consider separately the particular case of Lemma 2 when the second partition is obtained from the first by replacing one q-interval by several p-intervals so that the remainder is no greater than q, or one p-interval by a q-interval with the same condition. Put $w = [q/p] + 1$. Then in our particular case

$$\Delta \leq 4wql^{-1}. \tag{2}$$

Lemma 3. *Suppose we are given the numbers* γ, ρ, δ: $\gamma > 0$, $0 < \rho < 1$, $0 < \delta < 1$. *Suppose further that the semi-interval* $[0, l)$, $l \geq 2(p + q)$ *is subdivided into* N_p *p-intervals,* N_q *q-intervals and a remainder of length no greater than q. Assume that the semi-interval* $[l - \delta l, l]$ *contains at least* $[\gamma\delta l] + 1$ *p-intervals and* $[\gamma\delta l] + 1$ *q-intervals. If* $\eta_0 = \frac{1}{2}\rho(1 - \rho)$ *and*

$$|N_p(N_p + N_q)^{-1} - \rho| \leq \min(\eta_0, \eta_0\gamma\delta p), \tag{3}$$

then there exists another partition of $[0, l)$ *into* N'_p *(respectively* N'_q*) p- (respectively q-) intervals and a remainder of length no greater than q which differs from the original one only on* $[l - \delta l, l)$, *and which satisfies*

$$|N'_p(N'_p + N'_q)^{-1} - \rho| \leq 4wql^{-1}. \tag{4}$$

Proof. By (2), the replacement of only one p- or q-interval brings about a change of the relative frequency of no more than $4wql^{-1}$. Assume that the frequency of p-intervals was too large, i.e., $N_p(N_p + N_q)^{-1} > \rho$ and that $([\gamma\delta l] + 1)$ p-intervals on the semi-intervals $[l - \delta l, l)$ were replaced by the corresponding number of q-intervals. Then the new frequencies N^*_p, N^*_q satisfy the inequalities $N_p \geq N^*_p$, $N_q \leq N^*_q$ and

$$\begin{aligned} N_p(N_p + N_q)^{-1} - N^*_p(N^*_p + N^*_q)^{-1} &\geq N_p(N_p + N_q)^{-1} - N^*_p(N^*_p + N_q)^{-1} \\ &= N_q(N_p + N_q)^{-1}(N_p - N^*_p) \\ &\quad \times (N^*_p - N_q)^{-1} \\ &\geq \eta_0([\gamma\delta l] + 1)(N^*_p + N_q)^{-1}. \end{aligned}$$

The last inequality follows from the fact that $N_p(N_p + N_q)^{-1} \geq \rho - \eta_0 \geq \eta_0$. Further

$$\eta_0([\gamma\delta l] + 1)(N^*_p + N_q)^{-1} \geq \eta_0\gamma\delta l(N_p + N_q)^{-1} \geq \eta_0\gamma\delta p,$$

since $p(N_p + N_q) \leq l$. By (3), the new relative frequency satisfies $N^*_p(N^*_p + N^*_q)^{-1} \leq \rho$ and, by taking into consideration (2), we can replace the p-intervals successively one after the other, until inequality (4) becomes valid. The lemma is proved. □

Lemma 4. *If the semi-interval* $[0, l)$ *is subdivided in two ways, as in Lemma 2, and these two partitions differ only on a finite number of segments of total length no greater than* l_0, *then* $\Delta \leq 4wl_0 l^{-1}$.

Proof. Applying Lemma 1 and taking into consideration

$$|N_p - N'_p| \leq l_0 p^{-1}, \qquad |N_q - N'_q| \leq l_0 q^{-1},$$

we get

$$\Delta \leq 2ql^{-1}l_0(p^{-1} + q^{-1}) \leq 2ql^{-1}l_0(wq^{-1} + q^{-1}) \leq 4wl_0 l^{-1}.$$

The lemma is proved. □

Now we can define the parameters of the construction by induction. Recall that we started out with $0 < c_0 < f_0(x) < C_0$, $p < q$, $(w-1)p < q < wp$, $\eta_0 = \frac{1}{2}\rho(1-\rho)$. Choose ε, $0 \leq \varepsilon < 1$, so small that $g(\varepsilon) > q$, $\varepsilon < p$, and put

$$\begin{aligned} \varepsilon_i &= 2^{-i}\varepsilon \qquad (i \geq 1) \\ \gamma &= \tfrac{1}{12}\eta_0 q^{-1} \\ \eta_i &= \min(\tfrac{1}{12}\eta_0, 2^{-i-2}\eta_0\gamma p) \qquad (i \geq 1) \end{aligned} \tag{5}$$

Now choose natural numbers u_1, u_2 so that $r = u_1/(u_1 + u_2)$ implies $|r - \rho| < 2^{-3}\eta_1$. We shall say that the semi-interval $[0, \lambda_0)$ is partitioned quasi-r-periodically, if in this partition we first have u_1 p-intervals, then u_2 q-intervals, then again u_1 p-intervals, u_2 q-intervals, etc., until the right end point of the "block" thus defined becomes larger than $\lambda_0 - u_1 p - u_2 q$. After this point the partition may be arbitrary, as long as the remainder is not greater than q.

For sufficiently large J, any semi-interval $I = [0, \lambda_0)$ of length $\lambda_0 \geq J(u_1 p + u_2 q)$ has the following properties: the numbers N_p (respectively N_q) of p-intervals (respectively q-intervals) in any quasi-r-periodic partition of the semi-interval I satisfy the inequality

$$|N_p(N_p + N_q)^{-1} - \rho| < \eta_1. \tag{6}$$

Now fix $J \geq 2$ so that (6) holds and choose n_1 to meet the following conditions

$$c_1 = n_1 c_0 \geq J(u_1 p + u_2 q) \geq 2(p + q);$$

$$c_1 = n_1 c_0 > 4w[g(\varepsilon_2) + 2p + 2q]\eta_1^{-1}.$$

The number $C_1 = 2n_1 C_0$ will be the upper bound for the function $f_1(x)$. For the $n_1, n_2, \ldots, n_i$ that we have chosen, the number $c_i = n_1 \cdot n_2 \cdot \ldots \cdot n_i c_0$ is a lower bound for $f_i(x)$, the number $C_i = 2n_1 \cdot n_2 \cdot \ldots \cdot n_i C_0$—an upper bound for $f_i(x)$. Now we can choose the number n_{i+1} from the conditions

$$c_{i+1} > 2^{i+1} \cdot 3C_i, \tag{7}$$

$$c_{i+1} > 4w[g(\varepsilon_{i+2}) + q]\eta_{i+1}^{-1}, \tag{8}$$

$$c_{i+1} \geq 3 \cdot 2^{i+3} q \cdot \eta_0^{-1}. \tag{9}$$

These conditions together with (5) determine all the constants required to the construction.

Note that on the first step the semi-interval $[0, l)$ for $l = f_1(x) \geq c_1$ was subdivided quasi-r-periodically. Hence these partitions satisfy (6).

Now let us show that during the ith step ($i = 1, 2, \ldots$) the partition of the semi-interval $[0, f_i(x))$ into N_p p-intervals, N_q q-intervals and remainder not greater than q may be carried out so as to have

$$|N_p(N_p + N_q)^{-1} - \rho| < \eta_i \tag{10}$$

Since we have already proved this for $i = 1$, let us assume the validity of our statement for some i and consider the interval $[0, f_{i+1}(x))$, $x \in M_{i+1}$. Estimate the length l_0 of the part of the semi-interval on which the essential reconstructions were carried out. On each of the semi-intervals $[0, f_i(x))$, $[f_i(x), f_i(x) + f_i(T_i x)), \ldots$, whose lengths are greater than or equal to c_i, segments of length $\leq g(\varepsilon_{i+1}) + q$ were essentially reconstructed. Hence, if we denote as before $l = f_{i+1}(x)$, it follows that

$$\frac{l_0}{l} \leq \frac{g(\varepsilon_{i+1}) + q}{c_i}.$$

By Lemma 3, the new frequency $N'_p(N'_p + N'_q)^{-1}$ differs from the old one by $\Delta \leq 4w[g(\varepsilon_{i+1}) + q]c_i^{-1} \leq \eta_i$.

Taking into consideration (8), we get

$$|N'_p(N'_p + N'_q)^{-1} - \rho| < 2\eta_i = \min(\eta_0, 2^{-i-1}\eta_0 \gamma p).$$

Put $\delta = 2^{-i-1}$. By (7), at least a third of the semi-interval $[l - \delta l, l]$ consists entirely of semi-intervals of the form

$$\left[\sum_{k=0}^{n} f_i(T_i^k x), \sum_{k=0}^{n+1} f_i(T_i^k x)\right). \tag{11}$$

Denote by M_p (respectively M_q) the number of p- (respectively q-) intervals contained in the union of such semi-intervals. It follows from the induction hypothesis (10) that

$$M_p(M_p + M_q)^{-1} \geq \eta_0 \tag{12}$$

Since the total length of semi-intervals of the form (11) is no less than $\frac{1}{3} \cdot 2^{-i-1} l$ and the remainder of the partition on each such semi-interval is no greater than half its length, we have

$$(M_p + M_q)q \geq \tfrac{1}{3} 2^{-i-2} l.$$

Now (9) and (12) imply the inequality

$$M_p \geq \eta_0 \cdot \tfrac{1}{3} 2^{-i-2} l q^{-1} = 2\gamma\delta l \geq [\gamma\delta l] + 1.$$

Similarly it can be shown that $M_q \geq [\gamma\delta l] + 1$. Now we can apply Lemma 3 and pass to the new partition containing N_p'' (respectively N_q'') p- (respectively q-) intervals and differs from the previous one only on $[l - \delta l, l)$; for the new partition (by (4) and (9)), we have the estimate

$$|N_p''(N_p'' + N_p'')^{-1} - \rho| \leq 4wql^{-1} \leq 4wqc_{i+1}^{-1} < \eta_{i+1}.$$

This means that inequality (10) is valid at the $(i + 1)$th step; it is therefore proved for all $i \geq 1$.

It remains to show the convergence of the sets B_i.

Suppose D_i' consists of all points $(x, s) \in \overline{M}_{i+1}$ such that for some $n < \tau_{i+1}(x)$ we have

$$\sum_{k=0}^{n} f_i(T_i^k x) - g(\varepsilon_{i+1}) - 2q \leq s \leq \sum_{k=0}^{n} f_i(T_i^k x) + q.$$

Suppose further D_i'' consists of all $(x, s) \in \overline{M}_{i+1}$ such that either $0 \leq s \leq q$ or $s \geq f_{i+1}(x) - 2^{-i} f_{i+1}(x) - q$. If $(x, s) \notin D_i \overset{\text{dif}}{=} D_i' \cup D_i''$, then the set $\{T^t(x, s): |t| < q\}$ consists only of points on which the $(i + 1)$th partition was obtained from the ith one by a translation of distance no greater than ε_{i+1}.

Considering the fibers $\{(x, s): 0 \leq s < f_{i+1}(x)\}$ for a fixed x and applying Fubini's theorem, we get

$$\bar{\mu}(D_i') \leq [g(\varepsilon_{i+1}) + 3q]/c_i.$$

It follows from (7) that $c_i \geq 2^i c_0$, while (8) implies

$$[g(\varepsilon_{i+1}) + q]/c_i < \tfrac{1}{4} w^{-1} \eta_i.$$

Since $\sum_{i=1}^{\infty} \eta_i < \infty$, we obtain

$$\sum_{i=1}^{\infty} \bar{\mu}(D_i') < \infty. \tag{13}$$

Similarly from $\bar{\mu}(D_i'') \leq 2qc_{i+1}^{-1} + 2^{-i}$ we get

$$\sum_{i=1}^{\infty} \bar{\mu}(D_i'') < \infty. \tag{14}$$

Moreover, (5) implies

$$\sum_{i=1}^{\infty} \varepsilon_i < \infty. \tag{15}$$

Now put $B = \bigcup_{n=1}^{\infty} \bigcap_{i=n}^{\infty} B_i$.

It follows from (13), (14), and (15) that almost every trajectory of the flow $\{T^t\}$ possesses the following property: it remains for time α in the set B, then for time $p - \alpha$ or $q - \alpha$ in $\overline{M} \setminus B$, then again for time α in B, etc. Now we may conclude the proof of the theorem by standard arguments.

Suppose $M = \{\bar{x} \in \overline{M}_0 \colon T^t\bar{x} \in B \text{ for } 0 < t < \alpha\}$. The set M will be the base of the required special representation. Now let

$$f(x) = \inf\{t > 0 \colon T^t x \in M\}, \qquad x \in M,$$

and suppose the transformation $\tilde{T} \colon M \to M$ acts according to the formula $\tilde{T}x = T^{f(x)}x$. It follows from Lemma 1 that the partition of the space $\overline{M}$ into segments of trajectories between two successive appearances in the set M is regular. Therefore, as was shown in §3 (statement (B)), this partition induces a special representation of the flow $\{T^t\}$ constructed from the base automorphism $\tilde{T}$ and the function $f(x)$. Using inequality (10), it is easy to show that $f(x)$ assumes (mod 0) only the values p, q and

$$\mu(\{x \in M \colon f(x) = p\} = \tilde{\rho} \cdot \mu(\{x \in M \colon f(x) = q\}),$$

where μ is the invariant measure of the automorphism $\tilde{T}$. The theorem is proved. □

Part III

Spectral Theory of Dynamical Systems

Chapter 12

Dynamical Systems with Pure Point Spectrum

In this chapter we study an important class of dynamical systems—dynamical systems with pure point spectrum. Concerning the notions of the spectral theory of unitary operators used here see Appendix 2.

§1. General Properties of Eigen-values and Eigen-functions of Dynamical Systems

The spectra of unitary operators and one-parameter groups of unitary operators corresponding to dynamical systems possess a series of specific properties.

First consider an automorphism T of a measure space $(M, \mathfrak{S}, \mu)$. By $\Lambda_d(T)$ denote the set of eigenvalues of the unitary operator U_T adjoint to T. It is clear that $\Lambda_d(T)$ is a subset of the unit circle S^1 of the complex plane.

Theorem 1. *Suppose T is ergodic. Then*

(1) *the set $\Lambda_d(T)$ is a subgroup of the group S^1;*
(2) *every eigen-value is of multiplicity 1;*
(3) *the absolute value of every eigen-function is constant almost everywhere.*

Proof. We shall prove the statements of the theorem in converse order, starting from the last one. Suppose $\lambda \in \Lambda_d(T)$ and assume that the function $f_\lambda(x) \in L^2(M, \mathfrak{S}, \mu)$ satisfies

$$U_T f_\lambda(x) = f_\lambda(Tx) = \lambda \cdot f_\lambda(x)$$

almost everywhere and $f \neq 0 (\text{mod } 0)$. Then

$$U_T |f_\lambda(x)| = |f_\lambda(Tx)| = |\lambda| |f_\lambda(x)| = |f_\lambda(x)|$$

almost everywhere, i.e., $|f_\lambda(x)|$ is an invariant (mod 0) function. Since T is ergodic, we have $|f_\lambda(x)| = \text{const}$ almost everywhere.

It follows from the statement just proved above that

$$\frac{1}{f_\lambda(x)} \in L^2(M, \mathfrak{S}, \mu).$$

If two eigenfunctions $f_\lambda^1(x)$, $f_\lambda^2(x)$ correspond to the eigen-value $\lambda \in \Lambda_d(T)$, then the function $g(x) = f_\lambda^{(1)}(x) \cdot (f_\lambda^{(2)}(x))^{-1}$ has an absolute value which is constant almost everywhere and

$$g(Tx) = f_\lambda^{(1)}(Tx)(f_\lambda^{(2)}(Tx))^{-1} = \lambda f_\lambda^{(1)}(x) \cdot (\lambda f_\lambda^{(2)}(x))^{-1}$$

almost everywhere. It follows from the ergodicity of T that $g(x) = \text{const}$ almost everywhere, i.e.,

$$f^{(2)}(x) = \text{const} \cdot f^{(1)}(x)$$

almost everywhere.

Finally, if $\lambda_1, \lambda_2 \in \Lambda_d(T)$, $f_{\lambda_1}(x)$, $f_{\lambda_2}(x)$ are the corresponding eigenfunctions, then

$$U_T(f_{\lambda_1} f_{\lambda_2}) = U_T f_{\lambda_1} \cdot U_T f_{\lambda_2} = \lambda_1 \cdot \lambda_2 \cdot f_{\lambda_1} \cdot f_{\lambda_2},$$

$$U_T(f_{\lambda_1}/f_{\lambda_2}) = U_T(f_{\lambda_1})/U_T(f_{\lambda_2}) = \lambda_1 \lambda_2^{-1} \cdot f_{\lambda_1}/f_{\lambda_2},$$

i.e., $\lambda_1 \cdot \lambda_2 \in \Lambda_d(T)$, $\lambda_1 \lambda_2^{-1} \in \Lambda_d(T)$. The theorem is proved. □

It is useful at this point already to stress the basic reason which determines the specific properties of the spectrum described above. It consists in the following remarkable property of the operator U_T: if the functions f, $g \in L^2(M, \mathfrak{S}, \mu)$ are such that their product fg belongs to $L^2(M, \mathfrak{S}, \mu)$, then $U_T(fg) = U_T f \cdot U_T g$. In other words, if we take into consideration the fact that the Hilbert space $L^2(M, \mathfrak{S}, \mu)$ possesses an additional structure of partial multiplication, then the operator U_T preserves this structure, i.e., is an automorphism of this ring-like structure, often called a unitary ring.

Now consider an ergodic flow $\{T^t\}$ on the space $(M, \mathfrak{S}, \mu)$. By $\{U^t\}$ denote the adjoint one-parameter group of unitary operators, by $\Lambda_d(\{T^t\})$ the set of eigen-values of this group. We begin by proving the following lemma.

Lemma 1. *If $t_0 \in \mathbb{R}^1$ satisfies $e^{i\lambda t_0} \neq 1$ for all $\lambda \in \Lambda_d(\{T^t\})$, $\lambda \neq 0$, then the automorphism T^{t_0} is ergodic*

Proof. Represent $L^2(M, \mathfrak{S}, \mu)$ in the form

$$L^2(M, \mathfrak{S}, \mu) = H_d \oplus H_c,$$

where H_d is the subspace generated by the eigen-functions of the group $\{U^t\}$ and H_c is the subspace of the continuous spectrum. Then if $f \neq 0$ is an invariant function for U^{t_0}, i.e., $U^{t_0}f = f$, and $f = f_d + f_c$, where $f_d \in H_d$, $f_c \in H_c$, then $U^{t_0} f_d = f_d$, $U^{t_0} f_c = f_c$. The last relation implies $f_c = 0$. Now choose an orthogonal basis $\{f_\lambda\}$, $\lambda \in \Lambda_d(\{U^t\})$ in the subspace H_d of eigen-functions of the group $\{U^t\}$ and express the functions f_d, $U^{t_0}f_d$ in terms of this basis:

$$f_d = \sum_{\lambda \in \Lambda_d} c_\lambda f_\lambda, \qquad U^{t_0} f_d = \sum_{\lambda \in \Lambda_d} e^{it_0\lambda} c_\lambda f_\lambda.$$

The relation $U^{t_0} f_d = f_d$ implies $e^{it_0\lambda} c_\lambda = c_\lambda$ for all $\lambda \in \Lambda_d$. By the assumption of the lemma, $c_\lambda = 0$ for $\lambda \neq 0$, i.e., $f = \text{const}(\text{mod } 0)$. The lemma is proved. □

For the flow $\{T^t\}$ on the measure space $(M, \mathfrak{S}, \mu)$, let $\Lambda_d(\{T^t\})$ denote the set of eigen-values of the adjoint one parameter group $\{U^t\}$ of unitary operators.

Theorem 2. *Let $\{T^t\}$ be an ergodic flow; then*

(1) *the set $\Lambda_d(\{T^t\})$ is a subgroup of the additive group of real numbers;*
(2) *every eigen-value is of multiplicity* 1;
(3) *the absolute value of each eigen-function is constant almost everywhere.*

Proof. According to Lemma 1, we can find a t_0 such that the automorphism T^{t_0} is ergodic. Since the eigenfunctions of the flow $\{T^t\}$ are eigen-functions of the automorphism T^{t_0}, statement (3) follows from statement (3) of Theorem 1. The other two statements are proved just as in Theorem 1. Our theorem is proved. □

Suppose T^{t_0} is an ergodic automorphism of the flow $\{T^t\}$. Then the map sending each $\lambda \in \Lambda_d(\{T^t\})$ into $e^{it_0\lambda} \in \Lambda_d(T^{t_0})$ determines an isomorphism of the abelian groups $\Lambda_d(\{T^t\})$ and $\Lambda_d(T^{t_0})$. Let us prove the following consequence of this fact.

Corollary. *If for an ergodic automorphism T, we have $e^{2\pi ip/q} \in \Lambda_d(T)$ for some integer $q > 1$, p, q being relatively prime, the automorphism T cannot be included in a flow, i.e., there does not exist an ergodic flow $\{T^t\}$ for which T is metrically isomorphic to T^{t_0} for some $t_0 \in \mathbb{R}^1$.*

Proof. If such a flow exists then for some $\lambda \in \Lambda_d(\{T^t\})$, $\lambda \neq 0$, we would have $e^{it_0\lambda} = e^{2\pi ip/q}$. But $q\lambda \in \Lambda_d(\{T^t\})$, since $\Lambda_d(\{T^t\})$ is a group, and $e^{it_0 q\lambda} = 1$, which contradicts the fact that $\Lambda_d(\{T^t\})$ and $\Lambda_d(T)$ are isomorphic. The corollary is proved. □

§2. Dynamical Systems with Pure Point Spectrum. The Case of Discrete Time

Definition 1. The ergodic automorphism T is said to be an *automorphism with pure point spectrum* if the unitary operator U_T has a complete (in $L^2(M, \mathfrak{S}, \mu)$ system of orthogonal eigen-functions.

By Theorem 1, §1, the set $\Lambda_d(T)$ is a subgroup of the group S^1.

Theorem 1. *Suppose Λ is an arbitrary countable subgroup of the group S^1. Then there exists an automorphism T with pure point spectrum such that $\Lambda = \Lambda_d(T)$.*

The proof is based on certain facts from the theory of characters of commutative groups (see Pontrjagin [1]).

Consider the character group M of the subgroup Λ. Using the Pontrjagin duality theorem, we see that M is a commutative compact group. By μ denote the normalized Haar measure on M. Consider the identity map (inclusion) $g: \Lambda \to S^1$. Then g is a homomorphism of Λ into S^1 and is therefore a character of Λ, i.e., a point of the space M. By T_g denote the group translation on M, i.e., $T_g x = x + g$, $x \in M$. Let us show that T_g is an automorphism with pure point spectrum of the space M with invariant measure μ. Once again, by using the duality theorem, we see that the character group of the group M is isomorphic to Λ. For $\lambda \in \Lambda$ denote by χ_λ the character corresponding to λ. Then χ_λ is a function on M, $|\chi_\lambda| = 1$, and the family of functions χ_λ, $\lambda \in \Lambda$, constitutes an orthonormed basis in $L^2(M, \mu)$. We now have

$$U_T \chi_\lambda(x) = \chi_\lambda(Tx) = \chi_\lambda(x + g) = \chi_\lambda(x) \cdot \chi_\lambda(g) = \lambda \chi_\lambda(x),$$

since $\chi_\lambda(g) = \lambda$ (from the definition of g). The theorem is proved. □

The main result of the theory of automorphisms with pure point spectrum is the following theorem.

Theorem 2. *Suppose T_1, T_2 are ergodic automorphisms of the Lebesgue spaces $(M_1, \mathfrak{S}_1, \mu_1)$, $(M_2, \mathfrak{S}_2, \mu_2)$ with pure point spectrum, satisfying $\Lambda_d(T_1) = \Lambda_d(T_2)$. Then T_1 and T_2 are metrically isomorphic.*

Proof. Suppose

$$H_1 = L^2(M_1, \mathfrak{S}_1, \mu_1), \qquad H_2 = L^2(M_2, \mathfrak{S}_2, \mu_2).$$

The relation $\Lambda_d(T_1) = \Lambda_d(T_2)$ implies that for the unitary operators adjoint to U_1, U_2 there exists an isomorphism $\Phi: H_1 \to H_2$ of Hilbert spaces H_1, H_2, such that

$$U_2 \Phi = \Phi U_1, \qquad U_2^{-1} \Phi = \Phi U_1^{-1}.$$

This statement is valid for any pair of spectrally equivalent automorphisms. In the case of automorphisms with pure point spectrum we will show that such an isomorphism Φ may be generated by an isomorphism φ of the measure spaces $(M_1, \mathfrak{S}_1, \mu_1)$ and $(M_2, \mathfrak{S}_2, \mu_2)$, i.e., Φ acts according to the formula

$$\Phi(f(x)) = f(\varphi(x)), \qquad f \in L^2(M_1, \mathfrak{S}_1, \mu_1).$$

Suppose $f_\lambda^{(1)}, f_\lambda^{(2)}$ are normed eigen-functions for U_1, U_2 respectively and

$$\lambda \in \Lambda \overset{\text{def}}{=} \Lambda_d(T_1) = \Lambda_d(T_2).$$

Then $f_{\lambda_1}^{(1)} \cdot f_{\lambda_2}^{(1)}$ is an eigen-function of the operator U_1 with eigen-value $\lambda_1 \cdot \lambda_2$ (see Theorem 1, §1) so that

$$f_{\lambda_1}^{(1)} \cdot f_{\lambda_2}^{(1)} = a_{\lambda_1, \lambda_2}^{(1)} \cdot f_{\lambda_1 \cdot \lambda_2}^{(1)}, \qquad |a_{\lambda_1, \lambda_2}^{(1)}| = 1.$$

Thus any basis $\{f_\lambda^{(1)}, \lambda \in \Lambda\}$ gives rise to a function $a_{\lambda_1, \lambda_2}^{(1)}$ in two variables defined on $\Lambda \times \Lambda$. In a similar way, the function $a_{\lambda_1, \lambda_2}^{(2)}$ in two variables arises for the basis $\{f_\lambda^{(2)}\}$.

The functions $a_{\lambda_1, \lambda_2}^{(1)}, a_{\lambda_1, \lambda_2}^{(2)}$ possess the following properties:

(i) $|a_{\lambda_1, \lambda_2}^{(1)}| = |a_{\lambda_1, \lambda_2}^{(2)}| = 1;\ a_{1,1}^{(1)} = a_{1,1}^{(2)} = 1;$
(ii) $a_{\lambda_1, \lambda_2}^{(1)} = a_{\lambda_2, \lambda_1}^{(1)};\ a_{\lambda_1, \lambda_2}^{(2)} = a_{\lambda_2, \lambda_1}^{(2)};$
(iii) $a_{\lambda_1, \lambda_2}^{(1)} \cdot a_{\lambda_1 \lambda_2, \lambda_3}^{(1)} = a_{\lambda_1, \lambda_2 \lambda_3}^{(1)} \cdot a_{\lambda_2, \lambda_3}^{(1)};\ a_{\lambda_1, \lambda_2}^{(2)} \cdot a_{\lambda_1 \lambda_2, \lambda_3}^{(2)} = a_{\lambda_1, \lambda_2 \lambda_3}^{(2)} \cdot a_{\lambda_2, \lambda_3}^{(2)}$

The second property follows from commutativity and the third from associativity (of the multiplication).

Lemma 1. *Any function* a_{λ_1, λ_2}, $\lambda_1, \lambda_2 \in \Lambda$, *possessing properties* (i)–(iii), *is of the form*

$$a_{\lambda_1, \lambda_2} = d_{\lambda_1 \cdot \lambda_2} \cdot d_{\lambda_1}^{-1} \cdot d_{\lambda_2}^{-1}, \tag{1}$$

for some d_λ, $\lambda \in \Lambda$, $|d_\lambda| = 1$.

Proof. 1. The statement of the lemma may be reformulated as a purely algebraic statement on the triviality of a certain cohomology group. We shall give a direct proof by induction. Let us order the family of eigen-values in some way: $\lambda_0 = 1, \lambda_1, \lambda_2, \lambda_3, \ldots$. Denote by $\Lambda^{(r)}$ the subgroup of the group Λ generated by all λ_i, $0 \le i < r$. Put $d_1 = 1$ and assume that for a given r we have already constructed numbers d_λ, $\lambda \in \Lambda^{(r)}$ such that for all $\lambda_1, \lambda_2 \in \Lambda^{(r)}$ we have (1).

2. When we pass from r to $r+1$, we may assume that $\lambda = \lambda_{r+1} \notin \Lambda^{(r)}$: in the converse case $\Lambda^{(r+1)} = \Lambda^{(r)}$. In order to define d_λ, consider two cases:

(a) $\lambda^p \notin \Lambda^{(r)}$ for all integers $p \neq 0$; in this case we put $d_\lambda = 1$;
(b) $\lambda^p \in \Lambda^{(r)}$ for some integer $p \neq 0$; in this case let $h = \min\{p > 1: \lambda^p \in \Lambda^{(r)}\}$.

By the induction hypothesis d_{λ^h} is defined. It follows from (1) that for any p we must have the relation

$$a_{\lambda^p,\lambda} = d_{\lambda^{p+1}} \cdot d_{\lambda^p}^{-1} \cdot d_\lambda^{-1}. \tag{2}$$

Multiplying all these relations, with p varying from 0 to $h-1$, we obtain the equation for the definition of d_λ:

$$\prod_{p=0}^{h-1} a_{\lambda^p,\lambda} = d_{\lambda^h} \cdot (d_\lambda)^{-h}, \qquad d_\lambda^h = d_{\lambda^h}\left(\prod_{p=0}^{h-1} a_{\lambda^p,\lambda}\right)^{-1}.$$

In the role of d_λ take any of the h roots of this equation.

Then (in both cases (a) and (b)), use relation (2) in order to define d_{λ^p}, $0 \leq p < \infty$, consecutively

$$d_{\lambda^p} = (d_\lambda)^p \cdot \prod_{s=0}^{p-1} a_{\lambda^s,\lambda}. \tag{3}$$

For negative p define d_{λ^p} by the relation

$$a_{\lambda^p,\lambda^{-p}} = d_{\lambda^p} \cdot d_{\lambda^{-p}}.$$

3. Let us prove the relation

$$d_{\lambda^{p+q}} = a_{\lambda^p,\lambda^q} \cdot d_{\lambda^p} \cdot d_{\lambda^q}. \tag{4}$$

For the sake of simplicity we only consider the case $p \geq 0$; for $p < 0$ the argument is similar. Carry out an induction over p. First by putting $\lambda_2 = \lambda_3 = 1$, in condition (iii), we see that $a_{\lambda_1,1}^2 = a_{\lambda_1,1} \cdot 1$, i.e., $a_{\lambda,1} = 1$. This implies relation (4) for $p = 0$. Suppose (4) has been proved for some p. Using the expression (3) for d_{λ^p} we get

$$\prod_{s=0}^{p+q-1} a_{\lambda^s,\lambda} = a_{\lambda^p,\lambda^q} \cdot \prod_{s=0}^{p-1} a_{\lambda^s,\lambda} \cdot \prod_{s=0}^{q-1} a_{\lambda^s,\lambda}.$$

Thus in order to pass from p to $p+1$ it suffices to prove

$$a_{\lambda^{p+q},\lambda} \cdot a_{\lambda^p,\lambda^q} = a_{\lambda^{p+1},\lambda^q} \cdot a_{\lambda^p,\lambda}.$$

But this follows from (iii): in (iii) it suffices to put $\lambda_1 = 1$, $\lambda_2 = \lambda^p$, $\lambda_3 = \lambda^q$.

4. Now define $d_{\bar{\lambda}}$ for an arbitrary $\bar{\lambda} \in \Lambda^{(r+1)}$. The number $\bar{\lambda}$ is of the form

$$\bar{\lambda} = \lambda^n \cdot \mu, \tag{5}$$

where $\mu \in \Lambda^{(r)}$ and n is an integer. Put

$$d_{\bar{\lambda}} = a_{\lambda^n, \mu} \cdot d_{\lambda^n} \cdot d_\mu. \tag{6}$$

Since the representation of the number $\bar{\lambda}$ in the form (5) is not unique in general, we must prove that $d_{\bar{\lambda}}$ is well defined. If besides the relation (5), we have

$$\bar{\lambda} = \lambda^m \cdot \mu', \tag{7}$$

where $\mu' \in \Lambda^{(r)}$ and m is an integer, then it is easy to see that $(m - n)$ is a multiple of h, and it suffices to consider the case $m - n = h$. In this case $\mu = \mu'\lambda^h$, i.e., $d_\mu = a_{\mu', \lambda^h} \cdot d_{\mu'} \cdot d_h$. Substituting this expression into (6), we get

$$d_{\bar{\lambda}} = a_{\lambda^n, \mu} \cdot d_{\lambda^n} \cdot a_{\mu', \lambda^h} \cdot d_{\mu'} \cdot d_{\lambda^h}. \tag{8}$$

On the other hand (7) implies

$$d_{\bar{\lambda}} = a_{\lambda^{n+h}, \mu'} \cdot d_{\lambda^{n+h}} \cdot d_{\mu'}. \tag{9}$$

Taking into consideration (4), we see that the fact that the expressions (8) and (9) coincide reduces to the following relation:

$$a_{\lambda^n, \mu} \cdot a_{\mu', \lambda^h} = a_{\lambda^{n+h}, \mu'} \cdot a_{\lambda^n, \lambda^h}.$$

But this follows from (iii) when $\lambda_1 = \lambda^n$, $\lambda_2 = \lambda^h$, $\lambda_3 = \mu'$.

5. Now assume $\lambda_1, \lambda_2 \in \Lambda^{(r+1)}$, $\lambda_1 = \lambda^p \cdot \mu_1$, $\lambda_2 = \lambda^q \mu_2$, $\mu_1, \mu_2 \in \Lambda^{(r)}$. We have:

$$d_{\lambda_1} = a_{\lambda^p, \mu_1} \cdot d_{\lambda^p} \cdot d_{\mu_1}$$
$$d_{\lambda_2} = a_{\lambda^q, \mu_2} \cdot d_{\lambda^q} \cdot d_{\mu_2},$$
$$d_{\lambda_1 \lambda_2} = a_{\lambda^{p+q}, \mu_1 \mu_2} \cdot d_{\lambda^{p+q}} d_{\mu_1 \mu_2}$$

Put $\Pi = d_{\lambda_1 \lambda_2} d_{\lambda_1}^{-1} d_{\lambda_2}^{-1}$. It remains to prove that $\Pi = a_{\lambda_1, \lambda_2}$. By the induction hypothesis, $d_{\mu_1 \mu_2} \cdot d_{\mu_1}^{-1} d_{\mu_2}^{-1} = a_{\mu_1, \mu_2}$. Therefore

$$\Pi = a_{\lambda^{p+q}, \mu_1 \mu_2} \cdot d_{\lambda^{p+q}} \cdot a_{\lambda^p, \mu_1}^{-1} \cdot d_{\lambda^p}^{-1} \cdot a_{\lambda^q, \mu_2}^{-1} \cdot d_{\lambda^q}^{-1} \cdot a_{\mu_1, \mu_2}.$$

Using relation (4), we get

$$\Pi = a_{\lambda^{p+q}, \mu_1\mu_2} \cdot a^{-1}_{\lambda^p, \mu_1} \cdot a^{-1}_{\lambda^q, \mu_2} \cdot a_{\lambda^p, \lambda^q} \cdot a_{\mu_1, \mu_2}.$$

If in (iii) we put $\lambda_1 = \lambda^p$, $\lambda_2 = \lambda^q$, $\lambda_3 = \mu_1\mu_2$, then

$$a_{\lambda^{p+q}, \mu_1\mu_2} \cdot a_{\lambda^p, \lambda^q} = a_{\lambda^p, \lambda^q\mu_1\mu_2} \cdot a_{\lambda^q, \mu_1\mu_2} \tag{10}$$

further, putting $\lambda_1 = \mu_1$, $\lambda_2 = \mu_2$, $\lambda_3 = \lambda^q$ in (iii), we get

$$a_{\lambda^q, \mu_1\mu_2} \cdot a_{\mu_1, \mu_2} = a_{\lambda^q, \mu_2} \cdot a_{\lambda^q\mu_2, \mu_1}. \tag{11}$$

By using (10) and (11), we can write Π in the form

$$\Pi = a_{\lambda^p, \lambda^q\mu_1\mu_2} \cdot a_{\lambda^q\mu_2, \mu_1} \cdot (a_{\lambda^p, \mu_1})^{-1}.$$

Once again using (iii) for $\lambda_1 = \lambda^q\mu_2$, $\lambda_2 = \mu_1$, $\lambda_3 = \lambda^p$, we see that

$$\Pi = a_{\lambda^p\mu_1, \lambda^q\mu_2} = a_{\lambda_1, \lambda_2}.$$

The lemma is proved. □

Lemma 2. *The eigen-functions $\{f_\lambda\}$ of the ergodic automorphism T, $\lambda \in \Lambda_d(T)$, may be chosen so as to have $f_{\lambda_1} \cdot f_{\lambda_2} = f_{\lambda_1 \cdot \lambda_2}$ for all $\lambda_1, \lambda_2 \in \Lambda_d(T)$.*

Proof. Suppose $f_\lambda^{(1)}$ is an arbitrary family of eigenfunctions. Then we can find numbers a_{λ_1, λ_2}, $\lambda_1, \lambda_2 \in \Lambda_d(T)$, for which

$$f^{(1)}_{\lambda_1} \cdot f^{(1)}_{\lambda_2} = a_{\lambda_1, \lambda_2} \cdot f^{(1)}_{\lambda_1\lambda_2}.$$

From the previous lemma, we obtain $a_{\lambda_1, \lambda_2} = d_{\lambda_1\lambda_2} \cdot d^{-1}_{\lambda_1} d^{-1}_{\lambda_2}$. The functions $f_\lambda = d_\lambda f_\lambda^{(1)}$ satisfy the condition of the lemma. The lemma is proved. □

Returning to the problem of the metric isomorphism of the automorphisms T_1 and T_2, note that for any isomorphism Φ of the Hilbert spaces H_1, H_2 which commutes with the unitary operators U_1, U_2 we have

$$\Phi f_\lambda^{(1)} = c_\lambda f_\lambda^{(2)}, \qquad |c_\lambda| = 1.$$

If Φ is generated by the isomorphism φ of the measure spaces $(M_1, \mathfrak{S}_1, \mu_1)$, $(M_2, \mathfrak{S}_2, \mu_2)$, i.e.,

$$(\Phi h)(x_2) = h(\varphi^{-1}x_2), \qquad x_2 \in M_2,$$

then

$$\Phi(f_{\lambda_1}^{(1)} \cdot f_{\lambda_2}^{(1)}) = (f_{\lambda_1}^{(1)} \circ \varphi)(f_{\lambda_2}^{(1)} \circ \varphi) = c_{\lambda_1} c_{\lambda_2} f_{\lambda_1}^{(2)} f_{\lambda_2}^{(2)} = c_{\lambda_1} c_{\lambda_2} a_{\lambda_1, \lambda_2}^{(2)} f_{\lambda_1 \lambda_2}^{(2)}.$$

On the other hand, $f_{\lambda_1}^{(1)} f_{\lambda_2}^{(1)} = a_{\lambda_1, \lambda_2}^{(1)} f_{\lambda_1 \lambda_2}^{(1)}$ and therefore

$$\Phi(f_{\lambda_1}^{(1)} \cdot f_{\lambda_2}^{(1)}) = a_{\lambda_1, \lambda_2}^{(1)} c_{\lambda_1 \lambda_2} f_{\lambda_1, \lambda_2}^{(2)}$$

so that we get

$$a_{\lambda_1, \lambda_2}^{(1)} = c_{\lambda_1} \cdot c_{\lambda_2} \cdot c_{\lambda_1 \lambda_2}^{-1} a_{\lambda_1, \lambda_2}^{(2)}. \tag{12}$$

It turns out that the converse statement is also true.

Lemma 3. *Suppose that the map Φ is such that the relation* (12) *is valid for all $\lambda_1, \lambda_2 \in \Lambda$. Then there exists an isomorphism φ of Lebesgue spaces $(M_1, \mathfrak{S}_1, \mu_1)$, $(M_2, \mathfrak{S}_2, \mu_2)$ which generates Φ.*

Proof. It follows from (12) that

$$\begin{aligned}
\Phi(f_{\lambda_1}^{(1)} \cdot f_{\lambda_2}^{(1)}) &= a_{\lambda_1 \lambda_2}^{(1)} \Phi(f_{\lambda_1 \cdot \lambda_2}^{(1)}) = a_{\lambda_1, \lambda_2}^{(1)} \cdot c_{\lambda_1 \lambda_2} \cdot f_{\lambda_1 \cdot \lambda_2}^{(2)} \\
&= c_{\lambda_1} \cdot c_{\lambda_2} \cdot a_{\lambda_1, \lambda_2}^{(2)} f_{\lambda_1, \lambda_2}^{(2)} = c_{\lambda_1} \cdot f_{\lambda_1}^{(2)} \cdot c_{\lambda_2} \cdot f_{\lambda_2}^{(2)} \\
&= \Phi(f_{\lambda_1}^{(1)}) \cdot \Phi(f_{\lambda_2}^{(1)}).
\end{aligned}$$

Since the functions $\{f_\lambda^{(1)}\}$ constitute a basis in $L^2(M, \mathfrak{S}, \mu)$, the multiplicative property may be carried over to arbitrary bounded functions $f^{(1)}$, $f^{(2)} \in L^2(M, \mathfrak{S}, \mu)$ (the boundedness is needed to guarantee the inclusion $f^{(1)} \cdot f^{(2)} \in L^2(M, \mathfrak{S}, \mu)$):

$$\Phi(f^{(1)} \cdot f^{(2)}) = \Phi(f^{(1)}) \cdot \Phi(f^{(2)}). \tag{13}$$

Suppose, in particular, that χ_A is the indicator of a set A, $A \subset M$. Since $\chi_A^2 = \chi_A$, it follows from (13) that $[\Phi(\varphi_A)]^2 = \Phi(\chi_A)$, i.e., $\Phi(\varphi_A)$ is also the indicator of a certain set B, $B \subset M$. Since Φ is an isomorphism of Hilbert spaces, $\mu(A) = \mu(B)$.

Note that the multiplicative property is also valid for the inverse map Φ^{-1}, which also sends indicators into indicators. Thus we obtain an isomorphism ψ of the σ-algebras of measurable sets of the spaces $(M_1, \mathfrak{S}_1, \mu_1)$ and $(M_2, \mathfrak{S}_2, \mu_2)$ acting according to the formula $\psi(A) = B$.

Up to this point we have not used the fact that M_1, M_2 are Lebesgue spaces. Now let us use the fact that in Lebesgue spaces (see Appendix 1) every isomorphism of σ-algebras is generated by an isomorphism φ of the spaces themselves. The lemma is proved. □

Now we can conclude the proof of Theorem 2. It follows from Lemma 1 that

$$a^{(1)}_{\lambda_1, \lambda_2} = d^{(1)}_{\lambda_1 \lambda_2} \cdot (d^{(1)}_{\lambda_1})^{-1} \cdot (d^{(1)}_{\lambda_2})^{-1}, \qquad a^{(2)}_{\lambda_1, \lambda_2} = d^{(2)}_{\lambda_1 \lambda_2} \cdot (d^{(2)}_{\lambda_1})^{-1} \cdot (d^{(2)}_{\lambda_2})^{-2}.$$

Put

$$c_\lambda = d^{(1)}_\lambda \cdot (d^{(2)}_\lambda)^{-1}.$$

For such a choice of c_λ, we will have relation (12). Therefore, by Lemma 3, there is an isomorphism φ of measure spaces M_1 and M_2 satisfying $\varphi T_1 = T_2 \varphi$. The theorem is proved. □

Corollary. *Any ergodic automorphism of a Lebesgue space with pure point spectrum is metrically isomorphic to the group translation on the character group of the spectrum.*

§3. Dynamical Systems with Pure Point Spectrum. The Case of Continuous Time

The results of the previous section will now be carried over to the case of dynamical systems with continuous time.

Definition 1. The ergodic flow $\{T^t\}$ is said to be a flow with pure point spectrum if the adjoint group of unitary operators $\{U^t\}$ has a complete (in $L^2(M, \mathfrak{S}, \mu)$ system of orthogonal eigen-functions.

By Theorem 2, §1, the set $\Lambda_d(\{T^t\})$ is a subgroup of the additive group $\mathbb{R}^1$.

Theorem 1. *Suppose Λ is an arbitrary countable subgroup of the group $\mathbb{R}^1$. Then there exists a flow $\{T^t\}$ with pure point spectrum such that $\Lambda = \Lambda_d(\{T^t\})$.*

Proof. Just as in Theorem 1, §2, for the phase space M take the character group of the group Λ. Then M will still be a commutative compact group. Suppose μ is a normalized Haar measure on M. For every $t \in \mathbb{R}^1$ consider the homomorphism g_t of the group Λ into S^1: given by $g_t(\lambda) = e^{i\lambda t}$. Clearly $g_{t_1} \cdot g_{t_2} = g_{t_1 + t_2}$. In other words, $\{g_t\}$ is a one-parameter subgroup of M. Put $T^t x = x + g_t$, $x \in M$. It is easy to see that $\{T^t\}$ is the one-parameter subgroup of translations of the group M, preserving the measure μ.

If $\chi_\lambda \in M$ is the character corresponding by Pontrjagin duality to the point $\lambda \in \Lambda$, then

$$U^t \chi_\lambda = \chi_\lambda(T^t x) = \chi_\lambda(x + g_t) = \chi_\lambda(x) \cdot \chi_\lambda(g_t) = e^{i\lambda t} \cdot \chi_\lambda,$$

by the definition of g_t. Since the characters $\{\chi_\lambda\}$ constitute a complete orthogonal system in $L^2(M, \mathfrak{S}, \mu)$, we see that $\Lambda = \Lambda_d(\{T^t\})$. The theorem is proved. □

Now let us prove the main theorem on flows with pure point spectrum.

Theorem 2. *Suppose $\{T_1^t\}$, $\{T_2^t\}$ are ergodic flows on the Lebesgue spaces $(M_1, \mathfrak{S}_1, \mu_1)$, $(M_2, \mathfrak{S}_2, \mu_2)$ with pure point spectrum satisfying $\Lambda_d(\{T_1^t\}) = \Lambda_d(\{T_2^t\})$. Then $\{T_1^t\}$, $\{T_2^t\}$ are metrically isomorphic.*

Corollary. *Any ergodic flow on a Lebesgue space with pure point spectrum is metrically isomorphic to a flow generated by group translations along the one-parameter subgroup of the character group.*

Proof of Theorem 2. Denote by $\{T^t\}$ any of the flows $\{T_1^t\}$, $\{T_2^t\}$.

1. Suppose $f_\lambda(x)$ $(\lambda \in \Lambda(\{T^t\} = \Lambda))$ is an eigen-function of the adjoint group of operators $\{U^t\}$ corresponding to the eigen-value λ. The equality

$$U^t f_\lambda(x) = e^{i\lambda t} f_\lambda(x), \tag{1}$$

for each t holds for almost all x in the space $(M, \mathfrak{S}, \mu)$, but the set of all x where (1) holds depends on t in general. Put

$$A = \{(x, t) \in M \times \mathbb{R}^1 : U^t f_\lambda(x) \neq e^{i\lambda t} f_\lambda(x)\}.$$

Obviously $v(A) = 0$, where $v = \mu \times \rho$, and ρ is the Lebesgue measure the real line. By Fubini's theorem, there is a subset $N \subset M$, $\mu(N) = 0$, and sets $N_x \subset \mathbb{R}^1$, $\rho(N_x) = 0$ $(x \in M \backslash N)$ such that for $x \in M \backslash N$, $t \in \mathbb{R}^1 \backslash N_x$ we have (1).

Let us prove that if two points $x_1, x_2 \in M \backslash N$ are on the same trajectory of the flow $\{T^t\}$, i.e., $x_2 = T^{t_0} x_1$ for some $t_0 \in \mathbb{R}^1$, then we have the equality $f_\lambda(x_2) = e^{i\lambda t_0} f_\lambda(x_1)$. Consider the set $N_{x_1} - t_0 = \{t - t_0 : t \in N_{x_1}\}$. Clearly $\rho(N_{x_1} - t_0) = 0$ and therefore $\rho((N_{x_1} - t_0) \cup N_{x_2}) = 0$. Hence we can find a $\tau \in \mathbb{R}^1 \backslash ((N_{x_1} - t_0) \cup N_{x_2})$ so that we obtain

$$\begin{aligned} f_\lambda(x_2) &= e^{-i\lambda\tau} f_\lambda(T^\tau x_2) = e^{-i\lambda\tau} f_\lambda(T^{\tau + t_0} x_1) = e^{-i\lambda\tau} e^{i\lambda(\tau + t_0)} f_\lambda(x_1) \\ &= e^{i\lambda t_0} f_\lambda(x_1). \end{aligned}$$

Arguing further as in the proof of Lemma 1, §2, Chap. 1, we can show that the function f_λ may be changed on the set N only so that relation (1) will hold for all $x \in M$, $t \in \mathbb{R}^1$. For this changed function we retained the notation f_λ. Since λ ranges over only a countable set of values, we may assume that (1) holds also for all $\lambda \in \Lambda$.

2. As in the case of automorphisms, it follows from Lemma 1, §2, that there exists a complete orthonormed system $\{f_\lambda\}$ $\lambda \in \Lambda$ of eigen-functions for $\{U^t\}$ such that

$$f_{\lambda_1}(x) \cdot f_{\lambda_2}(x) = f_{\lambda_1+\lambda_2}(x), \qquad \lambda_1, \lambda_2 \in \Lambda. \tag{2}$$

Here again the equality is understood as that of elements of Hilbert space, i.e., it is valid almost everywhere in the space $(M, \mathfrak{S}, \mu)$.

Suppose E is the set of all $x \in M$ for which relation (2) does not hold for all $\lambda_1, \lambda_2 \in \Lambda$. It follows from Subsection 1 that the set E is invariant and, moreover, $\mu(E) = 0$. Hence we can eliminate the set E from M, assuming that relation (2) holds for all x, λ_1, λ_2.

3. Formula (2) shows that on a subset of full measure $M' \subset M$ the function $f_\lambda(x)$, viewed as a function of λ, is a character of the group Λ. Consider the map $\theta: M' \to G$, where G is the character group of the group Λ defined by the relation $\theta(x) = f_\lambda(x)$, and introduce the flow $\{S^t\}$ on G given by the formula

$$S^t g(\lambda) = e^{i\lambda t} g(\lambda), \qquad g(\lambda) \in G$$

with invariant Haar measure.

Let us show that $\{S^t\}$ is an ergodic flow with pure point spectrum. The fact that the spectrum of $\{S^t\}$ is a pure point spectrum is proved just as in Theorem 1 of the present section. Now let us prove the ergodicity.

Suppose $\chi_\lambda(g)$, $g \in G$ is a character of the group G corresponding by Pontrjagin duality to the element $\lambda \in \Lambda$; then

$$\chi_\lambda(S^t g) = \chi_\lambda(e^{i\lambda t} \cdot g_\lambda) = e^{i\lambda t} \cdot \chi_\lambda(g_\lambda),$$

which means that the elements of the group Λ constitute the family of eigen-values, while the χ_λ are the corresponding eigen-functions. But then 1 is a simple eigen-value of the group of unitary operators adjoint to $\{S^t\}$. Our statement is proved.

4. *The map θ is a metric isomorphism of the flows $\{T^t\}$ and $\{S^t\}$.* First let us show that the map θ is one-to-one (mod 0). To do this it suffices to prove that the partition ξ of the space M' into the level sets of all the functions f_λ is really the partition into separate points (mod 0). If this were not so, then by the measurability of the partition ξ (see Appendix 1) we would have a corresponding nontrivial σ-subalgebra $\mathfrak{S}(\xi) \subset \mathfrak{S}$. Then all the functions from the subspace spanning $\{f_\lambda\}$, $\lambda \in \Lambda$, would be measurable with respect to $\mathfrak{S}(\xi)$ and $\{f_\lambda\}$ would not be a complete system of functions in $L^2(M, \mathfrak{S}, \mu)$.

Further, it follows from (1) that $S^t\theta = \theta T^t$ for any t, so that θ maps M' into an invariant set of full measure for the flow $\{S^t\}$. Then the measure μ will be mapped into a certain measure ν on G, invariant with respect to $\{S^t\}$. Since $\{S^t\}$ is uniquely ergodic (see §1, Chap. 4), i.e., the only normalized

Borel measure for $\{S^t\}$ is the Haar measure, it suffices to check that θ is a Borel map. This will imply that the measure ν coincides with the Haar measure. Consider the open subsets $B \subset G$ of the following form

$$B = \{g(\lambda) \in G: |g(\lambda_i) - f_{\lambda_i}(x_0)| < \varepsilon, i = 1, 2, \ldots, n\},$$

for some $\varepsilon > 0$, $x_0 \in M'$, $\lambda_1, \ldots, \lambda_n \in \Lambda$.

Since the flow $\{S^t\}$ is minimal, $\theta(M')$ is dense in G and the sets B constitute a basis of open sets of the space G. But

$$\theta^{-1}(B) = \bigcap_{i=1}^{n} \{x \in M': |f_{\lambda_i}(x) - f_{\lambda_i}(x_0)| < \varepsilon\}.$$

Clearly $\theta^{-1}(B)$ is measurable, so that the inverse images of all the Borel sets are measurable, i.e., θ is Borel.

5. Since, according to the above, each of the flows $\{T_1^t\}$, $(T_2^t\}$ is metrically isomorphic to $\{S^t\}$, they are isomorphic to each other. The theorem is proved. □

Chapter 13

Examples of Spectral Analysis of Dynamical Systems

By the spectrum of a dynamical system we mean the spectrum of a unitary operator or group (semigroup) of unitary operators adjoint to the system on the invariant subset $L_0^2(M, \mathfrak{S}, \mu)$ of functions of zero mean. Two dynamical systems with the same spectrum are said to be *spectrally equivalent*. In this chapter we shall compute the spectrum of certain dynamical systems. The necessary facts from the spectral theory of unitary operators are provided in Appendix 2.

§1. Spectra of K-automorphisms

In this section we will show that all K-automorphisms are spectrally equivalent and have a countable Lebesgue spectrum (the definition is given in Appendix 2). We assume that the phase space $(M, \mathfrak{S}, \mu)$ is a Lebesgue space and that the measure μ is continuous. First we give a sufficient condition for the existence of countable multiplicity Lebesgue components in the spectrum of an automorphism. This condition is the existence for the automorphism T of an increasing σ-subalgebra, i.e., a σ-algebra $\mathfrak{S}^{(0)}$ such that $T\mathfrak{S}^{(0)} \supset \mathfrak{S}^{(0)}$, $T\mathfrak{S}^{(0)} \neq \mathfrak{S}^{(0)}$. Introduce the following notations. Suppose $\mathfrak{S}^{(n)}$ is the σ-algebra $T^n\mathfrak{S}^{(0)}$. Let $H^{(n)} = H^{(n)}(\mathfrak{S}^{(0)})$ be the Hilbert space $L_0^2(M, \mathfrak{S}^{(n)}, \mu)$. Put $H_n = H^{(n)} \ominus H^{(n-1)}$.

Theorem 1. *If the automorphism T possesses an increasing σ-algebra $\mathfrak{S}^{(0)}$, then the unitary operator U_T adjoint to T has the countable multiplicity Lebesgue spectrum on the subspace $\bigoplus_{n=-\infty}^{\infty} H_n$.*

The proof is based on the following general lemma.

Lemma 1. *The Hilbert space $H_1 = H^{(1)} \ominus H^{(0)}$ is infinite dimensional.*

The proof of the lemma shall be given later.

Proof of Theorem 1. Suppose $g_1, g_2, \ldots$ is an orthonormed basis in the space $H^{(1)} \ominus H^{(0)}$. Then the functions $g_k^{(n)} = U_T^n g_k$, $-\infty < n < \infty$, $1 \leq k < \infty$ constitute an orthonormed basis in the space

$$\bigoplus_{n=-\infty}^{\infty} [H^{(n)} \ominus H^{(n-1)}] = \bigoplus_{n=-\infty}^{\infty} H_n$$

and $U_T g_k^{(n)} = g_k^{(n+1)}$. The theorem is proved. □

Proof of Lemma 1. Since $H^{(1)} \neq H^{(0)}$ and $H^{(1)} \supset H^{(0)}$, there exists a function $f \in H^{(1)}$ for which $E[|f - E(f|\mathfrak{S}^{(0)})|^2|\mathfrak{S}^{(0)}| \neq 0$ on a set $C \in \mathfrak{S}^{(0)}$ of positive measure, where $E(f|\mathfrak{S}^{(0)})$ is the conditional expectation of f with respect to the σ-algebra $\mathfrak{S}^{(0)}$. Put

$$g(x) = \frac{f(x) - E(f|\mathfrak{S}^{(0)})}{\sqrt{E[|f - E(f|\mathfrak{S}^{(0)}|^2|\mathfrak{S}^{(0)}]}} \cdot \chi_C$$

(χ_C is the indicator of the set C). Then $E(g|\mathfrak{S}^{(0)}) = 0$ almost everywhere, $E(|g|^2|\mathfrak{S}^{(0)}) = 1$ for all conditions belonging to C. We may assume that the space $H^{(0)}(C) \subset H^{(0)}$ consisting of functions concentrated on C is infinite-dimensional. Take the infinite system of orthonormed uniformly bounded functions $e_1(x), e_2(x), \ldots \in H_0(C)$. Let us show that the functions $g_k(x) = e_k(x)g(x)$, $k = 1, 2, \ldots$ belong to H_1 and constitute an orthonormed system. The inclusion $g_k(x) \in H^{(1)}$ follows from $e_k(x) \in H^1$, $g(x) \in H^{(1)} |e_k(x)| \leq$ const. Further, for any bounded function $h(x) \in H^{(0)}$,

$$\int_M g_k(x)\overline{h(x)}\, d\mu = \int_M E[e_k(x)g(x)\overline{h(x)}|\mathfrak{S}^{(0)}]\, d\mu$$

$$= \int_M e_k(x)\overline{h(x)} \cdot E(g(x)|\mathfrak{S}^{(0)})\, d\mu = 0, \qquad \text{i.e.,} \quad g_k(x) \in H^{(1)} \ominus H^{(0)}.$$

Finally

$$\int_M g_i(x)\overline{g_j(x)}\, d\mu = \int_C e_i(x)g(x) \cdot \overline{e_j(x)}\overline{g(x)}\, d\mu$$

$$= \int_C e_i(x)\overline{e_j(x)} \cdot E(|g|^2|\mathfrak{S}^{(0)})\, d\mu = \int_M e_i(x)\overline{e_j(x)}\, d\mu = \begin{cases} 1 & \text{for } i = j, \\ 0 & \text{for } i \neq j. \end{cases}$$

The lemma is proved. □

Theorem 1 has the following consequence.

Corollary. *K-automorphisms have a countable Lebesgue spectrum.*

Proof. If for $\mathfrak{S}^{(0)}$ we take the σ-subalgebra corresponding to the K-partition, then, obviously, the subspace $\bigoplus_{n=-\infty}^{\infty} H_n(\mathfrak{S}^{(0)})$ coincides with $L_0^2(M, \mathfrak{S}, \mu)$. The corollary is proved. □

In particular, all Bernoulli automorphisms and all mixing Markov automorphisms have countable Lebesgue spectra.

§2. Spectra of Ergodic Automorphisms of Commutative Compact Groups

In this section we will show how a countable Lebesgue spectrum arises in dynamical systems of algebraic origin. Suppose $M = G$ is a commutative compact group, $T: M \to M$ its continuous algebraic automorphism. Then T is a metric automorphism of the space $(M, \mathfrak{S}, \mu)$, where μ is the normalized Haar measure on M and $\mathfrak{S}$ is the σ-algebra of Borel subsets of M.

Theorem 1. *Any ergodic automorphism T of a commutative compact group has a countable Lebesgue spectrum.*

Proof. 1. Suppose $\hat{G}$ is the character group of the group G. Since G is compact, it follows from Pontrjagin duality that the group $\hat{G}$ is countable and the elements of $\hat{G}$, viewed as functions on M, constitute an orthogonal basis in the space $L^2(M, \mathfrak{S}, \mu)$.

The unitary operator U_T leaves the set $\hat{G} \subset L^2(M, \mathfrak{S}, \mu)$ invariant. Indeed, if $\chi \in \hat{G}$, then

$$(U_T\chi)(x_1 \cdot x_2) = \chi(T(x_1x_2)) = \chi((Tx_1)(Tx_2)) = \chi(Tx_1) \cdot \chi(Tx_2)$$
$$= (U_T\chi)(x_1) \cdot (U_T\chi)(x_2),$$

i.e., $U_T\chi \in \hat{G}$. Therefore U_T induces a partition of $\hat{G}$ into trajectories (under the action of U_T).

2. Let us show that trajectory of each element $\chi \neq 1$ is infinite. Conversely, assume that $U_T^n\chi = \chi$, $\chi \neq 1$. Suppose $\varphi = \chi + U_T\chi + \cdots + U_T^{n-1}\chi$. Then $\varphi \in L^2(M, \mathfrak{S}, \mu)$, $U_T\varphi = \varphi$, and by the orthogonality of distinct characters $\varphi \neq \text{const} \pmod 0$, contradicting the ergodicity of T. It

follows from the statement just proved above that U_T has a homogeneous Lebesgue spectrum.

3. In this subsection we will show that the multiplicity of the spectrum is infinite, i.e., $\hat{G}$ falls apart into an infinite number of trajectories. Assuming the converse, let the number of trajectories be finite. Let us show that in this case $\hat{G}$ contains a finitely generated nontrivial subgroup invariant with respect to U_T and U_T^{-1}.

Take an element $\chi \in \hat{G}$, $\chi \neq 1$. Put $\chi_n = \prod_{i=0}^{n} U_T^i \chi$. Since the number of trajectories is finite, we can find integers $k, l, 0 \leq k \leq l$ such that χ_k and χ_l belong to the same trajectory, i.e., for some m we have

$$\prod_{i=0}^{k} U_T^i \chi = U_T^m \left(\prod_{i=0}^{l} U_T^i \chi \right) = \prod_{i=m}^{m+l} U_T^i \chi.$$

After carrying out the necessary cancellations, we obtain an identity of the form $\prod U_T^p \chi = 1$, where p ranges over a closed interval $p_1 \leq p \leq p_2$. Since a similar identity is valid for χ^{-1} as well, the subgroup generated by the element $\{U_T^p \chi\}$, $p_1 \leq p \leq p_2$ is the one we need.

4. Thus we can now assume that $\hat{G}$ is a commutative finitely generated group. It is well known that every such group may be represented in the form of a direct sum of a finite number of cyclic groups (finite or infinite). Hence $\hat{G}$ contains only a finite number of elements of finite order. Since every trajectory except the unit one under the action of U_T is infinite, while the order of elements is an invariant along the trajectory, all the elements of $\hat{G}$ (except 1) are of infinite order. Thus all the cyclic factors of the group $\hat{G}$ are infinite cyclic groups.

Let the *height* of the element $\chi \in \hat{G}$ be the largest n for which the equation $\chi = \varphi^n$ has a solution. In the direct product of a finite number of infinite cyclic groups, it is easy to see that elements of arbitrarily large height exist. But the height is an invariant along the trajectory, hence the number of trajectories cannot be finite. This contradiction proves our theorem.

Corollary. *Since an automorphism with homogeneous Lebesgue spectrum obviously is mixing, the properties of ergodicity and mixing (both weak and strong) are equivalent for automorphisms of commutative compact groups.*

Remark. One may have the impression that the appearance of countable Lebesgue spectra for K-automorphisms and ergodic automorphisms of commutative compact groups are caused by different factors. However, this is not so, since, as was shown by Rohlin ([9]), any ergodic automorphism of a commutative compact group is a K-automorphism.

§3. Spectra of Compound Skew Translations on the Torus and of Their Perturbations

Suppose M is the m-dimensional torus Tor^m, $m \geq 2$, with normalized Haar measure μ, T is a compound skew translation on it (see §2, Chap. 4), which, in cyclic coordinates $x_1, \ldots, x_m$, is written in the form

$$T(x_1, \ldots, x_m) = (x_1 + \alpha \,(\mathrm{mod}\ 1), x_2 + p_{21}x_1 \,(\mathrm{mod}\ 1), \ldots, x_m + p_{m,1}x_1 + \cdots + p_{m,m-1}x_{m-1} \,(\mathrm{mod}\ 1)), \tag{1}$$

where α is an irrational number and p_{ij}, $1 \leq j < i \leq m$ are integers. We will assume that $p_{k+1,k} \geq 1$ for $k = 1, 2, \ldots, m-1$.

Suppose $P = \|p_{ij}\|$ is a square matrix of order m whose elements for $j < i$ appear in (1) and vanish for $j \geq i$. By Q denote the matrix $P + E$.

Suppose H_k, $1 \leq k \leq m$ is the subspace of the Hilbert space $L^2(M, \mathfrak{S}, \mu)$ consisting of functions of the form $f(x_1, \ldots, x_k)$. Clearly $H_{k+1} \supset H_k$, and $U_T H_k = H_k$, $k = 1, \ldots, m$, where U_T is the unitary operator adjoint to T. Therefore $U_T(H_{k+1} \ominus H_k) = H_{k+1} \ominus H_k$.

Theorem 1. *The operator U_T has a pure point spectrum in the subspace H_1. In every subspace $H_{k+1} \ominus H_k$, $k = 1, 2, \ldots, m-1$ the operator U_T has a Lebesgue spectrum. The spectrum U_T in the subspace $H_m \ominus H_1$ is a countable Lebesgue spectrum.*

Proof. The functions

$$e_p(x) = \exp(2\pi i p x_1), \qquad p = 0, \pm 1, \pm 2, \ldots,$$

constitute an orthogonal basis in the subspace H_1 and each of them is an eigen-function for the operators U_T:

$$U_T[\exp(2\pi i p x_1)] = \exp(2\pi i p \alpha) \cdot \exp(2\pi i p x_1).$$

This implies the first statement of the theorem.

To prove the second one, consider the functions

$$e_s(x) = \exp[2\pi i(s, x)] = \exp\left(2\pi i \sum_{k=1}^{m} s_k x_k\right)$$

where s ranges over all possible vectors with integer coordinates $s =$

$(s_1, \ldots, s_m)$. Clearly, the functions $e_s(x)$ constitute an orthogonal basis in $L^2(M, \mathfrak{S}, \mu)$ and

$$\begin{aligned} U_T e_s(x) &= \exp(2\pi i s_1 \alpha) \cdot \exp[2\pi i(s, Qx)] \\ &= \exp(2\pi i s_1 \alpha) \cdot \exp[2\pi i(Q^* s, x)] \\ &= \exp(2\pi i s_1 \alpha) \cdot e_{Q^* s} x, \end{aligned}$$

where Q^* is the matrix adjoint to Q.

Hence, in order to prove the theorem, it suffices to show that every vector $s = (s_1, \ldots, s_m)$ with integer coordinates which has at least one nonvanishing coordinate $s_2, \ldots, s_m$ defines an infinite trajectory under the action of powers of the matrix Q^*, and that the number of such trajectories is infinite.

The last statement follows from the fact that for all integer coordinate vectors s_l of the form $s_l = (0, l, 0, \ldots, 0)$ and any n we have the relation

$$U_T^n e_{s_l} = \exp\left[2\pi i p_{21} \cdot \frac{n(n-1)}{2} \alpha\right] \cdot e_{(p_{21} nl, l, 0, \ldots, 0)},$$

which is easy to prove by induction. This relation also shows that the cyclic subspaces generated by the functions e_{s_l} for distinct l are two-by-two orthogonal.

Now let us show that the equality $(Q^*)^n s = s$ is not possible for any n, as soon as one of the coordinates $s_2, \ldots, s_m$ of the vector $s = (s_1, \ldots, s_m)$ is nonzero. We may assume that $n > 0$. Denote by $q_{ij}^{(n)}$ the elements of the matrix $(Q^*)^n$. It may be checked directly that $q_{ij}^{(n)} = 0$ for $i > j$; $q_{k,k}^{(n)} = 1$ for $k = 1, 2, \ldots, m$;

$$q_{k,k+1}^{(n)} = n \cdot q_{k,k+1}^{(1)} = n \cdot p_{k+1,k} \neq 0.$$

For the coordinates $s_1, \ldots, s_m$ of the vector s which satisfy the relation $(Q^*)^n s = s$, we have the following system of m equations

$$\begin{cases} s_1 + q_{12}^{(n)} s_2 + \quad \cdots \quad + q_{1m}^{(n)} s_m = s_1, \\ \qquad s_2 + q_{23}^{(n)} s_3 + \cdots + q_{2m}^{(n)} s_m = s_2, \\ \qquad\qquad\qquad\qquad \vdots \qquad\quad \vdots \\ \qquad\qquad\qquad\qquad s_m = s_m \end{cases}$$

i.e., we have a homogeneous system of $(m-1)$ equations

$$\begin{cases} q_{12}^{(n)} s_2 + \cdots + q_{1m}^{(n)} s_m = 0 \\ q_{23}^{(n)} s_3 + \cdots + q_{2m}^{(n)} s_m = 0 \\ \qquad\qquad \vdots \qquad\qquad \vdots \\ \qquad\qquad q_{m-1,m}^{(n)} s_m = 0 \end{cases}$$

in the unknowns $s_2, \ldots, s_m$. Since $q_{k,k+1}^{(n)} \neq 0$, $k = 1, \ldots, m-1$, this last system only possesses the trivial solution, which means that the relation $(Q^*)^n s = s$ is possible only for vectors of the form $s = (l, 0, \ldots, 0)$. The theorem is proved. □

Now let us show that in the example above the spectrum possesses a certain stability property. Consider the functions

$$h_1(x_1) \in H_1, \qquad h_2(x_1, x_2) \in H_2, \ldots, h_{m-1}(x_1, \ldots, x_{m-1}) \in H_{m-1}$$

such that all the h_k belong to $C^2(M)$ and

$$\frac{\partial h_k}{\partial x_k} + 1 > 0, \qquad k = 1, 2, \ldots, m-1. \tag{2}$$

Consider the transformation T_h of the torus M which acts according to the formula

$$\begin{aligned} T_h(x_1, \ldots, x_m) = (x_1 + \alpha \pmod 1),\ & x_x + p_{21}x_1 + h_1(x_1) \pmod 1, \ldots, x_m \\ & + p_{m1}x_1 + \cdots + p_{m,m-1}x_{m-1} \\ & + h_{m-1}(x_1, \ldots, x_{m-1}) \pmod 1) \end{aligned}$$

where the numbers α, p_{ij} are the same as above. Clearly T_h preserves the Haar measure μ, i.e., is an automorphism of the torus M. Denote by $U = U_{T_h}$ the unitary operator adjoint to T_h. Then $UH_k = H_k$, $k = 1, 2, \ldots, m$, and therefore $U(H_{k+1} \ominus H_k) = H_{k+1} \ominus H_k$.

Theorem 2. *The operator U in the subspace H_1 possesses a pure point spectrum. The spectrum of U in the subspace $H_m \ominus H_1$ is a countable Lebesgue spectrum.*

The proof is based on the following lemma.

Lemma 1. *For any function*

$$f, g \in H_k \cap C^2(M), \qquad 1 \leq k \leq m,$$

and any n we have

$$\left| \int_M g(x) \cdot U^n \frac{\partial f}{\partial x_k}\, d\mu \right| \leq \frac{\text{const}}{n}.$$

Proof of Lemma 1. Without loss of generality, we may assume that $k = m$. First we will prove the following relations

$$\frac{\partial}{\partial x_m}(U^n f) = U^n\left(\frac{\partial}{\partial x_m} f\right); \tag{3}$$

$$\frac{\partial}{\partial x_{m-1}} U^n f = U^n\left(\frac{\partial}{\partial x_{m-1}} f\right) + \left(U^n \frac{\partial f}{\partial x_m}\right) \times \left(np_{m,m-1} + \sum_{s=0}^{n-1} U^s\left(\frac{\partial}{\partial x_{m-1}} h_{m-1}\right)\right) \tag{4}$$

In order to obtain (3), note that for any n

$$U^n f = f(\ldots, x_m + h_{m-1}^{(n)}(x_1, \ldots, x_{m-1}))$$

where the dots denote the first $(m-1)$ arguments which depend only on the coordinates $x_1, \ldots, x_{m-1}$, while the function $h_{m-1}^{(n)}$ belongs to $H_{m-1} \cap C^2(M)$. The relation written above is easily proved by induction over n. It obviously implies (3).

Formula (4) will also be proved by induction over n. For $n = 1$ it follows from the definition of T_h that

$$\frac{\partial}{\partial x_{m-1}} Uf = U \frac{\partial f}{\partial x_{m-1}} + \left(U \frac{\partial f}{\partial x_m}\right)\left(p_{m,m-1} + \frac{\partial}{\partial x_{m-1}} h_{m-1}\right). \tag{5}$$

In order to pass from n to $(n+1)$, let us substitute the function Uf into (4) instead of f

$$\frac{\partial}{\partial x_{m-1}} U^{n+1} f = \frac{\partial}{\partial x_{m-1}} U^n(Uf) = U^n\left(\frac{\partial}{\partial x_{m-1}} Uf\right) + U^n\left(\frac{\partial}{\partial x_m} Uf\right) \cdot \left(np_{m,m-1} + \sum_{s=0}^{n-1} U^s \frac{\partial}{\partial x_{m-1}} h_{m-1}\right).$$

Using (3) and (5), we get

$$\frac{\partial}{\partial x_{m-1}} U^{n+1} f = U^{n+1} \frac{\partial f}{\partial x_{m-1}} + U^{n+1} \frac{\partial f}{\partial x_m}\left(p_{m,m-1} + U^n \frac{\partial}{\partial x_{m-1}} h_{m-1}\right) + \left(U^{n+1} \frac{\partial f}{\partial x_m}\right)\left(np_{m,m-1} + \sum_{s=0}^{n-1} U^s \frac{\partial}{\partial x_{m-1}} h_{m-1}\right).$$

This proves (4). Put

$$\Sigma_n = \left(np_{m,m-1} + \sum_{s=0}^{n-1} U^s \frac{\partial}{\partial x_{m-1}} h_{m-1}\right).$$

Then (4) implies

$$U^n \frac{\partial f}{\partial x_m} = \frac{1}{\Sigma_n} \frac{\partial}{\partial x_{m-1}} (U^n f) - \frac{1}{\Sigma_n} U^n \left(\frac{\partial f}{\partial x_{m-1}}\right),$$

so that

$$\int_M g \cdot U^n \left(\frac{\partial f}{\partial x_m}\right) d\mu = \int_M g \cdot \frac{1}{\Sigma_n} \cdot \frac{\partial}{\partial x_{m-1}} (U^n f)\, d\mu$$
$$- \int_M g \cdot \frac{1}{\Sigma_n} \cdot U^n \left(\frac{\partial f}{\partial x_{m-1}}\right) d\mu = I_1 - I_2.$$

In order to estimate I_1, integrate by parts:

$$I_1 = -\int \frac{\partial}{\partial x_{m-1}} \left(\frac{g}{\Sigma_n}\right) \cdot U^n f\, d\mu + \int g \cdot \frac{1}{\Sigma_n^2} \cdot \frac{\partial}{\partial x_{m-1}} \Sigma_n \cdot U^n f\, d\mu.$$

It follows from (2) and from the inequality $p_{m,m-1} \geq 1$ that $|\Sigma_n| \geq C_1 \cdot n$, where $C_1 > 0$ does not depend on n. Moreover, by (3)

$$\left|\frac{\partial}{\partial x_{m-1}} \Sigma_n\right| = \left|\sum_{s=0}^{n-1} \frac{\partial}{\partial x_{m-1}} \left(U^s \frac{\partial}{\partial x_{m-1}} h_{m-1}\right)\right|$$
$$= \left|\sum_{s=0}^{n-1} U^s \frac{\partial^2 h_{m-1}}{\partial x_{m-1}^2}\right| \leq C_2 \cdot n, \qquad C_2 > 0.$$

Hence

$$|I_1| \leq \left[\max\left|\frac{\partial g}{\partial x_{m-1}}\right| C_1^{-1} + \max|g| C_1^{-2} C_2 \max|f|\right] n^{-1}.$$

For I_2 we have immediately

$$|I_2| \leq \max|g| C_1^{-1} \max\left|\frac{\partial f}{\partial x_{m-1}}\right| n^{-1}.$$

The lemma is proved. □

Lemma 2. *The spectral type of any function of the form $\partial f/\partial x_k$, $f \in H_k \cap C^2(M)$ is absolutely continuous.*

The lemma follows immediately from Lemma 1 and the definition of an absolutely continuous spectrum (see Appendix 2).

Lemma 3. *The closure in $L^2(M, \mathfrak{S}, \mu)$ of the set of functions of the form $\partial f/\partial x_k$, $f \in H_k \cap C^2(M)$ is the subspace $H_k \ominus H_{k-1}$.*

The proof of this statement is easily obtained by developing the functions $f \in H_k \cap C^2(M)$ into Fourier series.

Proof of Theorem 2. The fact that the spectrum of the operator U in the subspace H_1 is discrete is proved just as in Theorem 1. It follows from Lemmas 2 and 3 that the maximal spectral type of the operator U in the subspace $H_m \ominus H_1$ is absolutely continuous with respect to Lebesgue measure. It remains to show that this spectral type is pure Lebesgue and that its multiplicity is infinite. To do this, consider an arbitrary function $g \in H_m \ominus H_1$, and the function $f = [\exp(2\pi i x_1)]g$. Then, for any n, we have

$$U^n f = U^n[\exp(2\pi i x_1)]U^n g = \exp[2\pi i n(x_1 + \alpha)]U^n g,$$

$$(U^n f, f) = \exp[(2\pi i n\alpha)](U^n g, g).$$

Therefore the spectral measure of the function f is the spectral measure of the function g rotated by the angle α. It therefore follows that the maximal spectral type is invariant with respect to the rotation by α and, α being irrational, is Lebesgue.

To prove that the Lebesgue spectrum is countable, introduce the subspaces $\Pi_{l,p}$, $1 \le l \le m$, $-\infty < p < \infty$, which are the closures of the sets of functions of the form

$$g(x_1, \ldots, x_{l-1})\exp(2\pi i p x_l), \qquad g \in H_{l-1}.$$

Then $U\Pi_{l,p} = \Pi_{l,p}$, and $\Pi_{l,p_1} \perp \Pi_{l,p_2}$ for $p_1 \ne p_2$. The previous argument shows that in each subspace $\Pi_{l,p}$ the spectrum is Lebesgue and the number of subspaces is infinite. The theorem is proved. □

§4. Examples of the Spectral Analysis of Automorphisms with Singular Spectrum

The theory of dynamical systems with pure point spectrum, developed in the previous chapter, for a long period of time gave grounds to hope that the structure of dynamical systems with continuous spectrum could be studied by similar means. However, it is now clear that this is not so. In this section we give two examples, meaningful in this respect, showing what new phenomena arise in the case of systems with continuous spectrum.

EXAMPLES

1. *The product of functions possessing a continuous spectrum may have a pure point spectrum.*

In the role of the measure space M take the direct product $S^1 \times Z_2$, where S^1 is the unit circle, which, as usual, will be identified with the semi-interval $0 \leq x < 1$; $Z_2 = \{1, -1\}$ is the group of square roots of 1, the measure μ on M being the product of the normalized Haar measures on S^1 and Z_2. The automorphism T is given by the formula

$$T(x, z) = (x + \alpha \,(\mathrm{mod}\ 1), g(x)z),$$

for $x \in S^1$, $z \in Z_2$; $g(x) = 1$ for $0 \leq x < 1/2$, $g(x) = -1$ for $1/2 \leq x < 1$. By $L_2(S^1)$ denote the subspace of the Hilbert space $L^2(M, \mu)$ consisting of functions which depend on x only. Put $H = L^2(M, \mu) \ominus L^2(S^1)$. If U_T is the unitary operator adjoint to T, then

$$U_T L^2(S^1) = L^2(S^1), \qquad U_T H = H.$$

It is also clear that in the subspace $L^2(S^1)$ the operator U_T has a pure point spectrum. Now assume that α is irrational. We will show that:

(1) in the space H the operator U_T has a continuous spectrum;
(2) if $f_1 \in H$, $f_2 \in H$ and $f_1 \cdot f_2 \in L^2(M, \mu)$, then $f_1 \cdot f_2 \in L^2(S_1)$; in other words the product of two functions possessing continuous spectra has a pure point spectrum.

Since $Z_2 = \{1, -1\}$, every function $f \in H$ may be written in the form

$$f(x, z) = z \cdot h(x), \qquad z \in Z, h(x) \in L^2(S^1).$$

Therefore the second statement is a consequence of the first.

To prove the first statement, assume that $f(x, z) = z \cdot h(x) \in H$ is an eigenfunction of the operator U_T with eigen-value λ, $|\lambda| = 1$. Then

$$f(x + \alpha)g(x) = \lambda f(x) \tag{1}$$

(the equality is understood as an equality between elements of Hilbert space). It follows from (1) that

$$\overline{f(x + \alpha + 1/2)} \cdot g(x + 1/2) = \bar{\lambda} \cdot \overline{f(x + 1/2)}. \tag{2}$$

Multiplying (1) by (2), we get

$$-f(x+\alpha)\overline{f(x+\alpha+1/2)} = f(x)\cdot\overline{f(x+1/2)},$$

i.e., the function $\varphi(x) = f(x)\cdot\overline{f(x+1/2)}$ is an eigen-function for the unitary operator U, adjoint to the rotation of the circle by the angle α, with eigen-value -1. This last situation is impossible, since the eigen-values of the operator U are of the form $e^{2\pi i n\alpha}$, $n = 0, \pm 1, \pm 2, \ldots$

2. *An automorphism with nonsimple finite multiplicity continuous spectrum.* The construction of this example will be split up into a series of steps.

(a) Suppose the space M_1 is the unit circle S^1 (once again identified with the semi-interval $0 \le x < 1$) with normalized Lebesgue measure and the transformation T_1 is the rotation of the circle by a certain angle β, i.e., $T_1 x = x + \beta \pmod 1$. The number β will be chosen later. Consider the function

$$F(x) = \begin{cases} 1 & \text{for } x \in [0, 1/4) \cup [1/2, 3/4), \\ 2 & \text{for } x \in [1/4, 1/2) \cup [3/4 \cup 1), \end{cases}$$

and construct the integral automorphism T_2 over the automorphism T_1 with function F. The phase space of the automorphism T_2 will be denoted by M_2. The points of the space M_2 are of the form $u = (x, y)$, where $x \in M_1$, $y = 1$ or 2.

Let us show that β may be chosen so that the unitary operator U_{T_2}, adjoint to the automorphism T_2, will have a continuous spectrum in the subspace of functions of zero mean. Suppose $f \in L^2(M_2)$ is an eigen-function for U_{T_2} with eigen-value $e^{2\pi i\lambda}$, $\lambda \in \mathbb{R}^1$. Then for almost all points $x \in M_1 \subset M_2$, we have

$$f(x+\beta) = e^{2\pi i\lambda F(x)} f(x). \tag{3}$$

On the circle $S^1 = M_1$, introduce the involution σ which acts according to the formula $\sigma(x) = (x + 1/2) \pmod 1$. Then the function f considered on M_1 may be represented, as any other function on M_1, in the form $f(x) = f_+(x) + f_-(x)$, where $f_+(\sigma x) = f_+(x)$ (such functions will be called *even* with respect to σ), $f_-(\sigma) = -f_-(x)$ (the function f_- is *odd* with respect to σ). It is obvious that the function F is even with respect to σ. Hence

$$f_+(x+\beta) = e^{2\pi i\lambda F(x)} f_+(x), \qquad f_-(x+\beta) = e^{2\pi i\lambda F(x)} f_-(x).$$

Since $f(x) \not\equiv 0 \pmod 0$, we either have $f_+(x) \not\equiv 0 \pmod 0$ or $f_-(x) \not\equiv 0 \pmod 0$. Put $\varphi(x) = f_+(x)$, $\theta = \lambda$ in the first case and $\varphi(x) = (f_-(x))^2$, $\theta = 2\lambda$—in the second. In both cases the function $\varphi(x)$ is even with respect to σ and satisfies the equation

$$\varphi(x+\beta) = e^{2\pi i\theta F(x)}\varphi(x).$$

Suppose $\psi(x) = \varphi(x/2)$, $\Phi(x) = F(x/2)$. Then $\psi(x)$ and $\Phi(x)$ may be viewed again as functions on the circle S^1 and

$$\psi(x + 2\beta) = e^{2\pi i\theta F(x)}\psi(x). \tag{4}$$

Lemma 1. *Suppose that, for the irrational number α, there exists a sequence of irreducible fractions $\{p_n/q_n\}$ such that q_n is odd and $q_n^2|p_n/q_n - \alpha| \to 0$ for $n \to \infty$. Suppose, moreover, that the function $g(x)$ is of the form*

$$g(x) = \begin{cases} \lambda_1 & \text{for } x \in [0, 1/2), \\ \lambda_2 & \text{for } x \in [1/2, 1), \end{cases}$$

where $|\lambda_1| = |\lambda_2| = 1$, $\lambda_1 \neq \lambda_2$. Then the equation $f(x + \alpha) = g(x)f(x)$ has no measurable solutions f, $f \not\equiv 0$.

The proof of the lemma shall be given later. Choose β so that the number $\alpha = 2\beta$ satisfies the assumptions of Lemma 1; then (4) has no nontrivial solution. This proves the continuity of the spectrum.

(b) Consider the space $M = M_2 \times Z_3$ where Z_3 is the group of cubic roots of 1 with normalized Haar measure. Points of the space M are of the form (u, z) where $u \in M_2$, $z \in Z_3$. Introduce the automorphism T of the space M by means of the formula

$$T(u, z) = (T_2 u, s(u)z),$$

where $u = (x, y) \in M_2$, $x \in M_1$, $y = 1$ or 2, and $s(u)$ depends only on the coordinate x and is determined by the formula

$$s(u) = \begin{cases} e^{2\pi i/3} & \text{for } x \in [0, 1/4) \cup [1/2, 3/4), \\ e^{4\pi i/3} & \text{for } x \in [1/4, 1/2) \cup [3/4, 1). \end{cases}$$

Let us show that the unitary operator U_T possesses a continuous spectrum in the subspace $L_0^2(M)$. To do this, introduce the subspaces $H_k \subset L^2(M)$ consisting of functions of the form $z^k f(u)$, $f \in L^2(M_2)$, $k = 0, 1, 2$. Each H_k is invariant with respect to U_T since

$$U_T(z^k f(u)) = s(u)z^k f(T_2 u) = z^k \cdot s(u) \cdot f(T_2 u).$$

Therefore, to prove the continuity of the spectrum, it suffices to establish the fact that the equation

$$s(u)f(T_2 u) = e^{2\pi i\lambda} f(u), \qquad \lambda \in \mathbb{R}^1, \lambda \neq 0 \tag{5}$$

has no measurable solution f, $|f| = 1$.

Extend the involution σ defined on M_1 to an involution on M_2 by means of the formula $\sigma(u) = \sigma(x, y) = (\sigma(x), y)$. Then $s(\sigma u) = s^2(u)$, $\sigma T_2 = T_2 \sigma$. Thererefore (5) implies

$$s^2(u) f(\sigma(T_2 u)) = e^{2\pi i \lambda} f(\sigma u). \tag{6}$$

Dividing (5) by (6), we get the relation

$$s(u) g(T_2 u) = g(u), \tag{7}$$

where $g(u) = f(\sigma u)/f(u)$. Putting

$$n(u) = s(u) \dots s(T_2^{F(x)-1} u),$$

we obtain by using (7)

$$n(u) g(T_2^{F(x)} u) = g(u). \tag{8}$$

Consider the last relation for the points $u = (x, 1) \in M_2$. The set of all such points can be naturally identified with the circle M_1. Denote by $T_2^{F(x)}$ the transformation which sends each point $u = (x, 1)$ into the point $T_2^{F(x)} u$. It is easy to check that $T_2^{F(x)}$ is a rotation of the circle M_1 by angle β, while the function $n(u)$ is of the form

$$n(u) = n(x, y) = \begin{cases} e^{2\pi i/3} & \text{if } x \in [0, 1/2), \\ e^{4\pi i/3} & \text{if } x \in [1/2, 1). \end{cases}$$

Therefore if the number β satisfies the assumptions of Lemma 1, then this lemma implies that equation (8) has no measurable solutions. The continuity of the spectrum of the automorphism T is proved.

(c) Let us prove that the spectrum of the automorphism T is of finite multiplicity. This follows from the fact that T may obviously be realized as an aperiodic interval exchange transformation of a finite number of intervals. In §2 of Chap. 5, we showed that any such transformation has only a finite number of ergodic normalized invariant measures. The arguments used there to prove this fact actually show that the spectrum of aperiodic interval exchange transformations is of finite multiplicity.

It remains to show that the spectrum of the automorphism T is not simple. To do this, note that the restrictions of the operator U_T to the subspaces H_1 and H_2 are unitarily equivalent. The equivalence is established by the unitary operator U_σ adjoint to the involution σ extended to the space M by means of the formula $\sigma(x, y, z) = (\sigma(x), y, z)$.

Proof of Lemma 1. Assume that, in contradiction to the statement of the lemma, the measurable function f, $f \not\equiv 0$ is a solution of our equation. Then $|f(x+\alpha)| = |f(x)|$, and since α is irrational, we have $|f| = \text{const} > 0$ almost everywhere. Therefore, $f(x+\alpha)/f(x) = g(x)$. Substituting the points $x, x+\alpha, \ldots, x+(q_n-1)\alpha$ for x in this equation, and multiplying all the relations obtained, we get

$$\frac{f(x+q_n\alpha)}{f(x)} = \prod_{r=0}^{q_n-1} g(x+r\alpha) \overset{\text{det}}{=} g_n(x).$$

Since $x + q_n\alpha \to x$ when $n \to \infty$, we have

$$\int_0^1 |g_n(x) - 1|\,dx = \text{const} \int_0^1 |f(x+q_n\alpha) - f(x)|\,dx \to 0 \tag{9}$$

for $n \to \infty$. Now put

$$g_n^{(1)}(x) = \prod_{r=0}^{q_n-1} g(x + r\cdot p_n/q_n).$$

Since p_n/q_n is an irreducible fraction, the point $x + r\cdot p_n/q_n$ ranges over all the values of the form

$$x + k/q_n \pmod 1, \qquad k = 0, 1, \ldots, q_n - 1.$$

It therefore follows that

$$g_n^{(1)}(x) = \begin{cases} \lambda_1^{[q_n/2]}\cdot\lambda_2^{[q_n/2]+1} & \text{if } \{q_n x\} < 1/2, \\ \lambda_1^{[q_n/2]+1}\cdot\lambda_2^{[q_n/2]}, & \text{if } \{q_n x\} \geq 1/2. \end{cases}$$

Therefore

$$\int_0^1 |g_n^{(1)}(x) - 1|\,dx = \tfrac{1}{2}|\lambda_1^{[q_n/2]}\cdot\lambda_2^{[q_n/2]+1} - 1|$$
$$+ \tfrac{1}{2}|\lambda_1^{[q_n/2]+1}\cdot\lambda_2^{[q_n/2]} - 1| \geq \frac{1}{2}\left|\frac{\lambda_2}{\lambda_1} - 1\right| > 0.$$

It follows from the assumptions of the lemma that $\int_0^1 |g_n(x) - g_n^{(1)}(x)|dx \to 0$ when $n \to \infty$. Therefore

$$\overline{\lim_{n\to\infty}} \int_0^1 |g_n(x) - 1|\,dx > 0,$$

which contradicts (9). The lemma is proved. □

§5. Spectra of K-flows

In this section we will prove a theorem which gives a sufficient condition for the existence of a countable Lebesgue component in the spectrum of a flow. In the case of K-flows, this theorem enables us to compute their spectra entirely.

Suppose that for the flow $\{T^t\}$ on the space $(M, \mathfrak{S}, \mu)$ there exists an increasing σ-subalgebra $\mathfrak{S}^{(0)}$ of the σ-algebra $\mathfrak{S}$, i.e., a σ-algebra such that $\mathfrak{S}^{(t)} \overset{\text{det}}{=} T^t\mathfrak{S}^{(0)} \supset \mathfrak{S}^{(0)}$ for $t > 0$ and $\mathfrak{S}^{(t)} \neq \mathfrak{S}^{(0)}$. This obviously implies $\mathfrak{S}^{(t_1)} \supset \mathfrak{S}^{(t_2)}$ for all $t_1, t_2, t_1 > t_2$. By $H^{(t)}$ denote the subspace of the space $L_0^2(M, \mathfrak{S}, \mu)$ consisting of functions which are measurable with respect to $\mathfrak{S}^{(t)}$. Then $H^{(t_1)} \supset H^{(t_2)}$ for $t_1 > t_2$. Suppose $\overline{H}$ is the smallest subspace of the space $L^2(M, \mathfrak{S}, \mu)$ containing all the $H^{(t)}$, $-\infty < t < \infty$, $\underline{H} = \bigcap_{-\infty<t<\infty} H^{(t)}$. Denote by H the subspace $\overline{H} \ominus \underline{H}$. According to von Neumann's theorem (see Appendix 2), the group of unitary operators $\{U^t\}$ adjoint to the flow $\{T^t\}$ has a homogeneous Lebesgue spectrum in the subspace H.

Theorem 1. *The group $\{U^t\}$ has a countable Lebesgue spectrum in the space H.*

The proof is based on the following simple lemma.

Lemma 1. *If for some $h \in L^2(M, \mathfrak{S}, \mu)$ we have the relation $(U^th, h) = 0$ for all t, $|t| \geq \text{const} > 0$, then in the cyclic subspace generated by the function h the group $\{U^t\}$ has a Lebesgue spectrum.*

The proof of this lemma will be given later.

We shall use a special representation of the flow $\{T^t\}$. According to Remark 3 after the proof of the theorem on special representations (see §3, Chap. 11), the flow $\{T^t\}$ possesses a special representation constructed from the automorphism T_1 of the Lebesgue space $(M', \mathfrak{S}', \mu')$ and the function F, $F \geq \text{const} > 0$ such that

(1) the automorphism T_1 has an increasing σ-subalgebra $\mathfrak{S}_0$, i.e.,

$$T_1\mathfrak{S}_0 \supset \mathfrak{S}_0 \text{ and } T_1\mathfrak{S}_0 \neq \mathfrak{S}_0;$$

(2) F is measurable with respect to the σ-algebra $\mathfrak{S}_0$;

(3) the minimal σ-subalgebra of the σ-algebra $\mathfrak{S}$ containing all the sets of the form $\bigcup_{0 \leq t \leq t_1} T^tC$, where $C \in \mathfrak{S}_0$ and $F(x) \geq t_1$ for almost all $x \in C$, coincides with $\mathfrak{S}^{(0)}$.

Using this special representation, we will show that there exists a sequence of functions $\{f_n\}$, $n = 1, 2, \ldots, f_n \in H$, such that the cyclic subspaces which they generate are pairwise orthogonal and the group $\{U^t\}$ has a Lebesgue spectrum in each such subspace. The statement of the theorem will then obviously follow.

Since $T_1\mathfrak{S}_0 \supset \mathfrak{S}_0$, $T_1\mathfrak{S}_0 \neq \mathfrak{S}_0$, we can find a function $h \in L^2(M', \mathfrak{S}', \mu')$ measurable with respect to $T_1\mathfrak{S}_0$ for which the conditional expectation $E(h|\mathfrak{S}_0)$ differs from h on a set of positive measure. Suppose $A \in \mathfrak{S}_0$ is such that $E((h - E(h|\mathfrak{S}_0))^2|\mathfrak{S}_0) > 0$ on it. Put

$$\psi(x) = \chi_A(x)\frac{h - E(h|\mathfrak{S}_0)}{\sqrt{E(|h - E(h|\mathfrak{S}_0)|^2|\mathfrak{S}_0)}},$$

where χ_A is the indicator of the set A. It is easy to check that

$$E(\psi(x)|\mathfrak{S}_0) = 0$$

almost everywhere; $E(|\psi|^2|\mathfrak{S}_0) = 1$ for almost all $x \in A$. Now take an arbitrary sequence of sets $A_n \in \mathfrak{S}_0$, $n = 1, 2, \ldots$ such that $A_i \cap A_j = \emptyset$ for $i \neq j$ and put

$$f_n(x, s) = \psi(x)\chi_{A_n}(x)\chi_{[0, \tau/3]}(s).$$

where $\tau = \inf F(x)$. Here (x, s) denotes points in the phase space of the special representation of the flow $\{T^t\}$.

Take an arbitrary t, $|t| < \tau/3$. Note that

$$(\operatorname{supp} U^t f_i) \cap (\operatorname{supp} f_j) = \emptyset$$

for $i \neq j$, where $\operatorname{supp} g = \{(x, s): g(x, s) \neq 0\}$. This follows from the fact that

$$\operatorname{supp} U^t f_i \subseteq (A_i \times [0, 2\tau/3))$$

when $t > 0$, while for $t < 0$,

$$\operatorname{supp} U^t f_i \subseteq (A_i \times [0, \tau/3)) \cup \{(x, s): x \in T_1^{-1}A_i, F(x) - \tau/3 \leq s < F(x)\}.$$

Therefore for such t the scalar products $(U^t f_i, f_j)$ vanish if $i \neq j$.

Now consider the scalar product $(U^t f_i, f_j)$ for $|t| \geq \tau/3$. We will assume that $t < 0$. The case $t > 0$ reduces to the previous one by means of the relation

$$(U^{-t} f_i, f_j) = (f_i, U^t f_j) = \overline{(U^t f_j, f_i)}.$$

First let us show that all the functions $U^t f_i$ are measurable with respect to the σ-algebra $\mathfrak{S}^{(0)}$. To do this note that if $C = C_1 \times [a, b]$ where $C_1 \in T_1\mathfrak{S}_0$, $[a, b] \subseteq [0, \tau/3)$, then $T^t C \in \mathfrak{S}^{(0)}$. This follows from condition (3). Since the functions f_i may be approximated by linear combinations of indicators of sets of the type C, it follows that $U^t f_i$ is measurable with respect to $\mathfrak{S}^{(0)}$. Now using the formula for the complete expectation, we may write

$$(U^t f_i, f_j) = E(E(U^t f_i \cdot \bar{f}_j | \mathfrak{S}_0)) = E(U^t f_i) E(\bar{f}_j | \mathfrak{S}_0).$$

But (1) implies that $E(\bar{f}_j | \mathfrak{S}_0) = 0$. Thus $(U^t f_i, f_j) = 0$ for all t, $|t| \geq \tau/3$. The theorem is proved. □

Proof of Lemma 1. Put $b(t) = (U^t h, h)$. By the Bohner–Khinchin theorem, the function $b(t)$ may be represented in the form $b(t) = \int_{-\infty}^{\infty} e^{i\lambda t}\, d\sigma_h(\lambda)$, where σ_h is a finite measure on $\mathbb{R}^1$. Since $b(t)$ is finite, it follows that $d\sigma_h(\lambda) = \rho(\lambda)\, d\lambda$, where $\rho(\lambda)$ may be extended to the whole complex plane so as to obtain an entire function there. Hence, for all real λ, the function $\rho(\lambda)$ vanishes on a set of points which is countable at most. The lemma is proved. □

Theorem 1 has the following consequence.

Corollary. *K-flows have countable Lebesgue spectra.*

Indeed, in the case of K-flows, the space H coincides with $L_0^2(M, \mathfrak{S}, \mu)$.

Chapter 14

Spectral Analysis of Gauss Dynamical Systems

Gauss dynamical systems were introduced in §2 of Chap. 8. There we constructed the real and complex subspaces $H_1^{(r)}$, $H_1^{(c)}$ of Hilbert space $L^2(M, \mathfrak{S}, \mu)$, where μ is the Gauss measure on the space M and, using them, obtained a necessary condition for the ergodicity of a Gauss dynamical system, consisting in the continuity of the spectral measure σ corresponding to the measure μ.

In this chapter we shall carry through a deeper study of the spectral properties of Gauss dynamical systems. We limit ourselves to the case of discrete time, i.e., to Gauss automorphisms. The results of this chapter may be carried over to Gauss systems in continuous time with obvious modifications.

We shall use the notions and notations introduced in §2, Chap. 8. Further, without special mention, we shall assume that the spectral measure σ is continuous.

§1. The Decomposition of Hilbert Space $L^2(M, \mathfrak{S}, \mu)$ into Hermite–Ito Polynomial Subspaces

Suppose T is a Gauss automorphism with phase space $(M, \mathfrak{S}, \mu)$. In this section we shall construct a special decomposition of Hilbert space $L^2(M, \mathfrak{S}, \mu)$ into subspaces invariant with respect to the unitary operator U_T adjoint to the automorphism T. The construction of this decomposition is intimately linked to the theory of Hermite–Ito polynomials of Gauss random variables. We begin by listing the necessary information concerning Hermite–Ito polynomials.

Suppose $y = f(x)$ is a real-valued random variable on the space M having a Gauss distribution with zero mean, i.e., for any α, $-\infty < \alpha < \infty$,

$$\mu(\{x \in M : f(x) < \alpha\}) = \frac{1}{\sqrt{2\pi b}} \int_{-\infty}^{\alpha} e^{-(u^2/2b)}\, du,$$

where $b = \int_M f^2\, d\mu$. Then all the powers $y^m = f^m(x)$, $m = 0, 1, 2, \ldots$ belong to $L^2(M, \mathfrak{S}, \mu)$. However, for distinct m, they are not orthogonal to each other in general.

Definition 1. The *mth Hermite–Ito polynomial* of the Gauss random variable $y = f(x)$ is the random variable $H_m(y) = y^m + a_1 y^{m-1} + \cdots + a_m$, for which

$$\int_M H_m(y) y^p \, d\mu = 0 \quad \text{for } p = 0, 1, 2, \ldots, m - 1.$$

In other words, $H_m(y)$ is an ordinary Hermite polynomial of degree m in a real variable into which the random variable y has been substituted in the role of the argument. Geometrically, $H_m(y)$ is the perpendicular constructed in $L^2(M, \mathfrak{S}, \mu)$ from the extremity of vector y^m to the subspace generated by all the $y^p, 0 \le p < m$. Clearly, $(H_{m_1}(y), H_{m_2}(y)) = 0$ for $m_1 \ne m_2$, and the subspace spanning all the $H_m(y)$, $m \ge 0$ coincides with the subspace of the space $L^2(M, \mathfrak{S}, \mu)$ of functions of the form $\varphi(y)$. Hermite–Ito polynomials (under the name of Vick polynomials) often appear in quantum field theory, where they have a different notation: $H_m(y) = \,:y^m:$. This notation is convenient and will be used below.

Now consider any finite set of Gauss random variables $y_1, \ldots, y_m \in H_1^{(r)}$.

Definition 2. *The Hermite–Ito polynomial* in the random variables $y_1, y_2, \ldots, y_m$ is the perpendicular constructed in $L^2(M, \mathfrak{S}, \mu)$ from the extremity of the vector $y_1 \cdot y_2 \cdot \cdots \cdot y_m$ to the subspace generated by all possible products $y'_1 \cdot y'_2 \cdot \cdots \cdot y'_p$, $p < m$ where all the y'_i belong to $H_1^{(r)}$.

The Hermite–Ito polynomial in the random variables $y_1, y_2, \ldots, y_m$ is denoted by $:y_1 y_2 \cdots y_m:$. By definition

$$\int :y_1 y_2 \cdots y_m: y'_1 y'_2 \cdots y'_p \, d\mu = 0 \tag{1}$$

for all $y'_1, y'_2, \ldots, y'_p \in H_1^{(r)}$. Note that among the vectors y_i and y'_j some may coincide. The following lemma establishes an important property of Hermite–Ito polynomials.

Lemma 1. *The Hermite–Ito polynomial* $:y_1 \cdots y_m:$ *coincides with the perpendicular lowered from the extremity of the vector* $y_1 \cdot y_2 \cdot \cdots \cdot y_m$ *to the subspace generated by all possible products* $y_{i_1} \cdot \cdots \cdot y_{i_p}$, $1 \le i_1, \ldots, i_p \le m$. *In other words, we can assume that the random variables* $y'_1, \ldots, y'_p$ *in the integral* (1) *are contained in the set of random variables* $y_1, \ldots, y_m$.

Proof. Denote the perpendicular mentioned in the statement of the lemma by $:\widetilde{y_1 \cdots y_m}:$. Any random variable $y' \in H_1^{(r)}$ may be uniquely written in the form $y' = \sum_{k=1}^m c_k y_k + z$ where $\int z y_k \, d\mu = 0$ for all k, $1 \le k \le m$. Therefore, if $y'_j = \sum c_{jk} y_k + z_j$, then the product $y'_1 \cdots y'_p$ may be written in the form

$$y'_1 \cdots y'_p = \sum P_l(y) Q_{p-l}(z)$$

where P_l is a polynomial of degree l in the variables $y_1, \ldots, y_m$ and Q_{p-l} is a polynomial of degree $p - l$ in the variables $z_1, \ldots, z_m$. For Gauss random variables, orthogonality implies independence. By the independence of $Q_{p-l}(z)$ from all the polynomials in $y_1, \ldots, y_m$, we shall have

$$\int :\widetilde{y_1 \cdots y_m}: P_l(y) Q_{p-l}(z) \, d\mu = \int :\widetilde{y_1 \cdots y_m}: P_l(y) \, d\mu \int Q_{p-l}(z) \, d\mu.$$

But $\int :\widetilde{y_1 \cdots y_m}: P_l(y) \, d\mu = 0$ by the definition of the random variable $\widetilde{y_1 \cdots y_m}$. The lemma is proved. □

Corollary. *If* $y_1, \ldots, y_m$ *are two-by-two orthogonal, then* $:y_1 \cdots y_m: = y_1 \cdot \cdots \cdot y_m$.

Now introduce the real and complex subspaces $H_m^{(r)}$ and $H_m^{(c)}$ of the space $L^2(M, \mathfrak{S}, \mu)$ as the closures of the sets of linear combinations of the vectors $:y_1 \cdot \cdots \cdot y_m:$ with real and complex coefficients respectively. Clearly $H_m^{(r)} \subset H_m^{(c)}$ and $H_{m_1}^{(c)} \perp H_{m_2}^{(c)}$ for $m_1 \ne m_2$. Moreover, each of the subspaces $H_m^{(r)}$, $H_m^{(c)}$ is invariant with respect to U_T.

The spaces $H_m^{(r)}$, $H_m^{(c)}$ may be constructed in the case of an arbitrary probability distribution in the space M. The next theorem sets off the Gaussian case from the others.

First we introduce the real Hilbert space $Q_m^{(r)}$, consisting of complex-valued functions $\varphi(\lambda) = \varphi(\lambda_1, \ldots, \lambda_m)$ defined for $\lambda_1, \ldots, \lambda_m \in S^1$ (i.e., for $\lambda = (\lambda_1, \ldots, \lambda_m) \in \mathrm{Tor}^m$), symmetric with respect to their variables and satisfying the relation $\varphi(-\lambda_1, \ldots, -\lambda_m) = \overline{\varphi(\lambda_1, \ldots, \lambda_m)}$ for which the norm $\| \;\; \|$ satisfies

$$\|\varphi\| \stackrel{\text{def}}{=} \left[\int_{\mathrm{Tor}^m} |\varphi(\lambda_1, \ldots, \lambda_m)|^2 \, d\sigma(\lambda_1) \cdots d\sigma(\lambda_m)\right]^{1/2} < \infty.$$ [1]

[1] The circle S^1, as usual, is identified with the semi-interval $[0, 1)$. Here and further in the chapter we sometimes write for brevity $(-\lambda_1, \ldots, -\lambda_m)$ instead of

$$(-\lambda_1) \pmod 1, \ldots, (-\lambda_m) \pmod 1).$$

Theorem 1. *For any $m \geq 1$, there exists an isometric map $\theta_m^{(r)}: Q_m^{(r)} \to H_m^{(r)}$ such that*

(1) $\theta_m^{(r)} Q_m^{(r)} = H_m^{(r)}$;
(2) *under the isomorphism $[\theta_m^{(r)}]^{-1}: H_m^{(r)} \to Q_m^{(r)}$ the operator U_T is mapped into the operator of multiplication by the function* $\exp\{2\pi i(\lambda_1 + \cdots + \lambda_m)\}$.

Proof. The map $\theta_1^{(r)}$ was constructed in Lemma 2, §2, Chap. 8. The argument in the case $m > 1$ will be split up into several steps.

1. We begin with some notations. By σ_m denote the measure $\sigma \times \sigma \times \cdots \times \sigma$ on the m-dimensional torus $\mathrm{Tor}^m = S^1 \times \cdots \times S^1$. Suppose S_m is the permutation group of the set of m elements. It acts in a natural way on the torus Tor^m and on the space of functions $\varphi(\lambda)$, $\lambda \in \mathrm{Tor}^m$. The action of a permutation $g \in S_m$ on the function φ is given by $(g\varphi)(\lambda) = \varphi(g\lambda)$. By the symmetrization of the function φ, we mean the function

$$S[\varphi] = \sum_{g \in S_m} \varphi(g\lambda).$$

Consider the group $\Gamma_m = S_m \times Z_2$, where $Z_2 = \{1, -1\}$ is the group of square roots of unity. Define the action of Γ_m on the torus Tor^m so that S_m acts as before, while the generator of the group Z_2 is the map $(\lambda_1, \ldots, \lambda_m) \to (-\lambda_1, \ldots, -\lambda_m)$.

Then the set

$$D = \{\lambda \in \mathrm{Tor}^m: \lambda_1 + \cdots + \lambda_m > m/2, \lambda_1 > \lambda_2 > \cdots > \lambda_m\},$$

will be the fundamental domain of the group Γ_m in the sense that $\sigma_m(\partial D) = 0$ and for any point $\lambda \in \mathrm{Tor}^m$ satisfying $\Gamma_m(\lambda) \cap \partial D = \varnothing$ the intersection $\Gamma_m(\lambda) \cap D$ reduces to one point. Indeed, by exchanging the coordinates of the point $\lambda = (\lambda_1, \ldots, \lambda_m)$ we can always (as soon as $\lambda_1 \neq \lambda_j$ for $i \neq j$) achieve the inequality $\lambda_1 > \lambda_2 > \cdots > \lambda_m$. Further, the point $\lambda_i \in [0, 1)$ is mapped into the point $(-\lambda_i) \pmod 1 = 1 - \lambda_i \in [0, 1)$ i.e., into the point symmetric to λ with respect to $1/2$. Therefore, by using the action of the generator of the group Z_2, we can achieve the inequality $\lambda_1 + \cdots + \lambda_m > m/2$. The relation $\sigma_m(\partial D) = 0$ follows from the fact that ∂D is the union of a finite number of hyperplanes on the m-dimensional torus of the form $\lambda_i = 0$ or $\lambda_i = \lambda_j$, while the σ_m-measure of each such hyperplane vanishes, since the measure σ is continuous.

2. We first define the map $\theta_m^{(r)}$ on some subset of functions $A \subset Q_m^{(r)}$. Suppose the $\Delta_k \subset S^1$, $k = 1, \ldots, m$, are Borel sets satisfying $\Delta_i^+ \cap \Delta_j^+ = \varnothing$ for $1 \leq i \neq j \leq m$, where $\Delta_k^+ = \Delta_k \cup (-\Delta_k)$ and $\Delta = \Delta_1 \times \cdots \times \Delta_m \subset D$. Put

$\chi^+_{\Delta_k}(\lambda_k) = \chi_{\Delta_k^+}(\lambda_k)$, $\chi^-_{\Delta_k}(\lambda_k) = i[\chi_{\Delta_k}(\lambda_k) - \chi_{-\Delta_k}(\lambda_k)]$. Fix a finite sequence $e = (e_1, \ldots, e_m)$ where e_i assumes the values ± 1 and introduce the function $\varphi \in Q_m^{(r)}$ by putting

$$\varphi(\lambda) = \sqrt{\frac{1}{m!}}\, S\left[\prod_{k=1}^{m} \chi_{\Delta_k}^{\operatorname{sgn} e_k}(\lambda_k)\right].$$

In the role of A now take the set of functions $\varphi(\lambda)$ of this type. Further put $\Phi(\Delta) = \theta_1^{(r)}\chi_\Delta(\lambda)$, $\Phi^+(\Delta) = \Phi(\Delta^+)$, $\Phi^-(\Delta) = i[\Phi(\Delta) - \Phi(-\Delta)]$.

Note that $\Phi^+(\Delta_k)$, $\Phi^-(\Delta_k)$ belong to the subspace $H_1^{(r)}$ and are therefore Gauss random variables. Further

$$(\Phi^+(\Delta), \Phi^-(\Delta)) = \int \chi_{\Delta^+}(\lambda)\chi_{\Delta^-}(\lambda)\, d\sigma(\lambda) = 0.$$

Therefore $\Phi^+(\Delta)$, $\Phi^-(\Delta)$ are independent. Now put

$$\theta_m^{(r)}(\varphi) = \prod_{k=1}^{m} \Phi^{\operatorname{sgn} e_k}(\Delta_k). \tag{2}$$

This expression is well defined, since the representation of the function φ in the form indicated above is unique.

Let us show that $\theta_m^{(r)}(\varphi) \in H_m^{(r)}$. It follows from the assumption $\Delta_i^+ \cap \Delta_j^+ = \varnothing$ that distinct factors in the right-hand side of (2) are orthogonal and therefore independent. But then, by the corollary to Lemma 1, we have: $:\theta_m^{(r)}(\varphi): = \theta_m^{(r)}(\varphi)$, i.e., $\theta_m^{(r)}(\varphi) \in H_m^{(r)}$. Further

$$\|\varphi\|^2 = \frac{1}{m!}\left\| S\left[\prod_{k=1}^{m} \chi_{\Delta_k}^{\operatorname{sgn} e_k}(\lambda_k)\right]\right\|^2 = \left\|\prod_{k=1}^{m} \chi_{\Delta_k}^{\operatorname{sgn} e_k}(\lambda_k)\right\|^2$$

$$= 2^m \left\|\prod_{k=1}^{m} \chi_{\Delta_k}(\lambda_k)\right\| = 2^m \prod_{k=1}^{m} \sigma(\Delta_k).$$

On the other hand, by the independence of the $\Phi^{\pm}(\Delta_k)$ for distinct k

$$\|\theta_m^{(r)}(\varphi)\|^2 = \left\|\prod_{k=1}^{m} \Phi^{\operatorname{sgn} e_k}(\Delta_k)\right\| = \prod_{k=1}^{m} \|\chi_{\Delta_k}^{\operatorname{sgn} e_k}(\lambda_k)\|^2 = 2^m \prod_{k=1}^{m} \sigma(\Delta_k).$$

The last two relations show that $\theta_m^{(r)}$ is an isometric map on the set A.

3. Consider k functions $\varphi_j \in A$, $1 \le j \le k$ whose supports are disjoint and, for any set of real numbers $a_1, \ldots, a_k$, put $\varphi(\lambda) = \sum_{j=1}^{k} a_j \varphi_j(\lambda)$. By $\tilde{A}$ denote the set of functions obtained in this way. Since the representation of a function

$\varphi \in A$ in the form indicated above is unique, we may extend $\theta_m^{(r)}$ by linearity to $\tilde{A}$ by putting

$$\theta_m^{(r)}(\varphi) = \sum_{j=1}^{k} a_j \theta_m^{(r)}(\varphi_j).$$

Then $\theta_m^{(r)}$ remains isometric, the supports of φ_j being pairwise disjoint.

4. Suppose $\Delta^{(j)} \subset D$, $1 \le j \le k$ are disjoint m-dimensional rectangles and c_j, $1 \le j \le k$ is a set of complex numbers. Put

$$\varphi(\lambda) = \sum_{j=1}^{k} \{c_j S[\chi_{\Delta^{(j)}}(\lambda)] + \bar{c}_j S[\chi_{-\Delta^{(j)}}(\lambda)]\},$$

and suppose C is the set of all functions of this form. Let us show that $\tilde{A} = \tilde{C}$.

Clearly $\tilde{A} \subseteq \tilde{C}$. Moreover, since the functions from $\tilde{A}$ and $\tilde{C}$ are invariant with respect to the action of the group Γ_m, it suffices to compare the restrictions of these functions to the fundamental domain D. Denote by $\tilde{A}_D$, $\tilde{C}_D$ the spaces obtained by restricting the functions from $\tilde{A}$, $\tilde{C}$ respectively to D. Then $\tilde{C}_D$, viewed as a real linear space, is generated by functions of the form

$$\varphi_\Delta^+ = \chi_\Delta, \qquad \varphi_\Delta^- = i\chi_\Delta, \qquad \Delta = \Delta_1 \times \cdots \times \Delta_m \subset D.$$

But for $\lambda \in D$

$$\varphi_\Delta^+(\lambda) = S\left[\prod_{k=1}^{m} \chi_{\Delta_k}^+(\lambda_k)\right] \in \tilde{A},$$

$$\varphi_\Delta^-(\lambda) = S\left[\chi_{\Delta_1}^-(\lambda_1) \prod_{k=2}^{m} \chi_{\Delta_k}^+(\lambda_k)\right] \in \tilde{A}.$$

Since $\tilde{A}$ is also a real linear space, our statement is proved.

5. In this subsection we will show that the closure of the set $\tilde{A}$ in the space $L^2(\mathrm{Tor}^m, \sigma_m)$ is the entire set $Q_m^{(r)}$.

Suppose $\varphi(\lambda) \in Q_m^{(r)}$. For any $\varepsilon > 0$ we can find a function

$$\tilde{\varphi}_\varepsilon(\lambda) \in L^2(\mathrm{Tor}^m, \sigma_m)$$

such that

(1) $\tilde{\varphi}_\varepsilon$ vanishes in an ε-neighborhood of the set $\mathrm{Tor}^m \setminus D$;
(2) $\int_D |\varphi - \tilde{\varphi}_\varepsilon|^2 \, d\sigma_m < \varepsilon$.

Further we can find a set of disjoint m-dimensional rectangles $\Delta^{(j)} \subset D$, $1 \le j \le k$ and complex numbers c_j, $1 \le j \le k$, such that

$$\left\| \tilde{\varphi}_\varepsilon(\lambda) - \sum_{j=1}^{k} c_j \chi_{\Delta^{(j)}}(\lambda) \right\| < \varepsilon.$$

Put

$$\varphi_\varepsilon = \sum_{j=1}^{k} \{c_j S[\chi_{\Delta^{(j)}}] + \bar{c}_j S[\chi_{-\Delta^{(j)}}]\}.$$

Then $\|\varphi - \varphi_\varepsilon\| < (1 + 2m!)\varepsilon$. It follows from the previous subsection that $\varphi_\varepsilon \in \tilde{A}$. Our statement is proved.

6. Since the map $\theta_m^{(r)}$ is isometric, it may be extended to the closure of $\tilde{A}$, i.e. (by the previous subsection) to $Q_m^{(r)}$. Let us show that

$$\theta_m^{(r)}(Q_m^{(r)}) = H_m^{(r)}.$$

It follows obviously from the construction that $\theta_m^{(r)}(Q_m^{(r)}) \subset H_m^{(r)}$. Further, the real linear envelope of the set of Hermite–Ito polynomials of the form

$$:\prod_{k=1}^{m} (c_k \Phi(\Delta_k) + \bar{c}_k \Phi(-\Delta_k):,$$

where Δ_k is a birational semi-interval of the form

$$\Delta_{p,q} = [p/2^q, (p+1)/2^q], \qquad 1 \le q < \infty, 0 \le p < 2^q$$

is dense in $H_m^{(r)}$. The relation

$$c\Phi(\Delta) + \bar{c}\Phi(-\Delta) = a\Phi^+(\Delta) + b\Phi^-(\Delta), \qquad a = \operatorname{Re} c, b = \operatorname{Im} c,$$

implies that the same linear envelope is generated by Hermite–Ito polynomials of the form $:\prod_{k=1}^{m} \Phi^{\pm}(\Delta_k):$. Therefore it suffices to prove that each $\Pi = :\prod_{k=1}^{m} \Phi^{\pm}(\Delta_k):$ belongs to $\theta_m^{(r)}(Q_m^{(r)})$.

Every birational semi-interval $\Delta_{p,q}$ may be represented as the union of semi-intervals $\Delta_{l,n}$ if n is large enough. Then Π may be represented as the sum of Hermite–Ito polynomials of the form $:\prod_{k=1}^{m} \Phi^{\pm}(\Delta_{p_k,n}):$. This corresponds to the decomposition of the m-dimensional rectangle $\Delta = \prod_{k=1}^{m} \Delta_k$ into nonintersecting m-dimensional rectangles $\prod_{k=1}^{m} \Delta_{p_k,n}$. Let us write Π in the form $\Pi = \Sigma_n' + \Sigma_n''$, where the sum in Σ_n' is taken over all the $p_1, \ldots, p_m$ for which $(\Delta_{p_1,n}, \ldots, \Delta_{p_m,n})$ is a family of nonintersecting intervals, while Σ_n'' is the sum over all the other families. For each family $(\Delta_{p_1,n}, \ldots, \Delta_{p_m,n})$ in Σ', the Gauss random variables $\Phi^{\pm}(\Delta_{p_k,n})$ are independent. Therefore in this case

$$:\prod_{k=1}^{m} \Phi^{\pm}(\Delta_{p_k,n}): = \prod_{k=1}^{m} \Phi^{\pm}(\Delta_{p_k,n}),$$

and

$$:\prod_{k=1}^{m} \Phi^{\pm}(\Delta_{p_k,n}): = \prod_{k=1}^{m} \Phi^{\pm}(\Delta_{p_k,n}) = \theta_m^{(r)}\left(\sqrt{\frac{1}{m!}}\, S\left[\prod_{k=1}^{m} \chi^{\pm}_{\Delta_{p_k,n}}(\lambda_k)\right]\right).^{1}$$

Thus $\Sigma_n' \in \theta_m^{(r)}(Q_m^{(r)})$. The necessary statement now follows from the fact (which we shall prove below) that $\|\Sigma_n''\|^2 = E(\Sigma_n''^2) \to 0$ when $n \to \infty$.

7. Now notice that the birational semi-intervals

$$[p'/2^n, (p' + 1)/2^n), \qquad [p''/2^n, (p'' + 1)/2^n)$$

are either disjoint or coincide. Consider a separate summand in Σ_n''. Let us enumerate the semi-interval which appear in it in increasing order of p_k: $\Delta_1 < \cdots < \Delta_{m'}$. Then $m' < m$, since at least one semi-interval appears twice. Let us write this summand in the form: $\prod_{i=1}^{m} (\Phi^+(\Delta_i))^{r_i^+}(\Phi^-(\Delta_i))^{r_i^-}$:. According to Lemma 1,

$$:\prod_{i=1}^{m'} (\Phi^+(\Delta_i))^{r_i^+}(\Phi^-(\Delta_i))^{r_i^-}: = \prod_{i=1}^{m'} :(\Phi^+(\Delta_i))^{r_i^+}(\Phi^-(\Delta_i))^{r_i^-}: = \Pi'.$$

Then $r_i^+ + r_i^- > 1$ for at least one i.

Now let us estimate (Σ_n'', Σ_n'') directly. It follows from the orthogonality of the Hermite–Ito polynomials that the scalar product of two distinct products of the Π' type vanishes. Therefore,

$$(\Sigma_n'', \Sigma_n'') = \sum_{\Pi'} (\Pi', \Pi') = \sum \prod_{i=1}^{m'} E[(:\Phi^+(\Delta_i))^{r_i^+}(\Phi^-(\Delta_i))^{r_i^-}:]^2.$$

Let us show that

$$E[:(\Phi^+(\Delta_i))^{r_i^+}(\Phi^-(\Delta_i))^{r_i^-}:]^2 \le \text{const}_m\, \sigma^{r_i^+ + r_i^-}(\Delta_i),$$

where the const_m depends only on m.

[1] In the notation $:\prod \Phi^{\pm}(\Delta_{p_k,n}):$ we implicitly assume that each factor is supplied with its own sign, the combination of signs being the same in the right-hand side as in the left-hand side.

Indeed, since $\Phi^+(\Delta)$ and $\Phi^-(\Delta)$ are independent and Gauss, we have

$$\begin{aligned} E[:(\Phi^+(\Delta_i))^{r_i^+}(\Phi^-(\Delta_i))^{r_i^-}:]^2 &= E[:(\Phi^+(\Delta_i))^{r_i^+}::(\Phi^-(\Delta_i))^{r_i^-}:]^2 \\ &= E[:(\Phi^+(\Delta_i))^{r_i^+}:]^2 E[:(\Phi^-(\Delta_i))^{r_i^-}:]^2 \\ &\le \mathrm{const}_{2r_i^+}\, \sigma^{r_i^+}(\Delta_i) \mathrm{const}_{2r_i^-}\, \sigma^{r_i^-}(\Delta_i) \\ &\le \mathrm{const}_m\, \sigma^{r_i^+ + r_i^-}(\Delta_i). \end{aligned}$$

But this is the required inequality. Therefore

$$\prod_{i=1}^{m'} E[:(\Phi^+(\Delta_i))^{r_i^+}(\Phi^-(\Delta_i))^{r_i^-}:]^2 \le \mathrm{const}_m\, \sigma_m(\Delta),$$

where $\Delta = \Delta_{p_1,n} \times \cdots \times \Delta_{p_m,n}$ is the original m-dimensional rectangle corresponding to the summand from Σ_m'' and const_m depends only on m. Thus

$$(\Sigma_n'', \Sigma_n'') \le \mathrm{const}_m\, \sigma_m(D_n),$$

where D_n is the union of the rectangles $\Delta_{p_1,n} \times \cdots \times \Delta_{p_m,n}$ which intersect hyperplanes of the type $\lambda_i = \lambda_j$, $\lambda_i = -\lambda_j$ on the m-dimensional torus. Since the measure σ is continuous, the σ_m-measure of each of these hyperplanes is 0. Therefore $\sigma_m(D_n) \to 0$ when $n \to \infty$ and our statement is proved.

8. It remains to show that the operator U_T is mapped by $\theta_m^{(r)}$ into the operator of multiplication by $\exp\{2\pi i \sum_{k=1}^m \lambda_k\}$. To do this we will show that

$$:\prod_{k=1}^m x(s_k): = \theta_m^{(r)}\left(\sqrt{\frac{1}{m!}}\, S\left[\exp\left\{2\pi i \sum_{k=1}^m \lambda_k s_k\right\}\right]\right). \tag{3}$$

Fix ε, $0 < \varepsilon < 1$. For any k, $1 \le k \le m$, construct the random variable

$$f_k^{(\varepsilon)}(x) = \sum a_{p,n}^{(k)} \Phi^{\pm}(\Delta_{p,n}) \in H_1^{(r)},$$

so as to have

$$\|x(s_k) - f_k^{(\varepsilon)}\| = \|e^{is_k\lambda} - \varphi_k^{(\varepsilon)}(\lambda)\| \le \varepsilon, \qquad \varphi_k^{(\varepsilon)}(\lambda) = \sum_p a_{p,n}^{(k)} \chi_{\Delta_{p,n}}^{\pm}(\lambda).$$

Put $\Phi_\varepsilon = \prod_{k=1}^m f_k^{(\varepsilon)}$. Then

$$\left\| \prod_{k=1}^m x(s_k) - \Phi_\varepsilon \right\| \le \mathrm{const}\ \varepsilon$$

and therefore

$$\left\| :\prod_{k=1}^{m} x(s_k): - :\Phi_\varepsilon: \right\| \leq \text{const } \varepsilon.$$

But

$$\Phi_\varepsilon = \sum a^{(1)}_{p_1, n} \cdot a^{(2)}_{p_2, n} \cdot \dots \cdot a^{(m)}_{p_m, n} \prod_{k=1}^{m} \Phi^{\pm}(\Delta_{p_k, n}).$$

Let us split up the last sum into two: $\Sigma = \Sigma' + \Sigma''$, $\Phi_\varepsilon = \Phi'_\varepsilon + \Phi''_\varepsilon$, where in Σ' we retain only the sum over those families $p_1, \dots, p_m$ for which

$$\prod_{k=1}^{m} \Delta_{p_k, n} \cap \partial D = \varnothing$$

and Σ'' contains all the other summands. As in the previous subsection, $\|\Sigma''\| \to 0$, when $n \to \infty$. Therefore, for n sufficiently large,

$$\left\| :\prod_{k=1}^{m} x(s_k): - :\Phi'_\varepsilon: \right\| \leq \text{const } \varepsilon.$$

Put $\varphi^{(\varepsilon)}(\lambda) = \varphi^{(\varepsilon)}(\lambda_1, \dots, \lambda_m) = \prod_{k=1}^{m} \varphi^{(\varepsilon)}_k(\lambda_k)$. Then

$$\left\| \exp\left(2\pi i \sum_{k=1}^{m} s_k \lambda_k\right) - \varphi^{(\varepsilon)}(\lambda) \right\|_{L^2(\mathrm{Tor}^m, \sigma_m)} \leq \text{const } \varepsilon,$$

therefore

$$\left\| \sqrt{\frac{1}{m!}}\, S\left[\exp\left(2\pi i \sum_{k=1}^{m} s_k \lambda_k\right)\right] - \sqrt{\frac{1}{m!}}\, S[\varphi^{(\varepsilon)}(\lambda)] \right\|_{Q^{(r)}_m} \leq \text{const } \varepsilon.$$

Further

$$\varphi^{(\varepsilon)}(\lambda) = \sum a^{(1)}_{p_1, n} \cdot a^{(2)}_{p_2, n} \cdot \dots \cdot a^{(m)}_{p_m, n} \prod_{k=1}^{m} \chi^{\pm}_{\Delta_{p_k, n}}(\lambda_k).$$

Split up this sum in the same manner as above into two sums Σ', Σ'', i.e., $\Sigma = \Sigma' + \Sigma''$, $\varphi^{(\varepsilon)} = (\varphi^{(\varepsilon)})' + (\varphi^{(\varepsilon)})''$. Then $\|(\varphi^{(\varepsilon)})''\| \leq \varepsilon$ for sufficiently large n, so that we have

$$\left\| \sqrt{\frac{1}{m!}}\, S\left[\exp\left(2\pi i \sum_{k=1}^{m} s_k \lambda_k\right)\right] - \sqrt{\frac{1}{m!}}\, S[(\varphi^{(\varepsilon)})'] \right\| \leq \text{const } \varepsilon.$$

Since $\theta_m^{(r)}$ is isometric

$$\left\| \theta_m^{(r)}\left(\sqrt{\frac{1}{m!}}\, S\left[\exp\left(2\pi i \sum_{k=1}^{m} s_k \lambda_k\right)\right]\right) - \theta_m^{(r)}\left(\sqrt{\frac{1}{m!}}\, S[(\varphi^{(\varepsilon)})']\right) \right\| \leq \text{const}\ \varepsilon.$$

But by the definition of $\theta_m^{(r)}$

$$\theta_m^{(r)}\left(\sqrt{\frac{1}{m!}}\, S[(\varphi^{(\varepsilon)})']\right) = :\Phi_\varepsilon':.$$

Thus we obtain

$$\left\| :\prod_{k=1}^{m} x(s_k): - \theta_m^{(r)}\left\{\sqrt{\frac{1}{m!}}\, S\left[\exp\left(2\pi i \sum_{k=1}^{m} s_k \lambda_k\right)\right]\right\} \right\| \leq \text{const}\ \varepsilon.$$

Since ε was arbitrary, formula (3) is proved.

In the space $Q_m^{(r)}$ consider the cylic group of operators $\{V^n\}$, $n \in Z$, where for $\varphi \in Q_m^{(r)}$ we have

$$(V^n \varphi)(\lambda) = \exp\left(2\pi i n \sum_{k=1}^{m} \lambda_k\right)\varphi(\lambda).$$

Let us show that the space $H_m^{(r)}$ is invariant with respect to U_T^n, $-\infty < n < \infty$ and $U_T^n = \theta_m^{(r)} V^n (\theta_m^{(r)})^{-1}$.

It obviously suffices to check the last equality. Put $f = :\prod_{k=1}^m x(s_k):$. Since $U_T^n f = :\prod_{k=1}^n x(s_k + n)$: it follows from the above that

$$\begin{aligned} U_T^n f &= \theta_m^{(r)}\left(\sqrt{\frac{1}{m!}}\, S\{\exp[2\pi i \sum (s_k + n)\lambda_k]\}\right) \\ &= \theta_m^{(r)}\left\{\exp(2\pi i n \sum \lambda_k)\sqrt{\frac{1}{m!}}\, S\left[\exp\left(2\pi i \sum_{k=1}^{m} s_k \lambda_k\right)\right]\right\} \\ &= \theta_m^{(r)} V^n (\theta_m^{(r)})^{-1} f. \end{aligned}$$

Thus our statement is valid for the function f. Since the linear envelope of such functions generates the entire space $H_m^{(r)}$, Theorem 1 is proved entirely. □

Now denote by $H^{(r)}$ the space of all the real-valued functions $f \in L^2(M, \mathfrak{S}, \mu)$.

Theorem 2. *The direct sum* $\oplus \sum_{m=0}^{\infty} H_m^{(r)}$ *coincides with* $H^{(r)}$. *Here* $H_0^{(r)}$ *denotes the one-dimensional space of real constants.*

Proof. For any family of integers $s_1, s_2, \ldots, s_p$ introduce the real subspace $H_{s_1,\ldots,s_p} \subset H^{(r)}$ consisting of functions of the form

$$f(x(s_1), \ldots, x(s_p)) \in L^2(M, \mathfrak{S}, \mu).$$

Since the closure of the sum of subspaces $H_{s_1,\ldots,s_p}$ coincides with $H^{(r)}$, it suffices to show that

$$H_{s_1,\ldots,s_p} \subset \bigoplus \sum_{m=0}^{\infty} H_m^{(r)}.$$

Suppose the random variables $y_i = x(s_i)$, $1 \le i \le p$ have a density of the joint probability distribution of the form

$$\psi(t_1, \ldots, t_p) = \frac{\sqrt{\det A}}{(2\pi)^{p/2}} \exp\{-\tfrac{1}{2}(At, t)\}$$

where A is a symmetric positive definite matrix and $t = (t_1, \ldots, t_p)$. Any function $f \in H_{s_1,\ldots,s}$ may be identified with a function $f(t_1, \ldots, t_p)$ for which

$$\int f^2(t_1, \ldots, t_p)\psi(t_1, \ldots, t_p)\, dt_1 \cdots dt_p < \infty.$$

By means of a linear change of variables we may put ψ in the form

$$\psi(t_1, \ldots, t_p) = \frac{1}{(2\pi)^{p/2}} \exp(-\tfrac{1}{2} \sum t_i^2)$$

For any indicator $\chi_\Delta(t)$ and any $\varepsilon > 0$, there is a polynomial $P_\varepsilon(t)$ satisfying

$$\frac{1}{\sqrt{2\pi}} \int_{-\infty}^{\infty} |\chi_\Delta(t) - P_\varepsilon(t)|^2 e^{-t^2/2}\, dt < \varepsilon.$$

This follows, for example, from the fact that ordinary Hermite polynomials (in a real variable) are dense in the Hilbert space of square integrable functions with respect to the measure $e^{-t^2/2}\, dt$.

This implies that for any indicator

$$\chi_\Delta(t_1, \ldots, t_p) = \prod_{i=1}^{p} \chi_{\Delta_i}(t_i),$$

there is a polynomial $P_\varepsilon(t_1, \ldots, t_p)$ such that

$$\frac{1}{(2\pi)^{p/2}} \int |\chi_\Delta(t) - P_\varepsilon(t)|^2 \exp(-\tfrac{1}{2} \sum t_i^2)\, dt < \varepsilon.$$

But the polynomial $P_\varepsilon(t_1, \ldots, t_p)$ can obviously be written as the sum of products of Hermite polynomials in one variable. Since to each such product corresponds an element of the space $\oplus \sum_{m=0}^{\infty} H_m^{(r)}$, our theorem is proved. □

Put $H_m^{(c)} = H_m^{(r)} + iH_m^{(r)}$, $Q_m^{(c)} = Q_m^{(r)} + iQ_m^{(r)}$, $m = 0, 1, \ldots$. Then $H_m^{(c)}$ is a (complex) subspace of the space $L^2(M, \mathfrak{S}, \mu)$. Clearly $H_{m_1}^{(c)} \perp H_{m_2}^{(c)}$ for $m_1 \neq m_2$. Define the map $\theta_m^{(c)}: Q_m^{(c)} \to H_m^{(c)}$ in the following way. Any function $\varphi \in Q_m^{(c)}$ may be written in the form

$$\varphi(\lambda) = \varphi_1(\lambda) + i\varphi_2(\lambda), \qquad \varphi_1(\lambda) = \tfrac{1}{2}[\varphi(\lambda) + \overline{\varphi(-\lambda)}],$$

$$\varphi_2(\lambda) = \frac{1}{2i}[\varphi(\lambda) - \overline{\varphi(-\lambda)}], \qquad \varphi_1, \varphi_2 \in Q_m^{(r)};$$

now put $\theta_m^{(c)}(\varphi) = \theta_m^{(r)}(\varphi_1) + i\theta_m^{(r)}(\varphi_2) \in H_m^{(c)}$. For any function $\varphi \in Q_m^{(c)}$, put

$$(V^n\varphi)(\lambda) = \exp\left(2\pi i \sum_{k=1}^{m} \lambda_k\right)\varphi(\lambda).$$

Theorem 3. *The subspaces $H_m^{(c)}$ are invariant with respect to the unitary operator U_T adjoint to the Gauss automorphism T. Moreover,*

(1) $\oplus \sum_{m=0}^{\infty} H_m^{(c)} = L^2(M, \mathfrak{S}, \mu)$; *here $H_0^{(c)}$ is the subspace of complex constants;*

(2) $\theta_m^{(c)}$ *is an isometric map between the spaces $Q_m^{(c)}$ and $H_m^{(c)}$ satisfying $U_T^n = \theta_m^{(c)} V^n (\theta_m^{(c)})^{-1}$.*

Theorem 3 immediately follows from Theorems 1 and 2.

§2. Ergodicity and Mixing Criteria for Gauss Dynamical Systems

In §2, Chap. 8, we show that the necessary condition for the ergodicity of a Gauss automorphism is the continuity of the spectral measure σ. Now we shall show that this condition is also sufficient. Here and in the following sections of Chap. 14 we shall use the notations introduced in §1.

Theorem 1. *If the spectral measure σ is continuous, then the space of eigenfunctions of the unitary operator U_T adjoint to the Gauss automorphism T consists of constants* (mod 0). *In other words, the automorphism T is ergodic and weak mixing.*

Proof. Suppose $f \in L^2(M, \mathfrak{S}, \mu)$ is an eigen-function of the operator U_T, i.e., $U_T f = e^{i\lambda_0} f$. We can represent f in the form $f = \sum_{m=0}^{\infty} f_m$, $f_m \in H_m^{(c)}$, $(f_{m_1}, f_{m_2}) = 0$ for $m_1 \neq m_2$. Since f is an eigen-function, while the space $H_m^{(c)}$ is invariant with respect to U_T, all the f_m are also eigen-functions with the same eigen-value $e^{i\lambda_0}$, i.e., $U_T f_m = e^{i\lambda_0} f_m$. Suppose $\varphi_m = (\theta_m^{(c)})^{-1} f_m$. By Theorem 3, §1, for $m \geq 1$, we shall have

$$\begin{aligned} V^n \varphi_m &= \exp\left(2\pi i n \sum_{k=1}^{m} \lambda_k\right) \varphi_m(\lambda_1, \ldots, \lambda_m) = V^n (\theta_m^{(c)})^{-1} f_m \\ &= (\theta_m^{(c)})^{-1} U_T^n f_m = (\theta_m^{(c)})^{-1} (e^{in\lambda_0} f_m) = e^{in\lambda_0} (\theta_m^{(c)})^{-1} f_m = e^{in\lambda_0} \varphi_m, \end{aligned}$$

i.e., $\exp(2\pi i n \sum_{k=1}^{m} \lambda_k) \varphi_m = e^{in\lambda_0} \varphi_m$, where the equal sign means equality of elements of the space $Q_m^{(c)}$. Therefore,

$$\begin{aligned} &\int \left| \left(2\pi i n \sum_{k=1}^{m} \lambda_k\right) \varphi_m(\lambda_1, \ldots, \lambda_m) - e^{in\lambda_0} \varphi_m(\lambda_1, \ldots, \lambda_m) \right|^2 d\sigma(\lambda_1) \cdots d\sigma(\lambda_m) \\ &\quad = \int \left| \exp\left(2\pi i n \sum_{k=1}^{m} \lambda_k\right) - e^{in\lambda_0} \right|^2 |\varphi_m(\lambda_1, \ldots, \lambda_m)|^2 \, d\sigma(\lambda_1) \cdots d\sigma(\lambda_m) = 0. \end{aligned}$$

Therefore it follows that $\varphi_m(\lambda_1, \ldots, \lambda_m) = 0$ almost everywhere outside of the set

$$D_{\lambda_0} = \{(\lambda_1, \ldots, \lambda_m) \colon \exp(2\pi i \textstyle\sum \lambda_k) = e^{i\lambda_0}\}.$$

By the continuity of the measure σ the measure σ_m of the set D_{λ_0} is 0. Therefore, for $m \geq 1$, we have $\varphi_m = 0$ almost everywhere with respect to the measure σ_m. In other words, $f = f_0$. The theorem is proved. □

It is just as easy now to find necessary and sufficient conditions for mixing.

Theorem 2. *In order that a Gauss automorphism T be a mixing, it is necessary and sufficient that the sequence*

$$b(n) = \int e^{2\pi i n \lambda} \, d\sigma(\lambda)$$

tends to zero when $|n| \to \infty$.

Proof. The necessity of the condition is obvious since $b(n) = (U_T^n f, f)$, where $f(x) = x(0)$. Let us prove its sufficiency.

Note that it suffices to establish the relation $\lim_{|n| \to \infty} (U_T^n f, f) \to 0$ where $f \in L_0^2(M, \mathfrak{S}, \mu)$ for some set F of functions f which is dense in $L_0^2(M, \mathfrak{S}, \mu)$.

Indeed, then for any function $f \in L_0^2(M, \mathfrak{S}, \mu)$, $\|f\| = 1$ and any $\varepsilon > 0$ we can find a function $f' \in F$, $\|f' - f\| < \varepsilon/3$ and a number $n_0(\varepsilon)$ such that

$$\|(U_T^n f', f'\| < \frac{\varepsilon}{3} \qquad \text{for all } n, |n| > n_0(\varepsilon).$$

For such n,

$$\begin{aligned}(U_T^n f, f)| &= |(U_T^n(f - f'), f) + (U_T^n f', f - f') + (U_T^n f', f')| \\ &\leq \|U_T^n(f - f')\| + \|f - f'\| + |(U_T^n f', f')| < \varepsilon.\end{aligned}$$

Therefore $\lim_{|n|\to\infty}(U_T^n f, f) \to 0$.

For the set F, take the family of all functions $\sum_{m=1}^N f_m$ where $f_m \in H_m^{(c)}$. By the orthogonality of $H_m^{(c)}$ (for distinct m), and their invariance with respect to U_T, we have

$$\left(U_T^n \sum_{m=1}^N f_m, \sum_{m=1}^N f_m\right) = \sum_{m=1}^N (U_T^n f_m, f_m).$$

It therefore suffices to consider the case $f = f_m$. Suppose $\varphi_m = (\theta_m^{(c)})^{-1} f_m$. Then by Theorem 3, §1,

$$(U_T^n f_m, f_m) = \int \exp\left(2\pi i n \sum_{k=1}^m \lambda_k\right) |\varphi_m(\lambda)|^2 \, d\sigma(\lambda_1) \cdots d\sigma(\lambda_m).$$

Denote by $\sigma^{(m)}$ the measure $\underbrace{\sigma * \sigma * \cdots * \sigma}_{m}$ (where $*$ is the convolution). We have

$$(U_T^n f_m, f_m) = \int e^{2\pi i n \alpha} \rho(\alpha) \, d\sigma^{(m)}(\alpha),$$

where $\rho(\alpha)$ is the conditional expectation of $|\varphi_m(\lambda)|^2$ under the condition $\sum_{k=1}^m \lambda_k = \alpha$. Since $\varphi_m \in Q_m^{(c)}$, we have $\int \rho(\alpha) \, d\sigma^{(m)}(\alpha) < \infty$. Now notice that

$$\int e^{2\pi i n \alpha} \, d\sigma^{(m)}(\alpha) = b^m(n) \to 0,$$

when $|n| \to \infty$ by the assumption of the theorem. Therefore the necessary statement follows from the generalized Riemann–Lebesgue lemma stated and proved below.

The Generalized Riemann–Lebesgue Lemma. *Suppose ν_0 is a finite measure on the circle and its Fourier coefficients $\int e^{2\pi i n\lambda}\, d\nu_0(\lambda)$ tend to zero when $|n| \to \infty$. Then, for any finite measure ν absolutely continuous with respect to ν_0 and possessing the density $\rho(\lambda) = d\nu(\lambda)/d\nu_0(\lambda)$, the Fourier coefficients $\int e^{2\pi i\lambda n} \times \rho(\lambda)\, d\nu_0(\lambda)$ tend to zero when $|n| \to \infty$.*

Proof. Arguing as in the proof of Theorem 2, we can show that it suffices to prove the lemma for some set of densities $\rho(\lambda)$ which is dense in the space $L^1(S^1, \nu_0)$. For such a set we can take the set of densities of the form $\rho(\lambda) = \sum a_k \chi_{\Delta_k}(\lambda)$, where each χ_{Δ_k} is the indicator of some interval.

In fact it suffices to consider the case $\rho(\lambda) = \chi_\Delta(\lambda)$. Let us find a sequence of trigonometric polynomials $P_r(\lambda)$ for which

$$\int |P_r(\lambda) - \chi_\Delta(\lambda)|\, d\nu_0(\lambda) \to 0 \quad \text{when } r \to \infty$$

For every P_r we obviously have

$$\int e^{2\pi i n\lambda} P_r(\lambda)\, d\nu_0(\lambda) \to 0 \quad \text{when } |r| \to \infty.$$

Therefore

$$\int e^{2\pi i n\lambda} \chi_\Delta(\lambda)\, d\nu_0(\lambda) \to 0 \quad \text{when } |n| \to \infty.$$

The lemma is proved. □

If σ is absolutely continuous, it follows from the usual Riemann–Lebesgue lemma that $b(n) \to 0$ when $|n| \to \infty$ and therefore in this case the Gauss automorphism T is mixing. There exist singular measures for which $b(n) \to 0$. But at the same time it is possible to construct continuous singular measures for which $b(n)$ does not tend to 0 when $|n| \to \infty$. This shows the existence of weak mixing Gauss automorphism which is not mixing.

§3. The Maximal Spectral Type of Unitary Operators Adjoint to Gauss Dynamical Systems

We shall now use the decomposition of Hilbert space $L^2(M, \mathfrak{S}, \mu)$ into subspaces $H_m^{(c)}$ of Hermite–Ito polynomials constructed in §1 in order to study the spectral properties of Gauss dynamical systems. Recall the following notation

$$\sigma_m = \underbrace{\sigma \times \sigma \times \cdots \times \sigma}_{m \text{ factors}}, \qquad \sigma^{(m)} = \underbrace{\sigma * \sigma * \cdots * \sigma}_{m \text{ factors}},$$

where σ is the spectral measure corresponding to the Gauss measure μ.

Lemma 1. *The maximal spectral type of a unitary operator U_T adjoint to a Gauss automorphism T in the subspace $H_m^{(c)}$ equals $\sigma^{(m)}$.*

Proof. Suppose $f \in H_m^{(c)}$ and $\varphi = (\theta_m^{(c)})^{-1}f$. Then for any n, we have

$$\begin{aligned}(U_T^n f, f) &= (V^n\varphi, \varphi)\\ &= \int \exp\left(2\pi i n \sum_{k=1}^{m} \lambda_k\right) |\varphi(\lambda_1, \ldots, \lambda_m)|^2 \, d\sigma_m\\ &= \int \exp(2\pi i n\alpha)\rho(\alpha)\, d\sigma^{(m)}(\alpha),\end{aligned}$$

where $\rho(\alpha)$ is the conditional expectation of $|\varphi(\lambda)|^2$ under the condition $\sum_{k=1}^{m} \lambda_k = \alpha$. It then follows that the spectral measure of the element f is absolutely continuous with respect to $\sigma^{(m)}$. For the function $f = \theta_m^{(c)}(1)$ this measure obviously equals $\sigma^{(m)}$. The lemma is proved. □

Theorem 1. *The maximal spectral type of the operator U_T is equal to the type of the measure*

$$e^{\sigma} = \delta + \sigma + \sigma^{(2)}/2! + \sigma^{(3)}/3! + \cdots,$$

where δ is a normalized measure supported at the point $\lambda = 0$.

Proof. The spectral type of δ is realized on the subspace of constants. Therefore the theorem follows from the invariance of the subspaces $H_m^{(c)}$ and from Lemma 1. □

For any measure σ on S^1 the measure $e^{\sigma} * e^{\sigma}$ is equivalent to the measure e^{σ}. If it were possible, for the measure e^{σ} to find a natural support Λ then this property would mean that $\Lambda + \Lambda = \Lambda$, where $\Lambda + \Lambda$ is the arithmetical sum of the set Λ with itself. Since for Gauss dynamical systems the spectral measure σ is symmetric, we see that e^{σ} is also symmetric, so that $-\Lambda = \Lambda$. In other words, the support Λ must be a subgroup of the group S^1. However, in the general case no natural notion of a support of a measure exists, and the fact that the maximal spectral type of the Gauss dynamical system is e^{σ} must be viewed as the analogue of the group property of spectra proved in Chap. 12 for dynamical systems with pure point spectra.

Theorem 2. *If the spectral measure σ is absolutely continuous with respect to the Lebesgue measure, then in the subspace $L_0^2(M, \mathfrak{S}, \mu)$ the unitary operator U_T adjoint to the Gauss automorphism T has the countable Lebesgue spectrum.*

The proof is based on the following lemma.

Lemma 2. *If the measure σ on S^1 is absolutely continuous with respect to the Lebesgue measure, then we can find a natural number m such that the measure $\sigma^{(m)}$ is equivalent to the Lebesgue measure.*

Proof of Lemma 2. Suppose $p(\lambda)$ is the density of the measure σ with respect to the Lebesgue measure, i.e.,

$$\frac{d\sigma(\lambda)}{d\lambda} = p(\lambda) \in L^1(S^1, d\lambda).$$

Then the density of the measure $\sigma^{(2)} = \sigma * \sigma$ equals

$$p_2(\lambda) = \int p(\lambda - \xi)p(\xi)\, d\xi.$$

It is easy to check that $p_2(\lambda)$ is a continuous function. Moreover, it is even, i.e., $p_2(-\lambda) = p_2(\lambda)$ since $p(\lambda)$ is an even function. This immediately implies that the function

$$p_4(\lambda) = \frac{d\sigma^4(\lambda)}{d\lambda} = \int p_2(\lambda - \xi)p_2(\xi)\, d\xi$$

is positive in some neighborhood Δ of the point $\lambda = 0$. But then $p_8(\lambda) = d\sigma^{(8)}(\lambda)/d\lambda$ is positive in the neighborhood $\Delta + \Delta = 2\Delta$, etc. For some k we shall obviously have $2^k\Delta \supset S^1$. Therefore, the measure $\sigma^{(m)}$, where $m = 2^k$, is equivalent to the Lebesgue measure. The lemma is proved. □

This lemma together with Theorem 1 implies that the maximal spectral type of the operator U_T in the subspace $L_0^2(M, \mathfrak{S}, \mu)$, i.e., the type of $\sigma + \sigma^{(2)}/2! + \sigma^{(3)}/3! + \cdots$ is Lebesgue, while the multiplicity function $n(\lambda)$ equals $+\infty$ almost everywhere. The theorem is proved. □

§4. Gauss Dynamical Systems with Simple Continuous Spectrum

In this section we shall construct a certain class of Gauss dynamical systems possessing a simple continuous spectrum.

Suppose Λ is a Borel subset of the circle S^1 without rational relations. This means that an equality of the form $n_1\lambda_1 + \cdots + n_k\lambda_k = 0$, where $\lambda_j \in \Lambda$, the n_j are integers, $1 \leq j \leq k$, and the sum is understood mod 1, is only possible if $n_1 = \cdots = n_k = 0$.

It is known that there exist perfect sets without rational relations (the proof of a stronger statement is given in Appendix 4).

Theorem 1. *A Gauss automorphism T whose spectral measure σ is continuous and supported on the set $\Lambda \cup (-\Lambda)$ where Λ is a set without rational relations has a simple continuous spectrum.*

Proof. Put

$$\Lambda_0 = \{0\}, \quad \Lambda_1 = \Lambda \cup (-\Lambda), \quad \Lambda_2 = \Lambda_1 + \Lambda_1, \quad \Lambda_m = \underbrace{\Lambda_1 + \cdots + \Lambda_1}_{m \text{ summands}},$$

(by the sum of sets we mean the set of all arithmetical sums of their elements, i.e., $\Lambda_m = \{\lambda_1 + \cdots + \lambda_m \colon \lambda_j \in \Lambda_j,\ 1 \le j \le m\}$). Then $\Lambda_{m_1} \cap \Lambda_{m_2} = \varnothing$ for $m_1 \neq m_2$, since in the converse case we would have a nontrivial relation between the points of the set Λ_1: $\lambda_1' + \cdots + \lambda_{m_1}' = \lambda_1'' + \cdots + \lambda_{m_2}''$, which in its turn implies the existence of a nontrivial relation between the points of Λ, which is not possible.

It follows from what we have proved above that the maximal spectral types of a unitary operator U_T in the spaces $H_m^{(c)}$ which are convolutions $\sigma^{(m)}$ are mutually singular for distinct m, the measure $\sigma^{(m)}$ being supported on the sets Λ_m.

Now let us show that the multiplicity of the spectral type $\sigma^{(m)}$ in the space $H_m^{(c)}$ equals 1. To do this, note that the measure σ_m is concentrated on the m-fold product $\Lambda_1 \times \Lambda_1 \times \cdots \times \Lambda_1$. Therefore, under the isomorphism $\theta_m^{(c)}$, we can consider only the symmetric functions $\varphi(\lambda_1, \ldots, \lambda_m)$ on $\Lambda_1 \times \cdots \times \Lambda_1$ (m times). But each such function may be written in the form

$$\varphi(\lambda_1, \ldots, \lambda_m) = \psi(\lambda_1 + \cdots + \lambda_m).$$

This follows from the fact that for any $\lambda \in S^1$, the equation $\lambda = \lambda_1 + \cdots + \lambda_m$ either has $m!$ solutions $(\lambda_1, \ldots, \lambda_m)$, $\lambda_i \in \Lambda_1$ which differ from each other by permutations (in the case $\lambda \in \Lambda_1 + \cdots + \Lambda_1$) or has no solutions whatsoever. Thus we have proved that the space $H_m^{(c)}$ is isomorphic to the space of functions $\psi(\lambda_1 + \cdots + \lambda_m) = \psi(\lambda)$ supported on $\Lambda_1 + \cdots + \Lambda_1$ (m summands), and square integrable with respect to the measure $\sigma^{(m)}$, while the operator U_T is mapped by this isomorphism into the operator of multiplication by $e^{2\pi i \lambda}$. This means that $H_m^{(c)}$ is a cyclic subspace for U_T and $\sigma^{(m)}$ is the maximal spectral type in this space. The theorem is proved. □

The following theorem is, in a certain sense, the converse to Theorem 1. In it we will prove that if a certain condition, stronger than the condition of absence of rational relations, is imposed on the set Λ, then every ergodic automorphism spectrally equivalent to the automorphism T (appearing in Theorem 1) is actually metrically isomorphic to it, i.e., is a Gauss automorphism. This shows that dynamical systems with continuous spectra of the type considered have properties similar to those of dynamical systems with pure point spectra (although the set Λ is continual, it is so poor that it retains certain properties of countable sets). We now pass to exact formulations.

Definition 1. A closed subset Λ of the circle S^1 is said to be a *Kronecker set*, if for any continuous map $\varphi\colon \Lambda \to S^1$ there is a sequence of integers n_s such that

$$\max_{\lambda \in \Lambda} |\varphi(\lambda) - \exp 2\pi i \lambda n_s| \to 0 \quad \text{for } s \to \infty.$$

It is easy to check that every Kronecker set is a set without rational relations. Indeed, in the converse case, the relation $\sum n_j \lambda_j = 0$, $\lambda_j \in \Lambda$, would imply

$$\exp(2\pi i n \textstyle\sum n_j \lambda_j) = \prod \exp(2\pi i n \lambda_j n_j) = 1 \quad \text{for all } n,$$

i.e., $\prod \varphi(n_j \lambda_j) = 1$ for any continuous map $\varphi\colon \Lambda \to S^1$, which is obviously false.

If Λ is a finite set, it will be a Kronecker set if and only if it has no rational relations. The proof of the existence of perfect Kronecker set is provided in Appendix 4.

Theorem 2. *Suppose T is an ergodic automorphism of the Lebesgue space $(M, \mathfrak{S}, \mu)$ and σ is a symmetric measure on S^1 concentrated on the set $\Lambda + (-\Lambda)$, where Λ is a Kronecker set. If the maximal spectral type of the operator U_T is e^σ and the multiplicity function equals* 1, *then T is metrically isomorphic to a Gauss automorphism with spectral measure σ.*

Proof. The statement of the theorem will follow from a more general statement, which we shall give in the form of another theorem.

Theorem 2′. *Suppose T is an ergodic automorphism of the Lebesgue space $(M, \mathfrak{S}, \mu)$ and the real-valued function $h \in L^2(M, \mathfrak{S}, \mu)$, has a continuous spectral measure σ supported on the set $\Lambda_1 = \Lambda \cup (-\Lambda)$, where Λ is a Kronecker set. Then the random variables $U_T^n h = h_n$ constitute a Gauss stationary random process with spectral measure σ.*

Proof of Theorem 2′. Suppose the coordinate λ of the circle S^1 varies in the limits $-\pi \le \lambda < \pi$, while the sum is understood mod 2π. Partition the semicircle $\{0 \le \lambda < \pi\}$ into semi-intervals

$$\Delta_j = [\lambda_{j-1}, \lambda_j), \qquad 1 \le j \le m < \infty, 0 = \lambda_0 < \lambda_1 < \cdots < \lambda_m = \pi,$$

where $\lambda_j \notin \Lambda_1$. Denote by $E(\Delta)$, $\Lambda \subset S^1$ the family of projection operators corresponding to U_T and put

$$y_j = \Phi(\Delta_j) = E(\Delta_j)h, \qquad 1 \le j \le m.$$

Moreover, we may assume that $\Lambda \subset \{0 \le \lambda < \pi\}$.

For an arbitrary family of integers $k = (k_1, \ldots, k_m)$ consider the random vector $y^{(k)} = (U_T^{k_1} y_1, U_T^{k_2} y_2, \ldots, U_T^{k_m} y_m)$. Let us show that the distribution of the vector $y^{(k)}$ does not depend on k.

Suppose $\varphi(\lambda) = \exp(ik_j\lambda)$ for $\lambda \in \Lambda_1 \cap \Delta_j$, $1 \le j \le m$. The function $\varphi(\lambda)$ is continuous on the set Λ. Therefore we can find a sequence of integers $\{n_q\}$ such that

$$\lim_{q\to\infty} \exp(in_q\lambda) = \varphi(\lambda),$$

uniformly on Λ. Therefore

$$\|U_T^{k_j}y_j - U_T^{n_q}y_j\|^2 = \int_{\Delta_j} |\varphi(\lambda) - \exp(in_q\lambda)|^2\, d\sigma(\lambda) \to 0,$$

when $q \to \infty$. In other words,

$$\lim_{q\to\infty} (U_T^{n_q}y_1, U_T^{n_q}y_2, \ldots, U_T^{n_q}y_m) = y^{(k)},$$

in the sense of the convergence in $L^2(M, \mathfrak{S}, \mu)$. Since T is an automorphism of the space M, the vectors on the left-hand side have a distribution which does not depend on q and coincides with the distribution of the vector $y = (y_1, y_2, \ldots, y_m)$. When the random variable in $L^2(M, \mathfrak{S}, \mu)$ converges the corresponding distributions weakly converge, hence we see that the distribution of the vector $y^{(k)}$ coincides with the distribution of the vector y.

Now let us show that for an arbitrary family of Borel subsets (of the complex plane) $B = (B_1, B_2, \ldots, B_m)$, we have the relation

$$\mu\{x: y_j \in B_j, 1 \le j \le m\} = \prod_{j=1}^{m} \mu\{x: y_j \in B_j\}. \tag{1}$$

The proof will be by induction.

Suppose that for any $k < m$ we have shown that

$$\mu(A_1 \cap A_2 \cap \cdots \cap A_k) = \mu(A_1)\cdot\mu(A_2)\cdot\cdots\cdot\mu(A_k),$$

where $A_j = \{x: y_j \in B_j\}$. Then it follows from the statement proved above that the measure

$$\mu(A_1 \cap A_2 \cap \cdots \cap A_k \cap T^r A_{k+1})$$

does not depend on r. By the Birkhoff–Khinchin theorem,

$$\frac{1}{N}\sum_{r=0}^{N-1} \mu(A_1 \cap A_2 \cap \cdots \cap A_k \cap T^r A_{k+1}) = \mu(A_1 \cap \cdots \cap A_k)\mu(A_{k+1}).$$

Therefore,

$$\mu(A_1 \cap A_2 \cap \cdots \cap A_k \cap A_{k+1}) = \mu(A_1 \cap \cdots \cap A_k)\mu(A_{k+1}) = \prod_{j=1}^{k+1} \mu(A_j),$$

which was to be proved.

For any λ, $-\pi \leq \lambda < \pi$, put $\Delta_\lambda = [-\pi, \lambda)$ and $y_\lambda = E(\Delta_\lambda)h = \Phi(\Delta_\lambda)$. This last equality defines a complex-valued random process which is stochastically continuous by the continuity of the measure σ. Since Λ is nowhere dense, (1) means that y_λ is a process with independent increments. The structure of processes with independent increments is well studied (see Doob [1]). Roughly speaking, every such process is obtained by putting together a continuous component, which is always Gaussian and a discontinuous component (independent of it), the discontinuous component being distributed according to the Poisson law or to a composition of Poisson laws. We will show that in our case the discontinuous part is absent. This follows from the fact that this part of the process should correspond to the eigenfunctions of the operator U_T. In the ergodic case the eigen-values are simple and do not depend (mod 0) on the realization y_λ. But this contradicts the stochastic continuity of y_λ.

Now let us give a more rigorous proof.

Consider the space $C^{(l)}$ of complex-valued functions y_λ defined on $[0, \pi)$, continuous from the left and possessing limits from the right, i.e., the limits

$$\lim_{\lambda' \to \lambda - 0} y_{\lambda'} = y_\lambda, \qquad \lim_{\lambda' \to \lambda + 0} y_{\lambda'} = y_{\lambda+0},$$

exist. It is possible to introduce a natural topology on this space, the so-called Skorokhod topology and consider the σ-algebra $\mathfrak{S}^{(c)}$ of Borel sets corresponding to it. The map $x \mapsto y_\lambda(x)$ defines a measure on the σ-algebra generated by cylindrical sets in the space of all the functions $y_\lambda(x)$. Denote this measure again by μ. The cylindrical sets and the entire σ-algebra generated by them is contained in $\mathfrak{S}^{(c)}$. We shall use the following statement (see Gihman and Skorokhod [1]).

For any stochastically continuous process with independent increments y_λ the measure μ may be extended uniquely to the entire σ-algebra $\mathfrak{S}^{(c)}$. The map $x \mapsto y_\lambda(x)$ sends almost every element x into an element of the space $C^{(l)}$.

Now let us show that in our case for almost every x we have the relation

$$y_{\lambda+0}(Tx) - y_\lambda(Tx) = e^{i\lambda}[y_{\lambda+0}(x) - y_\lambda(x)].$$

For any $\Delta = [\lambda', \lambda'') \subset [0, \pi)$ put

$$y_\Delta = y_{\lambda''} - \lambda_{\lambda'}, \qquad \lambda_\Delta = \tfrac{1}{2}(\lambda' + \lambda''), \qquad l_\Delta = \lambda'' - \lambda',$$

$$F_\Delta = \{x\colon |y_\Delta(Tx) - e^{-i\lambda_\Delta} y_\Delta(x)| > l_\Delta^{1/2}\},$$

$$G_\Delta = \{x\colon |y_\Delta(x)| > l_\Delta^{-1/2}\}.$$

Now by the Chebyshev inequality

$$\mu(F_\Delta) \le l_\Delta^{-1} \| y_\Delta(Tx) - e^{i\lambda_\Delta} y_\Delta(x) \|^2_{L^2(M,\mu)}$$

$$= l_\Delta^{-1} \int_\Delta |e^{i\lambda} - e^{i\lambda_\Delta}|^2 \, d\sigma(\lambda) \le l_\Delta \cdot \sigma(\Delta),$$

$$\mu(G_\Delta) \le l_\Delta \cdot \sigma(\Delta).$$

Suppose ξ_p, $1 \le p < \infty$ is a partition of the semi-interval $[0, \pi)$ into semi-intervals of length $\pi 2^{-p}$, and

$$E_n = \bigcup_{p=n}^{\infty} \bigcup_{\Delta \in \xi_p} F_\Delta \cup G_\Delta, \qquad n = 1, 2, \ldots.$$

We have

$$\mu(E_n) \le \sum_{p=n}^{\infty} \sum_{\Delta \in \xi_p} [\mu(F_\Delta) + \mu(G_\Delta)]$$

$$\le \sum_{p=n}^{\infty} 2^{1-p} \sum_{\Delta \in \xi_p} \sigma(\Delta) \le 2\sigma([-\pi, \pi)) \cdot 2^{-n}.$$

Since $\sum \mu(E_n) < \infty$, it follows from the Borel–Cantelli lemma that almost every point x belongs only to a finite number of sets E_n. Therefore, for almost every x and any $\lambda \in [0, \pi)$, denoting by $\Delta_p(\lambda)$ the elements of the partition ξ_p which contains λ, we get

$$\varlimsup_{p\to\infty} | y_{\Delta_p(\lambda)}(Tx) - e^{i\lambda} y_{\Delta_p(\lambda)}(x) |$$

$$\le \varlimsup_{p\to\infty} | y_{\Delta_p(\lambda)}(Tx) - e^{i\lambda_{\Delta_p(\lambda)}} y_{\Delta_p(\lambda)}(x) | + \varlimsup_{p\to\infty} l_{\Delta_p(\lambda)} | y_{\Delta_p(\lambda)} |$$

$$\le \lim_{p\to\infty} (2 l_{\Delta_p(\lambda)}^{1/2}) = 0.$$

For almost every x, we see that the $y_\lambda(x)$, as well as the $y_\lambda(Tx)$, are continuous from the left and have limits from the right for any λ. It therefore follows from the last relation that, for any λ, we have

$$y_{\lambda+0}(Tx) - y_\lambda(Tx) = e^{i\lambda}[y_{\lambda+0}(x) - y_\lambda(x)].$$

We can now conclude the proof of the theorem. For all $\varepsilon > 0$ and $\Delta \in [0, \pi)$, denote by $N_{\Delta,\varepsilon}$ the number of all those $\lambda \in \Delta$ for which $|y_{\lambda+0}(x) - y_\lambda(x)| \ge \varepsilon$. The random variable $N_{\Delta,\varepsilon}(x)$ is measurable and finite with probability 1, while, from what we have proved above, it follows that $N_{\Delta,\varepsilon}(Tx) = N_{\Delta,\varepsilon}(x)$ for almost all x. By the ergodicity of T we see that $N_{\Delta,\varepsilon}(x)$ is constant almost everywhere. Since $N_{\Delta,\varepsilon}$ is integer-valued and y_λ is stochastically continuous, we see that for sufficiently small Δ the relation $N_{\Delta,\varepsilon}(x) = 0$ holds almost everywhere. Since $N_{\Delta,\varepsilon}$ is additive with respect to Δ, it is clear that for almost every x and any Δ we have $N_{\Delta,\varepsilon}(x) = 0$. If $\varepsilon \to 0$, we see that for almost every x we have $N_{\Delta,\varepsilon}(x) = 0$ for all Δ and ε. Therefore the function $y_\lambda(x)$ is continuous with respect to λ with probability 1. The statement of Theorem 2 now follows from the following well-known theorem (see Doob [1]).

If a stochastically continuous process with independent increments has continuous realizations with probability 1, *then it is a Gaussian.*

Thus $y_\lambda(x)$ is a Gauss process and therefore the random variables

$$U^n h = \int e^{in\lambda}\, dy_\lambda = \int_{-\pi}^{\pi} e^{in\lambda}\, d\Phi(\lambda)$$

form a Gauss stationary process. Theorem 2′ is proved. □

Now let us conclude the proof of Theorem 2. Consider the element h whose spectral measure is σ. According to Theorem 2′, the sequence $\{U_T^n h, -\infty < n < \infty\}$ forms a Gauss stationary process. It follows from Theorem 1 that in the subspace spanning all possible products

$$U_T^{n_1} h \cdot U_T^{n_2} h \cdot \cdots \cdot U_T^{n_r} h,$$

the spectrum of the operator U_T is simple and the maximal spectral type equals e^σ. But this subspace coincides with the entire space $L^2(M, \mathfrak{S}, \mu)$. Theorem 2 is proved. □

§5. Gauss Dynamical Systems with Finite Multiplicity Spectrum

In this section we give a detailed study of the multiplicity function of a unitary operator U_T adjoint to a Gauss automorphism T. It is natural to begin its analysis in a single space $H_m^{(c)}$. Then the question reduces to the following problem in spectral analysis. Consider the Hilbert space of symmetric functions $\varphi(\lambda_1, \ldots, \lambda_m)$, where $\lambda_j \in S^1$, which are square integrable with respect to the measure σ_m, and the unitary operator

$$V\colon \varphi(\lambda_1, \ldots, \lambda_m) \to \exp\left(2\pi i \sum_{j=1}^{m} \lambda_j\right) \varphi(\lambda_1, \ldots, \lambda_m).$$

The operator V is unitarily equivalent to the restriction of the operator U_T to the invariant subspace $H_m^{(c)}$, therefore, by Lemma 1, §3, the maximal spectral type of the operator V is $\sigma^{(m)}$. The problem consists in finding the multiplicity function $n_m(\lambda)$ of the operator V.

We shall use the representation of the measure σ_m in the form $\sigma_m = \sigma^{(m)}\nu_m(\cdot|\alpha)$. Here, for almost every α (with respect to the measure $\sigma^{(m)}$), the measure $\nu_m(\cdot|\alpha)$ is a symmetric measure on the $(m-1)$-dimensional torus

$$\mathrm{Tor}_\alpha^{m-1} = \{(\lambda_1, \dots, \lambda_m) \in \mathrm{Tor}^m : \lambda_1 + \cdots + \lambda_m = \alpha\};$$

the sum, as usual, is understood mod 1. The equality written out above for the measure σ_m means that for any bounded measurable function $f(\lambda_1, \dots, \lambda_m)$ we have

$$\int f(\lambda_1, \dots, \lambda_m)\, d\sigma_m = \int d\sigma^{(m)}(\alpha) \int_{\sum \lambda_j = \alpha} f(\lambda_1, \dots, \lambda_m)\, d\nu_m(\lambda_1, \dots, \lambda_m | \alpha).$$

For almost every α (with respect to the measure $\sigma^{(m)}$), if the measure $\nu_m(\cdot|\alpha)$ is discrete, then it is concentrated in $p \cdot m!$ points, where $p \geq 1$ is an integer or $p = \infty$.

For $p \geq 1$ denote by $A_p^{(m)}$ the set of all $\alpha \in S^1$ such that the measure $\nu_m(\cdot|\alpha)$ is concentrated on the set N_α (consisting of $p \cdot m!$ points) which does not intersect any of the diagonals $\lambda_i = \lambda_j$, $i \neq j$. Suppose further $A_\infty^{(m)} = S^1 \setminus \bigcup_{p=1}^{\infty} A_p^{(m)}$; by $\sigma_p^{(m)}$ denote the restriction of the measure $\sigma^{(m)}$ to the set $A_p^{(m)}$, $p = 1, 2, \dots, \infty$.

Theorem 1. *The multiplicity function $n_m(\lambda)$ equals p on the set $A_p^{(m)}$.*

Proof. Let us assume that the coordinates $\lambda_j, j = 1, \dots, m$ on the torus Tor^m vary within the limits $0 \leq \lambda_j < 1$. For the fundamental domain of the group S_m (up to the boundary) we may take the set

$$D = \{(\lambda_1, \dots, \lambda_m) : \lambda_1 > \lambda_2 > \cdots > \lambda_m\}.$$

Suppose $\tilde{N}_\alpha = N_\alpha \cap D, 0 \leq \alpha < 1$. The set $\tilde{N}_\alpha$, for almost all $\alpha \in A_p^{(m)}$ (with respect to the measure $\sigma^{(m)}$), consists of p points and N_α is the symmetrization of $\tilde{N}_\alpha$.

First we shall assume that $p < \infty$. On the torus Tor^m introduce the coordinates $\alpha = \lambda_1 + \cdots + \lambda_m$ and $\beta_1, \dots, \beta_{m-1}$, which are the linear coordinates on each torus

$$\mathrm{Tor}_\alpha^{m-1} = \{(\lambda_1, \dots, \lambda_m) : \lambda_1 + \cdots + \lambda_m = \alpha\}.$$

This may be done since the fibration of the torus Tor^m into the tori $\mathrm{Tor}_\alpha^{m-1}$ is a Cartesian product. By $E_p^{(m)}$ denote the set of all $\lambda = (\lambda_1, \ldots, \lambda_m) \in D$ for which

$$\alpha = \sum_{j=1}^{m} \lambda_j \in A_p^{(m)}.$$

We shall now use the following fact from measure theory: there exist measurable sets $D_1, \ldots, D_p \subset D$, such that

(1) $\sigma_m(D_i) > 0$, $1 \le i \le p$;
(2) $D_i \cap D_j = \varnothing$ for $i \ne j$;
(3) $\sigma_m(\cup D_i) = \sigma_m(E_p^{(m)})$;
(4) for every $\alpha \in A_m^{(p)}$ the intersection $D_i \cap \mathrm{Tor}_\alpha^{m-1}$ reduces to a point belonging to the set $\tilde{N}_\alpha$.

Consider the subspaces H_i, $1 \le i \le p$ of the space $Q_m^{(c)}$ consisting of symmetric functions supported on the sets $D_i^{(s)}$ obtained by taking the symmetrization of D_i. Put $D^{(s)} = \bigcup_{i=1}^{p} D_i^{(s)}$. Since the D_i are disjoint, the H_i are pairwise orthogonal and (3) implies that $\oplus \sum_{i=1}^{p} H_i$ is the subspace of the space $Q_m^{(c)}$ consisting of functions supported on the symmetrizations of $E_p^{(m)}$. Every function from the subset H_i, $1 \le i \le p$, may be written in the form

$$\varphi(\lambda_1, \ldots, \lambda_m) = \psi(\lambda_1 + \cdots + \lambda_m) = \psi(\alpha).$$

Therefore the operator U_T in each H_i is unitarily isomorphic to the multiplication operator by $e^{2\pi i \alpha}$ in the space of functions $\psi(\alpha)$ on D with square integrable absolute values with respect to the measure $\sigma_p^{(m)}$. For $p < \infty$, the theorem is proved.

If $p = \infty$, then for sufficiently large N we have the inclusion $E_\infty^{(m)} \supset \bigcup_{i=1}^{N} D_i$, where the sets D_i possess the same properties as above. We then have $n_m(\lambda) \ge N$ for $\lambda \in A_\infty^{(m)}$. Since N was arbitrary the theorem is proved. □

Definition 1. A measure σ is said to *belong to the class D* if $\sigma^{(m)}(A_\infty^{(m)}) = 0$ for all $m \ge 1$.

A measure supported on a set without rational relations obviously belongs to the class D. If σ is a measure of class D, then for almost all λ (with respect to the measure $\sigma^{(m)}$) the equation $\lambda = \lambda_1 + \cdots + \lambda_m$ has a finite number of solutions. The inexactitude of this statement consists in that we have no natural notion of a support of a measure and therefore we cannot say to what set the λ_j belong. For measures of class D the multiplicity function of the spectrum of the operator U_T is almost everywhere finite on each subspace

$H_m^{(c)}$. The Gauss automorphisms for which the measure σ belongs to the class D are the closest ones to automorphisms with discrete spectrum.

Theorem 2. *Suppose $n(\lambda)$ is the multiplicity function of the unitary operator U_T corresponding to a Gauss ergodic automorphism T. Then either $n(\lambda) = 1$ for almost all λ with respect to a measure of maximal spectral type, or the function $n(\lambda)$ is unbounded.*

Proof. Suppose σ is the spectral measure of the automorphism T. If $\sigma^{(m)}(A_\infty^{(m)}) > 0$ for at least one m, then the statement of the theorem is obviously valid. If $\sigma^{(m)}(A_p^{(m)}) = 0$ for all $p > 1$ and all $m \geq 1$, then the statement of the theorem also holds. It remains to consider the case when $\sigma^{(m)}(A_p^{(m)}) > 0$ for some $m, p, m \geq 1, p > 1$. We will show that under this assumption the multiplicity function of the operator U_T restricted to the subspace $H_{2m}^{(c)}$ assumes values no smaller than p^2 on a set of positive measure. More precisely, we will construct p orthogonal cyclic subspaces in the space $H_m^{(c)}$ possessing the same spectral type and, by using them, p^2 orthogonal cyclic subspaces in the space $H_{2m}^{(c)}$, which shall also have the same spectral type. It is convenient to split up the argument into several steps.

1. Suppose A is a subset of the torus Tor^m. The set of $\lambda = (\lambda_1, \ldots, \lambda_m) \in \mathrm{Tor}^m$ for which there exists at least one $\tilde{\lambda} = (\tilde{\lambda}_1, \ldots, \tilde{\lambda}_m) \in A$ satisfying $\lambda_i = \tilde{\lambda}_i$ for at least one i, $1 \leq i \leq m$, will be referred to as the *cross* $\mathrm{Cr}(A)$ of the set A. The *symmetric cross* $\mathrm{Cr}^{(s)}(A)$ of the set A will be the symmetrization of $\mathrm{Cr}(A)$, i.e., the set of orbits of the symmetric group S_m which pass through $\mathrm{Cr}(A)$.

We shall assume that each coordinate λ_j varies within the limits $0 \leq \lambda_j < 1$. For the fundamental domain of the group S_m (up to the boundary) take the set

$$D = \{(\lambda_1, \ldots, \lambda_m) \in \mathrm{Tor}^m\colon \lambda_1 > \lambda_2 > \cdots > \lambda_m).$$

We shall construct $2p$ disjoint measurable sets $B_k^{(i)} \subset D$, $i = 1, 2$; $1 \leq k \leq p$ such that

(1) $\mathrm{Cr}^{(s)}(B_k^{(1)}) \cap B_l^{(2)} = \varnothing, \quad 1 \leq k, l \leq p;$

(2) there exist measurable sets $C_1, C_2 \subset S^1$, $C_1 \cap C_2 = \varnothing$, such that

$$B_k^{(1)} \subset \left\{(\lambda_1, \ldots, \lambda_m)\colon \sum_{j=1}^m \lambda_j \in C_1\right\};$$

$$B_k^{(2)} \subset \left\{(\lambda_1, \ldots, \lambda_m)\colon \sum_{j=1}^m \lambda_j \in C_2\right\} \quad \text{for all } k,\ 1 \leq k \leq p;$$

(3) there exist measurable subsets

$$C_1' \subset C_1, \qquad \sigma^{(m)}(C_1') > 0, \qquad C_2' \subset C_2, \qquad \sigma^{(m)}(C_2') > 0,$$

for which $\nu_m(B_k^{(1)}|\alpha) > \text{const} > 0$ for all $\alpha \in C_1'$ and any k, $1 \le k \le p$ and $\nu_m(B_k^{(2)}|\alpha) > \text{const} > 0$ for all $\alpha \in C_2'$ and any k, $1 \le k \le p$.

Here $\nu_m(\cdot|\alpha)$ is the conditional measure induced by the measure σ_m on the torus

$$\text{Tor}_\alpha^{m-1} = \{(\lambda_1, \ldots, \lambda_m) \in \text{Tor}^m \colon \lambda_1 + \cdots + \lambda_m = \alpha\}.$$

2. In this step we shall derive the necessary statement on the multiplicity of the spectrum from the existence of the sets $B_k^{(i)}$.

Suppose $\chi_k^{(i)}$ is the indicator of the set $B_k^{(i)} \cap C_i'$. Consider the cyclic subspaces $H_{k,l}$ generated by the functions

$$\chi_{k,l} = S^{(2m)}[\chi_k^{(1)}(\lambda_1, \ldots, \lambda_m)\chi_l^{(2)}(\lambda_{m+1}, \ldots, \lambda_{2m})], \qquad 1 \le k, l \le p.$$

We will show that the supports of the functions $\chi_{k,l}$ are disjoint, which will imply the orthogonality of the $H_{k,l}$ for different pairs (k, l).

Assume the converse, i.e., that the supports of $\chi_{\bar{k}\bar{l}}$ and $\chi_{\bar{\bar{k}}\bar{\bar{l}}}$ have a nonempty intersection. This means that there exist points

$$(\bar{\lambda}_1, \ldots, \bar{\lambda}_m) \in B_{\bar{k}}^{(1)}, \qquad (\bar{\lambda}_1', \ldots, \bar{\lambda}_m') \in B_{\bar{l}}^{(1)},$$

$$(\bar{\bar{\lambda}}_1, \ldots, \bar{\bar{\lambda}}_m) \in B_{\bar{\bar{k}}}^{(2)}, \qquad (\bar{\bar{\lambda}}_1, \ldots, \bar{\bar{\lambda}}_m) \in B_{\bar{\bar{l}}}^{(2)}$$

such that

$$\bar{\lambda} = (\bar{\lambda}_1, \ldots, \bar{\lambda}_m, \bar{\lambda}_1', \ldots, \bar{\lambda}_m'), \qquad \bar{\bar{\lambda}} = (\bar{\bar{\lambda}}_1, \ldots, \bar{\bar{\lambda}}_m, \bar{\bar{\lambda}}_1', \ldots, \bar{\bar{\lambda}}_m'),$$

implies that the coordinate of the point $\bar{\bar{\lambda}}$ may be rearranged so as to obtain $\bar{\lambda}$. This permutation must separately rearrange $\bar{\bar{\lambda}}_i$ and $\bar{\bar{\lambda}}_j'$, since in the converse case $B_{\bar{k}}^{(1)}$ would intersect $\text{Cr}^{(s)}(B_{\bar{l}}^{(2)})$. But such a permutation must be the identical one since all the $(\bar{\lambda}_1, \ldots, \bar{\lambda}_m)$, $(\bar{\lambda}_1', \ldots, \bar{\lambda}_m')$, $(\bar{\bar{\lambda}}_1, \ldots, \bar{\bar{\lambda}}_m)$, $(\bar{\bar{\lambda}}_1', \ldots, \bar{\bar{\lambda}}_m')$, are contained in the fundamental domain. Since at least one of the relations $\bar{k} \ne \bar{\bar{k}}$, $l \ne \bar{l}$ holds, we have $\bar{\lambda} \ne \bar{\bar{\lambda}}$. Thus the orthogonality of distinct $H_{k,l}$ is proved.

Now let us study the spectral type $\rho_{k,l}$ of each of the functions $\chi_{k,l}$. By the above this spectral type is dominated by the type of the measure $\sigma^{(2m)}$ and the density $p(\alpha) = (d\rho_{k,l}/d\sigma^{(m)})(\alpha)$ equals the conditional expectation of $|\chi_{k,l}|^2$ under the condition that the sum $\alpha = \sum_{j=1}^{2m} \lambda_j$ is fixed. Further, $\chi_{k,l} \ge 0$ and therefore this conditional expectation is bounded from below by the number $E(|\chi_k^{(1)}\chi_l^{(2)}|^2|\alpha)$. For any bounded measurable function $f \geq 0$ on S^1 we have

$$\int f(\alpha)E[(\chi_k^{(1)})^2(\chi_l^{(2)})^2|\alpha]\,d\sigma^{(2m)}(\alpha)$$

$$= \iint f(\lambda_1 + \cdots + \lambda_{2m})[\chi_k^{(1)}(\lambda_1, \ldots, \lambda_m)]^2[\chi_l^{(2)}(\lambda_{m+1}, \ldots, \lambda_{2m})]^2\,d\sigma_{2m}$$

$$= \iint f(\alpha_1 + \alpha_2)E[(\chi_k^{(1)})^2|\alpha_1]\,d\sigma^{(m)}(\alpha_1)E[(\chi_l^{(2)})^2|\alpha_2]\,d\sigma^{(m)}(\alpha_2)$$

$$\geq \text{const} \int_{C_1'}\int_{C_2'} f(\alpha_1 + \alpha_2)\,d\sigma^{(m)}(\alpha_1)\,d\sigma^{(m)}(\alpha_2). \tag{1}$$

Now note that the measure

$$\rho(\Delta) = \int_{C_1'}\int_{C_2'} \chi_\Delta(\alpha_1 + \alpha_2)\,d\sigma^{(m)}(\alpha_1)\,d\sigma^{(m)}(\alpha_2),$$

is absolutely continuous with respect to the measure $\sigma^{(2m)}$ and does not equal zero. In other words, there is a subset $\Delta_0 \subset S^1$ where the measures $\sigma^{(2m)}$ and ρ are equivalent. But together with (1) this means that the spectral type of each of the functions $\chi_{k,l}$ dominates the type of the restriction of the measure $\sigma^{(2m)}$ to the set Δ_0. Therefore the multiplicity function on Δ_0 is no less than p^2.

3. During the step we shall construct the necessary system of sets $B_k^{(1)}, B_l^{(2)}$.

On the torus Tor^m introduce the coordinates $\alpha = \lambda_1 + \cdots + \lambda_m$ and $\beta_1, \ldots, \beta_{m-1}$ which are the linear coordinates on every torus

$$\text{Tor}_\alpha^{m-1} = \{(\lambda_1, \ldots, \lambda_m) : \lambda_1 + \cdots + \lambda_m = \alpha\}.$$

Construct a sequence of partitions ξ_n of the torus Tor^m, where a generic element of each ξ_n is of the form

$$C_{k_0, k_1, \ldots, k_{m-1}}^{(n)} = \{(\alpha, \beta_1, \ldots, \beta_{m-1}) : k_0/2^n \leq \alpha < (k_0 + 1)/2^n,$$
$$k_i/2^n \leq \beta_i < (k_i + 1)/2^n,\ 1 \leq i \leq m - 1\}.$$

These partitions, which become finer and finer, converge to the partition into separate points. For sufficiently large n we can find numbers $k_0, k_j^{(l)}$, $1 \leq j \leq m - 1, l = 1, \ldots, p$ such that $C_{k_0, k_1^{(l)}, \ldots, k_{m-1}^{(l)}}^{(n)} \subset D$ and

$$\sigma_m(C_{k_0, k_1^{(l)}, \ldots, k_{m-1}^{(l)}})/\sigma_m(C_{k_0}) > \varepsilon_0, \qquad 1 \leq l \leq p$$

where ε_0 is a constant which does not depend on n and

$$C_{k_0} = \{\lambda \in \mathrm{Tor}^m: k_0/2^n \leq \sum \lambda_j < (k_0 + 1)/2^n\}.$$

Put $\bar{B}_l^{(1)} = C^{(n)}_{k_0, k_1^{(l)}, \ldots, k_{m-1}^{(l)}}$ and consider the sets $\mathrm{Cr}^{(s)}(\bar{B}_l^{(1)})$. When $n \to \infty$, the sum of measures of the sets $\mathrm{Cr}^{(s)}(\bar{B}_l^{(1)})$ tends to 0 in view of the continuity of the measure σ. Therefore we can find such a $\bar{k}_0$ and p families $\bar{k}_1^{(l)}, \ldots, \bar{k}_{m-1}^{(l)}$, $1 \leq l \leq p$, such that $C^{(n)}_{\bar{k}_0, \bar{k}_1^{(l)}, \ldots, \bar{k}_{m-1}^{(l)}} \cap \mathrm{Cr}^{(s)}(\bar{B}_l^{(1)}) = \varnothing$,

$$\sigma_m(C^{(n)}_{\bar{k}_0, \bar{k}_1^{(l)}, \ldots, \bar{k}_{m-1}^{(l)}})/\sigma_m(C^{(n)}_{\bar{k}_0}) > \varepsilon_0.$$

Put $\bar{B}_l^{(2)} = C^{(n)}_{\bar{k}_0 \bar{k}_1^{(l)}, \ldots, \bar{k}_{m-1}^{(l)}}$. Now suppose

$$\Omega^{(1)} = \bigcup_\alpha \mathrm{Tor}_\alpha^{m-1}, \Omega^{(2)} = \bigcup_\alpha \mathrm{Tor}_\alpha^{m-1},$$

where the sum is taken over all α such that

$$\nu_m(\bar{B}_l^{(1)} | \alpha) > \varepsilon_0^2, \qquad \nu_m(\bar{B}_l^{(2)} | \alpha) > \varepsilon_0^2,$$

in the first and second case, respectively.

The sets $B_l^{(1)} = \bar{B}_l^{(1)} \cap \Omega^{(1)}$, $B_l^{(2)} = \bar{B}_l^{(2)} \cap \Omega^{(2)}$ are the ones we need. The theorem is proved. □

Part IV

Approximation Theory of Dynamical Systems by Periodic Dynamical Systems and Some of its Applications

Chapter 15

Approximations of Dynamical Systems

The constructive theory of functions studies the relationship between the properties of functions and the speed of their approximations by functions of some particular fixed class. In a similar way, in ergodic theory we may study the dependence of various properties of dynamical systems on the rapidity of their approximations by the periodic dynamical systems which are simplest from some point of view. We shall see that many properties of dynamical systems are intimately related to the character of their approximations.

§1. Definition and Types of Approximations. Ergodicity and Mixing Conditions

Suppose T is an automorphism of the Lebesgue space $(M, \mathfrak{S}, \mu)$. We shall consider sequences of finite partitions $\{\xi_n\}$ of the space M and sequences of automorphisms $\{T_n\}$ such that T_n preserves the partition ξ_n. (The automorphism T_n *preserves the partition* ξ_n, if it sends every element of the partition ξ_n into an element of the same partition). The elements of the partition ξ_n will be denoted by $C_i^{(n)}$, $1 \leq i \leq q_n$. By $\mathfrak{S}(\xi_n)$ we denote the σ-algebra of subsets of the space M consisting (mod 0) of elements of the partition ξ_n. The notation $\xi_n \to \varepsilon$, when $n \to \infty$, where ε is the partition of M into separate points, means that for each $A \in \mathfrak{S}$ there is a sequence of sets $A_n \in \mathfrak{S}(\xi_n)$ such that $\mu(A_n \triangle A) \to 0$ when $n \to \infty$. Since the number of elements of the partition ξ_n is finite, the trajectory of each $C_i^{(n)}$ is finite, i.e., for some r_i, $1 \leq i \leq q_n$ we will have $T_n^{r_i} C_i^{(n)} = C_i^{(n)}$. For the sequel it is not important how $T_n^{r_i}$ interchanges the points within $C_i^{(n)}$, but it is convenient to assume that $T_n^{r_i} x = x$ for any point $x \in C_i^{(n)}$. By p_n we denote the order of T_n, i.e., the smallest natural number such that $T_n^{p_n} = Id$.

Definition 1. Suppose $f(n) \downarrow 0$. An automorphism T of the space $(M, \mathfrak{S}, \mu)$ possesses an *approximation of the first type by periodic transformations* (a.p.t. I) *with speed* $f(n)$, if there exists a sequence of partitions $\xi_n \to \varepsilon$ and a sequence of automorphisms T_n preserving ξ_n such that

$$\sum_{i=1}^{q_n} \mu(TC_i^{(n)} \triangle T_n C_i^{(n)}) < f(q_n), \qquad n = 1, 2, \ldots .$$

If for the sequences $\{\xi_n\}$, $\{T_n\}$, where T_n is a periodic automorphism of order p_n, we have the inequality

$$\sum_{i=1}^{q_n} \mu(TC_i^{(n)} \triangle T_n C_i^{(n)}) < f(p_n), \qquad n = 1, 2, \ldots$$

and $U_{T_n} \to U_T$ in the strong topology of operators in $L^2(M, \mathfrak{S}, \mu)$, then T *possesses an approximation of the second type by periodic transformations* (a.p.t. II) *with speed* $f(n)$.

If the automorphism T possesses a.p.t. I and T_n cyclically permutes the elements of ξ_n, then T is said to possess a *cyclic a.p.t. with speed* $f(n)$.

The following lemma immediately follows from the definition.

Lemma 1. *For any* $E \in \mathfrak{S}$ *and any natural* s *we have*

$$\mu(T^s E \triangle T_n^s E) \leq \sum_{i=0}^{s-1} \mu(T(T_n^i E) \triangle T_n^{i+1} E).$$

Proof. In view of the triangle inequality and the invariance of measure,

$$\mu(T^s E \triangle T_n^s E) \leq \mu(T^s E \triangle T^{s-1} T_n E)$$
$$+ \mu(T^{s-1} T_n E \triangle T^{s-2} T_n^2 E) + \cdots + \mu(T T_n^{s-1} \triangle T_n^s E)$$
$$= \sum_{i=0}^{s-1} \mu(T(T_n^i E) \triangle T_n^{i+1} E).$$

The lemma is proved. □

The Rohlin–Halmos lemma, proved in §4, Chap. 10, shows that any automorphism may be approximated by periodic ones. Clearly, the faster the automorphism T is approximated by periodic ones, the worse are its statistical properties, i.e., its ergodicity and mixing properties. On the other hand, a sufficiently good cyclic approximation, as we will see, guarantees the ergodicity of the automorphism. Let us give a precise formulation concerning the relationship of ergodicity, mixing and spectral properties with the speed of approximation.

Theorem 1. *Any automorphism* T *possesses a.p.t* I *with speed* $f(n) = a_n/\log n$, *where* a_n *is an arbitrary monotonic sequence of real numbers tending to infinity.*

Proof. It suffices to prove the theorem for an aperiodic T. The general case reduces to this particular one by decomposing into periodic and aperiodic parts.

By using the Rohlin–Halmos lemma for arbitrary n, we can find a set $A_n \in \mathfrak{S}$ such that the $T^k A_n$ are disjoint for $0 \leq k \leq n-1$ and

$$\mu\left(\bigcup_{k=0}^{n-1} T^k A_n\right) > 1 - \frac{1}{n}.$$

Define the approximating periodic automorphism T_n by putting

$$T_n x = \begin{cases} Tx, & \text{if } x \in \bigcup_{k=0}^{n-2} T^k A_n, \\ T^{-n+1} & \text{if } x \in T^{n-1} A_n, \\ x & \text{if } x \in M \setminus \bigcup_{k=0}^{n-1} T^k A_n. \end{cases}$$

Now let us construct the required sequence of finite partitions ξ_n. First, beginning with sufficiently large n, let us take an arbitrary sequence of finite partitions η_n, $\eta_n \to \varepsilon$ when $n \to \infty$, where the number k_n of elements of the partition η_n is so large that $k_n > 8 \log k_n$ and satisfies the inequality $2 \leq k_n < \min(a_n, n)$. Suppose ζ_n is the partition one element of which is $M \setminus A_n$ and all the others are of the form

$$A_n \cap C_{r_0} \cap T^{-1} C_{r_1} \cap \cdots \cap T^{-n+1} C_{r_{n-1}},$$

where C_{r_i}, $0 \leq i \leq n-1$ ranges over all elements of the partition η_n. Then $T^p \zeta_n$ (for $0 \leq p \leq n-1$) is the partition which subdivides $T^p A_n$ in the same way that ζ_n subdivides A_n and is degenerate on $M \setminus T^p A_n$, i.e., $M \setminus T^p A_n$ is an element of the partition $T^p \zeta_n$. Moreover, on each $T^p A_n$ the partition ζ_n is a refinement of η_n. The partition $\bigvee_{k=0}^{n-1} T^k \zeta_n$ coincides with ζ_p on each $T^p A_n$, $0 \leq p \leq n-1$ and $M \setminus \bigcup_{p=0}^{n-1} T^p A_n$ is an element of this partition. Finally put $\xi_n = \bigvee_{k=0}^{n-1} T^k \zeta_n$. Then, on the set $M \setminus \bigcup_{k=0}^{n-1} T^k A_n$, the partition ξ_n is a refinement of the partition η_n. Clearly, $\xi_n \to \varepsilon$ when $n \to \infty$.

The relation $T_n \xi_n = \xi_n$ follows immediately from the definition of T_n.

By construction, the number q_n of elements of the partition ξ_n is not greater than $nk_n^n + 1$. Therefore, using the inequality $n \leq 2^n - 1 \leq k_n^n - 1$, we get

$$q_n \leq n(k_n^n + 1) \leq (k_n^n - 1)(k_n^n + 1) \leq k_n^{2n}.$$

Hence $1/(2n) \leq \log k_n / \log q_n$. Finally

$$\sum_{i=1}^{q_n} \mu(TC_i^{(n)} \triangle T_n C_i^{(n)}) \leq 2\mu\left(M \setminus \bigcup_{k=0}^{n-2} T^k A_n\right)$$

$$\leq \frac{4}{n} \leq \frac{8 \log k_n}{\log q_n} < \frac{k_n}{\log q_n} < \frac{a_n}{\log q_n}.$$

The theorem is proved. $\square$

Theorem 2. *If the automorphism T possesses a cyclic approximation with speed $f(n) = \theta/n$, $\theta \geq 2$, then the number of its distinct invariant sets of positive measure is no greater than $\theta/2$. In other words, the number of ergodic components of the automorphism T is no greater than $\theta/2$.*

Proof. Assume, in contradiction to the statement of the theorem, that we can find $m > \theta/2$ disjoint sets $A_i \in \mathfrak{S}$, $\mu(A_i) > 0$, $1 \leq i \leq m$, such that $TA_i = A_i$ and $\bigcup_{i=1}^m A_i = M$ (mod 0). Let us obtain a contradiction from this assumption.

Let $\alpha = \min_{1 \leq i \leq m} \mu(A_i)$ and suppose δ is a fixed number, $0 < \delta < 1$. By the condition $\xi_n \to \varepsilon$, for sufficiently large n we can find sets $A_r^{(n)} \in \mathfrak{S}(\xi_n)$ such that

$$\sum_{r=1}^{m} \mu(A_r \triangle A_r^{(n)}) < \delta^2 \alpha. \tag{1}$$

This immediately implies that for every r there is an element $\tilde{C}_r^{(n)} = C_{i_r}^{(n)}$ of the partition ξ_n, $\tilde{C}_r^{(n)} \subset A_r^{(n)}$ for which

$$\mu(\tilde{C}_r^{(n)} \cap A_r) > \frac{1 - \delta}{q_n}.$$

Indeed, in the converse case we would have

$$\begin{aligned}\mu(A_r^{(n)} \cap A_r) &= \sum_{t: C_t^{(n)} \subset A_r^{(n)}} \mu(C_t^{(n)} \cap A_r) \\ &\leq (1 - \delta)\mu(A_r^{(n)}) < \mu(A_r)(1 - \delta)(1 + \delta) = \mu(A_r)(1 - \delta^2),\end{aligned}$$

i.e., $\mu(A_r^{(n)} \triangle A_r) \geq \delta^2 \cdot \mu(A_r) \geq \delta^2 \alpha$, contradicting (1).

Put $C = \tilde{C}_1^{(n)}$. Since the automorphism T_n cyclically permutes the elements of the partition ξ_n, we can find numbers k_j, $1 \leq j \leq m$ such that

$$0 < k_1 < k_2 < \cdots < k_m < q_n, \qquad T_n^{k_j} C = \tilde{C}_{r_j}^{(n)}$$

and $(r_1, r_2, \ldots, r_m)$ is a permutation of the numbers $(1, 2, \ldots, m)$. Denoting $k_{m+1} = q_n$ and $A_{r_{m+1}} = A_1$, we will have

$$\mu(T_n^{k_j} C \cap A_{r_j}) > (1 - \delta)/q_n. \tag{2}$$

This implies that for $j = 1, \ldots, m$

$$\mu(T_n^{k_{j+1}} C \cap A_{r_j}) = \mu(T_n^{k_{j+1} - k_j}(T_n^{k_j} C) \cap A_{r_j}) \leq \delta/q_n.$$

On the other hand, taking into consideration the fact that the A_{r_j} are invariant with respect to T, we get by using (2)

$$\begin{aligned}\mu(T^{k_{j+1}-k_j}(T_n^{k_j}C)\cap A_{r_j}) &= \mu(T^{k_{j+1}-k_j}(T_n^{k_j}C\cap A_{r_j}))\\ &= \mu(T_n^{k_j}C\cap A_{r_j}) > (1-\delta)/q_n.\end{aligned}$$

Therefore

$$\begin{aligned}&\mu(T^{k_{j+1}-k_j}(T_n^{k_j}C)\,\triangle\,T_n^{k_{j+1}-k_j}(T_n^{k_j}C))\\ &\quad = 2\mu(T^{k_{j+1}-k_j}(T_n^{k_j}C)\setminus T_n^{k_{j+1}-k_j}(T_n^{k_j}C)\\ &\quad \geq 2\mu((T^{k_{j+1}-k_j}(T_n^{k_j}C)\setminus T_n^{k_{j+1}-k_j}(T_n^{k_j}C))\cap A_{r_j})\\ &\quad = 2\mu((T^{k_{j+1}-k_j}(T_n^{k_j}C)\cap A_{r_j})\setminus T_n^{k_{j+1}-k_j}(T_n^{k_j}C)\cap A_{r_j})\\ &\quad \geq 2[\mu(T^{k_{j+1}-k_j}(T_n^{k_j}C)\cap A_{r_j}) - \mu(T_n^{k_{j+1}-k_j}(T_n^{k_j}C)\cap A_{r_j})]\\ &\quad \geq 2((1-\delta)/q_n - \delta/q_n) \geq (2-4\delta)/q_n. \qquad (3)\end{aligned}$$

Now putting $s = k_{j+1} - k_j$, $E = T_n^{k_j}C$ in Lemma 1, we will get

$$\begin{aligned}\mu(T^{k_{j+1}-k_j}(T_n^{k_j}C)\,\triangle\,T_n^{k_{j+1}-k_j}(T_n^{k_j}C) &\leq \sum_{i=0}^{k_{j+1}-k_j-1}\mu(T(T_n^{k_j+i}C)\,\triangle\,T_n(T_n^{k_j+i}C))\\ &= \sum_{i=k_j}^{k_{j+1}-1}\mu(T(T_n^iC)\,\triangle\,T_n(T_n^iC)).\end{aligned}$$

Adding these inequalities for all $j = 1, 2, \ldots$, we get, by using (3),

$$\begin{aligned}\frac{m(2-4\delta)}{q_n} &\leq \sum_{j=1}^{m}\sum_{i=k_j}^{k_{j+1}-1}\mu(T(T_n^iC)\,\triangle\,T_n(T_n^iC))\\ &= \sum_{i=0}^{q_n-1}\mu(TC_i^{(n)}\,\triangle\,T_nC_i^{(n)}) < \theta/q_n,\end{aligned}$$

and therefore $m \leq \theta/(2-4\delta)$. Since δ was arbitrary, the theorem is proved. □

Corollary. *If the automorphism T possesses a cyclic a.p.t. with speed θ/n, $\theta < 4$, then T is ergodic.*

Theorem 3. *If the automorphism T possesses an a.p.t.* II *with speed θ/n, where $\theta < 2$, then T is not mixing.*

Proof. We may assume that $\lim_{n\to\infty} p_n = \infty$ since in the converse case the condition $U_{T_n} \to U_T$ implies the periodicity of T.

Assume that T is a mixing automorphism. For an arbitrary integer $k \geq 2$, take the sets A_r, $\mu(A_r) = 1/k$, $1 \leq r \leq k$,

$$A_{r_1} \cap A_{r_2} = \varnothing \quad \text{for } r_1 \neq r_2,$$

where $\bigcup_{r=1}^k A_r = M$ (mod 0). In the case of mixing, we must have

$$\lim_{n\to\infty} \sum_{r=1}^{k} \mu(T^{p_n} A_r \cap A_r) = 1/k.$$

Now fix an arbitrary $\delta > 0$. For sufficiently large n, we can find sets $A_r^{(n)} \in \mathfrak{S}(\xi_n)$ such that

$$\mu(A_r \triangle A_r^{(n)}) < \delta/k, \qquad 1 \leq r \leq k.$$

Then

$$\begin{aligned}\sum_{r=1}^{k} \mu(T^{p_n} A_r \triangle A_r) \leq {} & \sum_{r=1}^{k} \mu(T^{p_n} A_r \triangle T^{p_n} A_r^{(n)}) \\ & + \sum_{r=1}^{k} \mu(T^{p_n} A_r^{(n)} \triangle A_r^{(n)}) + \sum_{r=1}^{k} \mu(A_r^{(n)} \triangle A_r) \qquad (4) \\ = {} & \Sigma_1 + \Sigma_2 + \Sigma_3.\end{aligned}$$

Clearly $0 \leq \Sigma_1 = \Sigma_3 < \delta$. Further, it follows from Lemma 1 that

$$\begin{aligned}\Sigma_2 &\leq \sum_{i=1}^{q_n} \mu(T^{p_n} C_i^{(n)} \triangle C_i^{(n)}) = \sum_{i=1}^{q_n} \mu(T^{p_n} C_i^{(n)} \triangle T_n^{p_n} C_i^{(n)}) \\ &\leq \sum_{i=1}^{q_n} \sum_{j=0}^{p_n - 1} \mu(T(T_n^j C_i^{(n)}) \triangle T_n(T_n^j C_i^{(n)})) \\ &= \sum_{j=0}^{p_n \;\; 1} \sum_{i=1}^{q_n} \mu(T(T_n^j C_i^{(n)}) \triangle T_n(T_n^j C_i^{(n)})).\end{aligned}$$

Since T_n^j acts as the permutation of elements of the partition ξ_n it follows that all the inner sums for a distinct j are equal to each other, so that

$$\Sigma_2 \leq p_n \sum_{i=1}^{q_n} \mu(T C_i^{(n)} \triangle T_n C_i^{(n)}) \leq \theta.$$

Returning to (4), we will have

$$\sum_{r=1}^{k} \mu(T^{p_n} A_r \triangle A_r) \leq 2\delta + \theta.$$

Therefore

$$\sum_{r=1}^{k} \mu(T^{p_n}A_r \cap A_r) = \sum_{r=1}^{k} \mu(A_r) - \frac{1}{2}\sum_{r=1}^{k} \mu(T^{p_n}A_r \triangle A_r) \geq 1 - \delta - \theta/2.$$

The left-hand side tends to $1/k$ when $n \to \infty$. Since k and δ were arbitrary, condition $\theta < 2$ yields a contradiction. The theorem is proved. □

§2. Approximations and Spectra

It is often possible to obtain definite information on the spectrum of an automorphism from the character of its approximation by periodic automorphisms. In this section we will prove a general theorem on the relationship between approximations and the multiplicity of the spectrum.

Theorem 1. *If the automorphism T possesses a cyclic a.p.t. with speed $f(n) = \theta/n$, $\theta < 2 - 2/m$, $m \geq 2$, then the multiplicity function $n(\lambda)$ of the spectrum of the operator U_T is no greater than $(m - 1)$.*

The proof is based on the following lemma from the spectral theory of unitary operators.

Lemma 1. *Suppose U is a unitary operator in a separable Hilbert space H, σ is the measure of the maximal spectral type for U, $n(\lambda)$ is the multiplicity function of U. If $n(\lambda) \geq m$ on a set $E \subset S^1$, $\sigma(E) > 0$, then we can find m orthogonal unit vectors $h^{(1)}, \ldots, h^{(m)}$ such that for any cyclic subspace $H' \subset H$ with respect to U and any m vectors of equal length $g^{(1)}, \ldots, g^{(m)} \in H'$, $\|g^{(i)}\| = a$, $1 \leq i \leq m$, we have the inequality*

$$\sum_{i=1}^{m} \|h^{(i)} - g^{(i)}\|^2 \geq m(1 + a^2 - 2a/\sqrt{m}).$$

Proof. According to the main theorem on the canonical form of the unitary operator (see Appendix 2), the space H may be decomposed into a continuous direct sum $H = \int_{S^1} \oplus H_\lambda \, d\sigma(\lambda)$ of Hilbert spaces H_λ, $\lambda \in S^1$ with respect to the measure σ, so that:

(1) $n(\lambda) = \dim H_\lambda$;
(2) every cyclic subspace H' with respect to U is of the form $H' = \int_{S^1} \oplus H'_\lambda \, d\sigma(\lambda)$, where $H'_\lambda \subset H_\lambda$ and $\dim H'_\lambda \leq 1$ for almost all λ with respect to the measure σ.

Suppose $\lambda \in E$ and $e_\lambda^{(1)}, \ldots, e_\lambda^{(m)}$ are the first m vectors of the basis in H_λ. Put $h^{(i)} = \{h_\lambda^{(i)}, \lambda \in S^1\}$, where $h_\lambda^{(i)} = (1/\sqrt{\sigma(E)})e_\lambda^{(i)}$ for $\lambda \in E$ and $h_\lambda^{(i)} = 0$ for $\lambda \notin E$, $1 \leq i \leq m$. Clearly $\|h^{(i)}\| = 1$, $(h^{(i)}, h^{(j)}) = 0$ for $i \neq j$.

By e'_λ denote the unit vector in H'_λ if $\dim H'_\lambda = 1$ and put $e'_\lambda = 0$ in the other cases. For arbitrary $g^{(i)} \in H'$, $\|g^{(i)}\| = a$, we have

$$\|h^{(i)} - g^{(i)}\|^2 = 1 + a^2 - 2\operatorname{Re}(h^{(i)}, g^{(i)}) \geq 1 + a^2 - 2|(h^{(i)}, g^{(i)})|,$$

$$\sum_{i=1}^{m} \|h^{(i)} - g^{(i)}\|^2 \geq m(1 + a^2) - 2\sum_{i=1}^{m} |(h^{(i)}, g^{(i)})|.$$

Write each vector $g^{(i)}$ in the form $g^{(i)} = \{g_\lambda^{(i)}, \lambda \in S^1\}$ where $g_\lambda^{(i)} = c_\lambda^{(i)} e'_\lambda$, $|c_\lambda^{(i)}| = \|g_\lambda^{(i)}\|_{H_\lambda}$. Then

$$\Sigma \overset{\text{def}}{=} \sum_{i=1}^{m} |(h^{(i)}, g^{(i)})| = \sum_{i=1}^{m} \left| \int_{S^1} (h_\lambda^{(i)}, g_\lambda^{(i)})\, d\sigma(\lambda) \right|$$

$$\leq \sum_{i=1}^{m} \int_E |(h_\lambda^{(i)}, g_\lambda^{(i)})|\, d\sigma(\lambda) = \frac{1}{\sqrt{\sigma(E)}} \int_E \sum_{i=1}^{m} |c_\lambda^{(i)}|\,|(e_\lambda^{(i)}, e'_\lambda)|\, d\sigma(\lambda).$$

Using the Cauchy–Buniakowski inequality, we get

$$\Sigma \leq \frac{1}{\sqrt{\sigma(E)}} \int_E (\Sigma |c_\lambda^{(i)}|^2)^{1/2} \left(\sum_{i=1}^{m} |(e_\lambda^{(i)}, e'_\lambda)|^2 \right)^{1/2} d\sigma(\lambda).$$

By the Bessel inequality, the second factor under the integral sign is no greater than 1. Therefore

$$\Sigma \leq \frac{1}{\sqrt{\sigma(E)}} \int_E \left(\sum_{i=1}^{m} |c_\lambda^{(i)}|^2 \right)^{1/2} d\sigma(\lambda) = \frac{1}{\sqrt{\sigma(E)}} \int_E \left(\sum_{i=1}^{m} \|g_\lambda^{(i)}\|^2 \right)^{1/2} \cdot d\sigma(\lambda)$$

$$\leq \frac{1}{\sqrt{\sigma(E)}} \left[\int_E \sum_{i=1}^{m} \|g_\lambda^{(i)}\|^2\, d\sigma(\lambda) \right]^{1/2} \left[\int_E 1 \cdot d\sigma(\lambda) \right]^{1/2}$$

$$= \left[\sum_{i=1}^{m} \int_E \|g_\lambda^{(i)}\|^2\, d\sigma(\lambda) \right]^{1/2} \leq \left[\sum_{i=1}^{m} \int_{S^1} \|g_\lambda^{(i)}\|^2\, d\sigma(\lambda) \right]^{1/2} = a\sqrt{m}.$$

Finally we obtain

$$\Sigma \|h^{(i)} - g^{(i)}\|^2 \geq m(1 + a^2) - 2a\sqrt{m}.$$

The lemma is proved. □

Proof of Theorem 1. By the corollary to Theorem 2, §1, the automorphism T is ergodic and, without loss of generality, we may assume that the invariant measure μ is continuous.

Suppose the multiplicity of the spectrum of the operator U_T on a set of positive measure of maximal spectral type is no less than m. Choose vectors

$h^{(1)}, \ldots, h^{(m)}$ as in the previous lemma and let $\delta > 0$. It follows from the condition $\xi_n \to \varepsilon$ that (for sufficiently large n) we may find functions

$$h_n^{(1)}, \ldots, h_n^{(m)} \in L^2(M, \mathfrak{S}, \mu), \qquad \|h_n^{(i)}\| = 1,$$

which are measurable with respect to the σ-algebra $\mathfrak{S}(\xi_n)$, and satisfy $\|h^{(i)} - h_n^{(i)}\| < \delta$, $1 \le i \le m$.

Take such an n and, for an arbitrary element C of the partition ξ_n, put $B_n' = \bigcap_{i=0}^{q_n-1} T^{-i}(T_n^i C)$. Clearly,

(1) $T^k B_n' \subseteq T^k C, \qquad 0 \le k \le q_n - 1,$

(2) $\mu(B_n') \ge \mu(C) - \dfrac{1}{2} \displaystyle\sum_{k=0}^{q_n-1} \mu(T T_n^k C \,\triangle\, T_n^{k+1} C) \ge \left(1 - \dfrac{\theta}{2}\right) \dfrac{1}{q_n}.$

By using the fact that T_n is cyclic, we can write the function $h_n^{(i)}$ in the form

$$h_n^{(i)}(x) = \sum_{k=0}^{q_n-1} b_{n,k}^{(i)} \chi_C(T_n^k x).$$

Notice that

$$1 = \|h_n^{(i)}\|^2 = \frac{1}{q_n} \sum_{k=0}^{q_n-1} |b_{n,k}^{(i)}|^2, \qquad \text{i.e.,} \quad \sum_{k=0}^{q_n-1} |b_{n,k}^{(i)}|^2 = q_n.$$

Suppose $B_n \subset B_n'$ is an arbitrary set satisfying $\mu(B_n) = (1 - (\theta/2))(1/q_n)$. Consider the cyclic subspace H' generated by the function χ_{B_n}. Put

$$g^{(i)} = \sum_{k=0}^{q_n-1} b_{n,k}^{(i)} \chi_{B_n}(T^k x).$$

Then

$$\|g^{(i)}\|^2 = \left(1 - \frac{\theta}{2}\right) \frac{1}{q_n} \sum_{k=0}^{q_n-1} |b_{n,k}^{(i)}|^2 = 1 - \frac{\theta}{2}.$$

Moreover

$$\begin{aligned} \|g^{(i)} - h_n^{(i)}\|^2 &= \left\| \sum_{k=0}^{q_n-1} b_{n,k}^{(i)} [\chi_C(T_n^k x) - \chi_{B_n}(T^k x)] \right\|^2 \\ &= \sum_{k=0}^{q_n-1} |b_{n,k}^{(i)}|^2 \|\chi_C(T_n^k x) - \chi_{B_n}(T^k x)\|^2 \\ &= \sum_{k=0}^{q_n-1} |b_{n,k}^{(i)}|^2 \mu(T_n^k C \setminus T^k B_n) = \frac{\theta}{2q_n} \sum_{k=0}^{q_n-1} |b_{n,k}^{(i)}|^2 = \frac{\theta}{2}. \end{aligned}$$

Therefore $\|h^{(i)} - g^{(i)}\| \le \sqrt{\theta/2} + \delta$, and by Lemma 1

$$m(\sqrt{\theta/2} + \delta)^2 \ge \sum_{i=1}^{m} \|h^{(i)} - g^{(i)}\|^2 \ge \sum_{i=1}^{m} \|h_n^{(i)} - g^{(i)}\|^2 - \sum_{i=1}^{m} \|h_n^{(i)} - h^{(i)}\|^2$$

$$\ge m(1 + 1 - \theta/2 - 2\sqrt{1 - \theta/2} \cdot 1/\sqrt{m}) - m\delta.$$

Since δ was arbitrary,

$$m\left(2 - \frac{\theta}{2} - 2\sqrt{1 - \theta/2} \cdot 1/\sqrt{m}\right) \le m\theta/2,$$

or

$$2 - \theta \le \sqrt{2}\sqrt{2 - \theta} \cdot 1/\sqrt{m}, \qquad \sqrt{m} \le \sqrt{2/(2 - \theta)},$$

$$m \le 2/(2 - \theta), \qquad \theta \ge 2 - 2/m$$

which contradicts the assumption. The theorem is proved. □

§3. An Application of Approximation Theory: an Example of an Ergodic Automorphism with a Spectrum Lacking the Group Property

By using approximation theory, we can construct examples of automorphisms with unexpected spectral properties. In this section we discuss the so-called group property of the spectrum of ergodic automorphisms.

According to Theorem 1, §1, Chap. 12, the point spectrum $\Lambda_d(T)$ for ergodic automorphisms T is a subgroup of the group S^1 i.e.,

(1) if $\lambda \in \Lambda_d(T)$, then $\lambda^{-1} \in \Lambda_d(T)$;
(2) if $\lambda_1, \lambda_2 \in \Lambda_d(T)$, then $\lambda_1\lambda_2 \in \Lambda_d(T)$.

This property of the spectrum, known as the group property, may be restated in a different way, so that in the new formulation it can, in principle, be applicable not only to a point spectrum. Namely, suppose T is an automorphism of the measure space $(M, \mathfrak{S}, \mu)$, U_T is the adjoint unitary operator and ρ is the maximal spectral type of U_T.

Consider the following two properties of the type ρ:

(1) the type ρ is symmetric; this means that for any measure σ of type ρ the measure σ' on S' defined by the relation $\sigma'(C) = \sigma(-C)$ where $-C = \{\lambda \in S^1 : -\lambda \pmod 1) \in C\}$ also belongs to type ρ (here we again identify S^1 with the semi-interval $0 \le \lambda < 1$);
(2) $\rho * \rho = \rho$; here by the type $\rho * \rho$ we mean the type of measures representable as convolutions of measures of the type ρ.

For discrete types, the properties (1′) and (2′) are respectively equivalent to the properties (1) and (2). It follows from (2′) that if ρ dominates a certain type which is absolutely continuous with respect to the Lebesgue measure, then ρ dominates the type of the Lebesgue measure. Indeed, in all the examples which have been considered up to now, the absolutely continuous type was always Lebesgue. Further, properties (1′) and (2′) hold for Gauss dynamical systems (see §3, Chap. 14).

Let us check that (1′) is valid for any automorphism. Indeed, suppose $f \in L^2(M, \mathfrak{S}, \mu)$ is of spectral measure σ, i.e.,

$$(U_T^n f, f) = \int f(T^n x)\overline{f(x)}\, d\mu(x) = \int_0^1 e^{2\pi i n\lambda}\, d\sigma(\lambda).$$

Then

$$(U_T^n \bar{f}, \bar{f}) = \int \overline{f(T^n x)} f(x)\, d\mu(x) = \int \overline{f(T^n x)\overline{f(x)}\, d\mu(x)}$$
$$= \int_0^1 e^{-2\pi i n\lambda}\, d\sigma(\lambda) = \int_0^1 e^{2\pi i n\lambda}\, d\sigma(-\lambda) = \int_0^1 e^{2\pi i n\lambda}\, d\sigma'(\lambda),$$

where $d\sigma'(\lambda) = d\sigma(-\lambda)$.

This shows that the set of spectral measures σ is symmetric, i.e., together with every measure σ this set contains the measure σ'. Thus (1) is proved.

Now let us show that (2′) is false in the general case.

An example of an ergodic automorphism with mixed spectrum not satisfying (2′)

Suppose the measure space $(M, \mathfrak{S}, \mu)$ is the direct product $M = S^1 \times Z_2$, where S^1 is the unit circle with Lebesgue measure ρ identified with the semi-interval $0 \le x < 1$, while $Z_2 = \{1, -1\}$ is the group of square roots of unity with normalized Haar measure. Points of the space M may be written in the form (x, y), $x \in S^1$, $y = \pm 1$. Any function $f \in L^2(M, \mathfrak{S}, \mu)$ is of the form $f(x, y) = f_1(x) + yf_2(x)$. Clearly $(f_1, yf_2) = 0$. Therefore, the previous relation means that

$$L^2(M, \mathfrak{S}, \mu) = H^+ \oplus H^-,$$

where

$$H^+ = \{f(x, y)\colon f(x, 1) = f(x, -1) \text{ almost everywhere}\},$$

$$H^- = \{f(x, y)\colon f(x, 1) = -f(x, -1) \text{ almost everywhere}\},$$

i.e., H^+ is the subspace of functions of the form $f(x)$, where H^- is the subspace of functions of the form $yf(x)$.

Suppose T is the rotation of the circle S^1 by a certain angle α, i.e., $Tx = x + \alpha \pmod 1$ and

$$w(x) = \begin{cases} -1 & \text{if } x \in [0, \beta), \\ 1 & \text{if } x \in [\beta, 1). \end{cases}$$

Consider the automorphism S of the space M acting according to the formula $S(x, y) = (Tx, w(x)y)$. It is clear that S is the skew product constructed from the automorphism T and the function w. The automorphism S may obviously be realized also as an interval exchange transformation.

The spaces H^+, H^- are invariant with respect to U_s and U_s possesses a pure point spectrum in the space H^+.

Lemma 1. *Suppose the numbers α, β satisfy the following conditions:*

(1) *α is irrational and there exists a sequence of irreducible fractions k_n/m_n such that $\lim_{n\to\infty} m_n^2 |k_n/m_n - \alpha| = 0$;*
(2) *there exists a constant c, $0 < c < 1$ such that for all integers r we have*

$$|r/m_n - \beta| > c/m_n.$$

Then the operator U_s has a continuous spectrum in H^-.

Lemma 2. *Assume that the automorphism T possesses a cyclic a.p.t. with speed $f(n)$, $nf(n) \to 0$ when $n \to \infty$ and $\{\xi_n\}$, $\{T_n\}$ are the corresponding sequences of partitions and approximating automorphisms. Suppose, moreover, that for every n there exists a set B_n consisting of d_n elements of the partition ξ_n, d_n is odd and $\rho([0, \beta) \triangle B_n) = o(1/q_n)$ where q_n is the number of elements of the partition ξ_n.*

Then $\lim_{n\to\infty} \|U^{q_n}f + f\| = 0$ for all $f \in H^-$.

Lemma 3. *Suppose σ is the normalized Borel measure on S^1 satisfying*

$$\int e^{2\pi i q_n \lambda}\, d\sigma(\lambda) \to -1$$

for some sequence $q_n \to \infty$. Then the measures σ and $\sigma * \sigma$ are singular with respect to each other.

Lemmas 1–3 shall be proved later. Choose the numbers α, β possessing properties (1) and (2) of Lemma 1. Assume in addition that there exist sequences of irreducible fractions $\{p_n/q_n\}$, $\{l_n/q_n\}$ such that the l_n are odd and

$$\lim_{n\to\infty} q_n^2 |p_n/q_n - \alpha| = 0, \qquad \lim_{n\to\infty} q_n |l_n/q_n - \beta| = 0.$$

It is easy to see that such pairs of numbers α, β exist. The assumptions of Lemma 2 will be satisfied if we take, for ξ_n, the partition of $[0, 1)$ into semi-intervals

$$C_i^{(n)} = [(i - 1)/q_n, i/q_n), \qquad 1 \le i \le q_n,$$

and, for T_n, the automorphism

$$T_n x = x + \frac{p_n}{q_n} (\text{mod } 1),$$

and then put $B_n = [0, l_n/q_n)$.

It follows from Lemma 1 that the operator U_s has a continuous spectrum in H^-. Take a unit vector $f \in H^-$ whose spectral measure σ is of the maximal spectal type of U_s in H^-. Then by Lemma 2

$$\int_0^1 e^{2\pi i q_n \lambda}\, d\sigma(\lambda) = (U^{q_n} f, f) \to -(f, f) = -1,$$

while Lemma 3 implies that $\sigma * \sigma$ is singular with respect to σ.

Suppose ν^+ is the maximal spectral type of U_s in H^+, which will of course be discrete, and ν^- is the maximal spectral type of U_s in H^-, which is continuous by Lemma 1. Then $\nu^+ + \nu^-$ is the maximum spectral type of U_s in $L^2(M, \mathfrak{S}, \mu)$ while

$$(\nu^+ + \nu^-) * (\nu^+ + \nu^-) = \nu^+ + \nu^+ * \sigma + \sigma * \sigma.$$

This type cannot coincide with $\nu^+ + \sigma$, since $\sigma * \sigma$ is singular with respect to σ, which was to be proved.

Proof of Lemma 1. If the vector $f = yg \in H^-$ is an eigenvector for U_s with eigenvalue λ, then for almost all x we have the relation $yg(x + \alpha)w(x) = \lambda y \cdot g(x)$. Put $w_1(x) = \lambda(w(x))^{-1}$. Then we have almost everywhere $g(x + \alpha) = w_1(x)g(x)$. This implies that the function $|g(x)|$ is invariant (mod 0) with respect to T. Since α is irrational, T will be ergodic and $|g(x)| = \text{const}$ almost everywhere. We will assume that $|g(x)| = 1$ almost everywhere. Then we can write $g(x + \alpha)/g(x) = w_1(x)$. Substitute the points $x + r\alpha$, $0 \le r < m_n$ for x in the last equality and multiply the obtained relations; then

$$\frac{g(x + m_n \alpha)}{g(x)} = \prod_{r=0}^{m_n - 1} w_1(x + r\alpha) \stackrel{\text{def}}{=} w_1^{(n)}(x).$$

Since $x + m_n\alpha$ tends to x when $n \to \infty$, we have

$$\int_0^1 \left| \frac{g(x + m_n\alpha)}{g(x)} - 1 \right|^2 dx = \int_0^1 |g(x + m_n\alpha) - g(x)|^2 \, dx$$
$$= \int_0^1 |w_1^{(n)}(x) - 1|^2 \, dx \to 0, \qquad (1)$$

when $n \to \infty$. Suppose

$$\tilde{w}_1^{(n)}(x) = \prod_{r=0}^{m_n - 1} w_1\left(x + r\frac{k_n}{m_n}\right), \qquad \beta = r_n/m_n + \theta_n/m_n.$$

By assumption $c < \theta_n < 1 - c$. It is easy to check that $\tilde{w}_1^{(n)}$ is of the form

$$\tilde{w}_1^{(n)}(x) = \begin{cases} \lambda_1^{r_n}\lambda_2^{m_n - r_n} & \text{if } \{q_n x\} \le \theta_n, \\ \lambda_1^{r_n+1}\lambda_2^{m_n - r_n - 1} & \text{if } \{q_n x\} > \theta_n \end{cases}$$

where $\lambda_1 \neq \lambda_2$. In other words, $\tilde{w}_1^{(n)}(x)$ assumes two values whose quotient is $\lambda_2/\lambda_1 \neq 1$, while each value is assumed on a set of measure greater than const > 0, i.e., $\|\tilde{w}_1^{(n)} - 1\| \ge$ const. Put $E_n = \{x\colon\ w_1^{(n)}(x) \neq \tilde{w}_1^{(n)}(x)\}$. It follows from the assumptions of the lemma that $\rho(E_n) \to 0$ when $n \to \infty$. Therefore

$$\overline{\lim_{n\to\infty}} \|w_1^{(n)} - 1\| \ge \text{const} > 0,$$

contradicting (1). The lemma is proved. □

Proof of Lemma 2. It suffices to prove the statement of the lemma for a dense set of functions $f \in H^-$. Therefore, we shall assume, in particular, that $|f| \le C$. By H_n^- denote the subspace of the space H^- consisting of functions of the form $yg(x)$, where g is constant mod 0 on each element of the partition ξ_n. For any function f, denote by f_n its orthogonal projection on H_n^-. If $|f| \le C$, then obviously $|f_n| \le C$. For any n we have

$$\|U_S^{q_n}f + f\| \le \|U_S^{q_n}f - U_S^{q_n}f_n\| + \|U_S^{q_n}f_n + f_n\| + \|f - f_n\|.$$

The condition $\xi_n \to \varepsilon$ implies that the first and last summands vanish when $n \to \infty$.

Suppose $f_n(x, y) = g_n(x)y$. Then

$$U_S^{q_n}f_n = f_n(S^{q_n}(x, y)) = g_n(T^{q_n}x)\prod_{k=0}^{q_n - 1} w(T^k x)y.$$

Put

$$E_n = \bigcup_{i=1}^{q_n} (TC_i^{(n)} \triangle T_n C_i^{(n)}), \qquad G_n = \bigcup_{j=0}^{q_n - 1} T^{-j} E_n,$$

$$F_n = \bigcup_{j=1}^{q_n} T^{-j}([0, \beta) \triangle B_n),$$

where $C_i^{(n)}$ are the elements of the partition ξ_n, $1 \leq i \leq q_n$.

If $x \in S^1 \setminus G_n$ then $g_n(T^{q_n} x) = g_n(x)$. If $x \in S^1 \setminus F_n$, then it follows from the fact that the approximation is cyclic that the number of those j, $0 \leq j \leq q_n - 1$, where $T_n^j x \in [0, \beta)$ is precisely equal to d_n, i.e., to the number of elements of the partition ξ_n which are contained in B_n. Therefore, for $x \in S^1 \setminus (F_n \cup G_n)$, we have

$$\prod_{j=0}^{q_n - 1} w(T^j x) = \prod_{j=0}^{q_n - 1} w(T_n^j x) = (-1)^{d_n} = -1.$$

Hence, if $(x, y) \in M \setminus ((F_n \cup G_n) \times \mathbb{Z}_2)$, then $U_S^{q_n} f_n = -f_n$. Further $\rho(F_n) \leq q_n \rho([0, \beta) \triangle B_n) \to 0$ when $n \to \infty$, $\rho(G_n) \leq q_n \rho(E_n) \to 0$ when $n \to \infty$, by assumption. Therefore

$$\|U_S^{q_n} f_n + f_n\|^2 = \int_{(F_n \cup G_n) \times Z_2} |U_S^{q_n} f_n + f_n|^2 \, d\mu \leq 4C^2 \rho(F_n \cup G_n) \to 0,$$

when $n \to \infty$. The lemma is proved. □

Proof of Lemma 3. The assumptions of the lemma imply that, when $n \to \infty$, we have

$$\int_0^1 e^{2\pi i q_n \lambda} \, d(\sigma * \sigma) = \left[\int_0^1 e^{2\pi i q_n \lambda} \, d\sigma(\lambda)\right]^2 \to 1.$$

We may assume that

$$\left| \int_0^1 e^{2\pi i q_n \lambda} \, d\sigma(\lambda) + 1 \right| \leq 2^{-n}, \qquad \left| \int_0^1 e^{2\pi i q_n \lambda} \, d(\sigma * \sigma) - 1 \right|^2 \leq 2^{-n},$$

by passing to subsequences if necessary.

Put

$$A_n = \{\lambda \in [0, 1): |1 + e^{2\pi i q_n \lambda}| < \sqrt{2}\}, \qquad \bar{A}_n = S^1 \setminus A_n,$$

$$B_n = \{\lambda \in [0, 1): |1 - e^{2\pi i q_n \lambda}| < \sqrt{2}\}, \qquad \bar{B}_n = S^1 \setminus B_n.$$

Then $A_n \cap B_n = \emptyset$. Moreover

$$\mathrm{Re}(1 + e^{2\pi i q_n \lambda}) \geq 0, \qquad \mathrm{Re}(1 - e^{2\pi i q_n \lambda}) \geq 0,$$

for all $\lambda \in S^1$, while for $\lambda \in \bar{A}_n$ we obviously have $\mathrm{Re}(1 + e^{2\pi i q_n \lambda}) \geq 1$. Hence

$$\begin{aligned} 2^{-n} &\geq \left| \int_0^1 (1 + e^{2\pi i q_n \lambda})\, d\sigma(\lambda) \right| \geq \left| \mathrm{Re} \int_0^1 (1 + e^{2\pi i q_n \lambda})\, d\sigma(\lambda) \right| \\ &= \int_0^1 \mathrm{Re}(1 + e^{2\pi i q_n \lambda})\, d\sigma(\lambda) \geq \int_{\bar{A}_n} \mathrm{Re}(1 + e^{2\pi i q_n \lambda})\, d\sigma(\lambda) \geq \sigma(\bar{A}_n). \end{aligned}$$

Therefore $\sigma(A_n) \geq 1 - 1/2^n$, $n = 1, 2, \ldots$. In a similar way, we get

$$(\sigma * \sigma)(B_n) \geq 1 - 1/2^n, \qquad n = 1, 2, \ldots.$$

Putting

$$A = \bigcup_{k=1}^{\infty} \bigcap_{n=k}^{\infty} A_n, \qquad B = \bigcup_{k=1}^{\infty} \bigcap_{n=k}^{\infty} B_n$$

we obtain $A \cap B = \emptyset$, $\sigma(A) = 1$, $(\sigma * \sigma)(B) = 1$. The lemma is proved. □

§4. Approximation of Flows

The definition of various types of approximations of a given automorphism by periodic ones given in §1 can be carried over to the case of flows in a natural way.

Definition 1. Suppose $g(u) \downarrow 0$ for $u \to \infty$. The flow $\{T^t\}$ on the Lebesgue space $(M, \mathfrak{S}, \mu)$ admits an *approximation by periodic transformations with speed* g, if we can indicate sequences of real numbers t_n, of partitions ξ_n of the space M into q_n measurable sets $C_i^{(n)} \subset M$ and of automorphisms S_n such that

(1) $\xi_n \to \varepsilon$;
(2) $S_n \xi_n = \xi_n$;
(3) $\sum_{i=1}^{q_n} \mu(T^{t_n} C_i^{(n)} \triangle S_n C_i^{(n)}) < g(q_n)$;
(4) if p_n is the order of the automorphism S_n (as a permutation of the sets $C_i^{(n)}$), then $p_n t_n \to \infty$ when $n \to \infty$.

If we have in addition:

(5) S_n cyclically permutes the set $C_i^{(n)}$, then the flow $\{T^t\}$ is said to possess a *cyclic approximation with speed* g.

The results of §1 may be easily carried over to the case of flows. We will not provide the corresponding formulations, but will prove two theorems on the relationship between approximations and the spectra of flows in a form which will be convenient for us in Chap. 16.

Theorem 1. *If the flow $\{T^t\}$ possesses a cyclic approximation with speed $g(u) = o(1/u)$, then in the space $L^2(M, \mathfrak{S}, \mu)$ we have the strong convergence of operators $U^{q_n t_n} \to E$, where E is the identity operator in $L^2(M, \mathfrak{S}, \mu)$.*

Proof. It suffices to show that $\|U^{q_n t_n} f - f\| \to 0$ when $n \to \infty$ for bounded functions $f \in L^2(M, \mathfrak{S}, \mu)$. Suppose $|f| \le C$. Denote by H_n the subspace of functions from $L^2(M, \mathfrak{S}, \mu)$ which are constant on elements of the partition ξ_n. If f_n is a projection of the element $f \in L^2(M, \mathfrak{S}, \mu)$ on H_n, then $\|f - f_n\| < \delta$ for all $\delta > 0$ when $n > n_0(\delta)$. For such n, we can write

$$\|U^{q_n t_n} f - f\| \le \|f(T^{q_n t_n} x) - f_n(T^{q_n t_n} x)\| + \|f_n(T^{q_n t_n} x) - f_n(x)\| + \|f_n(x) - f(x)\| = \Sigma_1 + \Sigma_2 + \Sigma_3.$$

Clearly $\Sigma_1 = \Sigma_3 < \delta$. To estimate Σ_2, note that $f_n(x) = f_n(S_n^{q_n} x)$, so that

$$\Sigma_2 = \|f_n(T^{q_n t_n} x) - f_n(S_n^{q_n} x)\|.$$

Further, if

$$x \notin E_n \overset{\text{def}}{=} \bigcup_{i=1}^{q_n} [T^{q_n t_n} C_i^{(n)} \triangle S_n^{q_n} C_i^{(n)}],$$

then $f_n(T^{q_n t_n} x) = f_n(S_n^{q_n} x)$.

By Lemma 1 in §1,

$$\mu(E_n) \le q_n \sum_{i=1}^{q_n} \mu(T^{t_n} C_i^{(n)} \triangle S_n C_i^{(n)}) \le q_n g(q_n).$$

Taking into consideration the inequality

$$|f_n(T^{q_n t_n} x) - f_n(S_n^{q_n} x)| \le 2C \quad \text{for all } x \in M,$$

we get $\Sigma_2 \le 2C\sqrt{q_n g(q_n)}$. Finally

$$\|U^{q_n t_n} f - f\| \le 2\delta + 2C\sqrt{q_n g(q_n)} \to 0,$$

when $n \to \infty$. The theorem is proved. □

Corollary 1. *If the flow $\{T^t\}$ possesses a cyclic approximation with speed $g(u) = o(1/u)$, then it is not mixing.*

Indeed, if $\{T^t\}$ were mixing, then for any function $f \in L_0^2(M, \mathfrak{S}, \mu)$ we would have $(U^t f, f) \to 0$, when $t \to \infty$. On the other hand, if $f \not\equiv 0$, then Theorem 1 implies $(U^t f, f) \to (f, f) \neq 0$. This contradiction proves the required statement.

Corollary 2. *If the flow $\{T^t\}$ possesses a cyclic approximation with speed $g(u) = o(1/u)$, then the maximal spectral type of the adjoint group of unitary operators $\{U^t\}$ is singular with respect to the Lebesgue measure.*

Indeed, in the converse case we could find a function $f \in L^2(M, \mathfrak{S}, \mu)$, $f \neq 0$ whose spectral type σ_f is absolutely continuous, i.e.,

$$(U^t f, f) = \int_{-\infty}^{\infty} e^{it\lambda}\, d\sigma_f = \int_{-\infty}^{\infty} e^{it\lambda} \varphi(\lambda)\, d\lambda,$$

where $\varphi(\lambda) = d\sigma_f(\lambda)/d\lambda \in L^1(\mathbb{R}^1, d\lambda)$. Since the Fourier transform of a function belonging to $L^1(\mathbb{R}^1, d\lambda)$ tends to zero at infinity, it follows that $(U^t f, f) \to 0$, contradicting Theorem 1.

Theorem 2. *If the flow $\{T^t\}$ possesses a cyclic approximation with speed $g(u) = o(u^{-2})$, then the spectrum of the adjoint group of unitary operators $\{U^t\}$ is simple.*

Proof. Suppose $\chi_n(x)$ is the indicator of the set $C_1^{(n)}$, H_n is the subspace of the space $L^2(M, \mathfrak{S}, \mu)$ consisting of functions constant on elements of the partition ξ_n. It follows from (1) that $(f, H_n) \to 0$ when $n \to \infty$ for any function $f \in L^2(M, \mathfrak{S}, \mu)$.

By H_n' denote the cyclic subspace generated by the function χ_n, i.e., the closure of the set of vectors of the form $\sum_k c_k U^{t_k} \chi_n$. By definition, the group $\{U^t\}$ has a simple spectrum in H_n'.

Let us show that $\operatorname{dist}(f, H_n') \to 0$ when $n \to \infty$ for any function $f \in L^2(M, \mathfrak{S}, \mu)$. It suffices to consider bounded $f : |f| \le C$.

For any $\varepsilon > 0$, when n is sufficiently large, it is possible to find (since the approximation is cyclic) functions

$$f_n \in H_n, \qquad f_n = \sum_{k=0}^{q_n - 1} a_k^{(n)} \chi_n(S_n^{-k} x),$$

such that $\|f - f_n\| < \varepsilon$. We can assume that

$$|a_k^{(n)}| \le C, \qquad k = 0, 1, \ldots, q_n - 1.$$

Put $f'_n = \sum_{k=0}^{q_n-1} a_k^{(n)} \chi_n(T^{-kt_n}x)$. Then

$$\|f - f'_n\| \le \|f - f_n\| + \|f_n - f'_n\|,$$

$$\begin{aligned}\|f_n - f'_n\| &= \left\| \sum_{k=0}^{q_n-1} a_k^{(n)} \chi_n(S^{-k}x) - \sum_{k=0}^{q_n-1} a_k^{(n)} \chi_n(T^{-kt_n}x) \right\| \\ &\le \sum_{k=0}^{q_n-1} |a_k^{(n)}| \, \|\chi_n(S_n^{-k}x) - \chi_n(T^{-kt_n}x)\| \\ &\le C \sum_{k=0}^{q_n-1} \|\chi_n(S_n^{-k}x) - \chi_n(T^{-kt_n}x)\|.\end{aligned}$$

Further

$$\begin{aligned}\|\chi_n(S_n^{-k}x) - \chi_n(T^{-kt_n}x)\|^2 &= \mu(S_n^k C_1^{(n)} \triangle T^{kt_n} C_1^{(n)}) \\ &\le \sum_{i=0}^{k-1} \mu(S_n(S_n^i C_i^{(n)}) \triangle T^{t_n}(S_n^i C_1^{(n)})) \\ &\le \sum_{i=0}^{q_n-1} \mu(S_n(S_n^i C_1^{(n)}) \triangle T^{t_n}(S_n^i C_1^{(n)})) \\ &\le g(q_n) = \delta_n q_n^{-2},\end{aligned}$$

where $\delta_n \to 0$ when $n \to \infty$. Therefore

$$\|f_n - f'_n\| \le C q_n \sqrt{\delta_n q_n^{-2}} = C\sqrt{\delta_n} \to 0 \qquad \text{when } n \to \infty,$$

i.e., $\operatorname{dist}(f, H'_n) \to 0$ when $n \to \infty$. It follows from the main theorem on the canonical form of unitary operators (see Appendix 2) that if the group $\{U^t\}$ has a simple spectrum on each of the subspaces H'_n and $\operatorname{dist}(f, H'_n) \to 0$ when $n \to \infty$ for all $f \in L^2(M, \mathfrak{S}, \mu)$, then the group $\{U^t\}$ has a simple spectrum on the entire space $L^2(M, \mathfrak{S}, \mu)$. The theorem is proved. □

Putting together the statements of Theorem 2 and Corollaries 1 and 2 to Theorem 1, we may formulate the following statement:

Theorem 3. *If the flow $\{T^t\}$ possesses a cyclic approximation with speed $g(u) - o(u^{-2})$, then it is not mixing, the spectrum of the adjoint group $\{U^t\}$ is simple and the maximal spectral type of this group is singular with respect to the Lebesgue measure.*

Chapter 16

Special Representations and Approximations of Smooth Dynamical Systems on the Two-dimensional Torus

§1. Special Representations of Flows on the Torus

Suppose the space M is the two-dimensional torus $\mathrm{Tor}^2 = \mathbb{R}^2/\mathbb{Z}^2$ with cyclic coordinates (u, v) and Lebesgue measure $du\,dv$. Consider the system of differential equations

$$\frac{du}{dt} = A(u, v), \qquad \frac{dv}{dt} = B(u, v) \tag{1}$$

on it, with right-hand sides of class C^r, $r \geq 2$. This system satisfies the existence and uniqueness conditions and we may therefore introduce the one-parameter group $\{T^t\}$ of translations along its solutions.

In this section we begin the study of ergodic properties of flows $\{T^t\}$ which arise in this way. We will assume that the flow $\{T^t\}$ possesses an absolutely continuous invariant measure λ with density $P(u, v) \in C^5(\mathrm{Tor}^2)$, $P(u, v) > 0$.

According to Liouville's theorem (see §2, Chap. 2)

$$\frac{\partial}{\partial u}(PA) + \frac{\partial}{\partial v}(PB) = 0. \tag{2}$$

Also assume that $A^2 + B^2 > 0$, i.e., the system (1) has no fixed points. Put

$$\lambda_1 = \iint_{\mathrm{Tor}^2} PA\,du\,dv, \qquad \lambda_2 = \iint_{\mathrm{Tor}^2} PB\,du\,dv.$$

Clearly $\lambda_1(\lambda_2)$ is the mean velocity of motion along the axis $u(v)$. The ergodic properties of the flow $\{T^t\}$ are intimately related to the properties of the number $\lambda = \lambda_1/\lambda_2$.

Lemma 1. *If λ is rational or if at least one of the numbers λ_1, λ_2 is equal to zero, then the flow $\{T^t\}$ is not ergodic.*

Proof. We may consider the system of equations (1) not on the torus $\mathrm{Tor}^2 = \mathbb{R}^2/\mathbb{Z}^2$ but on the plane $\mathbb{R}^2$ by extending the functions $A(u, v)$, $B(u, v)$ by periodicity to the entire plane. The density of the invariant measure $P(u, v)$ shall also be viewed as a periodic function on $\mathbb{R}^2$ with period 1 along u, v. Then (2) means that there exists a function $H(u, v)$ such that

$$\frac{\partial H}{\partial u} = PA, \qquad \frac{\partial H}{\partial v} = -PB,$$

and therefore

$$\frac{d}{dt} H(T^t(u, v)) = \frac{\partial H}{\partial u}\frac{du}{dt} + \frac{\partial H}{\partial v}\frac{dv}{dt} = 0,$$

i.e., H is a first integral for the system of equations (1) on the plane. The function H is not necessarily periodic, but since $\partial H/\partial u$, $\partial H/\partial v$ are periodic functions, H is of the form $H(u, v) = c_1 u + c_2 v + h(u, v)$, where $h(u, v)$ is now periodic with period 1 along u, v. The constants c_1, c_2 may be computed explicitly:

$$\begin{aligned} c_1 = H(u + 1, v) - H(u, v) &= \int_0^1 \frac{\partial H}{\partial u}\, du = \int_0^1 \int_0^1 \frac{\partial H}{\partial u}\, du\, dv \\ &= \iint PA\, du\, dv = \lambda_1, \end{aligned} \tag{3}$$

$$c_2 = H(u, v + 1) - H(u, v) = -\iint PB\, du\, dv = -\lambda_2. \tag{4}$$

Thus $H(u, v) = \lambda_1 u - \lambda_2 v + h(u, v)$. If $\lambda_1 = \alpha p$, $\lambda_2 = \alpha q$, where p, q are integers, $\alpha \neq 0$, then the function

$$\Phi(u, v) = \exp[2\pi i \alpha^{-1} H(u, v)]$$

is of period 1 along u, v, and therefore may be viewed as a well-defined function on the torus Tor^2 invariant with respect to $\{T^t\}$. Moreover, we clearly have $\Phi(u, v) \not\equiv \text{const}$. If precisely one of the numbers λ_1, λ_2 vanishes, say, $\lambda_2 = 0$, $\lambda_1 \neq 0$, then the function

$$\Phi(u, v) = \exp\left[\frac{2\pi i}{\lambda_1} H(u, v)\right] \not\equiv \text{const}.$$

will be invariant. If $\lambda_1 = \lambda_2 = 0$, then $H(u, v) = h(u, v) \not\equiv \text{const}$ is an invariant function on the torus. The lemma is proved. $\square$

We shall now assume that $\lambda = \lambda_1/\lambda_2$ is irrational. In this case we will prove the following theorem.

Theorem 1. *If λ is irrational, then the flow $\{T^t\}$ is metrically isomorphic to the special flow constructed from the automorphism T_1 of rotation of the circle S_1 by a certain irrational angle α, where α is of the form*

$$\alpha = \frac{m\lambda + n}{p\lambda + q}, \quad m, n, p, q \in \mathbb{Z}, \qquad \det \begin{Vmatrix} m & n \\ p & q \end{Vmatrix} = \pm 1,$$

and a function $F: S^1 \to \mathbb{R}^1$, $F \in C^5(S^1)$.

We shall begin the proof of Theorem 1 by proving the following three lemmas which are of intrinsic interest.

Lemma 2. *On the torus* Tor^2 *there exists a closed non-self-intersecting curve Γ of class C^∞ which is not tangent to the trajectories of the system* (1) *at any point.*

Proof. Together with the system (1) on the torus consider the system of equations

$$\frac{du}{dt} = -B(u, v), \qquad \frac{dv}{dt} = A(u, v), \tag{5}$$

whose trajectories at each point are orthogonal to those of the system (1). If the system (5) possesses a closed integral curve, then our statement is proved: it suffices to smooth out such a curve in order to obtain the required curve of class C^∞. In the converse case, take an arbitrary point $a \in \mathrm{Tor}^2$ and construct the integral curve $Q = \{T^t a: -\infty < t < \infty\}$ of the system (5). Put $a_n = T^n a$, $n = 1, 2, \ldots$. The sequence $\{a_n\}$ has a limit point $c \in \mathrm{Tor}^2$ and we may assume, by passing to subsequences if necessary, that

$$\lim_{n \to \infty} a_n = c.$$

From the theorem on the existence of solutions for systems of differential equations of type (5) and from the theorem on the smooth dependence of the solutions on the initial data, it follows that there exists a neighborhood $O(c) \subset \mathrm{Tor}^2$ of the point c such that

(1) every arc $\widehat{a_n a_{n+1}}$ of the trajectory Q possesses points outside of $O(c)$;
(2) vectors of the vector field corresponding to the system (5) at any two points of the neighborhood $O(c)$ form angles which are no greater than $\pi/8$.

Consider the segment $L = [c_1, c_2] \subset O(c)$ with mid-point c such that the direction of the vector $\overrightarrow{c_1c_2}$ forms an angle of $\pi/4$ with the direction of the vector field (5) at the point c. For all sufficiently large n, there is an arc $\widehat{a_nb_n}$ of the curve Q, $\widehat{a_nb_n} \subset O(c)$ such that $b_n \in L$ (the motion along this arc from a_n to b_n possibly corresponds to a decrease rather than an increase of the parameter t on Q). We may assume that such an arc exists for all $n = 1, 2, \ldots$. By property (1) of the neighborhood $O(c)$ all the points $b_1, b_2, \ldots$ are geometrically distinct and $\lim_{n\to\infty} b_n = c$. Now define two natural numbers $n_1, n_2, n_1 > n_2$ so that the point b_{n_1} lies between c and b_{n_2} on the segment L.

Suppose b_0 is the first intersection point of the arc $\widehat{b_{n_1}b_{n_2}}$ of the curve Q with the semi-interval $(b_{n_1}, b_{n_2}] \subset L$ (possibly $b_0 = b_{n_1}$). Taken together, the arc $\widehat{b_{n_2}b_0}$ of the curve Q and the segment $[b_0, b_{n_2}] \subset L$ form a simple closed curve $\tilde{\Gamma} \subset \mathrm{Tor}^2$. At each point of the curve $\tilde{\Gamma}$, except b_0, b_{n_2}, the curvature continuously depends on the point. At the points b_0, b_{n_2}, there exist tangents from the left and from the right, and the tangent from the side of the curve Q corresponds to the direction of the vector field (5), i.e., forms an angle of $\pi/2$ with the direction of the vector field (1), while the tangent on the side of the segment $[b_0, b_{n_2}]$, by property (2) of the neighborhood $O(c)$, forms an angle no greater than $\pi/4 + \pi/8 = 3\pi/8$ with the direction of the vector field (5), i.e., an angle greater than $\pi/8$ with the direction of the vector field (1). Since the direction of the vector field (1) on the segment $[b_0, b_{n_2}]$ changes no more than by $\pi/8$, we see that, smoothing the curve $\tilde{\Gamma}$ in the neighborhood of the points b_0, b_{n_2}, we shall obtain the required curve Γ. The lemma is proved. □

The curve Γ constructed in this lemma will be known as the *Siegel curve* for the system of differential equations (1).

Now suppose Γ is an arbitrary smooth closed curve on the torus $\mathrm{Tor}^2 = \mathbb{R}^2/\mathbb{Z}^2$ and $\mathbf{\Gamma}$ is a curve on the plane $\mathbb{R}^2$ which covers Γ. We may assume that the coordinates on the plane are chosen so that the origin of coordinates in $\mathbb{R}^2$ corresponds to some point $p \in \Gamma$. When we go around Γ in a fixed direction from the point p back to the same point, the corresponding motion in the plane will be from the point $(0, 0)$ to some point (p, q) and, since Γ is closed, the numbers p, q are integers. To curves Γ which are homotopic to zero corresponds the pair $p = 0, q = 0$.

Lemma 3. *Suppose Γ is a smooth curve without self-intersections which lies on the torus* Tor^2 *and is not homotopic to zero. Then the numbers p, q are relatively prime.*

Proof. Assume the converse, i.e., assume that $p = km$, $q = kn$, where k, m, n are integers, $k > 1$. Consider the infinite curve $\mathbf{\Gamma}$ on $\mathbb{R}^2$ which covers Γ. It divides the plane into two domains D_1 and D_2. Suppose $\mathbf{\Gamma}_1$ is the curve obtained from Γ by translation along the vector (m, n). Under the covering $\mathbb{R}^2 \to \mathrm{Tor}^2$ the curve Γ_1 is also mapped onto Γ. The curves Γ and Γ_1 do not intersect, since otherwise Γ would have a self-intersection point. Hence Γ_1 is entirely contained in one of the domains D_1, D_2, say in D_1. Then the curve Γ_2 obtained from Γ_1 by translation along the vector (m, n) is also contained in D_1. Similarly, D_1 contains the curves Γ_3, Γ_4, etc. But the curve Γ_k passes through the point $(p, q) \in \Gamma$, i.e., $\Gamma \cap \Gamma_k \neq \varnothing$. The contradiction thus obtained proves the lemma. □

Now notice that the Siegel curve Γ for the system (1) cannot be homotopic to zero. Indeed, in the converse case any of its inverse images in the plane $\mathbb{R}^2$ would have a nonzero vector field index. But then inside this curve there must be a point at which the vector field vanishes, while by assumption no such points exist.

Lemma 4 (The Return Lemma). *Suppose Γ is the Siegel curve for the system (1). Then for any point $p \in \Gamma$ there is a number $t > 0$ such that $T^t p \in \Gamma$.*

Proof. Assume the converse. Suppose the point $p \in \Gamma$ satisfies $T^t p \notin \Gamma$ for all $t > 0$. Then, taking into consideration the fact that the curve Γ is transversal to the vector field (1), we see that for all $t > 0$ the distance from $T^t p$ to the curve Γ is greater than some positive constant. Clearly the trajectory $\{T^t p\}$ is not closed. Therefore we can find a limit point q for the positive semi-trajectory $\{T^t p\colon t > 0\}$. Moreover, $T^t q \notin \Gamma$ for all $t \geq 0$, since otherwise the positive semi-trajectory of the point p would also intersect Γ. Suppose L is a small segment with centre at the point q which does not intersect Γ and is transversal to the vector field (1). On it we can always find a point q_0 of the form $q_0 = T^{t_0} p$, $t_0 > 0$.

Define by induction a sequence of points $q_1, q_2, \ldots, q_i, \ldots \in L$ in the following way: if the point q_{i-1}, $i \geq 1$ has already been defined, then q_i is the first intersection point of the positive semi-trajectory $\{T^t q_{i-1}\colon t > 0\}$ with the segment L. Obviously, such points q_i exist and $q_i = T^{t_i} q_0$ where $t_i \uparrow \infty$ when $i \to \infty$.

Let us prove that the points q_i on the segment L are situated in order of increasing numbers, namely $q_i \in [q, q_{i-1}]$, $i = 1, 2, \ldots$. In view of the fact that $\lim_{i\to\infty} q_i = q$, it suffices to prove that for any $n \geq 0$ there are no points of the form q_k, $k \geq 0$ on the segment $[q_n, q_{n+1}]$. Suppose C is the simple closed curve consisting of the segment $[q_n, q_{n+1}] \subset L$ and the arc of the trajectory of the point p from q_n to q_{n+1} which we shall denote by $\widehat{q_n q_{n+1}}$. Since the curves C and Γ do not intersect, they divide the torus into at least two domains. Suppose D_1 is the one which contains points of the form $T^\varepsilon q_{n+1}$ for small $\varepsilon > 0$. The semi-trajectory $\{T^t q_{n+1}\colon t > 0\}$ is entirely contained in the domain D_1, since it cannot intersect Γ and $\widehat{q_n q_{n+1}}$, while the vector field on the segment $[q_n q_{n+1}]$ is directed towards D_1 (here we make use of the fact that L is small). Therefore $q_k \notin [q_n q_{n+1}]$ for $k > n+1$. A similar argument works for $k < n$.

Now consider the trajectory $\{T^t q\}$ of the point q; let us prove that it is not closed. If it were a closed curve C', then the nonintersecting curves C' and Γ would divide the torus into two domains D'_1, D'_2.

Since no trajectory of the flow intersects C', while the vector field on Γ points everywhere to the same side, e.g., towards the domain D'_1, we see that for all $t > 0$ the domain $T^t D'_1$ is strictly contained in the domain D'_1, which contradicts the invariance of the measure $d\mu = P\,du\,dv$, $P > 0$. Thus the trajectory $\{T^t q\}$ is not closed. Therefore there exists a point $r \in \mathrm{Tor}^2$ which is the limit point of the positive semi-trajectory $\{T^t q : t > 0\}$.

Consider a small segment N with centre at the point r, transversal to the vector field (1). As before, define points $r_1, r_2, \ldots$ where $r_j = T^{s_j} q$ is the jth intersection point of the positive semi-trajectory of q with the segment N. As before, notice that $\lim_{j \to \infty} r_j = r$ and the points $r_1, r_2, \ldots$ are situated on the segment N in order of increasing numbers.

Since $q_i \to q$, for $j = 1, 2, 3, \ldots$ we can define (for sufficiently large i) numbers $s_{ij} > 0$ such that

$$T^{s_{ij}}(q_i) \in N, \quad \lim_{i \to \infty} s_{ij} = s_j, \quad \lim_{i \to \infty} T^{s_{ij}}(q_i) = r_j, \qquad j = 1, 2, 3, \ldots$$

Fix a k so large that the point r_2 is situated on the segment N between $T^{s_{k,1}}(q_k)$ and $T^{s_{k,3}}(q_k)$. Then the point $T^{s_{n,2}}(q_n)$ is between $T^{s_{k,1}}(q_k)$ and $T^{s_{k,3}}(q_k)$ for all sufficiently large n. But

$$T^{s_{k,1}}(q_k) = T^{s_{k,1}+t_k}(q_0), \qquad T^{s_{k,3}}(q_k) = T^{s_{k,3}+t_k}(q_0),$$
$$T^{s_{n,2}}(q_n) = T^{s_{n,2}+t_n}(q_0).$$

For sufficiently large n we will have

$$s_{n,2} + t_n > s_{k,1} + t_k, \qquad s_{n,2} + t_n > s_{k,3} + t_k,$$

therefore the intersection points of the positive semi-trajectories $\{T^t q_0 : t > 0\}$ with the segment N are not situated in order of increasing numbers, which contradicts the previous arguments. This contradiction proves the lemma. □

Lemma 4 has the following consequence.

Corollary. *Any trajectory of the flow $\{T^t\}$ intersects the Siegel curve Γ.*

Proof. Suppose $M' \subset M$ is a set invariant with respect to $\{T^t\}$ and consisting of those trajectories which intersect Γ. To each point $p \in \Gamma$, according to Lemma 4, we may assign, the number $f(p) > 0$ such that $T^{f(p)}(p)$ is the first return point of the semi-trajectory $\{T^t p: t > 0\}$ to Γ. Clearly $f(p)$ is a continuous function of p. Hence any point $q \in M'$ may be represented in the form $q = T^s p$ where $p \in \Gamma$, $0 \leq s \leq f(p)$.

The couples (p, s) correspond bijectively and bicontinuously to points $q \in M'$, if pairs of the form $(p, f(p))$ and $(T^{f(p)}(p), 0)$ are identified.

This means that the set M' is closed and homeomorphic to the two-dimensional torus. But any such set on the torus M coincides with the torus itself. The corollary is proved. □

Now we pass to the proof of Theorem 1. The proof shall be split up into several steps.

1. For the Siegel curve Γ of the system (2) introduce the parameter x proportional to arc length counted off from some point $p_0 \in \Gamma$ in a fixed direction, and normalized in such a way that x varies in the interval $[0, 1]$. Introduce the function $f(x)$, $0 \leq x \leq 1$,

$$f(x) = \inf\{t > 0;\ T^t p \in \Gamma\},$$

where the point $p \in \Gamma$ corresponds to the value of the parameter x.

By Lemma 4, $f(x)$ is finite for all x, $0 \leq x \leq 1$. Since $f(0) = f(1)$ the function f may be viewed as a continuous function on the circle $S_1 = \{x: 0 \leq x < 1\}$. It follows from the C^5-smoothness of the curve Γ and the C^5-smoothness of the trajectories of the system (1) that $f \in C^5(S^1)$.

2. Suppose $R: S^1 \to S^1$ is a diffeomorphism of class C^5 of the circle S^1 acting according to the formula $Rx = x'$, where x, x' are the values of the parameter corresponding to the points $p, p' \in \Gamma$ and $p' = T^{f(x)}p$. Consider the set

$$\overline{M} = \{(x, s) \in S^1 \times \mathbb{R}^1 : 0 \leq s \leq f(x)\},$$

in which points of the form $(x, f(x))$ and $(Rx, 0)$ are identified. Then $\overline{M}$ is homeomorphic to the two-dimensional torus. Between the points of the set $\overline{M}$ and those of the torus $\mathrm{Tor}^2 = \{(u, v): 0 \leq u, v < 1\}$ there exists, by the corollary to Lemma 4, a natural one-to-one map which sends (x, s) into $T^s p$, where p is the point of the curve Γ corresponding to the parameter value x. But $\overline{M}$ and Tor^2 possess smooth manifold structures of class C^5, while the C^5-smoothness of the curve Γ and of the functions A, B implies that the map indicated above is a C^5-diffeomorphism. The action of the flow $\{T^t\}$ is then mapped into the action of the special flow $\{\overline{T}^t\}$ constructed from the diffeomorphism R of the circle S^1 and the function $f(x)$, while the density of the invariant measure $P(u, v)$ is mapped into a function $\overline{P}(x, s) \in C^5(\overline{M})$, where the measure $d\bar{\mu} = \overline{P}(x, s)\, dx\, ds$ is invariant with respect to $\{\overline{T}^t\}$.

3. Carry out a C^5-smooth change of time in the flow $\{\bar{T}^t\}$ so that each point $x_0 = (x_0, 0) \in \bar{M}$ runs through the segment $\{(x_0, s): 0 \le s \le f(x_0)\}$ in time 1. In other words, introduce a new flow $\{\tilde{T}^t\}$ (on the space $\bar{M}$) whose trajectories coincide with the trajectories of $\{\bar{T}^t\}$, while the velocity of motion at the point (x, s) is defined by some positive function $h(x, s) \in C^5(\bar{M})$ satisfying

$$\int_0^{f(x_0)} \frac{ds}{h(x_0, s)} = 1,$$

for any $x_0 \in S^1$. Clearly, such a function h exists. Moreover, it follows from the transversality of the curve Γ that $f(x) \ge \text{const} > 0$, so that we can require in addition that $h(x, s) \ge \text{const} > 0$. By the Liouville theorem, the flow $\{\tilde{T}^t\}$ has the invariant measure $d\tilde{\mu} = [\bar{P}(x, s)/h(x, s)]\, dx\, ds$. Since

$$\tilde{\mu}(\bar{M}) = \iint_{\bar{M}} \frac{\bar{P}(x, s)}{h(x, s)}\, dx\, ds < \infty,$$

we may assume without loss of generality that the measure $\tilde{\mu}$ is normalized.

Now define the invariant measure v for the diffeomorphism R of the circle S^1. Suppose $\pi: \bar{M} \to S^1$ is the natural projection, i.e., $\pi(x, s) = x$. For any Borel set $E \subseteq S^1$, put

$$v(E) = \tilde{\mu}(\pi^{-1}E) = \iint_{\pi^{-1}E} \frac{\bar{P}(x, s)}{h(x, s)}\, dx\, ds.$$

The invariance of the measure v with respect to R follows from the invariance of the measure $\tilde{\mu}$ with respect to the flow $\{\tilde{T}^t\}$.

Introduce the function $y = \varphi(x) = v([0, x])$, $x \in S^1$. It follows from the definition of the measure v that $\varphi \in C^5$. We may assume that y is a new coordinate of the point $x \in S^1$. Let us show that the transformation R will be (for this coordinate) simply the rotation T_1 of the circle by a certain angle α: $T_1 y = y + \alpha \pmod 1$. Indeed, if

$$\varphi^{-1}(y) = x, \qquad \varphi^{-1}(T_1 y) = R(x),$$

then

$$\begin{aligned} T_1 y = v([0, R(x)]) &= \pm v([0, R(0)]) + v([R(0), R(x)]) \\ &= \pm v([0, R(0)]) + v([0, x]) = y \pm v([0, R(0)]). \end{aligned}$$

Our statement is proved and $\alpha = \pm v([0, R(0)])$.

Now consider the function $F(y) = f(\varphi^{-1}(y))$. Clearly $F \in C^5(S^1)$ and the original flow $\{T^t\}$ is metrically isomorphic to the special flow constructed from the rotation automorphism T_1 of the circle by the angle α, and the function F.

4. It remains to establish the relation between the rotation angle α and the number $\lambda = \lambda_1/\lambda_2$. This will be done as follows.

On one hand, we will show that for the original system of equations (1) there exists a C^5-smooth change of variables

$$u' = \psi'_1(u, v), \qquad v' = \psi'_2(u, v),$$

after which the trajectories of the system (1) (when we pass from the torus $\mathbb{R}^2/\mathbb{Z}^2$ to the plane $\mathbb{R}^2$) become the straight lines

$$u' = \alpha v' + \text{const.} \tag{6}$$

On the other hand, there is another C^5-smooth change of variables

$$u'' = \psi''_1(u, v), \qquad v'' = \psi''_2(u, v),$$

after which the trajectories become straight lines

$$u'' = \tilde{\lambda} v'' + \text{const,} \tag{7}$$

where $\tilde{\lambda} = (m\lambda - n)/(-p\lambda + q)$, and m, n, p, q are integers,

$$\det \begin{Vmatrix} m & n \\ p & q \end{Vmatrix} = 1.$$

Moreover, one of the coordinate axes in the coordinates (u', v') and (u'', v'') will be the same, e.g., $\psi'_1(u, v) = \psi''_1(u, v)$ and the origin will be the same. This will imply that when we pass from the coordinates (u', v') to the coordinates (u'', v'') the family of lines (6) is transformed into the family of lines (7). From this we will deduce in turn that $\alpha = \pm\tilde{\lambda}$.

5. We now pass to the description of the corresponding changes of variables.

(a) First assign to each point $q \in \text{Tor}^2$ the coordinates (θ, τ), $0 \le \theta, \tau < 1$ according to the following rule. The coordinate τ is defined by the relation $\tau = \inf\{t \ge 0\colon \tilde{T}^{-t}q \in \Gamma\}$. To determine the coordinate θ, consider the point $p = \tilde{T}^{-\tau}q \in \Gamma$ and put $\theta = y + \alpha\tau \pmod 1$, where y is the coordinate of the point p introduced in step 3. By construction, any trajectory of the flow $\{T^t\}$ will be of the form $\theta - \alpha\tau \equiv \text{const}$.

(b) Now for any point $q \in \text{Tor}^2$ introduce the coordinates (ξ, τ), $0 \le \xi, \tau < 1$, the coordinate τ being the same as before.

In order to construct the coordinate ξ, we will have to extend the coordinate τ from the torus Tor^2 to the covering plane $\mathbb{R}^2$.

Suppose (p, q) is the pair of numbers determining the homotopy type of the Siegel curve Γ and $\boldsymbol{\Gamma}$ is the covering curve in $\mathbb{R}^2$. By Lemma 3, we can find a pair of integers m, n such that

$$\det \begin{Vmatrix} m & n \\ p & q \end{Vmatrix} = 1.$$

The parallelogram Π in $\mathbb{R}^2$ spanning the vectors (m, n) and (p, q) generates the entire integer lattice. This means that we can define a real-valued function $\boldsymbol{\tau}(u, v)$ of class C^5 on Π such that

$$[\boldsymbol{\tau}(u, v)] \pmod 1 = \tau(u \pmod 1, v \pmod 1).$$

Then we can extend $\boldsymbol{\tau}(u, v)$ to the entire plane so as to have the relations

$$\begin{aligned} \boldsymbol{\tau}(u + m, v + n) - \boldsymbol{\tau}(u, v) &\equiv 1, \\ \boldsymbol{\tau}(u + p, v + q) - \boldsymbol{\tau}(u, v) &\equiv 0. \end{aligned} \tag{8}$$

Now consider the function $H(u, v)$ introduced in the proof of Lemma 1. It follows from (3) and (4) that

$$\begin{aligned} H(u + m, v + n) - H(u, v) &= m\lambda_1 - n\lambda_2 \stackrel{\text{def}}{=} C_1. \\ H(u + p, v + q) - H(u, v) &= p\lambda_1 - q\lambda_2 \stackrel{\text{def}}{=} C_2. \end{aligned} \tag{9}$$

Putting

$$\boldsymbol{\xi}(u, v) = -\frac{C_1}{C_2}\boldsymbol{\tau}(u, v) + \frac{1}{C_2}H(u, v),$$

and taking into consideration (8) and (9), we will get

$$\boldsymbol{\xi}(u + m, v + n) - \boldsymbol{\xi}(u, v) = 0, \qquad \boldsymbol{\xi}(u + p, v + q) - \boldsymbol{\xi}(u, v) = 1,$$

and since the parallelogram Π generates the entire lattice, $\xi = \boldsymbol{\xi}(u, v) \pmod 1$ may be viewed as a function defined on the torus.

In order to prove that the pair (ξ, τ) plays the role of the coordinates of the point $q \in \mathrm{Tor}^2$, it suffices to show that the function $H(u, v)$ assumes distinct values on different trajectories of the flow $\{T^t\}$ (on the plane).

Indeed, if we can find two trajectories with the same value of H, then we could, by the corollary to Lemma 4, join them by a segment of the curve

Γ. By Rolle's theorem, we could find a point $p \in \Gamma$ where the derivative of the function H in the direction of Γ vanishes. Since $dH/dt \equiv 0$, while Γ is a transversal curve, the point p would have to be a fixed point of the system (1), contradicting the assumption. Thus (ξ, τ) are the coordinates of a point $q \in \mathrm{Tor}^2$. By changing ξ by a constant, we may assume that the origin of coordinates was the same as that for the coordinates (θ, τ). Since H is the first integral, the trajectories of the system (1) in the coordinates (ξ, τ) will be the straight lines

$$\xi - \tilde{\lambda}\tau \equiv \text{const}, \quad \text{where } \tilde{\lambda} = -\frac{C_1}{C_2} = \frac{m\lambda_1 - n\lambda_2}{-p\lambda_1 + q\lambda_2} = \frac{m\lambda - n}{-p\lambda + q}.$$

Note that

$$\det \begin{Vmatrix} m & -n \\ -p & q \end{Vmatrix} = \det \begin{Vmatrix} m & n \\ p & q \end{Vmatrix} = 1.$$

To conclude the proof it remains to show that $\alpha = \pm\tilde{\lambda}$. To do this, consider the change of variables on the torus which sends the coordinates (θ, τ) into the coordinats (ξ, τ) and is the composition of the two change of variables described above. Then the family of lines $\theta - \alpha\tau \equiv$ const will be mapped into the family of lines $\xi - \tilde{\lambda}\tau \equiv$ const, the set $\tau = 0$ is mapped as a set into itself, while the origin stays put. Therefore any point of the form $\theta = k\alpha$, $\tau = 0$ (k is an integer) is mapped into a point of the form $\xi = k\tilde{\lambda}$, $\tau = 0$. Now the irrationality of $\tilde{\lambda}$ implies that the axis $\tau = 0$ is mapped into itself linearly. Hence, taking into consideration the fact that the coordinates are cyclic, i.e., are viewed mod 1, we have the relation $\xi = \pm\theta$ on the axis $\tau = 0$. Since the point $\theta = \alpha$, $\tau = 0$ corresponds to the point $\xi = \tilde{\lambda}$, $\tau = 0$, we have $\alpha = \pm\tilde{\lambda}$. The Theorem is proved. □

§2. Dynamical Systems with Pure Point Spectrum on the Two-dimensional Torus

The theorem proved in the previous section enables us to study the spectral properties of flows $\{T^t\}$ on the two-dimensional torus Tor^2 given by equations of the form

$$\frac{du}{dt} = A(u, v), \qquad \frac{dv}{dt} = B(u, v).$$

As in §1, we assume that the flow $\{T^t\}$ has no fixed points and there exists an absolutely continuous invariant measure with density $P(u, v) \in C^5(\mathrm{Tor}^2)$.

It turns out that the spectral properties of the flow $\{T^t\}$ are intimately related to arithmetical properties of the number

$$\lambda = \iint PA\, du\, dv \left[\iint PB\, du\, dv\right]^{-1}.$$

Theorem 1. *Suppose λ satisfies*

$$|\lambda - p/q| \geq \text{const}/q^4, \quad \text{const} > 0 \tag{1}$$

for all integers $p, q, q \neq 0$. Then

(1) *the flow $\{T^t\}$ is metrically isomorphic to the special flow constructed from the rotation automorphism T_1 of the circle S^1 by a certain angle α and the constant function, where α is the same as in Theorem* 1, §1;

(2) *the group of unitary operators $\{U^t\}$ adjoint to the flow $\{T^t\}$ has a pure point spectrum consisting of numbers of the form* const$(k + l\lambda)$ *where $-\infty < k, l < \infty$, (k, l being integers).*

Proof. In view of Theorem 1, §1, we may assume that $\{T^t\}$ is a special flow constructed from the automorphism T_1 and the functions $f(x) \in C^5(S^1)$, $f(x) \geq \text{const} > 0$. Hence it suffices to prove the following two statements:

(A) Suppose there exists a function $g \in C(S^1)$ which is the solution of the equation

$$f(x) - g(x) + g(x + \alpha) \equiv \beta, \tag{2}$$

where $\beta = \int_0^1 f(x)\, dx$. Then statements (1) and (2) of Theorem 1 hold.

(B) In the hypotheses of Theorem 1, a solution $g \in C(S^1)$ of equation (2) exists.

The meaning of equation (2) is clear: it is the equation for the curve

$$G_0 = \{(x, g(x))\} \subset M, \tag{3}$$

such that the trajectories of each of its points returns on it after the same period of time β. Here M is the phase space of the special flow:

$$M = \{(x, s): x \in S^1, 0 \leq s < f(x)\}.$$

Statement (A) follows from this remark in the case when we have inclusion (3), i.e., $0 \leq g(x) < f(x)$. In the general case the last inequality does not necessarily hold, but we will show how to avoid this difficulty.

The proof of statement (A) will be split up into several steps.

1. Consider the strip

$$\Pi = \{(x, s): 0 \le x \le 1, -\infty < s < \infty\},$$

and the nonintersecting sets $\Pi_n \subset \Pi$, $-\infty < n < \infty$, n being an integer:

$$\Pi_n = \left\{(x, s) \in \Pi: \sum_{k=0}^{n-1} f(x + k\alpha) \le s < \sum_{k=0}^{n} f(x + k\alpha)\right\}, \quad n \ge 0,$$

$$\Pi_n = \left\{(x, s) \in \Pi: -\sum_{k=1}^{-n} f(x - k\alpha) \le s < -\sum_{k=1}^{-n-1} f(x - k\alpha)\right\}; \quad n < 0.$$

It follows from the condition $f(x) \ge \text{const} > 0$ that $\Pi = \bigcup_{n=-\infty}^{\infty} \Pi_n$. The space M where the flow $\{T^t\}$ acts coincides with Π_0.

Let us define the map $\chi: \Pi \to M$ in the following way: if $z = (x, s) \in \Pi_n$, then

$$\chi(z) = \chi(x, s) = \left(x + n\alpha, s - \sum_{k=0}^{n-1} f(x + k\alpha)\right), \quad n \ge 0,$$

$$\chi(z) = \chi(x, s) = \left(x + n\alpha, s + \sum_{k=1}^{-n} f(x - k\alpha)\right), \quad n < 0.$$

Two points $z_1 = (x_1, s_1)$, $z_2 = (x_2, s_2) \in \Pi$ will be considered equivalent (notation $z_1 \sim z_2$) if $\chi(z_1) = \chi(z_2)$.

Consider the space $\boldsymbol{M}$ whose points are equivalence classes of points from Π. We shall denote by $(\boldsymbol{x}, \boldsymbol{s})$ the equivalence class containing the point (x, s). Introduce the measure $\boldsymbol{\mu}$ into the space $\boldsymbol{M}$ by putting $\boldsymbol{\mu}(\chi^{-1}A) \overset{\text{def}}{=} \mu(A)$ where $A \subset M$ is a measurable set while μ is the invariant measure for the special flow $\{T^t\}$. The action of the flow $\{T^t\}$ is also naturally carried over to $\boldsymbol{M}$:

$$\boldsymbol{T}^t(\boldsymbol{x}, \boldsymbol{s}) \overset{\text{def}}{=} \chi^{-1}[T^t\chi(\boldsymbol{x}, \boldsymbol{s})].$$

It is easy to check that the following formula

$$\boldsymbol{T}^t(\boldsymbol{x}, \boldsymbol{s}) = (\boldsymbol{x}, \boldsymbol{s} + \boldsymbol{t})$$

holds and that the measure $\boldsymbol{\mu}$ is invariant with respect to $\{\boldsymbol{T}^t\}$.

2. In the strip Π consider the graph G_0 of the function

$$G_0 = \{(x, g(x)), 0 \le x < 1\},$$

and put $G = \chi^{-1}(G_0)$. Let us prove that $G = \bigcup_{m=-\infty}^{\infty} G_m$ where

$$G_m = \{(x, g(x) + m); 0 \leq x < 1\}.$$

This equality follows directly from the next relation, which in turn follows from equation (2):

$$(x, g(x) + m\beta) \sim (x + \alpha, g(x + \alpha) + (m - 1)\beta), \tag{4}$$

where $x \in S^1$, $-\infty < m < \infty$, m being an integer.

Let us derive formula (4). Suppose $(x, g(x) + m\beta) \in \Pi_n$ and, for example, $n \geq 0$; the case $n < 0$ is similar. Then

$$\sum_{k=0}^{n-1} f(x + k\alpha) \leq g(x) + m\beta < \sum_{k=0}^{n} f(x + k\alpha). \tag{5}$$

By equation (2),

$$g(x) + m\beta = g(x + \alpha) + f(x) + (m - 1)\beta. \tag{6}$$

It follows from (5) and (6) that

$$\sum_{k=1}^{n-1} f(x + k\alpha) \leq g(x + \alpha) + (m - 1)\beta < \sum_{k=1}^{n} f(x + k\alpha),$$

i.e.,

$$\sum_{k=0}^{n-2} f((x + \alpha) + k\alpha) \leq g(x + \alpha) + (m - 1)\beta < \sum_{k=0}^{n-1} f((x + \alpha) + k\alpha),$$

if $n \geq 1$ and

$$-f((x + \alpha) - \alpha \leq g(x + \alpha) + (m - 1)\beta < 0,$$

if $n = 0$. This means that

$$(x + \alpha, g(x + \alpha) + (m - 1)\beta) \in \Pi_{n-1}.$$

Therefore

$$\chi(x + \alpha, g(x + \alpha) + (m - 1)\beta) = (x + \alpha + (n - 1)\alpha,$$

$$g(x + \alpha) + (m - 1)\beta - \sum_{k=0}^{n-2} f((x + \alpha) + k\alpha)).$$

On the other hand,

$$\chi(x, g(x) + m\beta) = \left(x + n\alpha, g(x) + m\beta - \sum_{k=0}^{n-1} f(x + k\alpha)\right).$$

By (6) the right-hand sides of the last two equalities are the same, so that formula (4) is proved.

3. Since the set G consists of entire equivalence classes, it may be considered as a subset of the space $\boldsymbol{M}$. Denote this subset by $\boldsymbol{G}$.

Consider the partition ξ of the space $\boldsymbol{M}$ each element of which is a semi-interval of the trajectory of the flow $\{\boldsymbol{T}^t\}$ from one appearance in the set $\boldsymbol{G}$ (inclusive) to a next one (not inclusive). It is easy to verify that the partition is a regular partition to the flow $\{\boldsymbol{T}^t\}$. (see §3, Chap. 11). This partition, as was shown in §3, Chap. 11, induces a special representation of the flow $\{\boldsymbol{T}^t\}$ and therefore of the flow $\{T^t\}$ which is metrically isomorphic to it. Suppose $\boldsymbol{T}_1$ is the base automorphism of the special representation. Then, using formula (6) for $m = 1$, we obtain

$$\boldsymbol{T}_1(\boldsymbol{x}, g(\boldsymbol{x})) = (\boldsymbol{x}, g(\boldsymbol{x}) + \boldsymbol{\beta}) = (\boldsymbol{x} + \boldsymbol{\alpha}, g(\boldsymbol{x} + \boldsymbol{\alpha})).$$

Therefore $\boldsymbol{T}_1$ may be identified with the automorphism $\tilde{T}_1$ which acts on the graph G_0 of the function $g(x)$ according to the formula

$$\tilde{T}_1(x, g(x)) = (x + \alpha, g(x + \alpha)).$$

This automorphism in its turn is metrically isomorphic to the automorphism T_1. Since the limiting function of the special representation, by construction, is identically equal to β, statement (1) of Theorem 1 is proved.

4. Statement (2) directly follows from (1). Indeed, the flow $\{T^t\}$ may now be identified with the flow given on the torus Tor^2 by the system of differential equations

$$\frac{du}{dt} = \frac{1}{\beta}, \qquad \frac{dv}{dt} = \frac{\alpha}{\beta},$$

i.e., the flow acting according to the formula

$$T^t(u, v) = (u + t/\beta, v + \alpha t/\beta).$$

The eigen-functions of the group of unitary operators $\{U^t\}$ adjoint to this flow are the exponents

$$\varphi_{k,l}(u, v) = \exp[2\pi i(ku + lv)],$$

where $k, l < \infty$, k, l are integers. Then

$$U^t \varphi_{k,l} = \exp\left[2\pi i t \frac{k + l\alpha}{\beta}\right] \varphi_{k,l}.$$

Thus the spectrum of the group $\{U^t\}$ consists of numbers of the form $\lambda_{k,l} = (1/\beta)(k + l\alpha)$. Since $\alpha = (m\lambda + n)/(p\lambda + q)$, the $\lambda_{k,l}$ may be expressed in the form

$$\lambda_{k,l} = \frac{1}{\beta}\left[k + l\frac{m\lambda + n}{p\lambda + q}\right] = \frac{1}{\beta(p\lambda + q)}[(kq + ln) + (kp + lm)\lambda]. \quad (7)$$

The relation

$$\det \begin{Vmatrix} m & n \\ p & q \end{Vmatrix} = \pm 1$$

implies that any number

$$\lambda_{k',l'} = \frac{1}{\beta(p\lambda + q)}(k' + l'\lambda), \qquad -\infty < k', l' < \infty,$$

where k', l' are integers may be represented in the form (7). Thus our statement is proved entirely. □

In the sequel, we shall need a simple lemma; for its statement we shall call numbers λ, satisfying the condition (1) of Theorem 1, poorly approximable.

Lemma 1. *If λ is poorly approximable, then $\alpha = (m\lambda + n)/(p\lambda + q)$, where m, n, p, q are integers and*

$$\det \begin{Vmatrix} m & n \\ p & q \end{Vmatrix} = \pm 1$$

is also poorly approximable.

The proof of the lemma shall be given later.

Proof of Statement B. Expand the function f into the Fourier series

$$f(x) = \sum_{n=-\alpha}^{\infty} f_n \exp(2\pi i n x), \quad (8)$$

and look for the solution g also in the form of a Fourier series

$$g(x) = \sum_{n=-\infty}^{\infty} g_n \exp(2\pi i n x). \tag{9}$$

It follows from (9) that

$$\begin{aligned} g(x+\alpha) &= \sum_{n=-\infty}^{\infty} g_n \exp[2\pi i n(x+\alpha)] \\ &= \sum_{n=-\infty}^{\infty} g_n \exp(2\pi i n\alpha)\exp(2\pi i n x). \end{aligned} \tag{10}$$

Noticing that the coefficient f_0 equals β, rewrite equation (2) by using (8), (9), and (10) in the form

$$\sum_{n\neq 0} [f_n - g_n + g_n \exp(2\pi i n\alpha)]\exp(2\pi i n x) \equiv 0,$$

we get

$$g_n = \frac{f_n}{1-\exp(2\pi i n\alpha)}, \quad n \neq 0.$$

Since the solution g is defined up to a constant, the coefficient g_0 is arbitrary. Since $f \in C^5(S^1)$, we have $|f_n| \leq \text{const}/n^5$. On the other hand, by Lemma 1, $\{n\alpha\} \geq \text{const}\ n^{-3}$ (the figure brackets denoting fractional parts). Thus

$$|1-\exp(2\pi i n\alpha)| = |1-\exp(2\pi i\{n\alpha\})| \geq \text{const}_1 \cdot n^{-3}.$$

Hence $|g_n| \leq \text{const}_2 |n^2$, so that the series

$$\Sigma\, g_n \exp(2\pi i n x) = g(x)$$

converges at every point and represents a continuous function. Statement B is proved. □

It remains to prove Lemma 1. The proof will be by *reductio ad absurdum.* Suppose α is not poorly approximable. Briefly we shall say that α is well approximable. The number λ may be expressed in terms of α rationally: $\lambda = (q\alpha - n) \times (m - p\alpha)^{-1}$. Therefore

$$\frac{p}{q}\lambda = -\frac{qp\alpha - pn}{qp\alpha - qm} = -\frac{qp\alpha - qm + qm - pn}{qp\alpha - qm} = -1 + \frac{pn - qm}{qp\alpha - qm}. \tag{11}$$

It is clear that if α is well approximable, then so is the number $\gamma = qp\alpha - qm$. Let us show that then $1/\gamma$ is also well approximable. Indeed, we can find a sequence of fractions p_k/q_k such that

$$|\gamma - p_k/q_k| = \varepsilon_k/q_k^4, \quad \varepsilon_k \to 0.$$

Hence $|\gamma q_k - p_k| = \varepsilon_k/q_k^3$ and

$$\begin{aligned}|1/\gamma - q_k/p_k| &= |(p_k - \gamma q_k)/\gamma p_k| = \varepsilon_k/(\gamma p_k q_k^3) \\ &= \varepsilon_k/p_k^4 \cdot 1/\gamma \cdot (p_k/q_k)^3 \le \text{const } \varepsilon_k/p_k^4.\end{aligned}$$

According to formula (11), we have $p\lambda/q = [(pn - qm)/\gamma] - 1$, so that λ is well approximable. The contradiction thus obtained proves the lemma. The theorem is proved. □

In conclusion let us show that the flows on the torus considered in this section are in a certain sense the "typical case." Namely, for almost all numbers λ condition (1) of Theorem 1 holds.

Actually, a stronger statement is valid. Let us say that the number λ is normally approximable by rational numbers, if there exists such constants $C > 0$, $\varepsilon > 0$ that $|\lambda - p/q| \ge C/q^{2+\varepsilon}$ for all integers p, q, $q \ne 0$.

Lemma 2. *The set N of normally approximable numbers on the closed interval $[0, 1]$ has Lebesgue measure 1.*

Proof. Fix ε, C and put

$$A_q^{\varepsilon, C} = \left\{\lambda \in [0, 1] \colon \min_p |\lambda - p/q| < C/q^{2+\varepsilon}\right\}.$$

The set $A_q^{\varepsilon, C}$ is the union of intervals of length $2C/q^{2+\varepsilon}$ with centres at the points p/q, $1 < p < q$ and of semi-intervals of length $C/q^{2+\varepsilon}$ by the points 0 and 1. Therefore $\rho(A_q^{\varepsilon, C}) \le 2C/q^{1+\varepsilon}$ where ρ is the Lebesgue measure so that $\rho(\bigcup_q A_q^{\varepsilon, C}) < \infty$. In view of Borel–Cantelli lemma we have $\rho(N) = 1$. The lemma is proved. □

§3. Approximations of Flows on the Torus

We shall continue the study of flows on the two-dimensional torus in the case when they are given by differential equations of the form

$$\frac{du}{dt} = A(u, v), \qquad \frac{dv}{dt} = B(u, v).$$

In this section we will consider the case when the number

$$\lambda = \iint PA\,du\,dv\left[\iint PB\,du\,dv\right]^{-1}$$

is well approximable by rational numbers (here as before P is the density of the invariant measure).

By Theorem 1, §1, instead of a flow on the torus we may consider a special flow $\{T^t\}$ constructed from the rotation automorphism T_1 of the circle S^1 by an angle α and from the function $F \in C^5(S^1)$. The phase space of the flow $\{T^t\}$ is denoted, as always, by $(M, \mathfrak{S}, \mu)$. The number α is of the form $\alpha = (m\lambda + n)/(p\lambda + q)$ and by Lemma 1, §2, is well approximable if λ is.

Theorem 1. *If for the number α there is a sequence p_n/q_n of irreducible fractions such that*

$$r_n = q_n^4 |\alpha - p_n/q_n| \to 0 \quad \text{when } n \to \infty,$$

then the flow $\{T^t\}$ possesses a cyclic approximation with speed $g(u) = o(u^{-2})$.

Proof. 1. Put $\beta = \int_0^1 F(x)\,dx$. Consider the sequence of natural numbers $\{m_n\}$, $m_n \uparrow \infty$. Further $\{m_n\}$ will be chosen so as to satisfy certain conditions. Presently we only assume that $\delta_n \stackrel{\text{def}}{=} \beta/m_n < \min_x F(x)$.

Choose a sufficiently large n and assume that $\alpha > p_n/q_n$. The case $\alpha < p_n/q_n$ is similar. Put

$$A_n = \left\{z = (x, s) \in M : 0 \le x < \frac{1}{q_n} - \{q_n \alpha\},\ 0 \le s < \delta_n\right\}.$$

It follows from the hypothesis of the theorem and the assumption $\alpha > p_n/q_n$ that $\{\alpha q_n\} = o(q_n^{-3})$ and therefore $\mu(A_n) > 0$.

Put $Q_n = q_n m_n$; let us show that for $k = 0, 1, \ldots, Q_n - 2$ the sets $T^{k\delta_n} A_n$ are disjoint. Obviously it suffices to show that

$$A_n \cap T^{k\delta_n} A_n = \varnothing, \qquad 0 \le k \le Q_n - 2.$$

Choose a point $z = (x, s) \in A_n$. By definition of the special flow, we have

$$T^t z = T^t(x, s) = \left(x + r(t)\alpha,\ s + t - \sum_{k=0}^{r(t)-1} F(T_1^k x)\right),$$

where $r(t)$ is a non-decreasing integer-valued function which depends, of course, also on x. Suppose

$$\tilde{t} = \tilde{t}(z) = \min\{t : t \geq \delta_n, T^t z \in A_n\}.$$

Let us show that $r = r(\tilde{t}) \geq q_n$.

Clearly, $r \geq 1$. Consider the semi-intervals

$$\Delta_k = [\{k\alpha\}, \{k\alpha + q_n^{-1} - q_n\alpha\}, \qquad \Delta_k^{(n)} = [\{kp_n/q_n\}, \{(kp_n + 1)/q_n\}),$$
$$k = 0, 1, \ldots, q_n - 1.$$

The inequalities

$$\{kp_n/q_n\} < \{k\alpha\} < \{k\alpha + q_n^{-1} - q_n\alpha\}$$
$$= \{(kp_n + 1)/q_n + k(\alpha - p_n/q_n) - q_n(\alpha - p_n/q_n)\} < \{(kp_n + 1)/q_n\}$$

imply that $\Delta_k \subset \Delta_k^{(n)}$ so that the semi-intervals Δ_k, $0 \leq k \leq q_n - 1$ are disjoint. If we would have the inequality $r < q_n$, then the first coordinate of the point $T^{\tilde{t}}z$ would belong to the semi-interval $\Delta_r, r > 1$. Since $\Delta_r \cap \Delta_0 = \varnothing$ in this case we would have $T^{\tilde{t}}z \notin \Delta_0$, contradicting the definition of the number $\tilde{t}$. Thus $r \geq q_n$.

Now let us make use of the following lemma.

Lemma 1. *Suppose $F \in C^5(S^1)$ and $|\alpha - (p/q)| \leq 0, 1/q^2$. Then*

$$\left|\sum_{k=0}^{q-1} F(x + k\alpha) - q\beta\right| < C_1 q^2 \left|\alpha - \frac{p}{q}\right| + C_2 q^{-3} \stackrel{\text{def}}{=} \Sigma\,(p, q),$$

for certain constants C_1, C_2.

The proof of the lemma will be given later.

Put $\Sigma_n = \Sigma\,(p_n, q_n)$. Since the second coordinate of the point $T^{\tilde{t}}z$ vanishes, we have

$$\tilde{t} = \sum_{k=0}^{r(\tilde{t})-1} F(T_1^k x) - s \geq \sum_{k=0}^{q_n-1} F(T_1^k x) - s \geq q_n\beta - \Sigma_n - s.$$

But $q_n\beta = Q_n\delta_n$, $|s| < \delta_n$ and the assumption of the theorem implies $\Sigma_n = o(q_n^{-2})$. Now let us require the inequality $m_n < q_n^2/\beta$. In this case $\delta_n > 1/q_n^2$ and therefore $\Sigma_n < \delta_n$ for large n. Finally,

$$\tilde{t} > Q_n\delta_n - \delta_n - \delta_n = (Q_n - 2)\delta_n.$$

This inequality shows that

$$A_n \cap T^{k\delta_n} A_n = \varnothing \quad \text{for } k = 0, 1, \ldots, Q_n - 2.$$

2. Note that if $s > \delta_n - \Sigma_n$ for the point $z = (x, s)$, then $\tilde{t}(z) > (Q_n - 1)\delta_n$. Hence

$$\mu(A_n \cap T^{(Q_n - 1)\delta_n} A_n) \leq \mu(\{z = (x, s) \in A_n : \delta_n - \Sigma_n < s < \delta_n\}) \leq \Sigma_n / q_n \beta. \tag{1}$$

Let us estimate the measure $\mu(A_n \cap T^{Q_n \delta_n} A_n)$ from below. Put $E_n = A_n \setminus T^{Q_n \delta_n} A_n$,

$$F_n = \{(x, s) \in A_n : 0 \leq s < \Sigma_n \text{ or } \delta_n - \Sigma_n \leq s < \delta_n\}.$$

Clearly $\mu(F_n) \leq 2\Sigma_n / q_n \beta$. If $z = (x, s) \in E_n \setminus F_n$, we first write

$$T^{Q_n \delta_n} z = T^{Q_n \delta_n - \tau_n(x)}(T^{\tau_n(x)} z), \quad \text{where } \tau_n(x) = \sum_{k=0}^{q_n - 1} F(x + k\alpha).$$

Since $s < \delta_n < \min F(x)$, we have $T^{\tau_n(x)} z = (x + q_n \alpha, s)$. By Lemma 1,

$$|Q_n \delta_n - \tau_n(x)| = |q_n \beta - \tau_n(x)| \leq \Sigma_n.$$

Therefore $T^{Q_n \delta_n} z = (x + q_n \alpha, s + Q_n \delta_n - \tau_n(x))$. Since

$$T^{Q_n \delta_n} z \notin A_n = \Delta_0 \times [0, \delta_n), \qquad s + Q_n \delta_n - \tau_n(x) \in [0, \delta_n),$$

we have $x + q_n \alpha \notin \Delta_0$. This means that

$$E_n \setminus F_n \subseteq \{z = (x, s) \in A_n : q_n^{-1} - 2\{q_n \alpha\} \leq x < q_n^{-1} - \{q_n \alpha\}\},$$

i.e.,

$$\mu(E_n \setminus F_n) \leq \frac{1}{\beta} \{q_n \alpha\} = \frac{q_n}{\beta} \left| \alpha - \frac{p_n}{q_n} \right|.$$

Therefore

$$\mu(E_n) \leq \mu(E_n \setminus F_n) + \mu(F_n) \leq \frac{2\Sigma_n}{q_n \beta} - \frac{q_n}{\beta} \left| \alpha - \frac{p_n}{q_n} \right|,$$

so that

$$\mu(A_n \cap T^{Q_n \delta_n} A_n) \geq \mu(A_n) - \frac{2\Sigma_n}{q_n \beta} - \frac{q_n}{\beta} \left| \alpha - \frac{p_n}{q_n} \right|. \tag{2}$$

3. In this subsection we shall construct the necessary partitions ξ_n, the automorphisms S_n and the numbers t_n which appear in the definition of a cyclic approximation.

The elements $C_i^{(n)}$, $0 \le i \le Q_n - 1$ of the partition ξ_n will be defined in the following way. First put

$$\tilde{C}_i^{(n)} = T^{i\delta_n} A_n \quad \text{for } i = 0, 1, \ldots, Q_n - 2; \qquad \tilde{C}_{Q_n-1}^{(n)} = T^{(Q_n-1)\delta_n} A_n \setminus A_n.$$

In order to obtain $C_i^{(n)}$, add to $\tilde{C}_i^{(n)}$ a part of the set $M \setminus \bigcup_{i=0}^{Q_n-1} \tilde{C}_i^{(n)}$ so as to make the measure of the element $C_i^{(n)}$ equal to $1/Q_n$.

The automorphism S_n will be defined by the formula

$$S_n z = T^{\delta_n} z \quad \text{for } z \in \bigcup_{k=0}^{Q_n-2} T^{k\delta_n} A_n \cup [(T^{(Q_n-1)\delta_n} A_n \setminus A_n) \cap T^{-\delta_n} A_n];$$

for the other values of $z \in M$ extend S_n arbitrarily but so as to retain properties (2) and (5) in the definition of cyclic approximation. It is clear that this may be done. We now have

$$\begin{aligned}\sum_{i=0}^{Q_n-1} \mu(T^{\delta_n} C_i^{(n)} \triangle S_n C_i^{(n)}) &\le 2\mu\left(M \setminus \bigcup_{k=0}^{Q_n-1} T^{k\delta_n} A_n\right) \\ &\quad + 2\mu(T^{(Q_n-1)\delta_n} A_n \cap A_n) + \mu(A_n \triangle T^{Q_n\delta_n} A_n) \\ &\stackrel{\text{def}}{=} 2\mu_1 + 2\mu_2 + \mu_3.\end{aligned}$$

Taking into consideration the fact that the sets

$$T^{k\delta_n} A_n, \qquad k = 0, 1, \ldots, Q_n - 2$$

are disjoint, we may write

$$\mu_1 \le 1 - Q_n \mu(A_n) + \mu(A_n \cap T^{(Q_n-1)\delta_n} A_n).$$

Since $\mu(A_n) = (1/\beta)\delta_n(q_n^{-1} - \{q_n\alpha\})$, it follows from (1) that

$$\begin{aligned}\mu_1 &\le 1 - \frac{1}{\beta} Q_n \delta_n (q_n^{-1} - \{q_n\alpha\}) + \frac{\Sigma_n}{q_n\beta} \\ &= 1 - \frac{1}{\beta} q_n \beta (q_n^{-1} - \{q_n\alpha\}) + \frac{\Sigma_n}{q_n\beta} = q_n^2 \left|\alpha - \frac{p_n}{q_n}\right| + \frac{\Sigma_n}{q_n\beta}. \qquad (3)\end{aligned}$$

The number μ_2 was estimated in (1), while (2) implies

$$\mu_3 \leq \frac{4\Sigma_n}{q_n\beta} + \frac{2q_n}{\beta}\left|\alpha - \frac{p_n}{q_n}\right|. \tag{4}$$

Putting together inequalities (1), (3), and (4) and the estimate for Σ_n, we finally get

$$\sum_{i=0}^{Q_n-1} \mu(T^{\delta_n}C_i^{(n)} \triangle S_n C_i^{(n)}) \leq \frac{\text{const}}{\beta q_n^2}\left(r_n + \frac{1}{q_n}\right).$$

Now let us specify the choice of the sequence $\{m_n\}$. We shall require, together with the inequality $m_n < q_n^2/\beta$ which appeared previously, the following relations:

$$r_n/q_n^2 = o(1/Q_n^2), \qquad 1/q_n^3 = o(1/Q_n^2).$$

Then we shall have

$$\sum_{i=0}^{Q_n-1} \mu(T^{\delta_n}C_i^{(n)} \triangle S_n C_i^{(n)}) = o(1/Q_n^2).$$

4. It remains to show that $\xi_n \to \varepsilon$. Topologically, the space M is a torus and there is a natural metric ρ on it. Denote by diam A the diameter of the set $A \subset M$ computed by means of this distance.

First let us show that

$$\max_{0 \leq k \leq Q_n-2} \operatorname{diam}(T^{k\delta_n}A_n) \to 0 \quad \text{for } n \to \infty.$$

Suppose $z_1 = (x_1, s_1)$, $z_1' = (x_1, 0)$, $z_2 = (x_2, s_2)$, $z_2' = (x_2, 0) \in A_n$. Then

$$\begin{aligned}\rho(T^{k\delta_n}z_1, T^{k\delta_n}z_2) &\leq \rho(T^{k\delta_n}z_1, T^{k\delta_n}z_1') + \rho(T^{k\delta_n}z_1', T^{k\delta_n}z_2') + \rho(T^{k\delta_n}z_2', T^{k\delta_n}z_2)\\ &\leq 2\delta + \rho(T^{k\delta_n}z_1', T^{k\delta_n}z_2').\end{aligned}$$

It follows from the formula for the action of the special flow that for $0 \leq k \leq Q_n - 2$, we have

$$\rho(T^{k\delta_n}z_1', T^{k\delta_n}z_2') \leq \max_{1 \leq r \leq q_n}\left|\sum_{j=0}^{r-1} F(x_1 + j\alpha) - \sum_{j=0}^{r-1} F(x_2 + j\alpha)\right|.$$

Thus the required statement follows from the next lemma.

Lemma 2. *If $F \in C^1(S^1)$ then for irrational α we have*

$$\max_{\substack{1 \le x \le q \\ 0 \le x_1, x_2 \le q^{-1}}} \left| \sum_{j=0}^{r-1} [F(x_1 + j\alpha) - F(x_2 + j\alpha)] \right| \to 0 \qquad \text{when } q \to \infty.$$

The proof of this lemma will be given later. Put

$$B_n = \bigcup_{k=1}^{Q_n - 2} T^{k\delta_n} A_n.$$

Clearly $\mu(B_n) \to 1$ when $n \to \infty$. It follows from the statement proved above that

$$\max_{0 \le i \le Q_n - 1} \operatorname{diam}(C_i^{(n)} \cap B_n) \to 0 \quad \text{when } n \to \infty.$$

Lemma 3. *Suppose M is a compact topological space, μ is a normalized Borel measure on M, $B_n \subset M$ is a sequence of Borel sets and $\xi_n = \{C_i^{(n)}\}$, $1 \le i \le q_n$ is a sequence of finite partitions. Then if*

$$\mu(B_n) \to 1, \qquad \max_{1 \le i \le q_n} \operatorname{diam}(C_i^{(n)} \cap B_n) \to 0 \quad \text{when } n \to \infty,$$

we have $\xi_n \to \varepsilon$.

The application of Lemma 3 concludes the proof of the theorem. □

Thus it remains only to prove Lemmas 1–3.

Proof of Lemma 1. Put

$$F(x) = \sum_{-\infty}^{\infty} F_m \exp(2\pi i m x), \qquad F_q(x) = \sum_{|m| < q} F_m \exp(2\pi i m x)$$

Since $F \in C^5$, we have

$$|F_m| < \text{const } m^{-5}, \qquad |F(x) - F_q(x)| \le \text{const } q^{-4},$$

$$\left| \sum_{k=0}^{q-1} F(x + k\alpha) - \sum_{k=0}^{q-1} F_q(x + k\alpha) \right| \le \text{const } q^{-3}.$$

On the other hand,

$$\sum_{k=0}^{q-1} F_q(x) - q\beta = \sum_{\substack{|m| < q \\ m \ne 0}} F_m \frac{\exp(2\pi i q\alpha m) - 1}{\exp(2\pi i \alpha m) - 1} \exp(2\pi i m x).$$

Put $\alpha = p/q + v$. By assumption, $|v| \le 0,1q^{-2}$. Further $m\alpha = mpq^{-1} + mv$. The distance from mp/q to the nearest integer is no less than $1/q$ while $|mv| \le 0,1q^{-1}$. Therefore

$$\exp(2\pi im\alpha) - 1 > 0,1q^{-1},$$

$$\begin{aligned}|\exp(2\pi iq\alpha m) - 1| &= |\exp(2\pi imv) - 1| \le \text{const} \cdot mv \\ &= \text{const}\,|m\alpha - mpq^{-1}| \le \text{const}\; q|\alpha - p/q|.\end{aligned}$$

Therefore

$$\begin{aligned}\left|\sum_{k=0}^{q-1} F(x + k\alpha) - q\beta\right| &\le \left|\sum_{k=0}^{q-1} F_q(x + k\alpha) - \sum_{k=0}^{q-1} F(x + k\alpha)\right| \\ &\quad + \left|\sum_{k=0}^{q-1} F_q(x + k\alpha) - q\beta\right| \\ &\le \text{const}\; q^{-3} + \text{const}\; q^2|\alpha - p/q| \sum_{|m|<q} |F_m|.\end{aligned}$$

The lemma is proved. □

Proof of Lemma 2. Obviously we have

$$\left|\sum_{j=0}^{r-1} [F(x_1 + j\alpha) - F(x_2 + j\alpha)]\right| \le \max_x |F'(x)| \cdot r \cdot q^{-1}.$$

Put $\varphi(x) = F'(x)$. Then

$$\sum_{j=0}^{r-1} [F(x_1 + j\alpha) - F(x_2 + j\alpha)] = \int_{x_1}^{x_2} \sum_{j=0}^{r-1} \varphi(\xi + j\alpha)\, d\xi.$$

In view of the unique ergodicity of the rotation of the circle by an irrational angle α, we have

$$\frac{1}{n} \sum_{j=0}^{n-1} \varphi(x + j\alpha) \to \int_0^1 \varphi(\xi)\, d\xi = 0 \quad \text{for } n \to \infty$$

uniformly with respect to x. In other words,

$$\left|\sum_{j=0}^{n-1} \varphi(x + j\alpha)\right| \le \varepsilon_n \cdot n, \quad \text{where } \varepsilon_n \to 0 \text{ for } n \to \infty.$$

For any $\varepsilon > 0$, we can find an $n(\varepsilon)$ such that $|\varepsilon_n| < \varepsilon$ whenever $n > n(\varepsilon)$.

Suppose $q > (n(\varepsilon)\max|F'|)/\varepsilon$. Then, if $r > n(\varepsilon)$, we have

$$\sum_{j=0}^{r-1}[F(x_1 + j\alpha) - F(x_2 + j\alpha)] \le r \cdot \varepsilon_r |x_1 - x_2| \le \frac{r}{q}\varepsilon_r < \varepsilon.$$

When $r \le n(\varepsilon)$, we have

$$\begin{aligned}\left|\sum_{j=0}^{r-1}[F(x_1 + j\alpha) - F(x_2 + j\alpha)]\right| &\le |x_1 - x_2|\max|F'|r \\ &\le \frac{r}{q}\max|F'| \le \varepsilon.\end{aligned}$$

The lemma is proved. □

Proof of Lemma 3. Suppose $A \subset M$ is a Borel set and $\chi_A(x)$ is its indicator. For a given $\varepsilon > 0$ let us first find a continuous function $f(x)$ on M such that

$$\int_M |\chi_A(x) - f(x)|\,d\mu < \varepsilon/2.$$

The function f is bounded: $|f| \le C$.

Using the uniform continuity of f, choose a $\delta = \delta(\varepsilon) > 0$ such that

$$\sup_{\substack{\rho(x_1, x_2) < \delta \\ x_1, x_2 \in M}} |f(x_1) - f(x_2)| < \varepsilon/4,$$

where ρ is the distance on M.

Now choose a number n so large that

$$\mu(M \backslash B_n) < \frac{\varepsilon}{8C}, \qquad \max_{1 \le i \le q_n} \operatorname{diam}(C_i^{(n)} \cap B_n) < \delta.$$

In each of the sets $C_i^{(n)}$, $1 \le i \le q_n$ choose an arbitrary point $x_i^{(n)} \in C_i^{(n)}$ and define the function $\tilde{f}_n(x)$, measurable with respect to the partition ξ_n, by putting $\tilde{f}_n(x) = f(x_i^{(n)})$ for $x \in C_i^{(n)}$. Then

$$\int_M |f - \tilde{f}_n|\,d\mu = \int_{B_n} |f - \tilde{f}_n|\,d\mu + \int_{M \backslash B_n} |f - \tilde{f}_n|\,d\mu = I_1 + I_2.$$

Clearly we have

$$I_2 \le \max|f - \tilde{f}_n| \cdot \mu(M \backslash B_n) \le 2C\frac{\varepsilon}{8C} = \frac{\varepsilon}{4}.$$

Moreover,

$$I_1 = \sum_{i=1}^{q_n} \int_{C_i^{(n)} \cap B_n} |f - \tilde{f}_n| \, d\mu \le \frac{\varepsilon}{4} \sum_{i=1}^{q_n} \mu(C_i^{(n)} \cap B_n) \le \frac{\varepsilon}{4}.$$

Therefore $\int_M |f - \tilde{f}_n| \, d\mu \le \varepsilon/2$. Now define the function $f_n(x)$, also measurable with respect to ξ_n, in the following way: if $x \in C_i^{(n)}$, $1 \le i \le q_n$ and $\mu(A \cap C_i^{(n)}) \le \frac{1}{2}\mu(C_i^{(n)})$, put $f_n(x) = 0$, while if $\mu(A \cap C_i^{(n)}) > \frac{1}{2}\mu(C_i^{(n)})$ put $f_n(x) = 1$. It is easy to check that for every i, $1 \le i \le q_n$,

$$\int_{C_i^{(n)}} |f - f_n| \, d\mu \le \int_{C_i^{(n)}} |f - \tilde{f}_n| \, d\mu.$$

Therefore

$$\int_M |f - f_n| \, d\mu \le \int_M |f - \tilde{f}_n| \, d\mu \le \varepsilon/2,$$

so that

$$\int_M |\chi_A(x) - f_n(x)| \, d\mu \le \int_M |\chi_A(x) - f(x)| \, d\mu + \int_M |f(x) - f_n(x)| \, d\mu < \varepsilon.$$

The function f_n is the indicator of a certain set $A_n \in \mathfrak{S}(\xi_n)$ which obviously satisfies the inequality $\mu(A \triangle A_n) < \varepsilon$. Since ε was arbitrary, the lemma is proved. □

Corollary. Putting this theorem together with Theorem 3, §4, Chap. 15 and Theorem 1, §2, Chapter 16, we see that for any irrational λ the flow $\{T^t\}$ is not mixing, the spectrum of the adjoint group $\{U^t\}$ is simple and the maximal spectral type is singular with respect to the Lebesgue measure.

§4. Example of a Smooth Flow with Continuous Spectrum on the Two-dimensional Torus

In this section we shall construct an example of a flow $\{T^t\}$ on the torus Tor^2 given in cyclic coordinates (u, v) by the equations

$$\frac{du}{dt} = \frac{1}{F(u, v)}, \qquad \frac{dv}{dt} = \frac{1}{\lambda F(u, v)}, \tag{1}$$

and possessing a continuous spectrum.

The function $F(u, v) > 0$ may be chosen smooth and even real analytic with period 1 with respect to each variable; λ is an irrational number, $0 < \lambda < 1$. The flow $\{T^t\}$ has the invariant measure $d\mu = F(u, v)du\,dv$.

By Theorem 1, §2, the number λ must have good approximations. It therefore follows from results of the previous section and from Theorem 3, §4, in Chap. 15, that the flow $\{T^t\}$ is an example of a smooth dynamical system with simple continuous singular spectrum.

To construct the example, it is convenient to use a special representation of flows of the form (1). For the base, we choose the circle S^1 with equation $v = 0$. It is easy to check that the base automorphism will be the rotation R_λ of the circle S^1 by the angle λ: $R_\lambda u = u + \lambda \pmod 1$ and the return time function will be

$$f(u) = \lambda \int_0^1 F(u + \lambda\xi, \xi)d\xi. \tag{2}$$

In order to construct the required example in the form of a special flow, we must make sure that equation (2) has a solution with respect to F in the appropriate class of functions.

Lemma 1. *Suppose*

$$f(u) = \sum_{k=-\infty}^{\infty} f_k \exp(2\pi iku) \tag{3}$$

is a real analytic function with period 1 *and*

$$f_0 > \frac{\pi}{2} \sum_{k \neq 0} |f_k|. \tag{4}$$

Then equation (2) *has a solution in the class of real analytic functions* $F(u, v) > 0$ *of period* 1 *in each variable.*

Proof. Let us find $F(u, v)$ in the form of a Fourier series

$$F(u, v) = \sum_{k, l=-\infty}^{\infty} F_{kl} \exp[2\pi i(ku + lv)], \tag{5}$$

Substituting (3) and (4) in equation (2), we get

$$f_0 + \sum_{k \neq 0} f_k \exp(2\pi iku) = \lambda\Bigg[F_{00} + \sum_{k^2 + l^2 > 0} \times F_{kl} \frac{\exp[2\pi i(k\lambda + l)] - 1}{2\pi i(k\lambda + l)} \exp(2\pi iku)\Bigg]$$

Equating the Fourier coefficients of both sides of this relation, we get

$$f_0 = \lambda F_{00}, \qquad f_k = \sum_{l=-\infty}^{\infty} F_{kl} \frac{\exp[2\pi i(k\lambda + l)] - 1}{2\pi i(k\lambda + l)}, \qquad k \neq 0. \tag{6}$$

We shall find the coefficients F_{kl} satisfying relations (6) in the following form: for each k all the F_{kl}, except one, vanish; the only nonzero coefficient is F_{k,l_k}, where l_k is the integer nearest to the number $(-k\lambda)$. Thus we put

$$F_{00} = \frac{f_0}{\lambda}, \qquad F_{0l} = 0 \quad \text{for} \quad l \neq 0;$$

$$F_{k,l_k} = \frac{2\pi i(k\lambda + l_k) \cdot f_k}{\lambda\{\exp[2\pi i(k\lambda + l_k)] - 1\}},$$

$$F_{k,l} = 0 \quad \text{for} \quad k \neq 0, l \neq l_k.$$

Let us show that $F(u, v)$ (with these Fourier coefficients) is the required function. We have

$$|F_{k,l_k}| = \frac{|f_k|}{\lambda}\left|\frac{2\pi i(k\lambda + l_k)}{\exp[2\pi i(k\lambda + l_k)] - 1}\right| = \frac{|f_k|}{\lambda} \cdot \frac{\pi|k\lambda + l_k|}{|\sin \pi(k\lambda + l_k)|}.$$

Since $|k\lambda + l_k| \leq 1/2$, the inequality $|\sin \alpha| \geq 2\alpha/\pi$, $|\alpha| \leq \pi/2$, implies

$$|F_{k,l_k}| \leq \frac{\pi}{2\lambda}|f_k|. \tag{7}$$

It follows from the fact that $f(u)$ is analytic that $|f_k| \leq c_1 \exp(-c_2|k|)$, where $c_1, c_2 > 0$ do not depend on k. Since $|l_k| \leq \text{const}\,|k|$, we finally obtain $|F_{k,l}| \leq c_3 \exp[-c_4(|k| + |l|)]$, where $c_3, c_4 > 0$ do not depend on k, l. Hence $F(u, v)$ is analytic. From the relation $\overline{F_{k,l}} = F_{-k,-l}$, we conclude that $F(u, v)$ is a real function. Moreover, using (7), we get

$$F(u, v) \geq F_{00} - \sum_{k^2+l^2>0} |F_{k,l}| \geq \frac{1}{\lambda}\left(f_0 - \frac{\pi}{2}\sum_{k\neq 0}|f_k|\right) > 0.$$

The lemma is proved. □

Now suppose the functions $f(u)$, $F(u, v)$ are related by the equality (2). Then the special flow $\{T_\lambda^t\}$, constructed from the rotation automorphism R_λ of the circle S^1 by the angle λ and from the function $f > 0$, is metrically isomorphic to the flow $\{T^t\}$ given by equations (1). Therefore, by Lemma 1, it suffices to construct a flow $\{T_\lambda^t\}$ with continuous spectrum such that the function f satisfies condition (4).

Suppose $M = \{(x, s): 0 \le x < 1, 0 < s < f(x)\}$ is the space of this special flow, μ is a normalized invariant measure. Consider the special flow $\{T_\beta^t\}$ on M with base rotation automorphism R_β by some rational angle $\beta = p/q$, where p, q are relatively prime, $0 < p < q$. Denote by $\mathfrak{S}_q$ the algebra of subsets of M generated by all possible rectangles with sides of the form $x = k/q$, $y = l/q$ (k, l-integers) contained in M. The elementary sets in $\mathfrak{S}_q$ will be the squares with sides $1/q$. Clearly $\mathfrak{S}_q$ is generated by its elementary sets. The example is based on the following lemma.

Lemma 2. *Suppose*

$$f(x) = \sum_{k=-\infty}^{\infty} f_k \exp(2\pi i k x)$$

not a trigonometric polynomial, i.e., for an infinite set $Q \subset \mathbb{Z}$ of values of k, the coefficients f_k do not vanish.[1] *Then*:

(a) *if $q \in Q$, $q > 0$, then for any two sets $A, B \in \mathfrak{S}_q$ the following limit exists*

$$\lim_{t\to\infty} \mu(T_\beta^t A \cap B) \overset{\text{def}}{=} \mu_{A,B,q};$$

(b) $$\max_{A,B\in\mathfrak{S}_q} |\mu_{A,B,q} - \mu(A)\mu(B)| = \varepsilon(q) \to 0 \quad \textit{when } q \to \infty. \tag{8}$$

The proof of the lemma will be given later; now we shall show how it can be used to construct the required example. First we prove the following statement.

Lemma 3. *Suppose $f(x) > 0$ satisfies the assumptions of Lemma* 1. *Then there exists a sequence of irreducible fractions*

$$\beta_n = \frac{p_n}{q_n}, \qquad 0 < p_n < q_n, \qquad q_n \to \infty, \qquad n = 1, 2, \ldots,$$

a sequence of natural numbers $t_n \to \infty$ and a sequence of numbers δ_n, $0 < \delta_n < 1/2^n$, such that

(a) $$|\beta_{n+1} - \beta_n| < \delta_n; \tag{$*$}$$

[1] Since the function f is real, the set Q is symmetric with respect to zero.

(b) *for any β in the interval $|\beta - \beta_n| < \delta_n$, any natural numbers k_1, k_2, $1 \le k_1, k_2 \le n$, and any sets $A \in \mathfrak{S}_{q_{k_1}}$, $B \in \mathfrak{S}_{q_{k_2}}$, we have the inequality*

$$|\mu(T_\beta^{t_n} A \cap B) - \mu(A)\mu(B)| \le \frac{1}{2^n} + \varepsilon(q_n),$$

where $\varepsilon(q)$ is defined in (8).

The proof will be by induction on n. Choose a natural number $q_1 > 1$ such that $f_{q_1} \ne 0$, $1/q_1 < \min_x f(x)$, and also a number p_1, $1 \le p_1 < q_1$ relatively prime to q_1, Applying Lemma 2 for $q = q_1$, $\beta = \beta_1 = p_1/q_1$, let us find a natural number t_1 such that

$$|\mu(T_{\beta_1}^{t_1} A \cap B) - \mu(A)\mu(B)| \le \tfrac{1}{4} + \varepsilon(q_1)$$

for all $A, B \in \mathfrak{S}_{q_1}$. Then choose a δ_1, $0 < \delta_1 < \frac{1}{2}$, such that for all β, $|\beta - \beta_1| < \delta_1$, we have

$$|\mu(T_\beta^{t_1} A \cap B) - \mu(A)\mu(B)| < \tfrac{1}{2} + \varepsilon(q_1)$$

for all $A, B \in \mathfrak{S}_{q_1}$. This may be done since $\mu(T_\beta^{t_1} A \cap B)$ is continuous in β. Suppose the statement of the lemma has been proved for all numbers $\le n$. Let us prove it for $n + 1$. If q_{n+1} is sufficiently large, then it follows from the law of asymptotic distribution of prime numbers that there exists a p_{n+1}, $1 \le p_{n+1} < q_{n+1}$, relatively prime to q_{n+1} and such that

$$\left| \frac{p_{n+1}}{q_{n+1}} - \frac{p_n}{q_n} \right| < \delta_n. \tag{9}$$

Moreover, if q_{n+1} is sufficiently large, then for any set $A \in \mathfrak{S}_{q_k}$, $1 \le k \le n$, there is a set $\tilde{A} \in \mathfrak{S}_{q_{n+1}}$ such that

$$\mu(A \bigtriangleup \tilde{A}) < \frac{1}{2^{n+3}}. \tag{10}$$

By the assumptions of the lemma, we may choose $q_{n+1} > q_n$ so as to have (9) and (10) and also $f_{q_{n+1}} \ne 0$. Applying Lemma 3 for $q = q_{n+1}$, $\beta = \beta_{n+1} = p_{n+1}/q_{n+1}$, let us find a natural number $t_{n+1} > t_n$ such that

$$|\mu(T_{\beta_{n+1}}^{t_{n+1}} \tilde{A} \cap \tilde{B}) - \mu(\tilde{A})\mu(\tilde{B})| < \frac{1}{2^{n+2}} + \varepsilon(q_{n+1})$$

for all $\tilde{A}, \tilde{B} \in \mathfrak{S}_{q_{n+1}}$. Then choose a δ_{n+1}, $0 < \delta_{n+1} < 1/2^{n+1}$, so that for all β, $|\beta - \beta_{n+1}| < \delta_{n+1}$; we will have

$$|\mu(T_\beta^{t_{n+1}}\tilde{A} \cap \tilde{B}) - \mu(\tilde{A})\mu(\tilde{B})| < \frac{1}{2^{n+1}} + \varepsilon(q_{n+1}) \tag{11}$$

for all $\tilde{A}, \tilde{B} \in \mathfrak{S}_{q_{n+1}}$. By (10), for all sets $A \in \mathfrak{S}_{q_{k_1}}$, $B \in \mathfrak{S}_{q_{k_2}}$, $1 \le k_1, k_2 \le n$, we can find sets $\tilde{A}, \tilde{B} \cap \mathfrak{S}_{q_{n+1}}$ such that

$$\mu(A \triangle \tilde{A}) < \frac{1}{2^{n+3}}, \qquad \mu(B \triangle \tilde{B}) \le \frac{1}{2^{n+3}}. \tag{12}$$

It follows from (11) and (12) that

$$|\mu(T_\beta^{t_{n+1}}A \cap B) - \mu(A)\mu(B)| < \frac{1}{2^{n+1}} + \varepsilon(q_{n+1}).$$

Thus the inductive step has been carried out and the lemma is proved. □

Let us now assume that the function f satisfies the assumption of Lemma 2 and also relations (4). Consider the flow $\{T_\lambda^t\}$, where $\lambda = \lim_{n\to\infty} \beta_n$; the numbers β_n were defined in Lemma 3 and the limit exists by (∗). Let us prove that this flow possesses a continuous spectrum.

To do this, it suffices to show that for any measurable sets $A, B \subset M$ we have

$$\lim_{n\to\infty} \mu(T_\lambda^{t_n}A \cap B) = \mu(A)\mu(B). \tag{13}$$

Indeed, (13) implies

$$\lim_{n\to\infty} \int_M f(T^{t_n}x)g(x)\,d\mu = \int f\,d\mu \cdot \int g\,d\mu$$

for any two functions $f, g \in L^2(M, \mathfrak{S}, \mu)$ and, for an eigen-function $h \in L^2(M, \mathfrak{S}, \mu)$, $\int h\,d\mu = 0$, $h \ne 0$, with eigen-value λ, we must have the relation

$$\left| \int h(T^{t_n}x)\bar{h}(x)\,d\mu \right| = |e^{it_n\lambda}| \cdot \int |h|^2\,d\mu = \int |h|^2\,d\mu > 0,$$

i.e.,

$$\overline{\lim_{n\to\infty}} \left| \int h(T^{t_n}x)\bar{h}(x)\,d\mu \right| > 0.$$

If $A \in \mathfrak{S}_{q_{k_1}}$, $B \in \mathfrak{S}_{q_{k_2}}$, $1 \le k_1, k_2 < \infty$, then for sufficiently large n, by Lemma 2, we shall have

$$|\mu(T_\lambda^{t_n} A \cap B) - \mu(A)\mu(B)| \le \frac{1}{2^n} + \varepsilon(q_n).$$

Passing to the limit when $n \to \infty$, we obtain (13) for all such sets.

If

$$\bar{A} = \bigcup_i A_i, \qquad \bar{B} = \bigcup_j B_j, \tag{14}$$

where

$$A_i, B_j \in \bigcup_{k=1}^{\infty} \mathfrak{S}_{q_k}, \qquad A_{i_1} \cap A_{i_2} = \varnothing,$$

$$B_{j_1} \cap B_{j_2} = \varnothing, \qquad \text{for} \quad i_1 \neq i_2, j_1 \neq j_2,$$

then

$$\lim_{n\to\infty} \mu(T_\lambda^{t_n} \bar{A} \cap \bar{B}) = \sum_{i,j} \lim_{n\to\infty} \mu(T_\lambda^{t_n} A_i \cap B_j)$$
$$= \sum_{i,j} \mu(A_i)\mu(B_j) = \mu(\bar{A})\mu(\bar{B}).$$

But for all measurable sets $A, B \subset M$ and any $\varepsilon > 0$, we can find $\bar{A}$, $\bar{B}$ of the form (14) such that

$$\mu(A \triangle \bar{A}) < \varepsilon, \qquad \mu(B \triangle \bar{B}) < \varepsilon.$$

Then

$$\begin{aligned} |\mu(T_\lambda^{t_n} A \cap B) - \mu(A)\mu(B)| &\le |\mu(T_\lambda^{t_n} A \cap B) - \mu(T^{t_n} \bar{A} \cap \bar{B})| \\ &+ |\mu(T_\lambda^{t_n} \bar{A} \cap \bar{B}) - \mu(\bar{A})\mu(\bar{B})| + |\mu(\bar{A})\mu(\bar{B}) - \mu(A)\mu(B)| \\ &= \Sigma_1 + \Sigma_2 + \Sigma_3. \end{aligned}$$

It is easy to show that $|\Sigma_1| \le \text{const} \cdot \varepsilon$, $|\Sigma_3| \le \text{const}\ \varepsilon$, while $|\Sigma_2| < \varepsilon$, for sufficiently large n. Then (13) will be proved, establishing that the flow $\{T_\lambda^t\}$ has a continuous spectrum.

It remains to prove Lemma 2.

The proof will be split up into several steps.

1. First we pass to a different special representation of the flow $\{T_\beta^t\}$. Choose a number $q \in Q$ and, for the base M_1, take the segment

$\{(x, s) \in M : 0 \le x < 1/q, s = 0\}$. The return time of the point $x = (x, 0) \in M_1$ in the set M_1 equals

$$f_q(x) = \sum_{k=0}^{q-1} f\left(x + \frac{pk}{q}\right).$$

Since p, q are relatively prime, we have

$$f_q(x) = \sum_{k=0}^{q-1} f\left(x + \frac{k}{q}\right).$$

The space of the new special representation will again be denoted by $M : M = \{(x, s) : 0 \le x < 1/q, 0 \le s < f_q(x)\}$. The base automorphism is the identical automorphism of M_1. By carrying out the change of variables $x' = x$, $s' = s \cdot q/f_q(x)$, we will transform the space M into a rectangle for which the notation M will be preserved:

$$M = \left\{(x', s') : 0 \le x' < \frac{1}{q}, 0 \le s' < q\right\}.$$

In the sequel, we will only consider the variables x', s', and, for brevity, will omit the superscripts (primes). The action of the flow $\{T^t_\beta\}$ in the new variables can be written in the form

$$T^t_\beta(x, s) = \left(x, s + \frac{qt}{f(x)} \pmod q\right),$$

and the invariant measure is

$$d\mu = \frac{f_q(x)}{q \cdot I}\, dx\, ds,$$

where

$$I = \int_0^{1/q} f_q(x) dx = \int_0^1 f(x)\, dx.$$

This measure is a direct product,

$$\mu = \mu_x \times \mu_s, \qquad d\mu_x = \frac{f_q(x)}{q \cdot I}\, dx, \qquad d\mu_s = ds.$$

To algebra of sets $\mathfrak{S}_q$ in the rectangle M corresponds a certain algebra $\tilde{\mathfrak{S}}_q$, the elementary sets of $\tilde{\mathfrak{S}}_q$ are sets of the form

$$\left\{(x, s): 0 \leq x < \frac{1}{q}, g(x) \leq s < g(x) + \frac{1}{f_q(x)}\right\},$$

where the function $g(x)$ is real analytic. These sets will be referred to as elementary sets of the algebra $\tilde{\mathfrak{S}}_q$.

2. For every $x \in [0, 1/q)$, the velocity of the point $(x, s) \in M$ under the action of the flow $\{T_\beta^t\}$ does not depend on s and equals $V(x) = q/f_q(x)$. The function $V(x)$ is real analytic. Assume that on a certain open interval $(a, b) \subset [0, 1/q)$, the derivative $dV(x)/dx$ does not change its sign, say, $dV(x)/dx > 0$. Consider the set A of the form

$$A = \{(x, s) \in M: a < x < b, l \leq s < l + \delta\}$$

and the closed interval $\Delta_c \subset M: \Delta_c = \{(x, s) \in M: a \leq x \leq b, s = c\}$. For some t let us compute the measure $\mu_x(T_\beta^t A \cap \Delta_c)$. To do this, it is convenient to introduce the flow $\{\tilde{T}_\beta^t\}$ which acts in the strip

$$\Pi = \{(x, s): 0 \leq x < 1/q, -\infty < s < \infty\}$$

according to the formula

$$\tilde{T}_\beta^t(x, s) = \left(x, s + \frac{qt}{f_q(x)}\right).$$

One checks directly that

$$\mu_x(T_\beta^t A \cap \Delta_c) = \sum_{n=-\infty}^{\infty} \mu_x(\tilde{T}_\beta^t A \cap \Delta_c^{(n)}), \tag{15}$$

where

$$\Delta_c^{(n)} = \{(x, s) \in \Pi: a < x < b, s = c + nq\}.$$

The set $\tilde{T}_\beta^t A$ is of the form

$$\tilde{T}_\beta^t A = \{(x, s) \in \Pi: a < x < b, l + tV(x) \leq s < l + \delta + tV(x)\}.$$

Since $V(x)$ is monotonic increasing on (a, b) the set $E_n = \tilde{T}_\beta^t A \cap \Delta_n^{(n)}$ is an open interval

$$E_n = \{(x, s) \in \Pi: x_n^{(1)} < x < x_n^{(2)}, s = c + nq\}.$$

Note that

$$l + \delta + tV(x_n^{(1)}) = l + tV(x_n^{(2)}) = c + nq,$$

i.e.,

$$V(x_n^{(1)}) = \frac{c + nq - l - \delta}{t}, \qquad V(x_n^{(2)}) = \frac{c + nq - l}{t}. \tag{16}$$

Suppose $x = U(y)$ is the function inverse to $V(x)$, $x \in (a, b)$. Then

$$x_n^{(1)} = U\left(\frac{c + nq - l - \delta}{t}\right), \qquad x_n^{(2)} = U\left(\frac{c + nq - l}{t}\right).$$

By the Lagrange theorem

$$x_n^{(2)} - x_n^{(1)} = [V(x_n^{(2)}) - V(x_n^{(1)})] \cdot U'(\tilde{y}_n), \tag{17}$$

where $\tilde{y}_n \in [V(x_n^{(1)}), V(x_n^{(2)})]$. From (16) and (17), we get

$$x_n^{(2)} - x_n^{(1)} = \frac{\delta}{t} \cdot U'(\tilde{y}_n). \tag{18}$$

Further

$$\mu_x(E_n) = \frac{1}{I} \int_{x_n^{(1)}}^{x_n^{(2)}} \frac{f_q(x)dx}{q} = \frac{1}{I} \int_{x_n^{(1)}}^{x_n^{(2)}} \frac{dx}{V(x)} = \frac{1}{I} (x_n^{(2)} - x_n^{(1)}) \cdot \frac{1}{V(\tilde{x}_n)}, \tag{19}$$

where $\tilde{x}_n \in [x_n^{(1)}, x_n^{(2)}]$. Putting together (18) and (19), we obtain

$$\mu_x(E_n) = \frac{1}{I} \cdot \frac{\delta}{t} \cdot U'(\tilde{y}_n) \cdot \frac{1}{V'(\tilde{x}_n)} = \frac{\delta}{q \cdot I} \cdot \frac{U'(\tilde{y}_n)(q/t)}{V(\tilde{x}_n)}. \tag{20}$$

Since $V(x) \geq \text{const} > 0$, it follows in particular from (20) that

$$\mu_x(E_n) \leq \frac{\text{const}}{t}. \tag{21}$$

From (15) and (20) we conclude that

$$\mu_x(T_\beta^t A \cap \Delta_c) = \frac{\delta}{q \cdot I} \sum_{n=-\infty}^{\infty} \frac{U'(\tilde{y}_n) \cdot (q/t)}{V(\tilde{x}_n)}.$$

3. Suppose $N_t = \{n: V(a) < nq/t < V(b)\}$. Then

$$\left| \mu_x(T_\beta^t A \cap \Delta_c) - \frac{\delta}{qI} \sum_{n \in N_t} \frac{U'(\tilde{y}_n)(q/t)}{V(\tilde{x}_n)} \right| \to 0, \tag{22}$$

when $t \to \infty$. Indeed the last sum includes the measures of all intervals E_n such that the point $\tilde{T}_\beta^t(a, 0)$ is located below the line $s = qn$, while the point $\tilde{T}_\beta^t(b, 0)$ is above this line. This means that the sum differs from $\mu_x(T_\beta^t A \cap \Delta_c)$ by the measure of no more than two intervals E_n, i.e., by (21) is no greater than const/t. It follows from (16) that

$$\left| \sum_{n \in N_t} \frac{U'(\tilde{y}_n)(q/t)}{V(\tilde{x}_n)} - \sum_{n \in N_t} \frac{U'(nq/t)(q/t)}{nq/t} \right| \to 0, \tag{23}$$

when $t \to 0$. But the last sum is the integral sum for the integral

$$\int_{V(a)}^{V(b)} \frac{U'(y)}{y} dy = \int_{V(a)}^{V(b)} \frac{dU(y)}{y} = \int_a^b \frac{dx}{V(x)}. \tag{24}$$

Putting together (22), (23), and (24), we get

$$\lim_{t \to \infty} \mu_x(T_\beta^t A \cap \Delta_c) = \frac{\delta}{qI} \int_a^b \frac{dx}{V(x)} = \frac{\delta}{q^2 I} \int_a^b f_q(x)\, dx$$

$$= \frac{\delta}{q} \mu_x([a, b]) = \frac{1}{q} \mu(A).$$

4. Consider, together with the set A, the set B of the form

$$B = \{(x, s) \in M: a < x < b, l_1 \leq s < l_1 + \delta_1\}.$$

Then

$$\mu(T_\beta^t A \cap B) = \int_{l_1}^{l_1 + \delta_1} \mu_x(T_\beta^t A \cap \Delta_s)\, ds,$$

so that

$$\lim_{t \to \infty} \mu(T_\beta^t A \cap B)$$

$$= \int_{l_1}^{l_1 + \delta_1} \lim_{t \to \infty} \mu_x(T_\beta^t A \cap \Delta_s)\, ds = \frac{\delta_1}{q} \mu(A) = \frac{\mu(A)\mu(B)}{\mu(D)},$$

where $D = \{(x, s) \in M: a < x < b, 0 \leq s < q\}$.

5. In the sequel we shall need the following statement:

If $q \in Q$, then the function $f_q(x)$ is not an identical constant and

$$|f_q(x) - qI| \leq \text{const} \cdot q^{-4}, \tag{25}$$

The proof of this statement will be given in step 7.

Now suppose $A_1, A_2 \in \tilde{\mathfrak{S}}_q$ are elementary sets:

$$A_j = \left\{(x, s): 0 \leq x < \frac{1}{q}, g_j(x) \leq s < g_j(x) + \frac{1}{f_q(x)}\right\}, \qquad j = 1, 2.$$

Since the function $V(x) = q/f_q(x)$ is real analytic, it follows from the statement formulated above that the derivative $dV(x)/dx$ vanishes only at a finite number of points $b_1, \ldots, b_s \in [0, 1/q)$ and preserves its sign between these points.

Decompose $[0, 1/q)$ into smaller segments by means of the points $0 = q_0 < a_1 < \cdots < a_n = 1/q$ so that all the points $b_1, \ldots, b_s$ are contained among the points a_i, $0 \leq i \leq n$. Denote

$$K_i^{(j)} = \left\{(x, s): a_{i-1} \leq x < a_i, g_j(a_i) \leq s < g_j(a_i) + \frac{1}{f_q(a_i)}\right\}$$

$$K_1 = \bigcup_{i=1}^{n} K_i^{(1)}, K_2 = \bigcup_{i=1}^{n} K_i^{(2)}, \qquad 1 \leq i \leq n, j = 1, 2.$$

Since the functions $f_q(x)$, $g_1(x)$, $g_2(x)$ are analytic, by choosing a sufficiently fine decomposition we may require that

$$\mu(A_1 \triangle K_1) < \frac{1}{q^5}, \qquad \mu(A_2 \triangle K_2) < \frac{1}{q^5}. \tag{26}$$

By step 4, we have

$$\lim_{t \to \infty} \mu(T_\beta^t K_1 \cap K_2) = \sum_{i=1}^{n} \lim_{t \to \infty} \mu(T_\beta^t K_i^{(1)} \cap K_i^{(2)}) = \sum_{i=1}^{n} \frac{\mu(K_i^{(1)})\mu(K_i^{(2)})}{\mu(D_i)}, \tag{27}$$

where $D_i = \{(x, s) \in M: a_{i-1} \leq x < a_i\ 0 \leq s < q\}$. But

$$\mu(K_i^{(1)}) = \frac{1}{qI \cdot f_q(a_i)} \int_{a_{i-1}}^{a_i} f_q(x)dx.$$

Using (25), we obtain

$$\left|\mu(K_i^{(1)}) - \frac{1}{q^2 I^2}\int_{a_{i-1}}^{a_i} f_q(x)dx\right| \le \frac{\text{const}}{q^5}\int_{a_{i-1}}^{a_i} f_q(x)\,dx.$$

Since

$$\sum_{i=1}^{n}\int_{a_{i-1}}^{a_i} f_q(x)\,dx = 1, \qquad \sum_{i=1}^{n}\mu(K_i^{(1)}) = \mu(K_1),$$

we shall obtain, by taking the sum over i, the relation

$$\left|\mu(K_1) - \frac{1}{q^2 I}\right| \le \frac{\text{const}}{q^5}. \tag{28}$$

Further

$$\frac{\mu(K_i^{(2)})}{\mu(D_i)} = \frac{(1/qI)\cdot[1/f_q(a_i)]\int_{a_{i-1}}^{a_i} f_q(x)\,dx}{(1/qI)\cdot q\int_{a_{i-1}}^{a_i} f_q(x)\,dx} = \frac{1}{q f_q(a_i)}.$$

Using (25) once again, we get

$$\left|\frac{\mu(K_i^{(2)})}{\mu(D_i)} - \frac{1}{q^2 I}\right| \le \frac{\text{const}}{q^5}. \tag{29}$$

Therefore

$$\left|\sum_{i=1}^{n}\mu(K_i^{(1)})\frac{\mu(K_i^{(2)})}{\mu(D_i)} - \frac{\mu(K_1)}{q^2 I}\right| \le \frac{\text{const}}{q^5},$$

i.e., by (29)

$$\left|\sum_{i=1}^{n}\mu(K_i^{(1)})\frac{\mu(K_i^{(2)})}{\mu(D_i)} - \frac{1}{q^4 I^2}\right| \le \frac{\text{const}}{q^5}. \tag{30}$$

But it follows from (28) that

$$\left|\mu(K_1)\mu(K_2) - \frac{1}{q^4 I^2}\right| = \left|[\mu(K_1)]^2 - \frac{1}{q^4 I^2}\right| \le \frac{\text{const}}{q^5}. \tag{31}$$

Putting together (27), (30), and (31), we finally get

$$\left|\lim_{t\to\infty}\mu(T_\beta^t K_1 \cap K_2) - \mu(K_1)\mu(K_2)\right| \le \frac{\text{const}}{q^5}.$$

By (26), we also have the inequality

$$\left| \lim_{t\to\infty} \mu(T^t_\beta A_1 \cap A_2) - \mu(A_1)(\mu(A_2)) \right| \le \frac{\text{const}}{q^5}.$$

6. Now choose an arbitrary pair of sets $A, B \in \tilde{\mathfrak{S}}_q$. Suppose $A = \bigcup_i A_i$, $B = \bigcup_j B_j$ where A_i, B_j are elementary sets and $B_{j_1} \cap B_{j_2} = \varnothing$, $A_{i_1} \cap A_{i_2} = \varnothing$ for $i_1 \ne i_2, j_1 \ne j_2$. Since $\mu(A_i) = \mu(B_j) = 1/Iq^2$, each of the indices i, j ranges over no more than $q^2 I$ values. Note that

$$\mu(T^t_\beta A \cap B) = \sum_{i,j} \mu(T^t_\beta A_i \cap B_j),$$

therefore

$$\sum \overset{\text{def}}{=} \left| \lim_{t\to\infty} \mu(T^t_\beta A \cap B) - \mu(A)\mu(B) \right| \le \sum_{i,j} \left| \lim_{t\to\infty} \mu(T^t_\beta A_i \cap B_j) - \mu(A_i)\mu(B_j) \right|.$$

The number of summands in the last sum is no greater than $q^4 I^2$, and each of them, by step 5, is no greater than const/q^5. Therefore $\sum \le \text{const}/q$. The lemma is proved. □

7. It remains to prove the statement formulated in the beginning of step 5. Let us expand $f_q(x)$ into a Fourier series:

$$f_q(x) = \sum_{k=0}^{q-1} f\left(x + \frac{k}{q}\right) = \sum_{k=0}^{q-1} \sum_{h=-\infty}^{\infty} f_n \exp\left[2\pi i n\left(x + \frac{k}{q}\right)\right]$$

$$= \sum_{h=-\infty}^{\infty} \left[\sum_{k=0}^{q-1} \exp\left(\frac{2\pi i n k}{q}\right)\right] f_n \exp(2\pi i n x).$$

If $n \ne ql$, where l is an integer, the inner sum vanishes, but if $n = ql$, then the sum equals q. Hence

$$f_q(x) = f_0 q + \sum_{l\ne 0} f_{ql} \cdot q \cdot \exp(2\pi i q l x). \tag{32}$$

Since $f(x)$ is a real analytic function, we have $|f_{ql}| \le c_1 \exp(-c_2 q|l|)$, where $c_1, c_2 > 0$ and therefore

$$|f_{ql}| \le \text{const}(q|l|)^5. \tag{33}$$

Noticing that $f_0 = I$, we obtain from (32) and (33):

$$|f_q(x) - qI| \le \text{const} \sum_{l\ne 0} \frac{q}{q^5|l|^5} \le \text{const} \cdot q^{-4}.$$

Thus the proof of the continuity of the spectrum of the flow $\{T^t_\lambda\}$ is concluded.

Appendix 1

Lebesgue Spaces and Measurable Partitions

1. Here we list the main facts relating to Lebesgue spaces and their measurable partitions. The proofs may be found, for example, in Rohlin's article [3]. We begin by some auxiliary definitions.

Suppose $(M, \mathfrak{S}, \mu)$ is a space with normalized measure. The sets $A \in \mathfrak{S}$ are called measurable. The measure μ is assumed complete. This means that for any set $A \in \mathfrak{S}$, $\mu(A) = 0$, all its subsets $B \subset A$ are also measurable. For any family of measurable sets $\{B_\alpha\}$, we denote by $F(\{B_\alpha\})$ the Borel field generated by all the B_α.[1]

Definition 1. A countable system of measurable sets

$$\mathbb{B} = \{B_i, i \in I\},$$

is said to be a *basis* of the space M if:

(i) for any $A \in \mathfrak{S}$ there exists a set $C \in F(\mathbb{B})$ such that $C \subset A$, $\mu(C \setminus A) = 0$;

(ii) for any pair of points $x_1, x_2 \in M$, $x_1 \neq x_2$, there exists an $i \in I$ such that either $x_1 \in B_i$, $x_2 \notin B_i$ or $x_2 \in B_i$, $x_1 \notin B_i$.

Now let us introduce the notion of completeness of the space M with respect to the basis $\mathbb{B} = \{B_i\}$. Suppose $e_i = \pm 1$ and $B_i^{(e_i)} = B_i$ if $e_i = 1$, $B_i^{(e_i)} = M \setminus B_i$ if $e_i = -1$. To any family of numbers $\{e_i, i \in I\}$ corresponds the intersection $\bigcap_{i \in I} B_i^{(e_i)}$. By (ii), every such intersection contains no more than one point.

Definition 2. The space $(M, \mathfrak{S}, \mu)$ is said to be *complete* with respect to the basis $\mathbb{B}$ if all the intersections $\bigcap_{i \in I} B_i^{(e_i)}$ are not empty.

[1] That is, the minimal σ-algebra containing all the B_α. We did not use the term "σ-algebra" in the text here, since most of the σ-algebras which appear in this book contain, together with any set, all the sets which coincide with it (mod 0), while $F(\{B_\alpha\})$ does not possess this property.

Definition 3. The space $(M, \mathfrak{S}, \mu)$ is called *complete* (mod 0) with respect to the basis $\mathbb{B}$ if M may be included as a subset of full measure into a certain measure space $(\overline{M}, \overline{\mathfrak{S}}, \bar{\mu})$ which is complete with respect to its own basis $\overline{\mathbb{B}} = \{\overline{B}_i, i \in I\}$ satisfying $\overline{B}_i \cap M = B_i$ for all $i \in I$.

It turns out that a space which is complete (mod 0) with respect to one of its own basis is also complete (mod 0) with respect to any other basis.

The following definition is the main one.

Definition 4. The space $(M, \mathfrak{S}, \mu)$ which is complete (mod 0) with respect to one of its basis is said to be a *Lebesgue space.*

The meaning of this definition is the following. On one hand, the notion of a Lebesgue space is so wide that it includes practically all the measure spaces used in the applications. In particular, any complete separable metric space in which the measure is defined on a σ-algebra generated by the open sets (the Borel σ-algebra) is a Lebesgue space. The direct product of a finite or countable family of Lebesgue spaces is also Lebesgue.

On the other hand, Lebesgue spaces possess many nice properties which are not valid in the general case. Let us mention one of them. Any automorphism T of the measure space $(M, \mathfrak{S}, \mu)$ induces a certain map (isomorphism) S of the σ-algebra $\mathfrak{S}$ onto itself according to the formula $S(A) = TA$, $A \in \mathfrak{S}$. For a Lebesgue space the converse statement is valid: any isomorphism of the σ-algebra of measurable sets is generated in this way by a certain automorphism of the space itself. Further in this appendix we will assume that $(M, \mathfrak{S}, \mu)$ is a Lebesgue space.

2. A partition of the space $(M, \mathfrak{S}, \mu)$ is, by definition, any family $\xi = \{C\}$ of nonempty disjoint measurable subsets C such that $\bigcup_{C \in \xi} C = M$. If $\bigcup_{C \in \xi} C = M$(mod 0), then ξ is referred to as a partition (mod 0). Speaking of partitions and operations on them, we often mean partitions (mod 0) without mentioning this explicitly. One naturally defines inequalities between partitions: $\xi \preccurlyeq \eta$ (ξ is not finer than η) if each element $C_\xi \in \xi$ is the union of a certain number of elements $C_\eta \in \eta$. The notations $\xi = \eta$, $\xi \preccurlyeq \eta$ are also used in the case when they only hold (mod 0). The sets $A \in \mathfrak{S}$ which are the unions of elements $C_\xi \in \xi$ are called *measurable* with respect to ξ (or ξ-*sets*).

The main role in the study of Lebesgue spaces is played by the notion of measurable partition.

Definition 5. The partition ξ is said to be *measurable*, if there exists a countable system of sets $\mathbb{B} = \{B_i, i \in I\}$ which are measurable with respect to ξ such that for all $C_1, C_2 \in \xi$ we can find an $i \in I$ such that either $C_1 \subset B_i, C_2 \not\subset B_i$ or $C_2 \subset B_i, C_1 \not\subset B_i$.

The following statement holds: the quotient space of a Lebesgue space by a measurable partition ξ, i.e., the space M/ξ whose points are the elements C_ξ of the partition ξ, is also a Lebesgue space.

If $\{\xi_\alpha\}$ is a system of measurable partitions, then by its product $\xi = \bigvee_\alpha \xi_\alpha$ we mean the measurable partition ξ which is entirely defined by the following properties:

(i) $\xi \succcurlyeq \xi_\alpha$ for any α;
(ii) if $\xi' \succcurlyeq \xi_\alpha$ for all α and ξ' is measurable, then $\xi' \succcurlyeq \xi$.

The intersection $\eta = \bigwedge_\alpha \xi_\alpha$ is, by definition, the measurable partition η which is well defined by the following properties:

(i) $\eta \preccurlyeq \xi_\alpha$ for any α;
(ii) if $\eta' \preccurlyeq \xi_\alpha$ for all α and η is measurable, then $\eta' \preccurlyeq \eta$.

3. There exists a natural one-to-one correspondence between the measuable partitions ξ of a Lebesgue space $(M, \mathfrak{S}, \mu)$ and the complete σ-subalgebras of the σ-algebra $\mathfrak{S}$, i.e., such σ-algebras $\mathfrak{M} \subset \mathfrak{S}$ that the measure μ restricted to $\mathfrak{M}$ is complete. Namely, assign to the partition ξ the complete σ-algebra $\mathfrak{S}(\xi) \subset \mathfrak{S}$ consisting of the sets $A \in \mathfrak{S}$ which coincide (mod 0) with one of the sets which is measurable with respect to ξ. It turns out that every complete σ-algebra $\mathfrak{M} \subset \mathfrak{S}$ is of the form $\mathfrak{M} = \mathfrak{S}(\xi)$ for some $\xi = \xi(\mathfrak{M})$ and the map $\mathfrak{M} \to \xi(\mathfrak{M})$ is one-to-one. To operations on partitions correspond operations of the corresponding σ-subalgebras. Namely, if $\{\xi_\alpha\}$ is a system of measurable partitions, then

$$\mathfrak{S}\left(\bigvee_\alpha \xi_\alpha\right) = \bigvee_\alpha \mathfrak{S}(\xi_\alpha), \qquad \mathfrak{S}\left(\bigwedge_\alpha \xi_\alpha\right) = \bigwedge_\alpha \mathfrak{S}(\xi_\alpha).$$

Here $\bigwedge_\alpha \mathfrak{S}(\xi_\alpha) = \bigcap_\alpha \mathfrak{S}(\xi_\alpha)$ is the set-theoretic intersection of the σ-algebras, while $\bigvee_\alpha \mathfrak{S}(\xi_\alpha)$ is the intersection of all the σ-algebras which contain all $\mathfrak{S}(\xi_\alpha)$.

4. The main property of measurable partitions is that the elements $C \in \xi$ of such partitions can themselves be transformed into spaces with measure μ_C, and these measures play the role of conditional probabilities.

Definition 6. *By a canonical system of conditional measures belonging to the partition* ξ, we mean a system of measures $\{\mu_C\}$, $C \in \xi$ possessing the following properties:

(i) μ_C is defined on some σ-algebra $\mathfrak{S}_C$ of subsets of the set C;
(ii) the space $(C, \mathfrak{S}_C, \mu_C)$ is Lebesgue;

(iii) for any $A \in \mathfrak{S}$ the set $A \cap C$ belongs to $\mathfrak{S}_C$ for almost all $C \in M/\xi$, the function $\mu_C(A \cap C)$ is measurable on M/ξ and

$$\mu(A) = \int_{M/\xi} \mu_C(A \cap C)\, d\mu.$$

Every measurable partition possesses a canonical system of conditional measures. This system is unique (mod 0) in the sense that any other system $\{\mu'_C\}$ coincides with it for almost all $C \in M/\xi$. Conversely, if some partition possesses a canonical system of conditional measures, then it is measurable.

Appendix 2

Relevant Facts from the Spectral Theory of Unitary Operators

1. Suppose H is a separable complex Hilbert space, U a unitary operator in H, i.e., U is a linear continuous invertible operator such that $U^* = U^{-1}$. The complex number λ is said to be an eigen-value of the operator U if there exists a vector $h \in H$ which satisfies $Uh = \lambda h$. Then h is called an eigen-vector of the operator U with eigen-value λ. Sometimes, to stress the relationship between λ and h, we shall denote this vector by h_λ. Since

$$\begin{aligned}\lambda(h_\lambda, h_\lambda) &= (Uh_\lambda, h_\lambda) = (h_\lambda, U^*h_\lambda) = (h_\lambda, U^{-1}h_\lambda) = \overline{(U^{-1}h_\lambda, h_\lambda)}\\ &= \overline{\lambda^{-1}}(h_\lambda, h_\lambda),\end{aligned}$$

we have $\lambda\bar{\lambda} = 1$, i.e., $|\lambda| = 1$.

The family of all eigen-values $\Lambda_d(U)$ of the operator U is known as the discrete (point) spectrum of this operator. If $\lambda_1, \lambda_2 \in \Lambda_d(U)$, $\lambda_1 \neq \lambda_2$, then $(h_{\lambda_1}, h_{\lambda_2}) = 0$. The smallest closed subspace generated by all the h_λ is denoted by $H_d(U)$. The orthogonal complement to it $H \ominus H_d(U)$ is denoted by $H_c(U)$. The subspaces $H_d(U)$ and $H_c(U)$ may be described in purely "ergodic" terms as well. Namely, consider the limits

$$\lim_{n\to\infty} \frac{1}{2n+1} \sum_{k=-n}^{n} |(U^k h, h)| = m(h).$$

Then $H_d(U)$ consists of all vectors h for which $m(h) = \|h\|^2$, while $H_c(U)$ consists of all vectors h satisfying $m(h) = 0$.

For an arbitrary $h \in H$, consider the sequence

$$b_n(h) = (U^n h, h), \qquad -\infty < n < \infty.$$

This sequence is positive definite:

$$\Sigma\, c_{k_1} \bar{c}_{k_2} b_{k_1 - k_2}(h) = \left(\sum c_k U^k h, \sum c_k U^k h\right) \geq 0$$

for any finite sequence $\{c_k\}$. Therefore by the Bochner–Khinchin theorem there exists a (unique) measure σ_h (on the circle S^1) such that

$$b_n(h) = \int \exp[2\pi i n\lambda]\, d\sigma_h(\lambda).$$

This measure is known as the spectral measure of the vector h. If $h = h_{\lambda_0}$, $\lambda_0 \in \Lambda_d(U)$, then σ_h is the measure concentrated at the point λ_0. In a more general way, for an arbitrary $h \in H_d(U)$ the corresponding measure σ_h is discrete. Conversely, for any $h \in H_c(U)$ the measure σ_h is continuous. These properties may also be used to define $H_d(U)$ and $H_c(U)$.

For any vector $h \in H$ consider the sequence of vectors $U^n h$, $-\infty < n < \infty$ and the subspace $C(h)$ which is the closure of the set of finite linear combinations $\sum c_k U^k h$. Clearly, $C(h)$ is the smallest closed subspace containing the vector h and invariant with respect to U. The subspace $C(h)$ is said to be the cyclic subspace generated by the vector h. We have the following theorem.

Theorem 1. *The space $C(h)$ may be isomorphically mapped onto the Hilbert space $L^2(S^1, \mathfrak{S}, \sigma_h)$, where $\mathfrak{S}$ is the Borel σ-algebra of subsets of the circle so that the action of the operator U becomes the multiplication by* $\exp(2\pi i\lambda)$. *If $\varphi \in L^2(S^1, \mathfrak{S}, \sigma_h)$ corresponds to the vector $h_1 \in C(h)$ under this isomorphism, the measure σ_{h_1} is absolutely continuous with respect to σ_h and we have $d\sigma_{h_1}/d\sigma_h = |\varphi|^2$. For a vector $h_1 \in C(h)$ the relation $C(h_1) = C(h)$ holds if and only if $|\varphi(\lambda)| > 0$ almost everywhere with respect to the measure σ_h.*

It follows from this theorem that the set of vectors h_1 satisfying $C(h_1) = C(h)$ corresponds to the set of finite measures equivalent to the measure σ_h. The class of equivalent finite measures will be referred to as the spectral type. Using this terminology, we can say that every cyclic subspace $C(h)$ is characterized by its spectral type.

A spectral type is called absolutely continuous (with respect to the Lebesgue measure) if the measure σ_h is absolutely continuous. If the measure σ_h is equivalent to the Lebesgue measure, then the spectral type is said to be Lebesgue. It follows from Theorem 1 that in the case of a Lebesgue type there is a vector h_1 in the space $C(h)$ such that σ_{h_1} is the Lebesgue measure. In this case $(U^n h_1, U^m h_1) = 0$ for $n \neq m$ and the vectors $U^n h_1$, $-\infty < n < \infty$ constitute an orthogonal basis in the subspace $C(h)$. This shows that the Lebesgue spectral type is, in a certain sense, the opposite of the discrete spectral type, i.e., of the spectral type of a discrete measure. If σ_h is singular with respect to the Lebesgue measure, then the spectral type is known as singular.

If the spectral type of a certain cyclic subspace is discrete (continuous, Lebesgue, etc.) then we sometimes say that the spectrum of the operator U possesses discrete (continuous, Lebesgue, etc.) components. If the spectrum of the operator U possesses discrete as well as continuous components, then we say that U has a mixed spectrum.

By using the notion of spectral type we may state the main theorem on the canonical form of a unitary operator. We shall give two equivalent formulations.

The Main Theorem on the Canonical Form of a Unitary Operator

First Formulation. For any unitary operator U there exists a sequence of vectors h_i, $i = 1, 2, \ldots$ such that

(1) $C(h_i) \perp C(h_j)$ for $i \neq j$, $H = \bigoplus_{i=1}^{\infty} C(h_i)$;
(2) $\sigma_{h_{i+1}} \preccurlyeq \sigma_{h_i}$, $i = 1, 2, \ldots$;[1]
(3) *for any sequence of vectors h_i', $i = 1, 2, \ldots$ satisfying conditions* (1) *and* (2) *we have $\sigma_{h_i'} \approx \sigma_{h_i}$ for $i = 1, 2, \ldots$.*

Second Formulation. For any unitary operator U and any $n = 1, 2, \ldots, \infty$ we can find a Borel subset A_n of the circle S^1 and a sequence of vectors

$$h_{n,k} \in H, \quad k = 1, 2, \ldots, n (k = 1, 2, \ldots \text{ for } n = \infty),$$

such that

(1) $A_{n_1} \cap A_{n_2} = \varnothing$ *for* $n_1 \neq n_2$, $\bigcup_n A_n = S^1$
(2) $C(h_{n,k}) \perp C(h_{n_1,k_1})$ *for* $(n, k) \neq (n_1, k_1)$, $\bigoplus_n \bigoplus_{k=1}^{n} C(h_{n,k}) = H$;
(3) $\sigma_{h_{n,k}} = \sigma_{h_{n,l}} \overset{\text{def}}{=} \sigma^{(n)}$ *for* $1 \leq k, l \leq n$ *and* $\sigma^{(n)}(S^1 \setminus A_n) = 0$.

For any other sequence $h_{n,k}'$ satisfying the above conditions, we have $\sigma_{h_{n,k}'} \approx \sigma_{h_{n,k}}$.

The spectral type of the measure σ_{h_1} appearing in the first formulation of the main theorem is known as the *maximal spectral type* of the operator U. The spectral measure of any element $h \in H$ is absolutely continuous with respect to measures of maximal spectral type. The function $n(\lambda)$ defined on S^1 by the relation $n(\lambda) = n$ for $\lambda \in A_n$ (see the second formulation) is known as the *multiplicity function*, while the spectral type of the measure $\sigma^{(n)}$ is a *spectral type of multiplicity n*.

If the entire space H is a cyclic subspace for some vector $h \in H$, we say that the operator U has a simple spectrum. If $\sigma^{(n)} = 0$ when $n > n_0 (n_0 < \infty)$, we say that U has a finite multiplicity spectrum. If there exists an n_0 such that $\sigma^{(n)} = 0$ for $n \neq n_0$, we say that U has a homogeneous spectrum of multiplicity n_0. In particular, for $n_0 = \infty$ we have a homogeneous countable multiplicity spectrum. If $\sigma^{(n_0)}$ is equivalent to the Lebesgue measure on S^1,

[1] The notation $\sigma' \preccurlyeq \sigma''$ for two measures σ', σ'' means that the measure σ' is absolutely continuous with respect to the measure σ''; the notation $\sigma' \approx \sigma''$ means that $\sigma' \preccurlyeq \sigma''$ and $\sigma'' \preccurlyeq \sigma'$.

then we speak of a homogeneous Lebesgue spectrum. In this case it is possible to choose a sequence $h_{n_0,k}$ so that the vectors

$$U^p h_{n_0,k}, \qquad -\infty < p < \infty, 1 \le k \le n_0 \quad \text{for } n_0 < \infty, \text{ and } 1 \le k < \infty \quad \text{for } n_0 = \infty$$

constitute an orthogonal basis in H. When $n_0 = \infty$ we have a countable multiplicity Lebesgue spectrum (countable Lebesque spectrum). The existence of a sequence of vectors $h_1, h_2, \ldots$ such that $U^p h_i$, $-\infty < p < \infty$, $i = 1, 2, \ldots$ constitutes an orthogonal basis in the space H is a necessary and sufficient condition for homogeneous Lebesgue spectra. The number of vectors in this sequence equals the multiplicity of the spectrum.

The sequence of spectral types of measures σ_{h_i} which appears in the first formulation of the main theorem constitutes a system of unitary invariants of the operator U. In a similar way, a system of unitary invariants is constituted by the maximal spectral type and the multiplicity function. Each of these systems is complete in the sense that two unitary operators U_1, U_2 possessing identical systems of unitary invariants are unitarily equivalent, i.e., there is a unitary operator V such that $U_2 = V^{-1}U_1V$.

Now we give another important corollary from the main theorem on the canonical form of unitary operator. The statement of this corollary uses the notion of continuous direct sum of Hilbert spaces (see Fomin and Naimark [1]).

Suppose U is a unitary operator in the Hilbert space H. The space H may be decomposed into a direct continuous sum

$$H = \int \bigoplus H_\lambda \, d\sigma(\lambda)$$

with respect to the measure σ of maximal spectral type, so that

(1) *the multiplicity function satisfies* $n(\lambda) = \dim H_\lambda$;
(2) *any cyclic (with respect to U) subspace H' is of the form $H' = \int \oplus H'_\lambda \, d\sigma(\lambda)$ where H'_λ is a subspace of the space H_λ and* $\dim H'_\lambda$ *equals* 0 *or* 1 *for almost all λ with respect to the measure σ.*

2. The notions described above can be carried over directly to the case of continuous time, i.e., to one-parameter groups of unitary operators.

If $\{U^t\}$, $-\infty < t < \infty$ is a continuous one-parameter group of unitary operators in a separable Hilbert space H, then the set $\Lambda_d(\{U^t\})$ consists of all $\lambda \in \mathbb{R}^1$ for which we can find a vector $h_\lambda \in H$ satisfying $U^t h_\lambda = e^{i\lambda t} h_\lambda$. The smallest closed subspace containing all the h_λ is denoted by

$$H_d(\{U^t\}); \qquad H_c(\{U^t\}) \stackrel{\text{def}}{=} H \ominus H_d(\{U^t\}).$$

The cyclic subspace corresponding to the vector $h \in H$ is the closure of the set of all possible finite linear combinations $\sum c_s U^{t_s}h$. As before, this subspace is denoted by $C(h)$. The spectral measure σ_h is defined by means of the relation

$$(U^t h, h) = \int_{-\infty}^{\infty} e^{it\lambda}\, d\sigma_h(\lambda), \qquad -\infty < t < \infty.$$

The measure σ_h in this case is defined on the Borel σ-algebra $\mathfrak{S}$ of the line $\mathbb{R}^1$.

A result similar to Theorem 1 is valid; in it appears the isomorphism between the subspace $C(h)$ and the Hilbert space $L^2(\mathbb{R}^1, \mathfrak{S}, \sigma_h)$ which transforms the action U^t into multiplication by $e^{it\lambda}$. The main theorem on the canonical form of a unitary operator can be carried over to the case under consideration with obvious modifications.

3. We shall now state and prove von Neumann's important theorem giving conditions for a group of unitary operators $\{U^t\}$ to have a homogeneous Lebesgue spectrum.

Theorem 2. *The following conditions are equivalent*:

(1) *the group* $\{U^t\}$ *has homogeneous Lebesgue spectrum*;
(2) *there exists a subspace* $H^{(0)}$ *of the space* H *such that*

 (i) $U^t H^{(0)} \supset H^{(0)}$ *for* $t > 0$;
 (ii) *the minimal closed subspace containing all the* $U^t H^{(0)}$, $-\infty < t < \infty$ *coincides with* H;
 (iii) $\bigcap_{-\infty < t < \infty} U^t H^{(0)} = \{0\}$;

(3) *there exists a continuous one-parameter group* $\{V^t\}$ *of unitary operators related to the group* $\{U^t\}$ *by the following commutation relations*

$$U^t V^s = e^{-ist} V^s U^t, \qquad -\infty < s, t < \infty$$

(here e^{-ist} *denotes the operator of multiplication by* e^{-ist}*).*

Proof. 1. Let us prove that (1) $\Rightarrow$ (2). Consider the vectors $h_1, h_2, \ldots \in H$ of maximal spectral type whose cyclic subspaces $C(h_i)$ are two-by-two orthogonal and $\bigoplus_i C(h_i) = H$. Suppose $H^{(0)}$ is the subspace generated by the vectors of the form $U^t h_i$ for all $t \le 0$, $i = 1, 2, \ldots$. It may be checked directly that conditions (i), (ii), and (iii) hold in this case.

2. Let us prove the implication (2) $\Rightarrow$ (3). Suppose E_p is the orthogonal projection operator on the subspace $U^p H^{(0)}$, $-\infty < p < \infty$. Then $\{E_p\}$ is a spectral family of projection operators. It is easy to check that $E_p U^t = U^t E_{p-t}$ for all p, t.

Let us form the one-parameter group of unitary operators $\{V^s\}$, $V^s = \int e^{ips}\,dE_p$. For any element $h \in H$, we have

$$V^sU^th = \int e^{ips}\,dE_p(U^th) = \int e^{ips}\,d(E_pU^th)$$

$$= \int e^{ips}\,d(U^tE_{p-t}h) = e^{its}U^t\int e^{i(p-t)s}\,dE_{p-t}h = e^{its}U^tV^sh,$$

i.e., (3) is proved.

3. Let us prove (3) $\Rightarrow$ (1). Suppose $h \in H$ is a vector whose spectral measure σ_h with respect to the group $\{U^t\}$ is measure of maximal spectral type. Note that the spectral measure of the vector V^sh with respect to the group $\{U^t\}$ is obtained from the spectral measure σ_h by a translation by s. Hence all the translations of the measure σ_h are absolutely continuous with respect to σ_h. It is well known that this is possible only when σ_h is equivalent to the Lebesgue measure.

Suppose $n(\lambda)$ is the multiplicity function for the group $\{U^t\}$. Assume that $n(\lambda) \geq n_0$ on a set of positive measure of maximal spectral type. This means that the measure $\sigma^{(n_0)}$ is nonzero (for the group $\{U^t\}$). Take the vectors $h_{n_0,k}$, $k = 1, 2, \ldots, n_0$ and the corresponding cyclic subspaces $C(h_{n_0,k})$ with respect to the $\{U^t\}$ for which $\sigma_{h_{n_0,k}} = \sigma^{(n_0)}$. Suppose the measure $\sigma^{(n_0)}$ is concentrated on the set A_{n_0}. Then the cyclic subspaces

$$C(V^sh_{n_0,k}), \qquad k = 1, 2, \ldots, n_0,$$

are orthogonal and their spectral measures are obtained by a translation by s of the measure $\sigma^{(n_0)}$. This means that the measures $\sigma_{V^sh_{n_0,k}}$, $k = 1, 2, \ldots, n_0$ are concentrated on the set $A_{n_0} + s$, i.e., for any s we have $n(\lambda) \geq n_0$ for almost all $\lambda \in (A_{n_0} + s)$. Since s was arbitrary, we see that $n(\lambda) \geq n_0$ almost everywhere on $\mathbb{R}^1$. Therefore the function $n(\lambda)$ is a constant (mod 0). The theorem is proved. $\square$

Appendix 3

Proof of the Birkhoff–Khinchin Theorem

The statement of this theorem is given in §2, Chap. 1. We shall carry out the proof for endomorphisms only.

The proof is based on an important lemma, sometimes called maximal ergodic theorem. First we introduce some notations. Put

$$s_n(x) = s_n(x, f) = \sum_{k=0}^{n-1} f(T^k x), \qquad x \in M, n = 1, 2, \ldots, s_0(x) \equiv 0.$$

$$A = A(f) = \left\{x \in M: \sup_{n \geq 0} s_n(x) > 0\right\}.$$

Lemma 1.

$$\int_A f(x)\, d\mu \geq 0.$$

The proof of the lemma shall be given later, while now we shall use it to deduce the ergodic theorem.

For all rational $a, b, a < b$ denote

$$E_{a,b} = \left\{x \in M: \underline{\lim}\, \frac{1}{n} s_n(x) < a < b < \overline{\lim}\, \frac{1}{n} s_n(x)\right\}.$$

Obviously $E_{a,b} \in \mathfrak{S}$ and for the proof of the existence of the limit $\lim(1/n)s_n(x)$ it suffices to show that $\mu(E_{a,b}) = 0$ for all a, b. Fix a, b, put $E = E_{a,b}$, and consider the function

$$g(x) = \begin{cases} f(x) - b & \text{for } x \in E, \\ 0 & \text{for } x \notin E. \end{cases}$$

Applying Lemma 1 to this function, we get

$$\int_{A(g)} g(x)\, d\mu \geq 0, \tag{1}$$

where

$$A(g) = \left\{x \in M: \sup_{n \geq 1} \frac{1}{n} s_n(x; g) > 0\right\} = \left\{x \in M: \sup \frac{1}{n} s_n(x, f) > b\right\}.$$

Clearly $A(g) \supset E$. It follows from the invariance of the set E that $s_n(x; g) \equiv 0$ for $x \notin E$, i.e., $A(g) \subset E$. Therefore $A(g) = E$, we can rewrite (1) in the form

$$\int_E f(x)\, d\mu \geq b\mu(E). \tag{2}$$

In a similar way, consider the function

$$g'(x) = \begin{cases} a - f(x) & \text{for } x \in E, \\ 0 & \text{for } x \notin E; \end{cases}$$

we then get

$$\int_E f(x)\, d\mu \leq a\mu(E). \tag{3}$$

It follows from (2) and (3) that $\mu(E) = 0$. Thus the limit $\bar{f}(x) = \lim_{n\to\infty} (1/n)s_n(x; f)$ exists almost everywhere.

Since for any n we have

$$\int_M \frac{1}{n} s_n(x; f)\, d\mu \leq \frac{1}{n} \int \sum_{k=0}^{n-1} |f(T^k x)|\, d\mu = \int_M |f(x)|\, d\mu < \infty,$$

it follows that $\bar{f} \in L^1(M, \mathfrak{S}, \mu)$. It remains to prove the relation

$$\int_M \bar{f}\, d\mu = \int_M f\, d\mu.$$

Consider the set $C_{a,b} = \{x \in M: a < \bar{f}(x) < b\}$. As before, we can use Lemma 1 to obtain

$$a\mu(C_{a,b}) \leq \int_{C_{a,b}} f\, d\mu \leq b\mu(C_{a,b}).$$

Moreover, obviously

$$a\mu(C_{a,b}) \leq \int_{C_{a,b}} \bar{f}\, d\mu \leq b\mu(C_{a,b})$$

Hence

$$\left| \int_{C_{a,b}} \bar{f}\, d\mu - \int_{C_{a,b}} f\, d\mu \right| \le (b - a)\mu(C_{a,b}).$$

Now fix a natural number q, and consider all possible pairs a, b of the form

$$a = p/2^q,\ b = (p + 1)/2^q; \qquad p = 0, \pm 1, \pm 2, \dots$$

We may assume without loss of generality that

$$\mu(\{x: f(x) = p/2^q\}) = \mu(\{x: \bar{f}(x) = p/2^q\}) = 0.$$

Then

$$\left| \int_M \bar{f}\, d\mu - \int_M f\, d\mu \right| \le \sum_{p=-\infty}^{\infty} \left| \int_{C_{p/2^q,(p+1)/2^q}} \bar{f}\, d\mu - \int_{C_{p/2^q,(p+1)/2^q}} f\, d\mu \right|$$

$$\le \sum_{p=-\infty}^{\infty} \frac{1}{2^q} \mu(C_{p/2^q,(p+1)/2^q}) = \frac{1}{2^q}.$$

When $q \to \infty$, we get $\int_M \bar{f}\, d\mu = \int_M f\, d\mu$. The theorem is proved. □

Proof of Lemma 1. For any $k \ge 0$,

$$s_k(Tx) = s_{k+1}(x) - f(x).$$

Fix $n \ge 1$ and take the maximum over $k = 0, 1, \dots, n - 1$ of both sides of this equality. First, for the left-hand side we can write

$$\begin{aligned}\max_{0 \le k \le n-1} s_k(Tx) &= \max_{0 \le k \le n-1} (0, f(Tx), \dots, f(Tx) + f(T^2x) + \cdots + f(T^n x))\\ &= \max_{0 \le k \le n-1} (f(x), f(x) + f(Tx), \dots, f(x) + f(Tx) + \cdots\\ &\quad + f(T^n x)) - f(x).\end{aligned}$$

For the right-hand side write

$$\begin{aligned}\max_{0 \le k \le n-1} (s_{k+1}(x) - f(x)) &= \max_{0 \le k \le n-1} (f(x), f(x) + f(Tx), \dots, f(x) + \cdots\\ &\quad + f(T^n x)) - f(x)\\ &= \max_{0 \le k \le n-1} (0, f(Tx), \dots, f(Tx) + \cdots + f(T^n x)).\end{aligned}$$

Introducing the notations

$$\Phi_n(x) = \max(0; s_1(x), \ldots, s_n(x)),$$
$$\Phi_n^*(x) = \max(s_1(x), \ldots, s_n(x)),$$

we now get from (4)

$$\Phi_{n+1}^*(x) - f(x) = \Phi_n(Tx).$$

Hence, $f(x) = \Phi_{n+1}^*(x) - \Phi_n(Tx) \geq \Phi_n^*(x) - \Phi_n(Tx)$. Suppose $A_n = \{x \in M : \Phi_n(x) > 0\}$. Then

$$\int_{A_n} f(x)\, d\mu \geq \int_{A_n} \Phi_n^*(x)\, d\mu - \int_{A_n} \Phi_n(Tx)\, d\mu.$$

But for $x \in A_n$ we have the relation $\Phi_n^*(x) = \Phi_n(x)$ and for $x \notin A_n$ the relation $\Phi_n(x) = 0$. Therefore

$$\int_{A_n} \Phi_n^*(x)\, d\mu = \int_{A_n} \Phi_n(x)\, d\mu = \int_M \Phi_n(x)\, d\mu.$$

Moreover, since Φ_n is non-negative,

$$\int_{A_n} \Phi_n(Tx)\, d\mu \leq \int_M \Phi_n(Tx)\, d\mu.$$

Finally we obtain

$$\int_{A_n} f(x)\, d\mu \geq \int_M \Phi_n(x)\, d\mu - \int_M \Phi_n(Tx)\, d\mu = 0.$$

The last equality follows from the fact that T preserves the measure μ. Now let n tend to ∞. The set A_n will obviously tend to A in the sense that $\mu(A_n \triangle A) \to 0$. Hence when $n \to \infty$ we shall obtain the required inequality. The lemma is proved. □

Appendix 4

Kronecker Sets

Suppose $S^1 = \{z: |z| = 1\}$ is the unit circle S^1 viewed as a commutative compact group with respect to multiplication. To each point $z = e^{i\lambda} \in S^1$ assign its argument λ belonging to the semi-interval $[-\pi, \pi)$. The continuous characters of the group S^1 are of the form

$$\chi_n(z) = \chi_n(e^{i\lambda}) = e^{in\lambda}, \qquad -\infty < n < \infty$$

where n is an integer. The character group X of the group S^1 is obviously isomorphic to the additive group of integers $\mathbb{Z}$.

Definition. *A subset $K \subset S^1$ is said to be a Kronecker set* if for any continuous map $\varphi: K \to S^1$ there is a sequence $\{n_r\}$, $r = 1, 2, \ldots$ of integers such that

$$\sup_{z \in K} |\varphi(z) - \chi_{n_r}(z)| \to 0 \quad \text{for } r \to \infty.$$

The definition above may be generalized in a natural way to arbitrary compact commutative groups.

We shall now give an example of a perfect Kronecker set on the circle S^1. The construction of the example will be split up into several steps.

1. Suppose $E \subset S^1$, $E = \{z_1, \ldots, z_m\}$, where $z_j = e^{i\alpha_j}$, $1 \leq j \leq m$ is a finite set without rational relations, i.e., a set such that the equality $\sum_{j=1}^{m} r_j \cdot \alpha_j = 0 (\text{mod } 2\pi)$, where the r_j are rational, is possible only when $r_1 = r_2 = \cdots = r_m = 0$. Let us prove that E is a Kronecker set.

Suppose the map $\varphi: E \to S^1$ is given by the formula $\varphi(z_j) = w_j$, where $w_j = e^{i\beta_j}$, $1 \leq j \leq m$. Consider the translation automorphism T of the m-dimensional torus $\text{Tor}^m = S^1 \times \cdots S^1$ which, in the cyclic coordinates $(\lambda_1, \ldots, \lambda_m)$, acts according to the formula

$$T(\lambda_1, \ldots, \lambda_m) = (\lambda_1 + \alpha_1 (\text{mod } 2\pi), \ldots, \lambda_m + \alpha_m (\text{mod } 2\pi))$$

By Theorem 1, §1, Chap. 4, the automorphism T is minimal. Hence, in particular, there is a sequence $\{n_r\}$ of integers satisfying $T^{n_r}x = y$, where x is the point with coordinates $\lambda_1 = \cdots = \lambda_m = 0$, y is the point with coordinates $\lambda_1 = \beta_1, \ldots, \lambda_m = \beta_m$. This means that the relations

$$\lim_{r\to\infty} e^{in_r\alpha_j} = \varphi(e^{i\alpha_j}), \qquad j = 1, \ldots, m$$

hold, i.e., E is a Kronecker set.

2. For any $r = 0, 1, \ldots$ let us construct:

(a) a family of open sets $K_j^{(r)} \subset S^1$, $1 \le j \le 2^r$ such that

$$\operatorname{diam} K_j^{(r)} < \frac{1}{r}, \qquad \overline{K}_i^{(r)} \cap \overline{K}_j^{(r)} = \varnothing \quad \text{for } i \neq j;$$

(b) a family of points $z_j^{(r)} = e^{i\lambda_j^{(r)}} \in K_j^{(r)}$, $1 \le j \le 2^r$;
(c) a finite set $F^{(r)} \subset X$ of characters of the group S^1 satisfying the following conditions:

(1) for any sequence of points $(z_1, \ldots, z_{2^r})$, $z_j \in S^1$ there exists a character $e^{in_r\lambda} \in F^{(r)}$ such that

$$|\exp[in_r\lambda_j^{(r)}] - z_j| < 1/r, \qquad 1 \le j \le 2^r;$$

(2) for any character $\chi \in F^{(r)}$ we have the inequality $|\chi(z) - \chi(z_j^{(r)})| < 1/r$ for all $z \in \overline{K}_j^{(r)}$, $1 \le j \le 2^r$.

This construction will be carried out by induction. For $r = 0$ choose for $K_1^{(0)}$ any arc of the circle and pick a point $z_1^{(0)} = e^{i\lambda_1^{(0)}}$ on it so that $\lambda_1^{(0)}$ is incommeasurable with π. From compactness, there is a finite set $F^{(0)} \subset X$ for which condition (1) will hold for $r = 0$. Condition (2) is trivial when $r = 0$.

Suppose the construction has been carried out for all numbers less than r; let us carry it through for r. Suppose W_{2j-1}, W_{2j} are disjoint open subsets of $K_j^{(r-1)}$, $1 \le j \le 2^{r-1}$. It follows from step 1 that there exists a sequence of points $z_j^{(r)} \in W_j$, $1 \le j \le 2r$ which forms a Kronecker set. It follows from compactness that there exists a finite set $F^{(r)} \subset X$ for which (1) holds. For $K_j^{(r)}$ take such a small neighborhood of the point $z_j^{(r)}$ that (2) holds for it. Thus the construction is carried out for all r.

3. Put $K = \bigcap_{r=0}^{\infty} \bigcup_{j=1}^{2^r} \overline{K}_j^{(r)}$. Clearly K is a perfect set. Suppose $\varphi: K \to S^1$ is a continuous map. For sufficiently large r_0 we can extend the map φ to the set

$$K^{(r_0)} = \bigcup_{j=1}^{2^{r_0}} \overline{K}_j^{(r_0)},$$

so that the extended map is still continuous. In particular, the value $\varphi(z_j^{(r)})$ will be defined for $r \geq r_0$, $1 \leq j \leq 2^r$.

Take an $\varepsilon > 0$ and choose an r, $r \geq r_0$, $r > 3/\varepsilon$ such that

$$|\varphi(z) - \varphi(z_j^{(r)})| < \varepsilon/3,$$

for $z \in \overline{K}_j^{(r)}$, $1 \leq j \leq 2^r$. Find a character $\chi = e^{in_r\lambda} \in F^{(r)}$ satisfying

$$|\chi(z_j^{(r)}) - \varphi(z_j^{(r)})| < \varepsilon/3, \qquad 1 \leq j \leq 2^r.$$

Since $|\chi(z) - \chi(z_j^{(r)})| < 1/r$ for $z \in \overline{K}_j^{(r)}$, $1 \leq j \leq 2^r$, we have $|\varphi(z) - \chi(z)| < \varepsilon$ for all $z \in \overline{K}_j^{(r)}$, $1 \leq j \leq 2^r$, and therefore for all $z \in K$. Since ε was arbitrary, this means that K is a Kronecker set.

Bibliographical Notes

Chapter 1

One of the first books on ergodic theory was Hopf's monograph [1]. Let us also mention von Neumann's fundamental paper [1] and its continuation, the article by Halmos and von Neumann [1]. An important review article on ergodic theory is due to Rohlin [5]. Other monographs of general character on ergodic theory known to us are the following: Halmos [4], Jacobs [1], Billingsley [1]; and the later books by Arnold and Avez [1], Friedman [1], Brown [1], Sinai [9] and [12], and Walters [1]. Among the more specialized books note Parry [1], Smorodinsky [1], Ornstein [2], and Shields [1]. Further we give a list of references more or less related to the text. This list makes no claim to completeness. An important bibliography of works in ergodic theory may be found in the review article by Vershik and Yuzvinsky [1] and Katok, Sinai, and Stepin [1].

Measure spaces, considered in this book are Lebesgue spaces. In the form generally accepted now the theory of such spaces was constructed in Rohlin's article [3]. A different approach is in von Neumann's paper [1]. Concerning the relationship between the notions of measurable flows and continuous flows, see Rohlin [5], Vershik [3], and Maruyama [2]. The first proofs of the Birkhoff–Khinchin ergodic theorem were given in the works of Birkhoff [2], Khinchin [2], and Kolmogorov [1]. The shortest proof of this theorem given in Appendix 3 is due to Garsia [1].

The relationship between invariant functions and functions invariant (mod 0) was first discussed in detail in von Neumann's work [1]. Concerning the general theory of the decomposition of arbitrary dynamical systems into ergodic components, see von Neumann [1] and Rohlin [4]. Induced automorphisms first appeared in the work of Kakutani [1]. Lemma 1, §5, is due to Kac [1]. The notion of mixing goes back to Gibbs [1]. Concerning weak mixing, see Hopf [1]. The definition of multiple mixing was given by Rohlin [6]. K-mixing is intimately related to the notion of K-system, introduced by Kolmogorov in the paper [3], where such systems were called quasi-regular.

The group of unitary operators adjoint to a dynamical system was introduced in Koopman's work [1]. For the Wiener lemma, see Wiener [1].

For the von Neumann ergodic theorem, see von Neumann [2]. Theorem 1, §8, is the classical Bogoluboff–Kriloff theorem from [1].

Chapter 2

For the proof of Liouville's theorem see, for example, Arnold's book [3]. A series of problems relating to the theory of Hamiltonian systems is discussed in Moser's book [1]. The generalizations of geodesic flows described in Subsection 5, §2, were considered in Arnold's article [2]. The reduction of Hamiltonian systems to geodesic flows is carried out just as in Anosov's and Sinai's article [1]. For the Liouville theorem on integrable Hamiltonian systems, see Arnold's book [3]. For numerous examples of integrable geodesic flows on Riemann surfaces, see Kagan's book [1].

Systems of point vortices were the object of numerous studies. The system consisting of three vortices was studied in detail in Novikov's work [1].

Recently, interesting results for systems consisting of four vortices were obtained by Ziglin and for systems of an arbitrary number of vortices by Khanin.

The (L, A)-pair method, or the method of integration of dynamical systems by using the inverse problem of scattering theory, is an extremely important and rapidly developing direction in the theory of dynamical systems. This method was first applied for finding the solutions of the Korteveg–de Vries equation in the work of Kruskal, *et al.* (Gardner, Green, Kruskal, and Miura [1]). The (L, A)-pair method proper was introduced by Lax [1]. Important results here are due to Zakharov, Calogero, Manakov, Novikov, Faddeev, Shabat and others. The study of Toda lattice by this method was carried out by Manakov [1] and Flashka [1] and [2]. Integrable systems of one-dimensional particles with pairwise interaction were discovered by Moser [2] and Calogero [1].

Chapter 3

Translations on the torus were considered in Weyl's work [1] in connection with the problem of uniform distribution of fractional parts of various functions. Concerning the Lagrange problem, see Weyl [2], Jessen and Tornehave [1]. The notion of rotation number for homeomorphisms of the circle was introduced by Poincaré. In the works of Poincaré [1] and Denjoy [1], the topological classification of diffeomorphisms of the circle was constructed.

The example of a homeomorphism with a nowhere dense derived set is due to Denjoy [1].

For the Denjoy theorem, see Denjoy [1]. Our exposition of this theorem follows along the lines of Herman's work [1]. The example in §5 is due to Arnold [1]. Theorem 1 in §6 is due to Herman [1] and Katok whose proof is given in the text.

Chapter 4

Lemma 1, §1, is due to Stepin. Translations on commutative groups were introduced by von Neumann [1] who found conditions for their ergodicity. Theorem 1, §2, on unique ergodicity of skew translations on the torus is due to Fürstenberg [1]. Endomorphisms and automorphisms of commutative compact groups were first considered from the ergodic point of view in the works of Halmos [1] and Rohlin [6]. The theorem on the uniform distribution of periodic trajectories is one of the versions of the general theorem on the uniform distribution of periodic trajectories for the so-called hyperbolic dynamical systems, see Bowen [1]. Hopf's geometric proof of ergodicity can be found in his article [2] dealing with geodesic flows on manifolds of negative curvature.

The general construction of dynamical systems on homogeneous spaces of Lie groups was introduced in Fomin's and Gelfand's article [1], where, by using methods of the theory of infinite-dimensional representations of Lie groups, the spectrum of geodesic flows on surfaces of constant negative curvature was found. In the same paper the relationship between such geodesic flows and one-parameter subgroups of the group $SL(2, \mathbb{R})$ was discovered. Our study of the ergodicity and mixing of geodesic flows essentially follows Hopf's work [2], see also Mautner [1], and Auslander, Green, and Hahn [1].

Chapter 5

Theorem 1, §1, is due to Keane [1] and Zemlyakov. The results of §2 are due to Oseledets. Theorem 1, §3, is due to Katok, see Katok, Sinai, and Stepin [1], Chap. 4. The example in §4 is Satayev's [1]. A similar example was constructed by Keynes and Newton [1] and Keane [2]. The first example of minimal but not uniquely ergodic dynamical system was found by Markov and is described in Nemytsky's and Stepanov's book [1]. The study of the ergodicity of a certain class of interval exchange transformation is the subject of works by Keane and Rosy, as well as Chulayevsky [1].

Chapter 6

Systems of the billiard type were already studied by Hadamard [1] and Birkhoff [1]. The accurate construction of billiard systems given in §1 is, as far as we know, the first such construction. Theorem 2, §2, is due to Boldrigini, Keane, and Marchetti [1]. Concerning billiards in an ellipse see Birkhoff's book [1] where a reference to Leibnitz appears. Theorem 2, §3, is due to Birkhoff [1]. The existence of a set of positive measure consisting of caustics in the phase space of billiards in a convex domain was proved in Lazutkin's paper [1]. Lorentz gases were introduced in the work of Lorentz [1] and since then have been intensively studied in connection with many problems of statistical mechanics. Let us mention Hauge's interesting review article [1] on this topic. The reduction of systems of hard spheres to systems of the billiards type first appeared, as far as we know, in Krylov's book [1].

Chapter 7

The results of §§1–3 are due to Weyl [1]. The original proof of Theorem 1, §2, due to Weyl was based on certain estimates of trigonometric sums. The "ergodic proof" which we provide was proposed by Fürstenberg [1]. Certain applications of ergodic theory to number theory are given in Postnikov's work [1].

Concerning the decomposition of real numbers into continuous fractions and metrical properties of continued fractions, see Khinchin's book [1]. Theorem 3, §4, was essentially proved for the first time in Kuzmin's work [1] where an estimate of the speed of mixing for Gauss transformation was also obtained. The derivation of various properties of continued fractions by using the ergodic theorem can be found in Billingsley's book [1]. A wider class of decompositions of real numbers than the decomposition into continued fractions was considered by Renyi [1].

Ergodic properties of piecewise monotonic transformations of the interval were studied by many authors. The theorem on the existence of an absolutely continuous invariant measure for such transformations was proved by Kosyakin and Sandler [1] and Lasota and Yorke [1]. Our proof of Theorem 1, §4, follows along the lines of Adler's work [1].

Chapter 8

As far as we know, the relationship between ergodic theory and the theory of stationary random processes of probability theory was first pointed out by Kolmogorov and repeatedly accented in the 1930's. The proof of the ergodicity of Bernoulli automorphisms and Markov automorphisms is essentially a variant of Kolmogorov's famous "0–1 law," which he established for sequences of independent random variables in the book [5]. The actual terms "Bernoulli automorphism" and "Markov automorphism" first appeared when the entropy theory of dynamical systems was developed, see Part II and Rohlin's review [8]. The direct relationships between the theory of stationary random processes and ergodic theory is based on the notion of generating partition. Rohlin [8] showed that every aperiodic automorphism has a countable generating partition, i.e., is isomorphic (mod 0) to a stationary random process with a countable number of states. Krieger [1] showed that automorphisms with finite entropy possess finite generating partitions.

The isomorphism between group automorphisms of the torus and Markov automorphisms were established in Adler's and Weiss's work [1]. The notion of Markov partition for the so-called hyperbolic systems was introduced in the work of Sinai [7] and [8], and Bowen [2].

Ergodicity and mixing of Gauss dynamical systems were first studied by Fomin [1] and Maruyama [1].

Chapter 9

Dynamical systems corresponding to the motion of an infinite number of noninteracting particles were studied in probability theory, where a more general case was considered, i.e., the case when the particles, besides mechanical motion, are involved in purely random motion (in this connection see Doob's book [1] and the articles by Harris [1] and Dobrushin [1]). A more delicate study of the ergodic properties of an ideal gas in $\mathbb{R}^d$ is given in the article by Volkovyssky and Sinai [1].

The general notion of a Poisson suspension over an arbitrary dynamical system is due to Aisenman, Goldstein, and Leibowitz [1].

The construction of infinite dimensional dynamical systems appearing in statistical mechanics was considered by Lanford, III, O.E. [1], [2], Sinai [10], [11], and Pressutti, Pulvirenti, and Tirozzi [1].

Chapter 10

Skew products of dynamical systems were introduced by Anzai [1]. The notion of equivalence of dynamical systems according to Kakutani [1] are intimately connected with induced and integral automorphisms. Recently, interest in these notions has grown again, see Weiss's review article [1] and the papers by Feldman [1] and Katok [3].

Properties of dynamical systems under change of time were studied by Grabar [1] and [2], Chacon [1], Friedman and Ornstein [1], Kochergin [1], and other authors.

Natural extensions of endomorphisms were introduced by Rohlin [7]. Theorem 1, §5, is due to Rohlin [5] and Halmos [3]. Concerning the sharpened form of the Rohlin–Halmos lemma, see the work of Thouvenot [1].

The notion of the entropy of a dynamical system first appeared in the works of Kolmogorov [3] and [4], which initiated the entropy theory of dynamical systems. The variant of the definition of entropy given in this book appears in Sinai's work [1]. The general entropy theory of dynamical systems is developed in Rohlin's review article [8] and the monographs by Billingsley [1], Sinai [9], and Smorodinsky [1]. Statements (3) and (4) in Theorem 2, §6, are due to Abramov [1]. A simple proof of these statements is given in Brown's book [1]. Theorem 3, §6, is also due to Abramov [2]. The proof given in the text was proposed by Pinsker. Theorem 4 in §6 is a very particular case of the general theorem, see Shannon [1], McMillan [1], and Breiman [1].

Theorem 1, §7, is due to Ornstein [1]. This important theorem gives a final solution of the isomorphism problem for Bernoulli automorphisms. The first nontrivial example of the metric isomorphism of Bernoulli automorphisms was constructed by Meshalkin [1]. Other examples are given in the works of Blum and Hanson [1] and Livshits [1]. Sinai in [5] obtained the result which means, in particular, that any two Bernoulli automorphisms with the same entropy are weakly isomorphic, i.e., each of them is metrically isomorphic to some factor automorphism of the other.

Ornstein constructed a profound and delicate theory enabling him to study the isomorphism problem for a still wider class of dynamical systems. His theory together with its applications is developed in the monograph [2].

The proof of Theorem 1, §7, given in the text is due to Keane and Smorodinsky [1]. The Keane–Smorodinsky construction not only gives a new proof of the Ornstein theorem but also yields the construction of a finitary isomorphism of Bernoulli automorphisms with the same entropy.

K-systems were introduced by Kolmogorov in his paper [3], where they were called quasi-regular. In the same paper some of their properties were indicated. A series of results of the same type can be found in Rohlin's works [7] and [8]. For Example 3, §8, see Volkovyssky and Sinai [1]. Theorem 1, §8, is a particular case of more general results due to Rohlin and Sinai [1]; see also Rohlin's review article [8]. Theorem 2, §8, was proved by Rudolph [1] for flows of finite entropy. The proof in the general case is contained in the works of Blanshard [1] and Gurevich [1].

Exact endomorphisms were introduced by Rohlin [7]. Theorem 4, §8, is a generalization of Rohlin's results, see also Adler [1].

Chapter 11

The first proof of the theorem on special representations of flows can be found in the works of Ambrose [1], and Ambrose and Kakutani [1]. Another proof, based on the theory of measurable partitions of Lebesgue spaces, is contained in Rohlin's review article [5]. In Example 3, §2, we deduce, by using the theory of special representations the well-known Rice formula from the theory of Gauss stationary processes, proved in its most general form by Bulinskaya [1]. Theorem 1, §4, in somewhat weaker form was proved by Rudolph [1]. The proof given in the text follows along the lines of Krengel's paper [1].

Chapter 12

The theory of dynamical systems with pure point spectrum was constructed by von Neumann [1]. An exposition of this theory is given in Halmos's book [4] and in Rohlin's review [5].

Chapter 13

The theorem on the countable Lebesgue spectrum of *K*-automorphisms is due to Kolmogorov [3]. Concerning Theorem 1, §2, see Halmos's book [4]. Rohlin [9] proved that every ergodic automorphism of a commutative compact group is a K-automorphism. Theorem 2, §3, was proved by Kushnirenko [1]. The results of §4 belong to Oseledets [1], those of §5 are due to Sinai [2]. The spectral properties of general dynamical systems were studied by Alekseev [1].

Chapter 14

The decomposition of Hilbert space into Hermite–Ito polynomial subspaces was constructed in the works of Ito [1] and [2]. In the same papers he carried out the spectral analysis of Gauss dynamical systems in several cases. In Fomin's paper [1] the properties of spectral types were studied. Theorem 1, §4, was repeatedly stated as a conjecture by Kolmogorov and was proved by Girsanov [1]. In this paper there is also a series of examples of Gauss dynamical systems with various spectral properties. Theorem 2, §4, under an additional restriction, was proved in Sinai's paper [4]. The proof in the general case is due to Foias and Stratila [1]. The results of §5 are due to Girsanov [1]. In the works of Vershik [1] and [2], it is proved that all Gauss systems with countable Lebesgue spectrum are metrically isomorphic. Perfect sets without rational relations were first constructed by Von Neumann. Concerning Kronecker sets see Rudin's book [1].

Chapter 15

The approximation of automorphisms of measure spaces by periodic automorphisms first appears in the works of Halmos [3] and Rohlin [2]. The general definitions of approximations given in §1 are due to Katok [1] and Stepin [1]. These definitions were somewhat generalized and improved by Schwartzbauer [1]. Another approach to the study of approximations of automorphisms by periodic ones was proposed by Vershik [4]. Problems of approximation theory are developed in the review article by Katok and Stepin [1]. The results of §2 and §3 are due to Stepin [2]. Theorem 2, §4, is essentially Oseledets' theorem carried over to the case of flows, see Katok [2].

Chapter 16

Concerning Theorem 1, §1, see Kolmogorov's paper [2]. Lemma 2, §1, is due to Siegel [1]. Our exposition in §1 is close to that in Sternberg's book [1]. The results of §2 are due to Kolmogorov [2]. Our exposition is close to the one in Sinai's book [9]. Theorem 1, §3, is due to Katok [2]. The existence of smooth flows with continuous spectrum on the two-dimensional torus was mentioned by Kolmogorov [2]. The construction presented in §4 is due to Shklover [1].

Bibliography

Abramov, L. M.
[1] Entropy of an induced automorphism. Dokl. Acad. Sci. USSR **128**, no. 4, 647–650 (1959).
[2] On the entropy of flows. Dokl. Acad. Sci. USSR **128**, no. 5, 873–876 (1959).

Adler, R.
[1] *F*-expansions revisited. Lecture Notes in Mathematics. No. 318, 1–5. New York–Berlin–Heidelberg: Springer-Verlag (1973).

Adler, R. and Weiss, B.
[1] Entropy a complete metric invariant for automorphisms of the torus. Proc. Nat. Acad. Sci. USA **57**, no. 6, 1573–1576 (1967).

Aisenman, M., Goldstein, S., and Lebowitz, J.
[1] Ergodic properties of infinite systems. Lecture Notes in Physics. Vol. 38, 112–143 (1975).

Alekseev, V. M.
[1] The existence of a bounded function of maximal spectral type. Vestnik Moscow Univ. no. 5, 13–15 (1958).

Ambrose, W.
[1] Representation of ergodic flows. Ann. Math. **42**, 723–739 (1941).

Ambrose, W. and Kakutani, S.
[1] Structure and continuity of measurable flows. Duke Math. J. **9**, 25–42 (1942).

Anosov, D. V. and Sinai, Ya. G.
[1] Certain smooth ergodic systems. Uspehi Mat. Nauk **22**, no. 5, 107–172 (1967).

Anzai, H.
[1] Ergodic skew product transformations on the torus. Osaka Math. J. **3**, no. 1, 83–99 (1951).

Arnold, V. I.
[1] Small denominators. I. On maps of the circle onto itself. Izv. Acad. Sci. USSR Ser. Mat. **25**, 1, 21–86 (1961).
[2] Some remarks on flows of linear elements and frames. Dokl. Acad. Sci. USSR **138**, no. 2, 255–257 (1961).
[3] Mathematical Methods in Classical Mechanics. Moscow; Nauka 1974. New York–Berlin–Heidelberg: Springer-Verlag 1978.

Arnold, V. I. and Avez, A.
[1] Problèmes Ergodiques de la Mecanique Classique. Paris: Gauthier-Villars 1967.

Auslander, L., Green, L., and Hahn, F.
[1] Flows on Homogeneous Spaces. Princeton, N.J.: Princeton University Press 1963.

Billingsley, P.
[1] Ergodic Theory and Information. New York: Wiley 1965.

Birkhoff, G. D.
[1] Dynamical Systems. New York: 1927.
[2] Proof of the ergodic theorem. Proc. Nat. Acad. Sci. USA **17**, 656–660 (1931).

Blanchard, F.
[1] Partition extrêmales de flots d'entropie infinie. Z. Wahrscheinlichkeitstheorie, **36**, no. 2, 129–136 (1976).

Blum, J. and Hanson, D.
[1] On the isomorphism problem for Bernoulli schemes. Bull. Amer. Math. Soc. **63**, 221–223 (1963).

Bogoluboff, N. N., and Kriloff, N. M.
[1] La théorie générale de la mesure dans son application à l'étude des systèmes dynamiques de la mécanique non-linéaire. Ann. Math. **38**, 65–113 (1937).

Boldrigini, C., Keane, M., and Marchetti, F.
[1] Billiards in polygons. Ann. of Probab. **6**, no. 4, 532–540 (1978).

Bowen, R.
[1] Periodic points and measures for axiom A diffeomorphisms. Trans. Amer. Math. Soc. **154**, 377–397 (1971).
[2] Markov partitions for axiom A diffeomorphisms. Amer. J. Math., **92**, no. 3, 725–747 (1970).

Breiman, L.
[1] The individual ergodic theorem of information theory. Ann. Math. Stat., **28**, 809–811 (1957). (Correction: Ann. Math. Stat., **31**, 809–810 (1960).

Brown, J. R.
[1] Ergodic Theory and Topological Dynamics. New York: Academic Press 1976.

Bulinskaya, E. V.
[1] On the mean number of intersections of levels by stationary Gauss processes. Probab. Theory Appl. **6**, no. 4, 474–478 (1961).

Calogero, F.
[1] Exactly solvable one-dimensional many-body problem. Roma 1975 (Preprint).

Chacon, R.
[1] Change of velocity in flows. J. Math. Mech. **16**, no. 5, 417–431 (1966).

Chulaevski, V. A.
[1] Cyclic approximations of interval exchange transformations. Uspehi Mat. Nauk, **34**, no. 2, 215–216 (1979).

Denjoy, A.
[1] Sur les courbes défines par les équations différentielles à la surface du tore. J. Math. Pures et Appl. **11**, 333–375 (1932).

Dobrushin, R. L.
[1] On the Poisson law for the distribution of particles in space. Ukr. Mat. J. **8**, no. 2, 127–134 (1956).
[2] Description of a random field by means of conditional probabilities and its regularity conditions. Probab. Theory Appl. **13**, no. 4, 44–57 (1968).

Doob, J.
[1] Stochastic Processes. New York: Wiley 1953.

Feldman, J.
[1] New K-automorphisms and a problem of Kakutani. Israel J. Math. **24**, N1, 16–38 (1976).

Flashka, H.
[1] Toda lattice, I. Phys. Rev. B9, 1924–1925 (1974).
[2] Toda lattice, II. Progr. Theor. Phys. **51**, 703–716 (1974).

Foias, C. and Stratila, S.
[1] Ensembles de Kroneker dans la théorie érgodique. C. R. Acad. Sci. Paris **267**, 20, A166–A168 (1967).

Fomin, S. V.
[1] Normal dynamical systems. Ukr. Mat. J. **2**, no. 2, 25–47 (1950).

Fomin, S. V. and Gelfand, I. M.
[1] Geodesic flows on manifolds of constant negative curvature. Uspehi Mat. Nauk **7**, no. 1, 118–137 (1952).

Fomin, S. V. and Naimark, M. A.
[1] Continuous direct sums of Hilbert spaces. Uspehi Mat. Nauk **10**, no. 2, 111–142 (1955).

Friedman, N.
[1] Introduction to Ergodic Theory. Princeton, N.J.: Van Nostrand–Reinhold 1970.

Friedman, N. and Ornstein, D.
[1] Ergodic transformations induce mixing transformations. Advan. Math. **10**, no. 1, 147–163 (1973).

Fürstenberg, M.
[1] Strict ergodicity and transformation of the torus. Amer. J. Math. **83**, no. 4, 573–601 (1961).

Gardner, C., Green, J., Kruskal, M., and Miura, R.
[1] A method for solving the Korteveg–de Vries equation. Phys. Rev. Lett. **19**, 1095–1097 (1967).

Garsia, A. M.
[1] A simple proof of Eberhard Hopf's maximal ergodic theorem. J. Math. Mech. **14**, no. 3, 381–382 (1965).

Gibbs, J.
[1] Elementary Principles in Statistical Mechanics. New Haven: Yale University Press 1902.

Girsanov, I. V.
[1] On spectra of dynamical systems generated by Gauss stationary processes. Dokl. Acad. Sci. USSR **119**, no. 5, 851–853 (1958).

Grabar, M. I.
[1] On changes of time in dynamical systems. Dokl. Acad. Sci. USSR **169**, no. 2, 250–252 (1966).
[2] On changes of time in dynamical systems. Dokl. Acad. Sci. USSR **169**, no. 3, 431–433 (1966).

Gurevich, B. M.
[1] Perfect partitions for ergodic flows. Funct. Anal. Appl. **11**, no. 3, 20–23 (1977).

Hadamard, J.
[1] Non-Euclidean Geometry in the Theory of Automorphic Functions (Russian translation from French). Moscow: Gostekhizdat 1951.

Halmos, P.
[1] On automorphisms of compact groups. Bull. Amer. Math. Soc. **49**, 619–624 (1943).
[2] In general a measure-preserving transformation is not mixing. Ann. Math. **45**, 776–782 (1944).
[3] Approximation theories for measure-preserving transformations. Trans. Amer. Math. Soc. **55**, no. 1, 1–18 (1944).
[4] Lectures on Ergodic Theory. Tokyo: 1953.

Halmos, P. and Neumann J. von
[1] Operator methods in classical mechanics, II. Ann. Math. **43**, 332–350 (1942).

Harris, T.
[1] Diffusion with collisions between particles. J. Appl. Probab. **2**, no. 2, 323–338 (1965).

Hauge, E.
[1] What one can learn from Lorentz Models. Lecture Notes in Physics. **31**, 337–367 (1974).

Herman, M.
[1] Sur la conjugation différentiable des difféomorphismes du cercle à des rotations. Paris. Publ. Math. **49**, 5–233 (1979).

Hopf, E.
[1] Ergodentheorie, Berlin: 1937.
[2] Statistik der geodätischen Linien in Mannigfaltigkeiten negativer Krümmung. Leipzig Ber. Verhandl. Sächs. Akad. Wiss. **91**, 261–304 (1939).

Ito, K.
[1] Multiple Wiener integral. J. Math. Soc. Japan. **3**, 157–169 (1951).
[2] Complex multiple Wiener integral. Japan J. Math. **22**, 63–86 (1952 (1953)).

Jacobs, K.
[1] Neure Methoden and Ergebnisse der Ergodentheorie. Berlin–Göttingen–Heidelberg: Springer-Verlag, 1960.

Jessen, B. and Tornhave, H.
[1] Mean motions and almost periodic functions. Acta Math. **77**, 137–279 (1945).

Kac, M.
[1] On the notion of recurrence in discrete stochastic processes. Bull. Amer. Math. Soc. **53**, 1002–1010 (1947).

Kagan, V. F.
[1] Foundations of the Theory of Surfaces, Volume I. Moscow: Gostekhizdat 1947.

Kakutani, S.
[1] Induced measure-preserving transformations. Proc. Imp. Acad. Tokyo **19**, 635–641 (1943).

Katok, A. B.
[1] Entropy and approximations of dynamical systems by periodic transformations. Funct. Anal. Appl. **1**, no. 1, 75–85 (1967).
[2] Spectral properties of dynamical systems with an integral invariant on the torus. Funct. Anal. Appl. **1**, no. 4, 46–56 (1967).
[3] Change of time, monotonic equivalence and standard dynamical systems. Dokl. Acad. Sci. USSR **223**, no. 4, 789–792 (1975).

Katok, A. B., Sinai, Ya. G., and Stepin, A. M.
[1] Theory of dynamical systems and general transformation groups with invariant measure. In: Mathematical Analysis, Vol. 13, 129–262. Moscow: Itogi Nauki, Viniti Editions 1975.

Katok, A. B. and Stepin, A. M.
[1] Approximations in ergodic theory. Uspehi Mat. Nauk **22**, no. 5, 81–106 (1967).

Keane, M.
[1] Interval exchange transformations. Math. Z. **141**, 25–31 (1975).
[2] Non-ergodic interval exchange transformations. Israel J. Math. **26**, no. 2, 188–196 (1977).

Keane, M. and Smorodinsky, M.
[1] Bernoulli schemes of the same entropy are finitary isomorphic. Ann. Math. **109**, no. 2, 397–406 (1979).

Keynes, H., Newton, D.
[1] A "minimal" non-uniquely ergodic interval exchange transformation. Math. Z. 148, 101–105 (1976).

Khinchin, A. Ya.
[1] Continued Fractions. Moscow: Nauka 1978.
[2] Zur Birckhoffs Lösung des Ergodenproblems. Math. Ann. 485–488 (1932).

Kochergin, A. V.
[1] Change of time in flows and mixing. Izv. Acad. Sci. USSR, Ser. Mat. **37**, no. 6, 1275–1298 (1973).

Kolmogorov, A. N.
[1] A simplified proof of the Birkhoff–Khinchin theorem, Uspehi Mat. Nauk no. 5, 52–56 (1938).
[2] On dynamical systems with an integral invariant on the torus. Dokl. Acad. Sci. USSR **93**, no. 5, 763–766 (1953).
[3] A new metric invariant of transitive dynamical systems and Lebesgue space automorphisms. Dokl. Acad. Sci. USSR **119**, no. 5, 861–864 (1958).
[4] On entropy per unit time as a metric invariant of automorphisms. Dokl. Acad. Sci. USSR **124**, no. 4, 754–755 (1959).
[5] The foundations of probability theory. Moscow: Nauka 1974.

Koopman, B. O.
[1] Hamiltonian systems and transformations in Hilbert space. Proc. Nat. Acad. Sci. USA **17**, no. 5, 315–318 (1931).

Kosiakin, A. A. and Sandler, E. A.
[1] Ergodic properties of a certain class of piecewise smooth transformations of the segment. Izv. VUZov, Mat., no. 3, 32–40 (1972).

Krengel, U.
[1] On Rudolph's representation of aperiodic flows. Ann. Inst. Poincaré **B12**, no. 4, 319–338 (1976(1977)).

Krieger, W.
[1] On entropy and generators of measure-preserving transformations. Trans. Amer. Math. Soc. **119**, no. 2, 453–464 (1970).

Krylov, N. S.
[1] Works on the foundations of statistical physics. Acad. Sci. USSR Editions, 1950.

Kushnirenko, A. G.
[1] Spectral properties of certain dynamical systems with polynomial divergence. Vestnik Moscow Univ., Ser. Mat.-Mech., no. 1, 101–108 (1974).

Kuzmin, R. O.
[1] On a problem of Gauss. Dokl. Acad. Sci. USSR, Ser. A. 375–380 (1928).

Lanford, III. O. E.
[1] Classical mechanics of one-dimensional systems of infinitely many particles. Commun. Math. Phys. **9**, 169–181 (1969); **11**, 257–292 (1969).
[2] Time evolution of large classical systems. Lecture Notes in Physics. Vol. 38, 1975.

Lanford, III. O. E. and Ruelle, D.
[1] Observables at infinity, and states with short range correlations in statistical mechanics. Commun. Math. Phys. **13**, 194–215 (1969).

Lasota, A. and Yorke, J.
[1] On the existence of invariant measures for piecewise monotonic transformations. Trans. Amer. Math. Soc. **186**, 481–488 (1973).

Lax, P.
[1] Integrals of nonlinear equations of evolution and solitary waves. Commun. Pure Appl. Math. **21**, 467–490 (1968).

Lazutkin, V. F.
[1] Existence of caustics for the billiards problems in a convex domain. Izv. Acad. Sci. USSR, Ser. Mat. **37**, no. 186–216 (1973).

Livshits, A. N.
[1] On the isomorphism problem of Bernoulli schemes. Probab. Theory Appl. **19**, no. 2, 409–416 (1974).

Lorentz, H.
[1] The motion of electrons in metallic bodies. Proc. Amsterdam Acad. **7**, 438, 585, 604 (1905).

Manakov, S. V.
[1] On complete integrability and stochastisation in discrete dynamical systems. J. Exp. Theor. Phys. **67**, no. 2, 543–555 (1974).

Maruyama, G.
[1] The harmonic analysis of stationary stochastic processes. Mem. Fac. Sci. Kyusyu Univ., Ser. A. **4**, no. 1, 45–105 (1948).
[2] Transformations of flows. J. Math. Soc. Japan. 18, no. 3, 303–330 (1966).

Mautner, F.
[1] Geodesic flows on symmetric Riemann spaces. Ann. Math. 65, no. 3, 416–431 (1957).

McMillan, B.
[1] The basic theorems of information theory. Ann. Math. Stat. **24**, 196–219 (1953).

Meshalkin, L. D.
[1] A case of Bernoulli scheme isomorphism. Dokl. Acad. Sci. USSR, **128**, no. 1, 41–44 (1959).

Moser, Ju.
[1] Lectures on Hamiltonian Systems. New York: Courant Institute of Mathematical Science 1968.
[2] Three integrable Hamiltonian systems connected with isospectral deformations. Advan. Math. **16**, no. 2, 197–220 (1975).

Nemytski, V. V. and Stepanov, V. V.
[1] Qualitative Theory of Differential Equations. Moscow–Leningrad: Gostekhizdat 1947.

Neumann, J., von
[1] Zur Operatorenmethode in der Klassischen Mechanik. Ann. Math. **33**, 587–642 (1932).
[2] Proof of the quasi-ergodic hypothesis. Proc. Nat. Acad. Sci. USA, **18**, 70–82 (1932).

Novikov, E. A.
[1] Dynamics and statics of vortex systems, J. Exp. Theor. Phys. **68**, no. 5, 1868–1882 (1975).

Ornstein, D.
[1] Bernoulli shifts with the same entropy are isomorphic. Advan. Math. **4**, no. 3, 337–352 (1970).
[2] Ergodic Theory, Randomness, and Dynamical Systems. New Haven and London: Yale University Press 1974.

Oseledets, V. I.
[1] On the spectrum of ergodic automorphisms. Dokl. Acad. Sci. USSR, **168**, no. 5, 1009–1011 (1966).

Parry, W.
[1] Entropy and Generators in Ergodic Theory. New Haven and London: Yale University Press 1966.

Pinsker, M. S.
[1] Dynamical systems with completely positive and zero entropy. Dokl. Acad. Sci. USSR. **133**, no. 5, 1025–1026 (1960).

Poincaré, H.
[1] Mémoire sur les courbes définies par une équation différentielle. I, II, III, IV. J. Math. Pures et Appl. **7**, 3, 375–422 (1881); **8**, 3, 251–286 (1882); **1**, 4, 167–244 (1885); **2**, 4, 151–217 (1886).

Pontriagin, L. S.
[1] Continuous Groups. Moscow: Nauka 1973.

Postnikov, A. G.
[1] Ergodic questions of residue theory and the theory of Diophantine approximations. Trudy V. A. Steklov Mat. Inst. **82**, 3–111 (1966).

Pressutti, E., Pulvirenti, M., and Tirozzi, B.
[1] Time evolution of infinite classical systems with singular, long range, two body interactions. Commun. Math. Phys. **47**, 81–95 (1976).

Renyi, A.
[1] Representations of real numbers and their properties. Acta Math. Sci. Hung. **8**, 477–493 (1957).

Rohlin, V. A.
[1] Unitary rings. Dokl. Acad. Sci. USSR. **59**, no. 4, 643–646 (1948).
[2] The general measure-preserving transformation is mixing. Dokl. Acad. Sci. USSR. **60**, no. 3, 349–351 (1948).
[3] On the main notions of measure theory. Mat. Sb. **67**, no. 1, 107–150 (1949).
[4] On the decomposition of dynamical systems into transitive components. Mat. Sb. **67**, no. 2, 235–249 (1949).
[5] Selected problems in the metric theory of dynamical systems. Uspehi Mat. Nauk. **30**, no. 2, 57–128 (1949).
[6] On the endomorphisms of compact commutative groups. Izv. Acad. Sci. USSR, Ser. Mat. **13**, 323–340 (1949).
[7] Exact endomorphisms of Lebesgue spaces. Izv. Acad. Sci. USSR, Ser. Mat. **25**, 499–530 (1961).
[8] Lectures on the entropy theory of transformations with invariant measure. Uspehi Mat. Nauk. **22**, no. 5, 3–56 (1967).
[9] Metric properties of endomorphisms of compact commutative groups. Izv. Acad. Sci. USSR, Ser. Mat. **28**, 4, 867–874 (1964).

Rohlin, V. A., and Sinai, Ya. G.
[1] Construction and properties of invariant measurable partitions. Dokl. Acad. Sci. USSR. **141**, no. 5, 1038–1041 (1961).

Rudin, W.
[1] Fourier Analysis on Groups. New York: 1962.

Rudolph, D.
[1] A two-valued step-coding for ergodic flows. In: Proceedings of the International Conference on Dynamic Systems in Mathematical Physics, Rennes, Sept. 14–21, 1975.

Ruelle, D.
[1] Statistical mechanics. New York: Benjamin 1969.

Sataev, E. A.
[1] On the number of invariant measures of flows on orientable surfaces. Izv. Acad. Sci. USSR, Ser. Mat. **39**, no. 4, 860–878 (1975).

Schwartzbauer, T.
[1] A general method for approximating measure preserving transformations. Proc. Amer. Math. Soc. **24**, no. 3, 643–648. (1970).

Shannon, C.
[1] A mathematical theory of communication. Bell Syst. Tech. J. **27**, 379–423, 623–656 (1948).

Shields, P.
[1] The Theory of Bernoulli Shifts. Chicago: University of Chicago Press 1973.

Shklover, M. D.
[1] On classical dynamical systems on the torus with continuous spectrum. Izv. VUZov, Mat. no. 10, 113–124 (1967).

Siegel, C.
[1] Note on differential equations on the torus. Ann. Math. **46**, no. 3, 423–428 (1945).

Sinai, Ya. G.
[1] On the notion of entropy of dynamical systems. Dokl. Acad. Sci. USSR, **124**, no. 4, 768–771 (1959).
[2] Dynamical systems with countable multiplicity Lebesgue spectrum, I. Izv. Acad. Sci. USSR, Mat. **25**, no. 6, 899–924 (1961).
[3] On the properties of spectra of dynamical systems. Dokl. Acad. Sci. USSR. **150**, no. 6, 1235–1237 (1963).
[4] On higher order spectral measures of ergodic stationary processes. Probab. Theory. Appl. **8**, no. 4, 463–469 (1963).
[5] On the weak isomorphism of transformations with invariant measure. Mat. Sb. **63**, no. 1, 23–42 (1964).
[6] Classical Dynamical Systems with Countable Multiplicity Lebesque Spectrum. Izv. Acad. Sci. USSR Mat. 30, no. 1, 15–68 (1966).
[7] Markov partitions and C-diffeomorphisms. Funct. Anal. Appl. **2**, 3, 64–89 (1968).
[8] Construction of Markov partitions. Funct. Anal. Appl. **2**, no. 3, 70–80 (1968).
[9] Introduction to ergodic theory. Erevan Univ. Editions, 1973. English translation Princeton University Press 1977.
[10] Construction of dynamics in one-dimensional systems of statistical mechanics. Theor. Math. Phys. no. 11, 248–258 (1972).
[11] Construction of cluster dynamics for dynamical systems of statistical mechanics. Vestnik Moscow Univ., Ser. Mat. Mech. **29**, 152–158 (1974).
[12] Theory of Dynamical Systems. Part I. Ergodic Theory. Warsaw: Warsaw University Press 1969.

Smorodinsky, M.
[1] Ergodic Theory, Entropy. Lecture Notes in Mathematics, no. 214. Berlin–Heidelberg–New York: Springer-Verlag. 1971.

Sternberg, Sh.
[1] Celestial mechanics, Parts I and II. New York–Amsterdam: Benjamin, 1969.

Stepin, A. M.
[1] Spectrum and approximation of metric automorphisms by periodic transformations. Funct. Anal. Appl. **1**, no. 2, 77–80 (1967).
[2] The relation between approximation and spectral properties of metric automorphisms. Mat. Zametki. **13**, no. 3, 403–409 (1973).

Thouvenot, J. P.
[1] Quelques proprietes des systèmes dynamiques qui se décomposent en un produit, de deux systemes dont l'un et un schema de Bernoulli. Israel J. Math. **21**, 2–3, 177–207 (1975).

Vershik, A. M.
[1] On the theory of normal dynamical systems. Dokl. Acad. Sci. USSR. **144**, no. 1, 9–12 (1962).
[2] On the spectral and metric isomorphism of some normal dynamical systems. Dokl. Acad. Sci. USSR, **144**, no. 2, 255–257 (1962).

[3] Measurable realization of continuous automorphism groups of the unitary ring. Izv. Acad. Sci. USSR, Ser. Mat., **29**, no. 1, 127–136 (1965).

[4] Four definitions of the scale of an automorphism. Funct. Anal. Appl. **7**, no. 3, 1–17 (1973).

Vershik, A. M. and Yuzvinsky, S. A.

[1] Dynamical systems with invariant measure. In: Mathematical Analysis, 133–187 (1967). Moscow: Itogi Nauki, Viniti Editions 1969.

Volkovyssky, K. L. and Sinai, Ya. G.

[1] Ergodic properties of an ideal gas with an infinite number of degrees of freedom. Funct. Anal. Appl. **5**, 19–21 (1971).

Walters, P.

[1] Ergodic theory. Introductory lectures. Lecture Notes in Mathematics, no. 458. Berlin–Heidelberg–New York: Springer-Verlag 1975.

Weiss, B.

[1] Equivalence of measure preserving transformations (Preprint).

Weyl, H.

[1] Über der Gleichverteilung von Zahlen mod. 1. Math. Ann. **77**, 313–352 (1916).

[2] Mean Motion, I, II. Amer. J. Math. **60**, 889–896 (1938). **61**, 143–148 (1939).

Wiener, N.

[1] Generalized harmonic analysis. Acta Math. **55**, 117–258 (1930).

Index

U

V